Materials Science and Engineering: Novel Developments

Materials Science and Engineering: Novel Developments

Edited by
Reece Hughes

WILLFORD PRESS
www.willfordpress.com

Published by Willford Press,
118-35 Queens Blvd., Suite 400,
Forest Hills, NY 11375, USA

ISBN: 978-1-68285-609-3

Cataloging-in-Publication Data

Materials science and engineering : novel developments / edited by Reece Hughes.
    p. cm.
Includes bibliographical references and index.
ISBN 978-1-68285-609-3
1. Materials science. 2. Materials. 3. Materials--Technological innovations.
4. Engineering. I. Hughes, Reece.
TA403 .M38 2019
620.11--dc23

For information on all Willford Press publications
visit our website at www.willfordpress.com

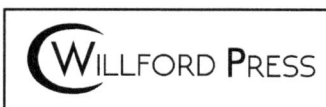

WILLFORD PRESS

Contents

Permissions

List of Contributors

Index

Preface

Every book is a source of knowledge and this one is no exception. The idea that led to the conceptualization of this book was the fact that the world is advancing rapidly; which makes it crucial to document the progress in every field. I am aware that a lot of data is already available, yet, there is a lot more to learn. Hence, I accepted the responsibility of editing this book and contributing my knowledge to the community.

The interdisciplinary field of materials science and engineering studies the structure, production and properties of materials. It incorporates the principles of physics, chemistry and engineering to advance the understanding of metallurgy, nanotechnology and biomaterials. Materials science and engineering has diverse applications in the areas of mineralogy, ceramics, microfabrication, forensic analysis, etc. Metals, polymers and ceramics are some of the primary materials studied under this field. The advancements in this field have resulted in the development of better materials like biomaterials and nanomaterials. This book is a compilation of topics that discuss the most vital concepts and emerging trends in the area of materials science and engineering. It includes some of the vital pieces of work being conducted across the world on various aspects related to this field. This book is appropriate for students seeking detailed information in this area as well as for experts.

While editing this book, I had multiple visions for it. Then I finally narrowed down to make every chapter a sole standing text explaining a particular topic, so that they can be used independently. However, the umbrella subject sinews them into a common theme. This makes the book a unique platform of knowledge.

I would like to give the major credit of this book to the experts from every corner of the world, who took the time to share their expertise with us. Also, I owe the completion of this book to the never-ending support of my family, who supported me throughout the project.

Editor

Application of Gettering Process on the Improvement of the Structural and Mineralogical Properties of Tunisian Phosphate Rock

Daik R[1]*, Lajnef M[2], Amor SB[1], Elgharbi S[3], Meddeb H[1], Abdessalem K[4], Férid M[3] and Ezzaouia H[1]

[1]Photovoltaic Laboratory Research and Technology Centre of Energy, Borj-Cedria Science and Technology Park, BP 95, 2050 Hammam-Lif, Tunisia
[2]Sfax Preparatory Engineering Institut, Route Menzel Chaker, 0.5 km, BP 1172, 3080 Sfax, Tunisia
[3]National Research Center in Sciences of Materials, Borj-Cedria Science and Technology Park, B.P. 95 Hammam-Lif, 2050, Tunisia
[4]L3M, Department of Physics, Faculty of Sciences of Bizerte, 7021 Zarzouna, Tunisia

Abstract

The present study deals with the effects of gettering process on the structural and mineralogical composition of Tunisian phosphate rock. The treated samples were characterized to investigate the variation of physical structure and chemical composition as compared to the reference phosphate rock. The quantitative analysis of the impurities concentration before and after gettering treatment using energy-dispersive (EDX) reveals a significant reduction of impurity concentration (more than 75%) such as Al, Si, S, Na, and Mg. Scanning electron microscopy (SEM) shows that gettering process promoted structural alterations of phosphate rock sample due to fusion of impurities. The XRD patterns show that the chief mineral constituent of treated sample is only fluorapatite, while those in the reference ore were calcite, dolomite, quartz and carbonate-fluorapatite. FT-IR characterization show a disappearance of the bands related to calcite at 714 cm^{-1} as well as B carbonate situated at 1430 cm^{-1}, 1458 cm^{-1} after gettering treatment. This result is in good correlation with Raman analysis.

Keywords: Phosphate rock; Gettering; Impurities; Structure

Introduction

Gafsa which is located in the south of Tunisia is one of the largest phosphate producers in the word (more than 10 million tons per year since the early nineties) [1]. The phosphate rock is used to manufacture phosphate fertilizers and industrial products and, also the only significant global resource of phosphorus used in animal feed supplements, food preservatives, anti-corrosion agents, cosmetics, fungicides, ceramics, water treatment and metallurgy [2]. The rock is composed essentially of the apatite group in association with a wide assortment of accessory minerals mainly fluorides, carbonates (calcite and/or dolomite), clays, quartz, silicates, metal oxides as well as organic matters and trace impurities such as U, REEs (rare earth elements), Cd, As, V, Cr, Zn, Cu, Ni, etc., which can be harmful for several application at certain concentration [3-6].

Gettering process typically consists on a combined a rapid thermal treatment (RTP) followed by a chemical etching after the growth of a porous layer in order to reduce impurities amount and to enhance the phosphate quality. The rapid thermal treatment aims to migrate the impurities to the boundaries surface where they undergone an elimination process by a chemical attack.

The gettering treatment is an effective process to eliminate these impurities which was already applied in our laboratory for the purification of silica [7]. Although, characterization and quantification of the impurities contained in Tunisian phosphate is well established, there are not many reports about their elimination and phosphate purification [5,8-12]. In the present work, we aim not only to improve the structural and mineralogical properties but also to eliminate the majority of these impurities contained in Tunisian phosphate by gettering treatment. The changes in physical structure and chemical composition of the samples after gettering treatment have been investigated by using X-ray powder diffraction (XRD), scanning electron microscopy (SEM/EDX), FT-IR, and Raman spectroscopies. Transmission electron microscope (TEM) micrographs were performed to inspect the morphological properties after treatment.

Materials, Procedure and Methods

Materials

Phosphate rock samples used in this study was obtained from the phosphate deposits from the Metlaoui basins located in the south of Tunisia. It was crushed, ground, and then sieved, the fraction in the range between 180 µm and 600 µm was used. This fraction was crushed by a jaw breaker, reaching a dimension of to 180 µm. Another manual grinding is performed using an agate mortar in order to increase specific surface area.

Procedure

The experimental procedure consists in two steps:

First step (formation of porous layer): the porous layer of phosphate rock is formed by CAVP technique (Chemical Attack in the Vapor Phase) when the sample is exposed to an acidic vapor composed of 64% HNO_3, 20% CH_3COOH and 16% HF. The vapor phase etching is performed under heating at 45°C for 60 minutes. The objective of growing of porous layer on the grain surface is to increase the specific surface layer, thus the impurities can be removed easily.

Second step (gettering process): the sample of porous phosphate is introduced in the rapid thermal furnace (RTP) at a fixed temperature 900°C for 45 minutes under a flow of oxygen. In order to remove the impurities from the samples, the thermally treated porous phosphate undergoes four iterative etchings: 1 g of the former sample is etched

*Corresponding author: Daik R, Photovoltaic Laboratory Research and Technology Centre of Energy, Borj-Cedria Science and Technology Park, BP 95, 2050 Hammam-Lif, Tunisia, E-mail: daik.ridha.crten@gmail.com

with 20 ml of diluted solution of CP4 (3 ml HCl + 1 ml HNO_3 were dissolved in 996 ml of deionizer water). The mixed solution undergoes a stirring for 3 minutes. The treated phosphate in the final phase is separated from the obtained solution by a filtration system of 0.54 mesh diameter. The solid remaining was washed, dried during 1 hour at 100°C then weighed with a precision. For simplification, we have noted (RP) the reference phosphate sample and (TP) phosphate after gettering treatment (treated phosphate).

Methods

X-Ray diffraction were performed using X'PERT Pro Philips analytical diffractometer operating at wavelength Kα copper (λ = 1.5418 nm) and the obtained results were analyzed using the software X'PertisHigh Score Plus.

The IR spectra were recorded using a Nicolet 560 spectrometer; samples pelletized using a pressein potassium bromide (KBr) to 2 mg of product 300 mg of KBr. Registration is realized in the range between 4000 cm^{-1} and 400 cm^{-1}.

Raman shift were recorded with micro-Raman spectroscopy (Jobin Yvon Horibra LABRAMHR) in 400 cm^{-1}-1100 cm^{-1} range. The excitation source was 632.8 nm line of He-Ne laser. The microstructure of samples was characterized by transmission (Technai G2) electron microscopy. The chemical composition was determined by energy diversive X-ray EDX analysis. For the TEM sample preparation, we employed the ultrasound vibration method [13]. The samples were immersed in ethanol solution and ultrasound vibration was applied to separate precipitates from the phosphate. After that, the precipitates were carefully extracted in the solution and picked up using TEM copper meshes with carbon film coatings.

Results and Discussion

XRD characterization

A powder X-ray diffraction (XRD) analysis was used to determine the crystalline phases of the Tunisian natural phosphates rocks before and after getting process. The XRD patterns of the treated phosphate rocks as well as the raw material are illustrated in Figure 1.

The main minerals in reference phosphate rocks (RP) are carbonate-fluorapatite (2θ: 25.99°; 28.17°; 29.44°; 31.97°; 33.24°; 34.15°; 40.31°; 42.48°; 44.48°; 47.01°; 49.55°; 50.81°; 53.09°) (JCPDS 00-021-0141),

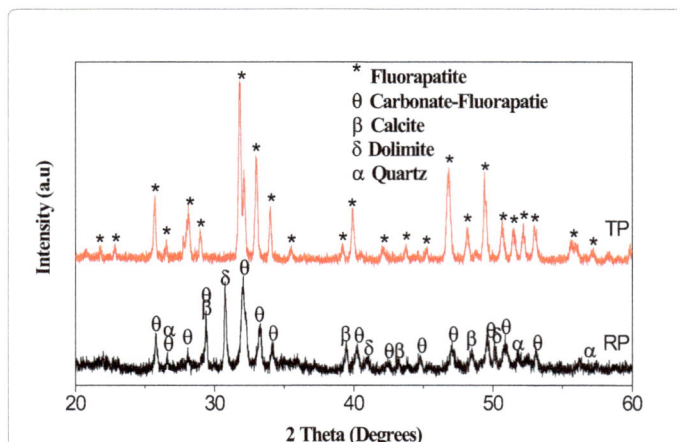

Figure 2: FT-IR spectrum of Tunisian phosphate rock before (RP) and after treatment (TP).

quartz SiO_2 (2θ: 26.54°; 51.9°; 56.25°) (JCPDS 01-080-2146), carbonates which are in the form of dolomite CaMg $(CO_3)_2$ (2θ : 30.7°, 41.06°, 50.09°) (JCP"DS 01-073-2409) and calcite $CaCO_3$ (2θ : 29.44°, 39.42°, 43.18° and 48.46°) (JCPDS 01-072-1652). Calcite and quartz were the main gangue minerals in the Tunisian phosphate rock. Concerning the treated sample (TP), as expected from Figure 1, the major crystalline phase is hexagonal fluorapatite (FAp) $(Ca_{10}(PO_4)6F_2)$, space group $P6_3/m$ (JCPDS 01-079-1459). The highest intensity near 33° confirms the fluorapatite behavior of the treated sample [2,7,9,10,14,15].

Calcite and quartz diffraction lines are disappeared as a result of gettering process, also carbonate-fluorapatite has changed to fluorapatite because carbonates are decomposed by rapid thermal treatment. Therefore, it was proved in others works that rapid thermal treatment at 900°C leads to a phosphate with relatively higher P_2O_5 and CaO contents and a disappearance of organic matter [16,17]. In this work CaO formed after rapid thermal treatment was eliminated by chemical attack in vapor phase (ACPV).

XRD pattern of treated phosphate compared with reference sample (Figure 1) shows a good resolution of the peaks and a decrease of the width at half maximum which proves an amelioration of crystalinity after gettering process.

X-ray diffraction analysis indicates that certain level of impurities were removed during gettring process of phosphate rock. However, the improvement of the phosphate quality of treated sample depends on the mass percentages of the remaining impurities, notably the quartz, calcite and dolomite. In previous work, it was demonstrated also that the reduction of phosphate impurities was associated with some structural changes in the apatite [16].

FT-IR characterization

Figure 2 shows the FT-IR spectra of Tunisian phosphate before and after gettering treatment in the region of 4000 cm^{-1}-400 cm^{-1}. From this figure, we can observe that the gettering process had a remarkable an important effect on the vibrational bands intensity and its positions; also we can note the appearance and disappearance of some pics.

The FTIR spectrum indicates that the reference phosphate rock spectrum shows that the characteristic absorption bands corresponds to the carbonate fluorapatite [7,17,18]. The symmetric υ1 (stretching) mode assigned to PO_4^{3-} is represented by a single band at 966 cm^{-1}. The υ2 (bending) mode of phosphate groupment is located at 474 cm^{-1}. The strong absorption band at 1044 cm^{-1} ascribed to asymmetric υ3

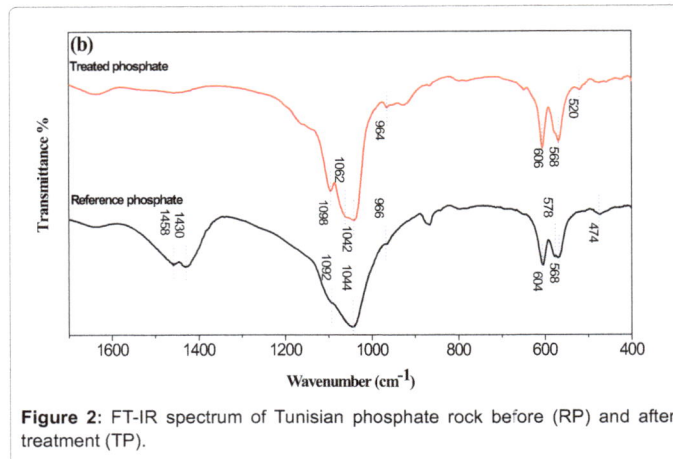

Figure 1: Diffractogramms of the reference phosphate (RP) and the treated phosphate (TP).

mode. The asymmetric υ4 mode is splited in three bands: 568, 578 and 604 cm^{-1}. The two bands at 1430 cm^{-1} and 1458 cm^{-1} were assigned to υ2 vibration of CO_3^{2-} group located in the B site of apatite (carbonate substituting phosphate) [17]. The spectrum of main component of the phosphate rock reference sample is in a good agreement with published IR spectra of apatite [19-21].

FT-IR spectra of the TP sample illustrated in Figure 2 reveals that the vibrational bands of treated phosphate were clearly observed compared with reference phosphate rock. The absorption peaks located at 1098 cm^{-1} and 1062 cm^{-1} originated from asymmetrical stretching υ3 of PO_3^{-4} and the peaks localized at 568 cm^{-1} and 606 cm^{-1} were attributed to bending modes υ4 of PO_3^{-4}. While the symmetric stretching modes υ1 and υ2 of PO_3^{-4} were also observed at around 964 cm^{-1} and 520 cm^{-1} respectively [21].

Moreover, after gettering process, the band positions and their intensities are slightly affected and we observe a change in the number of phosphate bands, the treated phosphate indicates that the bands at 520 cm^{-1} corresponding to υ2 strongly shifted from 474 cm^{-1} to 520 cm^{-1}. Concerning the shift, it can be due to the variation repulsion potential of the contracted or dilated crystal lattice which is confirmed by XRD analysis [7,22]. The positions of υ4 and υ1 modes didn't change but an important increase of intensity was marked. The υ3 asymmetric mode was degenerated in tow distiguitched peacks at 1042 cm^{-1} and 1062 cm^{-1}. The appearance of the two distinct peaks is due to the presence of different P-O distances in the crystal.

Besides, a considerable reduction in the absorption of carbonate bending is shown clearly after gettering treatment. In fact, we remark a disappearance of the bands related to calcite at 714 cm^{-1} as well as B carbonate situated at 1430 cm^{-1}, 1458 cm^{-1}. This implies that carbonate and calcite substitutions induce vacancies at the OH sites, and we assume that thermal treatment is responsible of the total decomposition of carbonate bands and intensities decreases [23]. Thus, the results indicate that mixture acids can be used to reduce calcium carbonate in low-grade calcareous phosphate rock as it improves the degree of beneficiation [24].

RAMAN characterization

Raman scattering is a sensitive tool for studying the phosphate material because it gives direct structural evidence qualitatively related to the different components in the material. Figure 3 shows the Raman spectra for RP and TP. From this figure, we can't observe any vibrational mode for reference sample (RP). This is due to the fact that Raman bands are completely overlapped by the fluorescence background originated from organic matter, metal compouned and rare earth existing in natural phosphate rock (RP) [25-29]. As a result of this overlap, we can't differentiate between the different vibrational modes.

For the TP, Raman spectra shows obviously the different vibrational modes of phosphates groupement after Gettering process. The strongest Raman active υ1 of PO_3^{-4} mode appearance in the spectrium of the TP sample at 961 cm^{-1} [30,31].

To better clarify the vibrationnal modes existing in TP, deconvolution of the Raman spectrum were shown in Figures 4 and 5. The Raman spectrum of phosphate in the 125-300 cm^{-1} spectral range is illustrated in Figure 4. Raman bands are observed at 139, 169, 214, 234, 265 and 283 cm^{-1}. These bands are assigned to lattice vibrations as it was reported by many authers [11,18,32].

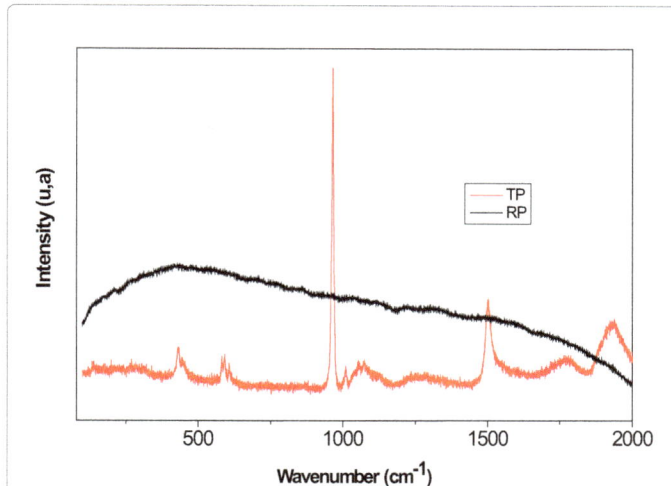

Figure 3: RAMAN spectrum of Tunisian phosphate rock before (RP) and after treatment (TP).

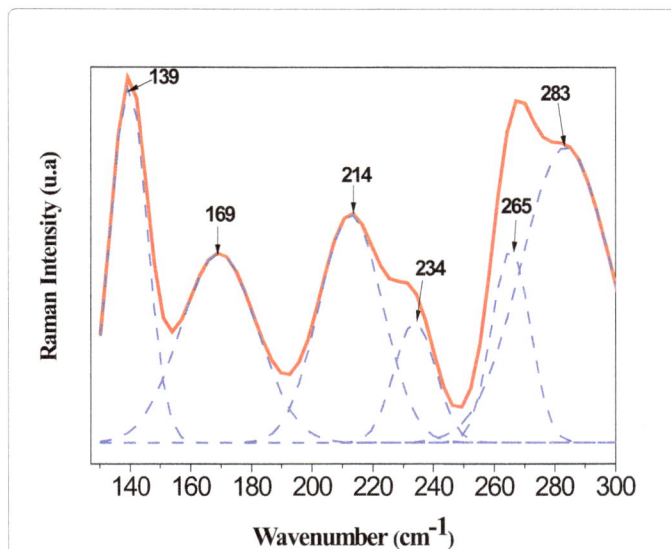

Figure 4: Raman spectrum of treated phosphate over the 125 cm^{-1}-300 cm^{-1} spectral range.

The Raman spectrum of treated phosphate over the 400-620 cm^{-1} spectral range is reported in Figure 6. This range is assigned to the vibration of υ2 and υ4 PO_4^{3-} bending modes. It was reported by S. Elgharbi and H. Lefires when they work about Tunisian phosphate rock that the Raman bands at 582, 591 and 607 cm^{-1} are assigned to υ4 PO_4^{3-} and the bands at 431 and 435 cm^{-1} are due to the υ2 PO_4^{3-} [12,13]. Then, the work reported by Karampasa about calcium phosphate confirmed very well the results above [29].

The Raman spectrum over the 850-1200 cm^{-1} range is reported in Figure 6. Similar intensity bands are found at 1056, 1100, 1116 cm^{-1} which are assigned to υ3 PO_4^{3-} antisymmetric stretching vibration, the three bandes are attributed to a pure fluorapatite [18]. Low intensity Raman band at 1009 cm^{-1} is attributed to υ1 PO_4^{3-} symmetric stretching mode [28,33,34].

From Raman analysis, we can conclude that the gettereing process in necessary to eliminate the impurities and organic matters which

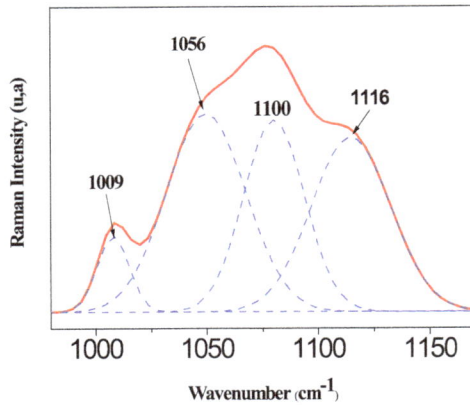

Figure 5: Raman spectrum of treated phosphate over the 850 cm^{-1}-1200 cm^{-1} spectral range.

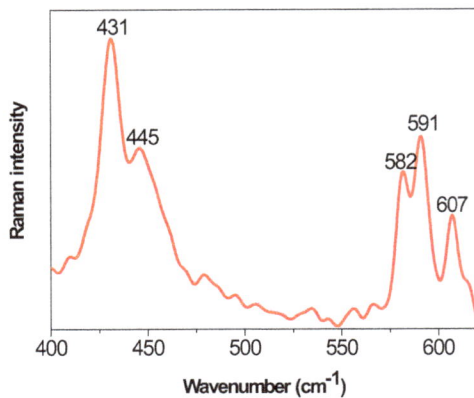

Figure 6: Raman spectrum of treated phosphate over the 400 cm^{-1} -620 cm^{-1} spectral range.

are the main causes of overlopped in reference phosphate rock, and consequently improves the structural properties.

TEM-EDAX characterization

The change in the physical structure of the treated sample in comparison with the raw ore was investigated by TEM. This scanning procedure consisted of looking for structure alterations, agglutination, porosity, morphology, compaction, and distribution, with qualitative and semi quantitative identification of elements [35].

It is shown in Figure 7 that the phosphate rock consists of two different particule phases with estimated sizes of 60 μm. Moreover, these phases are defined with tow portions which can be due to the accumulation of the impurities which escape the dispersion of the particles. The portion in light grey is formed by phosphorous rich components whereas the portion in dark grey are formed by calcium-rich components, which can be defined as $CaCO_3$, based on chemical analysis. No phosphorous was found in the carbonate parts. Carbonate-fluorapatite existing in the ore has only been observed in the parts with phosphorous-rich components. The surfaces of the parts with phosphate exhibit a compact structure with only little porosity.

The TEM micrographs of the treated sample by gettering process (porous phosphate treated 45 min at 900°C and eatching in mixture acid) is given in Figure 8 shows that the TP sample is formed by many crystals with baton forms. It seems that the RP sample was subdivided to many particles with different sizes. It was determined that the shrinkage and the cracks at the surrounding parts with phosphate occurring due to thermal and etching treatments [36]. The holes on the surfaces of the parts with phosphate prove that carbonate–fluorapatite was calcined and that the carbonate–fluorapatite changed to fluorapatite. This is due to the disappearance of the impurities which occupied interstitial sites, grain boundaries. Only the preponderant elements appear in the imagery which is confirmed by quantitative analysis. These results are in good agreement with the XRD analysis.

To get more insight of the composition of the RP and TP samples, Energy-dispersive X-ray (EDX) was used in many places of the sample area. The results were summarized in Table 1. From Figures 7 and 8, we noticed that the major elements before treatment are P, Ca, F in addition to the impurities such as Al, Si, S, Na, Mg…Whereas, after treatment only the P, Ca, F are presented with small traces S, Na and Si.

Table 1 shows the quantitative chemical composition. The analysis shows a homogeneous phase composed by P, Ca and F as being major elements consists mainly of fluorapatite. The chemical composition of phosphate rock shows that after treatement process, it changes to a rather poor in magnesium, in silica and metal such as Al, Fe.

Moreover, the electron microprobe analysis of samples allows us to

Element	RP	TP
Ca	26.27	49.03
P	10.92	28.8
F	3.4	13.43
Si	37.62	7.73
Al	9.08	1.53
Mg	8.87	1.85
Fe	2.31	0.06
S	1.38	0.51
Na	1.12	0.06
K	0.28	0.07
Ca/P	2.4	1.7

Table 1: Atomic percent of reference phosphate (RP) and treated phosphate (TP).

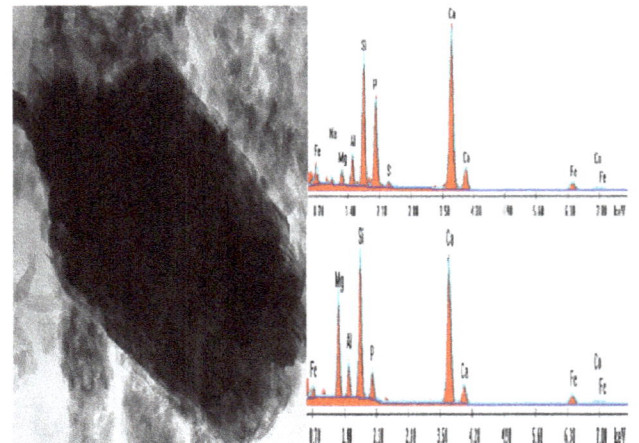

Figure 7: TEM and EDX analysis of reference phosphate rock.

Application of Gettering Process on the Improvement of the Structural and Mineralogical Properties...

5

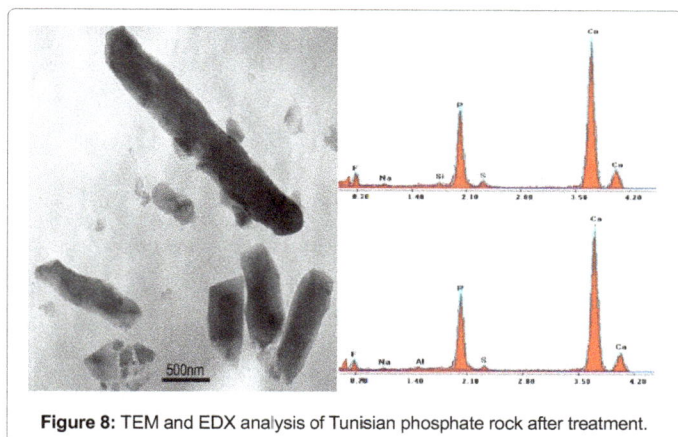

Figure 8: TEM and EDX analysis of Tunisian phosphate rock after treatment.

evaluate the ratio Ca/P. Compared to the reference phosphate are 2.4, treated phosphate (TP) is become 1.7 which is closer to the theoretical Ca/P molar ratio of pure FAP: 1.67. However, this proves the presence of carbonate-FAP and calcite in reference sample and the presence of calcium oxide in excess [28]. This difference in composition may take place by incorporation of ion present in the site of PO_4^{2-} group. In our case these elements are F^-, Na^+, CO_3^{2-} known to be incorporated into the network of the apatite.

Conclusion

A marked change on the properties of Tunisian phosphate rock was observed following the gettering process. The experimental results in this study suggest a significant improvement in the structure as well as the composition of the treated phosphate rock. Therefore, we consider that gettering process is not only a promising way to eliminate the impurities but also it enhances the use of phosphate rock in many fields.

References

1. Popescu IC, Filip P, Humelnicu D, Humelnicu I, Scott TB, et al. (2013) Removal of uranium (VI) from aqueous systems by nanoscalezero-valent iron particles suspended in carboxy-methyl cellulose. Journal of Nuclear Materials 443: 250-255.

2. Cevik U, Baltas H, Tabak A, Damla N (2010) Radiological and chemical assessment of phosphate rocks in some countries. Journal of Hazardous Materials 182: 531-535.

3. Zendah H, Khattech I, Jemal M (2013) Thermochemical and kinetic studies of the acid attack of "B" type carbonate flucrapatites at different temperatures (25-55°C). Thermochemica Acta 565: 46-51.

4. Mar SS, Okazaki M (2012) Investigation of Cd contents in several phosphate rocks used for the production of fertilizer. Microchemical Journal 104: 17-21.

5. Silva EF, Mlayah A, Gomes C, Noronha F, Charef A, et al. (2010) Heavy elements in the phosphorite from KalaatKhasba mine (North-western Tunisia): Potential implications on the environment and human health. Hazardous Materials 182: 232-245.

6. Guoa Z, Fanb J, Zhanga J, Kanga Y, Liua H, et al. (2015) Sorption heavy metal ions by activated carbons with well-developed microporosity and amino groups derived from Phragmites australis by ammonium phosphates activation. Journal of the Taiwan Institute of Chemical Engineers.

7. Marouan K, Messaoud H, Hatem E (2012) Purification of silicon powder by the formation of thin porous layer followed byphoto-thermal annealing. Nanoscale Research Letters 7: 444.

8. Koleva V, Petkova V (2012) IR spectroscopic study of high energy activated Tunisian phosphorite. Vibrational Spectroscopy 58: 125-132.

9. Garnit H, Bouhlel S, Baraca D, Chtara C (2012) Application of LA-ICP-MS to sedimentary phosphaticparticules from Tunisian phosphorite deposits: Insights from trace elemnts and REE into paleo-depositionelenvirenments. Chemie der Erde 72: 127-139.

10. Bachoua H, Othmani M, Coppel Y, Fatteh N, Debbabi M, et al. (2014) Structural and thermal investigations of a Tunisian natural phosphate rock. J Mater Environ Sci 5: 1152-1159.

11. Galai H, Sliman F (1994) Mineral characterization of the Oum El Khachebphosphorites.

12. Amini R, Vakili H, Ramezanzadeh B (2015) Studying the effects of poly (vinyl) alcohol on the morphology and anti-corrosion performance of phosphate coating applied on steel surface. Journal of the Taiwan Institute of Chemical Engineers.

13. Frost RL, Scholz R, Lopées A, Xi Y, Gobac ZZ (2013) Raman and infrared spectroscopic characterization of the phosphate mineral paravauxite $Fe^{2+}Al_2(OH)_2 \cdot 8H_2O$. Spectrochimica Acta A Molecular and Biomolecular Spectrocopy 116: 491-496.

14. Elgharbi S, Horchani-Naifer K, Ferid M (2015) Investigation of the structural and mineralogical changes of Tunisian phosphorite during calcinations. J Therm Anal Calorim 119: 265-271.

15. Lefires H, Medini H, Megriche A, Mgaidi A (2014) Dissolution of Calcareous Phosphate Rock from Gafsa (Tunisia) Using Dilute Phosphoric Acid Solution. International Journal of Nonferrous Metallurgy 3: 1-7.

16. Guo F, Li J (2010) Separation strategies for Jordanian phosphate rock with siliceous and calcareous gangues. International Journal of Mineral Processing 97: 74-78.

17. Fleet ME (2009) Infrared spectra of carbonate apatites: v_2-Region bands. Biomaterials 30: 1473-1481.

18. Antonakosa A, Liarokapis E, Leventouri T (2007) Theodora Leventouri, Micro-Raman and FTIR studies of synthetic and natural apatites. Biomaterials 28: 3043-3054.

19. Farmer VC (1974) The layer silicates. Mineralogical Society of London.

20. Incee DE, Johnston CT, Moudgil BM (1991) Fourier transformation infared spectroscopy study of adsorption of oleic acid on surfaces of apatite. J Angmuir 7: 1453-1457.

21. Wei W, Cui J, Wei Z (2014) Effects of low molecular weight organic acids on the immobilization of aqueous Pb(II) using phosphate rock and different crystallized hydroxyapatite. Chemosphere 105: 14-23.

22. Kolevo V, Stevo V (2013) Phosphate ion vibrations in dihydrogen phosphate salts of the type $M(H_2PO_4)_2 \cdot 2H_2O$ (M = Mg, Mn, Co, Ni, Zn, Cd): Spectra–structure correlations. Vibrational Spectroscopy 64: 89-100.

23. Fahamin A, Nasiri-Tabrizi B, Ebrahimi-Kahrizsangi R (2012) Synthesis of calcium phosphate-based composite nanopowders by mechanochemical process and subsequent thermal treatment. Ceramics International 38: 6729-6738.

24. Zafar ZI, Ashraf M (2007) Selective leaching kinetics of calcareous phosphate rock in lactic acid. Chemical Engineering Journal 131: 41-48.

25. Sheng Q, Sheng V, Liu S, Li W, Wang L, Tang C (2013) Enhanced broad band near-infrared luminescence and peak wavelength shit of Yb-Bi ions co-doped phosphate glasses containing. Journal of Luminescence 144: 26-29.

26. Wenyuan YU, Guanlai LI, Li Z (2010) Sonochemical synthesis and photoluminescence properties of rare-earth phosphate core/shell nanorods. Journal of Rare Earths 28: 171-175.

27. Babu SS, Babu P, Jayasankar CK, Sievers W, TroSter T, et al. (2007) Optical absorption and photoluminescence studies of Eu3+-doped phosphate and fluorophosphate glasses. Journal of Luminescence 126: 109-120.

28. Bandara AMTS, Senanayake G (2015) Leachability of rare-earth, calcium and minor metal ions from natural fluoroapatite in perchloric, hydrochloric, nitric and phosphoric acid solutions: Effect of proton activity and anion participation. Hydrometallurgy 153: 179-189.

29. Karampasa IA, Kontoyannis CG (2013) Characterization of calcium phosphates mixtures. Vibrational Spectroscopy 64: 126-133.

30. Nakamoto K (2009) Infrared and Raman spectra of inorganic and coordination compounds: part A: Theory and applications in inorganic chemistry (6th edn.). John Wiley & Sons inc, USA.

31. Williams Q, Knittle E (1996) Infrared and Raman spectra of $Ca_5(PO_4)3F_2$-fluorapatite at high pressures: compression-induced changes in phosphate site and Davydov splitting. J Phys Chem Solids 57: 417-422.

32. Bushiri MJ, Jayasree RS, Fakhfakh M, Nayar VU (2002) Raman and infrared spectral analysis of thallium niobyl phosphates: $Tl_2NbO_2PO_4$, $Tl_3NaNb_4O_9(PO_4)_2$ and $TlNbOP_2O_7$. Materials Chemistry and Physics 73: 179-185.

33. Frost RL, Lopez A, Scholz R, Xi Y, Belotti FM (2013) Infrared and Raman spectroscopic characterization of the carbonate mineral huanghoite and in comparison with selected rare earth carbonate. Journal of Molecular Structure 1051: 221-225.

34. Frost RL, López A, Xi Y, Cardoso LH, Scholz R (2014) A vibrational spectroscopic study of the phosphate mineral minyulite $KAl_2(OH,F)(PO_4)_24(H_2O)$ and in comparison with wardite. Spectrochimica Acta A Molecular and Biomolecular Spectroscopy 124: 34-39.

35. Francisco EAB, Prochnow LI, Motta de Toledo MC, Ferrari VC, Luís de Jesus S (2007) Thermal Treatment Of Aluminous Phosphates Of The Crandallite Group And Its Effect On Phosphorus Solubility. Sci Agric 64: 269-274.

36. Özer AK, Gülaboglu MS, Bayrakçeken S, Weisweiler W (2006) Changes in physical structure and chemical composition of phosphate rock during calcination in fluidized and fixed beds. Advanced Powder Technol 17: 481-494.

Low Velocity Impact of Filament-Wound Glass-Fiber Reinforced Composite Pipes

Khan Z[1]*, Naik MK[2], Al-Sulaiman F[3] and Merah N[1]

[1]Mechanical Engineering Department, King Fahd University of Petroleum and Minerals, Dhahran, KSA
[2]The Petroleum Institute, Abu Dhabi, UAE
[3]National Company of Mechanical Systems, Riyadh, Saudi Arabia

Abstract

Low velocity single-bounce impact tests have been conducted on filament-wound glass fiber reinforced/vinylester and glass fiber reinforced/epoxy composite pipes. An instrumented drop weight testing system was used for the impact testing. The tests were performed on 300 mm long sections of 150 mm diameter pipes having 6 mm wall thickness. The impact energy required to just initiate the damage in glass fiber reinforced/epoxy pipes was found to be larger than the energy needed for glass fiber reinforced/vinylester pipe. The load-time curves also reveal that vinyl ester-based pipes exhibit a ductile failure under impact, whereas, in the epoxy-based pipes the failure was rather brittle in nature.

Keywords: Low velocity impact; GFRP; Filament-wound pipes

Introduction

Although Glass Fiber Reinforced Plastic (GFRP) composites are known for high degree of tailor ability and many excellent chemical and mechanical properties, a major concern that limits the usage of GFRP composites is their low resistance to impact loading. Low velocity impacts can induce significant damage in the material in the form of matrix cracking, delamination, and fiber fracture. Very often these types of damages remain invisible to the naked eye, but may causes serious degradation in the otherwise excellent mechanical properties of the FRP composites and cause premature and unexpected failure. The response of composite materials to these impact loadings is complex, as it depends on the structural configuration as well as on the intrinsic material properties. Furthermore, it depends on the type of material, geometry, and velocity of the impactor. Each plays an important role in characterizing the overall effect of transverse impact.

Generally, impact with impactor speeds less than 100 m/s are classified as low velocity impact. But there are several other definitions of low velocity impact, with no universal agreement. Sometimes low velocity impact is used in the context of low energy impact, i.e., less than 136 J (100 ft-lb). Low velocity impact normally involves deformation of the entire structure during the contact duration of the impactor, and this situation is considered quasi-static with no consideration of the stress waves that propagate between the impactor and the boundary of the impacted component.

The effect of low-velocity impact damage on the FRP composites laminates and pipes has been studied by a number of researchers over the past several years [1-5]. It is well established that the impact damage occurs in two phases: fracture initiation phase, and fracture propagation phase where ratio of damage initiation energy to damage propagation energy is shown to be a function of material ductility as higher ductility exhibit higher initiation energy to propagation energy ratio than the more brittle ones [6]. Static and single-bounce low velocity (up to 10 m/s) drop weight impact tests on ± 55° filament-wound E-glass/epoxy resin pipes produces a two-part failure process of an elastic deformation followed by failure due to delamination initiation and local crushing [7]. In curved graphite/epoxy composite plates, low velocity impact can cause a dent formation on the impacted surface of the plate while cracking and ply separation occurs on the opposite surface [8]. In the typical load time plots the first load drop

indicates on set of matrix cracking [9] which corresponds to impact damage initiation and becomes the cause of subsequent delamination immediately along the top or bottom interface of the cracked layer [10].

The impact damage development in ± 55° filament wound glass/epoxy tubes of 55-mm ID and 6-mm thick tubes intended for underwater applications the mean damage threshold values were 3 to 4 J and subsequent through thickness delamination damage occurs at energies up to 7 J [11]. In glass, carbon, and aramid fabrics-reinforced composites, dome shaped fracture occurs on the front face as a result of the localized matrix crushing and fiber shearing, while the damage on the rear face shows a characteristic pattern of cracks in the fiber direction [12].

Though a large number low velocity impact studies on FRP composites have been carried out during the past twenty years, the work involving GFRP pipes have remained significantly limited. It is obvious that due to the increasing use of GFRP pipes in many diverse applications and due to the advent of new materials and processing techniques the area of understanding and characterization of impact behavior of GFRP pipes continues to draw considerable attention of the research and design communities. In the present investigation low velocity impact behavior has been investigated in two FRP pipes, the filament-wound glass fiber reinforced/vinylester and the glass fiber reinforced/epoxy composite pipes. Fiber reinforced polymers have captured a significant market as a material of preferred choice in a variety of structural applications around the globe.

Fiber reinforced polymer (FRP) composite materials show great potential for integration into the highway infrastructure. Typically, these materials have long and useful lives; are light in weight and easy

*Corresponding author: Khan Z, Mechanical Engineering Department, King Fahd University of Petroleum and Minerals, Dhahran, Saudi Arabia
E-mail: zukhan@kfupm.edu.sa

to construct; provide excellent strength-to weight characteristics; and can be fabricated for "made-to-order" strength, stiffness, geometry, and other properties. FRP composite materials may be the most cost-effective solution for repair, rehabilitation, and construction of portions of the highway infrastructure. FRP composite materials have a high strength-to-weight ratio and are generally not affected by the harsh highway environment (they do not corrode, and they have excellent fatigue resistance). These composite materials while offer a range of advantages in terms of excellent formability, high specific mechanical strength, better thermal resistance, excellent chemical and corrosion resistance, are at the same time quite susceptible to damage under impact loading. This impact damage can severely impair the otherwise excellent mechanical properties and often results in causing premature failure of the composite. The conventional materials, which are being replaced by the composite materials, have well-defined impact characteristics and the standards are well defined but the laminated composites are more susceptible to impact damages which are often internal and cannot be observed visually [13].

Material and Methods

The specimens used for impact testing were 300-mm long pipe section cut from commercially available filament-wound E-glass fiber reinforced vinyl ester and epoxy based pipes. Both types of pipes had internal diameters of 150 mm and wall thickness of 6 mm. The winding angle of all the pipes was ± 54.5° to the pipe axis. Glass fiber reinforced/vinyl ester pipes are referred as GFRV pipes, whereas, the glass fiber reinforced/epoxy pipes are referred as GFRE pipes.

Single bounced low velocity impact tests were carried out using an instrumented free falling drop weight impact test system (Dynatup 9250G, Instron Corp., USA). A 1.27 cm diameter spherical head steel tup was used as the impactor. Tests at different impact energies were performed by choosing suitable combinations of crosshead mass and drop height. The contact force was measured with a load transducer located between the cross head and hemispherical tup nose. Impact tests were carried out by varying mass and energy until the energy required to just initiate the impact damage and the energy required for total penetration were determined. The tests were then carried out at intermediate energies to examine the impact behavior of the pipe samples. A total of four impact energy levels, 6, 30, 70, and 100 J were investigated for the GFRV pipes, while the impact energy levels for the GFRE pipes were 12, 35, 80, and 110 J. Three impact tests were performed at each energy level. For low impact energy (up to 50 J) tests, the impactor mass of 10 kg was used, while a 25 kg mass was used for higher energy tests. The data used for microscopic evaluation in the study is 60X magnification power is used. The resolution of optical images used in the images is 600 dpi for various images.

Visual and optical inspections of the impact damaged pipes were performed after each test. With the drop height and weight known the data acquisition system provided the calculated values of maximum (peak) load, energy at maximum load, impact energy, deflection at maximum load, and impact velocities.

Results and Discussion

The results of the impact tests carried out at various impact energies for the GFRV and GFRE composite pipes are presented in the tabular form in Tables 1 and 2. The load-time, energy-time, and load-deflection histories for the two materials are shown in the Figures 1-6.

It is well known now that the load-time history can be divided into two distinct regions, a region of damage initiation and a region of

damage propagation. As the load increases during damage initiation phase, elastic strain energy is accumulated in the specimen and no gross failure takes place. However, failure on a micro-scale, for example, transverse matrix cracking, fiber micro-buckling, or debonding at the fiber-matrix interface is possible. These micro damage events are indicated by the pronounced fluctuations in the load-time curves. When a critical load is reached at the end of the initiation phase, the load monotonically decrease with time indicating damage propagation and the composite specimen may fail either by a tensile or a shear failure depending on the relative values of the tensile and inter-laminar shear strengths. At this point the fracture propagates either in a catastrophic manner, indicated by continual load drop or in a progressive manner by continuing to absorb energy at smaller loads, indicated by the load fluctuations in the load-time history. Energy to peak force is the energy that the specimen has absorbed up to the point of maximum load. The total penetration energy is thus the sum of the initiation energy i.e., energy to reach the peak point on the load-time trace and the energy consumed in the damage propagation. Deflection at peak load is the maximum deflection that the specimen experience during the impact loading. It is the deflection value at the point where the load-time curve reaches its peak.

Impact Energy (J)	Specimen Number	Peak Force (kN)	Deformation at Peak Force (mm)	Energy to Peak Force (J)	Total Penetration Energy (J)
6J	6.1	3.22	2.08	3.82	4.92
	6.2	3.29	1.47	2.39	4.97
	6.3	3.29	1.49	2.6	4.37
	Average	3.27	1.68	2.94	4.75
30J	30.1	5.47	7.08	27.39	28.1
	30.2	5.38	7.11	26.8	27.76
	30.3	5.01	7.42	26.15	28.36
	Average	5.28	7.2	26.78	28.07
70J	70.1	6.58	11.29	51.5	66.83
	70.2	6.57	11.54	51.23	66.9
	70.3	6.53	13.51	62.33	68.69
	Average	6.56	12.11	55.02	67.47
100J	100.1	6.57	16.59	79.1	96.22
	100.2	6.58	12.48	58.05	96.35
	100.3	6.57	14.4	44.69	95.13
	Average	6.57	14.49	60.61	95.9

Table 1: Results of the impact tests of GFRV composite pipes.

Impact Energy (J)	Specimen Number	Peak load (kN)	Deformation at Peak Load (mm)	Energy to Peak Load (J)	Total Penetration Energy (J)
12J	12.1	6.55	3.19	10.36	19.54
	12.2	6.55	3.7	12.91	19.63
	12.3	5.81	3.88	12.39	18.78
	Average	6.3	3.59	11.89	19.32
35J	35.1	6.72	3.02	10.1	31.18
	35.2	6.66	3.06	9.83	32.45
	35.3	6.54	3.01	9.66	31.52
	Average	6.64	3.03	9.87	31.72
80J	80.1	6.62	2.97	10.36	80.6
	80.2	6.66	3.19	10.73	80.25
	80.3	6.59	3.12	11.48	81.12
	Average	6.62	3.09	10.86	80.66
110J	110.1	6.62	6.11	28.66	114.47
	110.2	6.56	5.58	25.42	115.14
	110.3	6.58	5.1	23.02	115.3
	Average	6.59	5.59	25.7	114.97

Table 2: Results of the impact tests of GFRE composite pipes.

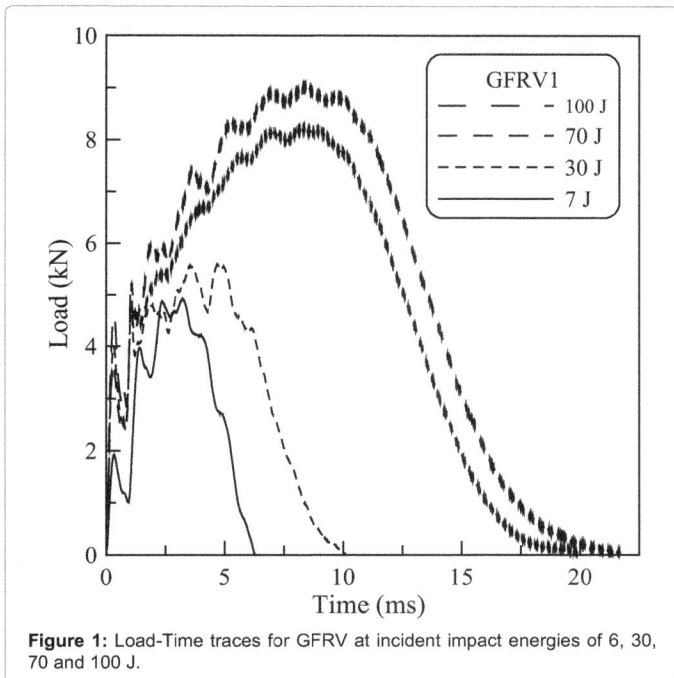

Figure 1: Load-Time traces for GFRV at incident impact energies of 6, 30, 70 and 100 J.

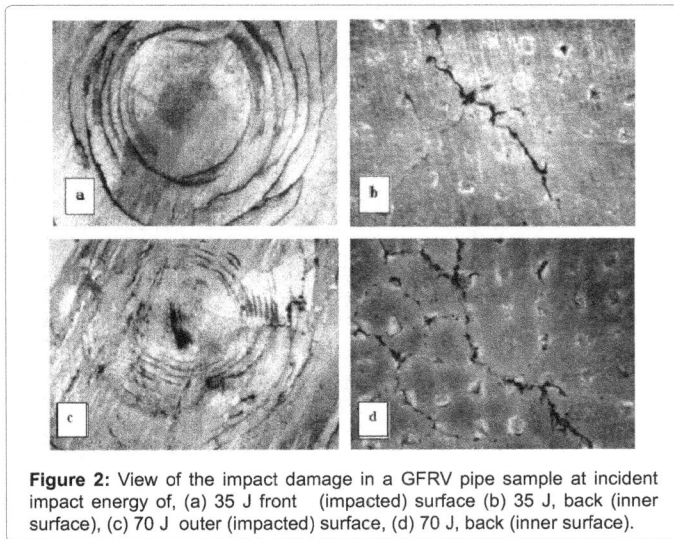

Figure 2: View of the impact damage in a GFRV pipe sample at incident impact energy of, (a) 35 J front (impacted) surface (b) 35 J, back (inner surface), (c) 70 J outer (impacted) surface, (d) 70 J, back (inner surface).

An examination of the load-time traces presented in Figure 1 shows that for low to intermediate incident energies of 6 J and 30 J the damage initiation phase predominates as indicated by the pronounced fluctuations in the load-time history.

The damage at 6 J remains mostly invisible on the impacted surface with no sign of cracking on the back (inner) surface. The impact damage however becomes extensive at higher incident energy levels. Figure 2 provides view of the front (impacted) and back (inner) surface of the specimen impacted with the incident energy of 35 J and 70 J.

It is evident from Figure 2a that at this energy level the specimen shows clear indentation (gross plastic deformation) on the front (impacted) surface and cracking on the back (inner) surface (Figure 2b). At 70 J the damage on the front (impacted) surface shows signs of cracking along around the indentation, while extensive cracking occurs on the back (inner) surface of the pipe sample.

For the GFRV pipes the total absorbed energy (which is the sum of energy absorbed to peak load and the energy absorbed after the peak load) also increases with an increase in incident impact energy as evidenced by Figure 3.

The energy up to the peak load is absorbed through elastic deformation and increases linearly with time, followed by the energy absorbed by the damage initiation and propagation events. This energy absorption increases non-linearly with time, which is suggestive of the fact that the damage events beyond the peak load occurs in an inelastic manner. The examination of the load-deflection traces for GFRV pipes tested at different incident energy levels (Figure 4) also reveal the existence of different stages of deformation and damage creation events.

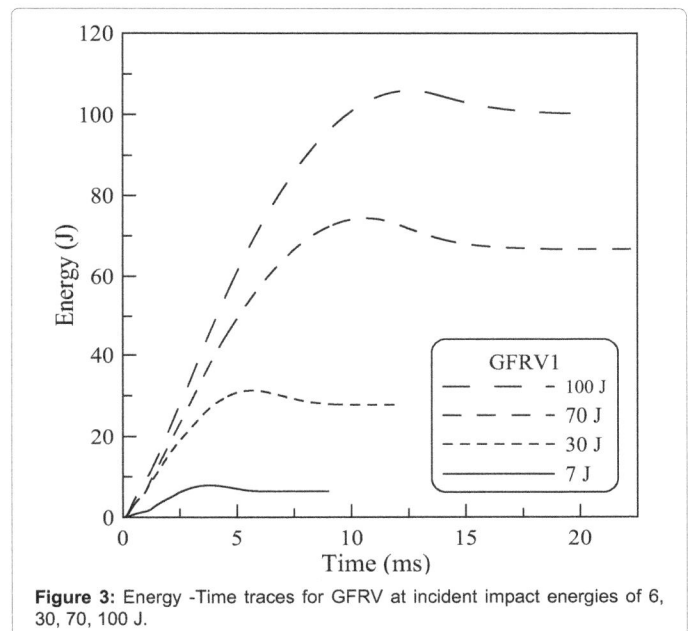

Figure 3: Energy -Time traces for GFRV at incident impact energies of 6, 30, 70, 100 J.

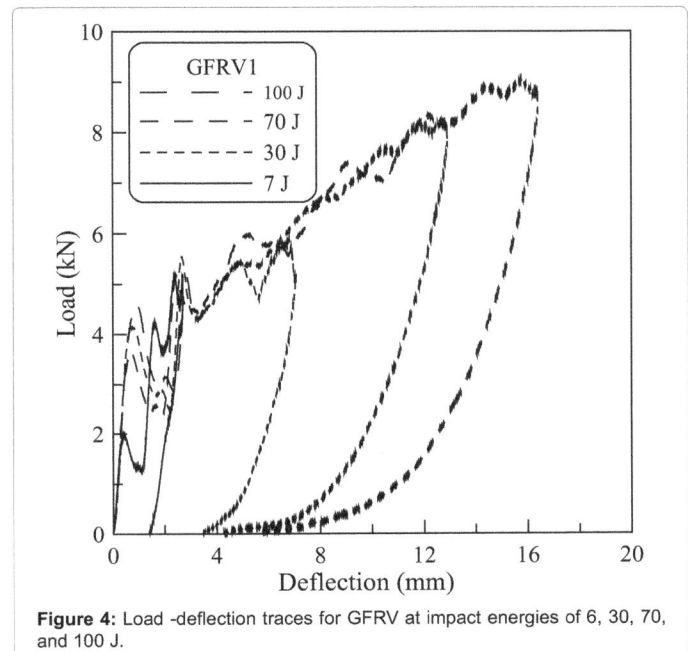

Figure 4: Load -deflection traces for GFRV at impact energies of 6, 30, 70, and 100 J.

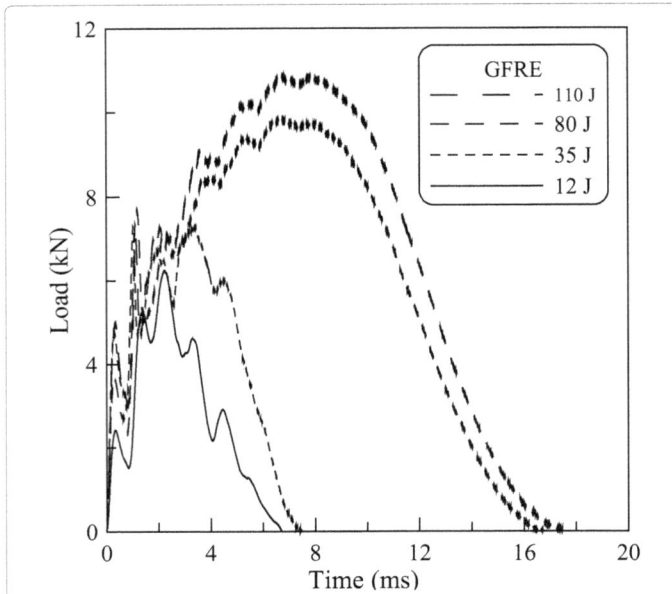

Figure 5: Load-Time traces for GFRE at impact energies of 12, 35, 80 and 110 J.

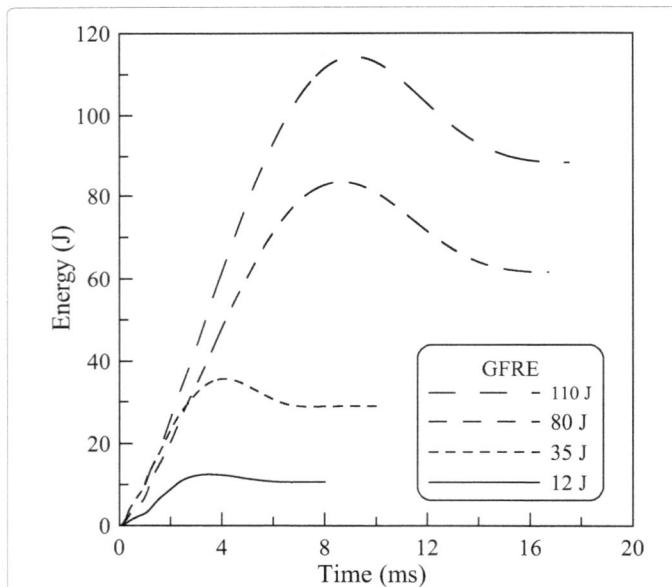

Figure 6: Energy -Time traces for GFRE at impact energies of 12, 35, 80 and 110 J.

The initial linear portion of the load-deflection traces points at the elastic response. This is followed by a portion of inelastic response with characteristic pronounced fluctuations in the load-deflection traces. Maximum specimen deflection is attained at the peak load and as the load begins to drop from its peak value the deflection also begins to reverse through elastic recovery and attains its minimum value at failure. The plateaus in the load-deflection traces for tests at incident energies of 70 and 100 J once again indicate that for full penetration impact events very large deflections are attained due to the physical passing of the impactor through the sample thickness.

The load-time traces from the impact tests at various incident energy levels for GFRE material are provided in Figure 5. It is evident

that for the GFRE pipes, the impact damage was also produced in a similar two stage damage initiation and damage propagation process.

The energy-time and the load-deflection for the GFRE pipe samples tested at different incident energy levels are displayed in Figures 6 and 7, respectively. These traces are approximately analogues to the traces observed for GFRV pipe samples and follow the same trends of increase in absorbed energy and deflection with increase in the incident impact energy as seen in GFRV material.

The impact damage areas for the GFRE pipes impacted at two incident energy levels of 35 and 80 J are shown in Figure 8. At 35 J the impacted surface shows clear indentation at the impacted surface (Figure 8a) and a network of very fine cracks on the back (inner) surface (Figure 8b) of the pipe sample. At 80 J large indentation has formed but the area surrounding the indentation remains much more intact (Figure 8c) than what was observed in the GFRV pipe samples where the indentation was associated with a much larger damage area with wide spread plastic deformation. The back (inner) surface of the GFRE pipe sample at this 80 J incident impact energy also show much less pronounced cracking (Figure 8d) than that observed in GFRV pipe samples impacted with incident energy of 80 J.

The back (inner) surface of the GFRE pipe sample at this 80 J incident impact energy also show much less pronounced cracking (Figure 8d) than that observed in GFRV pipe samples impacted with incident energy of 80 J. The data presented in Tables 1 and 2 summarizes the impact test results for the GFRV and GFRE pipes. Table 1 shows that for GFRV pipe the average peak loads at which the samples failed were 3.27 kN, 5.28 kN, 6.56 kN, and 6.57 kN for the impact energies of 6, 30, 70, and 100 J, respectively. This means that in GFRV pipes the failure occurs at increasingly higher impact loads as the incident impact energy levels are increased. Table 1 also show that for GFRV pipes, the average absorbed energy values (absorbed energy = the total penetration energy minus energy at peak load) for impact energies of 6, 30, 70, and 100 J were, 1.81, 1.29, 12.45, and 35.2 J, respectively. The average deflection at peak load also increase with increase in the incident impact energy level. For GFRV pipes these deflection values

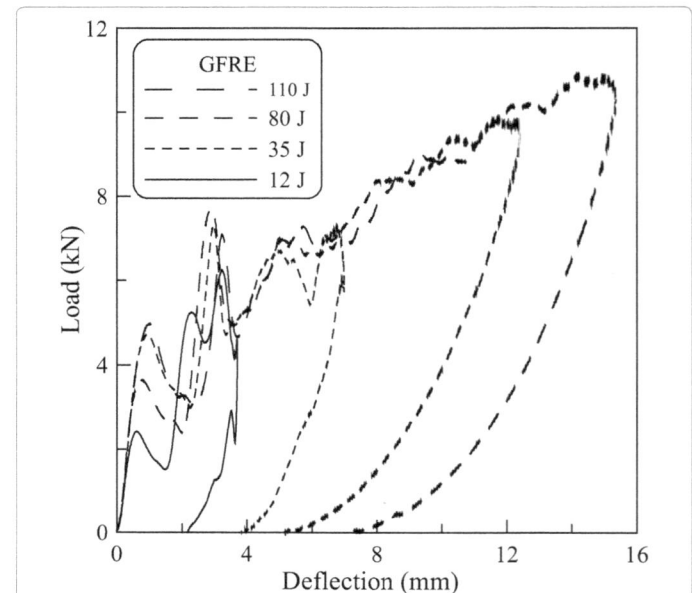

Figure 7: Load -deflection traces for GFRE at impact energies of 12, 35, 80 and 110 J.

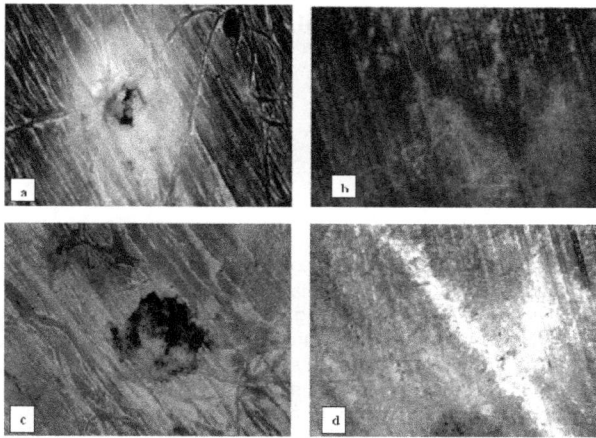

Figure 8: View of the impact damage in a GFRV pipe sample at incident impact energy of 35 J, (a) outer (impacted) surface (b) inner surface.

were 1.68, 7.2, 12.11, and 14.49 mm for incident energy levels of 6, 30, 70, and 100 J, respectively.

The GFRE pipes require higher incident impact energies for damage initiation and propagation. The damage initiation threshhold for GFRE was 12 J and the full penetration was achieved at 110 J as against the corresponding 6 and 100 J for GFRV pipes. Table 2 shows that for GFRE pipes the peak load was independent of the incident impact energy and for all the four energy levels the peak load remained essentially constant at approximately 6.5 kN.

The average absorbed energy for GFRE pipes increased with increase in the incident impact energy and was noted as 7.4, 21.9, 69.8, and 89.3 J for the incident impact energies of 12, 35, 80, and 110 J, respectively. The energy at all peak load values for the GFRE pipes at various impact energy levels were studied and the deflections at peak load values for GFRE pipe were substantially lower than those for the GFRV pipes. GFRV pipes were noted as 3.59, 3.03, 3.09, and 3.59 mm for the incident impact energy levels of 12, 35, 80, and 110 J. The lower the energy of peak load and deflection at peak load values for GFRE pipes in comparison to GFRV pipes indicates brittle nature of the epoxy matrix as compared to the relatively less brittle vinyl ester matrix.

Conclusion

Low velocity impact response of filament-wound Glass Fiber Reinforced/vinylester (GFRV) and Glass Fiber Reinforced/epoxy (GFRE) pipes have been examined using instrumented drop weight testing machine. From the impact response data and damage evaluation the following conclusions can be made like Load-time, energy-time, and load-deflection histories are indicative of the damage initiation and damage propagation. Two distinct responses to impact can be identified. The first response is elastic deformation during which no gross damage takes place. The second response is the initiation and propagation of major damage under elastic and plastic deformation. Energy up to the peak load is dissipated in elastic deformation followed by energy dissipation in major damage initiation and propagation. For GFRV pipes, the peak load increases with increase in the incident impact energy, while for the GFRE pipes the peak load remains essentially constant and could be considered independent of the magnitude of the incident impact energy. The energy to peak load and the deflection at peak load values for GFRE pipes were found to be substantially lower than the values observed for the GFRV pipes. The lower energy to peak

load and deflection at peak load values were indicative of the rather brittle nature of the epoxy matrix as compared to the relatively less brittle vinyl ester matrix.

Acknowledgement

The authors wish to acknowledge King Fahd University of Petroleum & Minerals, Dhahran, Saudi Arabia and Saudi Aramco for funding this research through project ME 2236.

References

1. Xiong DX, Wang TY, Liew JYR (2015) Experimental Investigation on the Behavior of Hollow GFRP Pipes Subjected to Transverse Impact. Advanced Materials Research 1110: 36-39.

2. Safri SNA, Sultan MTH, Yidris N, Mustapha F (2014) Low Velocity and High Velocity Impact Test on Composite Materials-A review. The International Journal of Engineering And Science 3: 50-60.

3. Hui-min D, Xue-feng A, Xiao-su Y, Li Y, Zheng-tao S, et al. (2015) Progress in Research on Low Velocity Impact Properties of Fibre Reinforced Polymer Matrix Composite. Journal of Materials Engineering 43: 89-100.

4. Zhang X (1998) Characterization of Filament wound GRP Pipes under Lateral Quasi-static and Low Velocity Impact Loads. University Of Aberdeen thesis, pp: 1-194.

5. Arif AFM, Malik MH, Al-Omari AS (2014) Impact Resistance of Filament Wound Composite Pipes: A Parametric Study. ASME 3: 1-7.

6. Nahas MN (1987) Radial Impact Strength of Fibre-Reinforce Composite Tubes. J of Mat Science 22: 657-662.

7. Alderson KL, Evans KE (1992) Low velocity transverse impact of filament-wound pipes: Part 1. Damage due to static and impact loads. Composite Structures 20: 37-45.

8. Ambur DR, Starnes JH (1998) Struct., Structural Dynamics, and Materials Conference and Exhibit, and AIAA/ASME/AHS Adaptive Structures Forum, Long Beach, CA.

9. Shyr TW, Pan YH (2003) Impact Resistance and Damage Characteristics Of Composite Laminates. Composite Structures 62: 193-203.

10. Aslan Z, Karakuzu R, Okutan B (2003) The Response of Laminated Composite Plates Under Low-Velocity Impact Loading. Composite Structures 59: 119-127.

11. Gning PB, Tarfaoui M, Collombet F, Riou L, Davies P (2005) Damage Development in Thick Composite Tubes Under Impact Loading and Influence on Implosion Pressure: Experimental Observations. Composites: Part B 36: 306-318.

12. Morais WA, Monteiro SN, Almeida JRM (2005) Effect of the Laminate Thickness on the Composite Strength to Repeated Low Energy Impacts. Composite Structures 70: 223-228.

13. Abrate S (1998) Impact on Composite Structures. Cambridge, Cambridge University Press, UK.

3

Effect of Size, Temperature, and Structure on the Vibrational Heat Capacity of Small Neutral Gold Clusters

Vishwanathan K* and Springborg M

Physical and Theoretical Chemistry, University of Saarland, Germany

Abstract

The vibrational heat capacity Cvib of a re-optimized neutral gold cluster was investigated at temperatures 0.5-300 K. The vibrational frequency of an optimized cluster was revealed by small atomic displacements using a numerical finite-differentiation method. This method was implemented using density-functional tight-binding (DFTB) approach. The desired set of system Eigen frequencies (3N -6) was obtained by diagonalization of the symmetric positive semi definite Hessian matrix. Our investigation revealed that the Cvib curve is strongly influenced by temperature, size, and structure and bond-order dependency. The effect of the range of interatomic forces is studied; especially the lower frequencies make a significant contribution to the heat capacity at low temperatures. In addition to that, we have exactly predicted the vibrational frequencies (ω_i) which occur between 0.55 to 370.72 cm-1, depending on the nanoparticle morphology at T=0 for small neutral gold clusters AuN=3-20. This result has been proved and confirmed by the size effect values. It was found that beside the particle size, geometric shape, defect structure and an increase in asymmetry of nanoparticles effects on heat capacity. Surprisingly, the Boson peaks are typically ascribed to an excess density of vibrational states for the small clusters. Finally, temperature dependencies of the vibrational heat capacities of the re-optimized neutral gold clusters have been studied for the first time.

Keywords: Gold atomic clusters; Density-functional tight-binding (DFTB) approach; Finite-differentiation approximation; Force constants (FCs); Vibrational density of states (VDOS); Vibrational heat capacity (Cvib); Boson Peaks (BP)

Introduction

The study of nanostructured materials exhibiting novel properties is one of the most fascinating fields of current research. Small nanomaterial's are of particular interest because of intriguing characteristics [1-3]. Nanoparticles with smaller dimensions may exhibit different properties in comparison with bulk material. The nanoparticles possess unique physic-chemical, optical and biological proper-ties which can be manipulated suitably for desired applications [4]. The advances in the field of nanoscience and nanotechnology has brought to fore the Nano sized inorganic and organic particles which are finding extensive applications as amendments in medicine and therapeutics, synthetic textiles and food packaging products [5]. The incorporation of engineered nanoparticles into household, personal care, consumer, and industrial products is increasing the exposure of humans and the ecosystems to these materials through production, transportation, storage, use, and disposal. Due to their small sizes, nano-materials (NMs) can enter into cells and interact with cell organelles and/or macromolecules and may thus disrupt the normal cellular functions [6-9]. Various NMs have been observed to show systemic effects when administered into systemic circulation, either intentionally for biomedical therapy or accidentally during environmental exposures, and may even cross the blood/brain barrier [6,8,10].

Clusters are well suited for a rapidly increasing number of applications and they have been an active eld of research for about a quarter of a century. Instead of reviewing all the literature, we refer the reader to the already existing review article by Baletto and Ferrando and also to the book by Wales [11,12]. Clusters can be viewed as solids at the Nano scale; yet molecular cluster chemistry and solid state chemistry have traditionally been considered as separate topics [13]. Nowadays, gold chemistry plays a very important role in Nano electronics and bio-nanoscience [14]. Particularly, gold clusters are of potential relevance to the Nano electronics industry and hence remain the subject of many experimental as well as theoretical studies. They contain edge atoms that have low coordination [15] and can adopt binding geometries that lead to a more reactive electronic structure [16]. Gold in the Nano regime, especially gold Nano crystals have shown size-sensitive reactive properties and are considered to be as promising chemical catalysts [17]. As one of the precious metals, gold has good corrosion resistance and extremely high stability and has been widely researched for biomedical applications [18].

The vibrational properties of clusters and small particles have been studied very intensively [19-26], and are vital for understanding and describing the atomic interactions in the cluster [27-32]. Thermal properties like heat capacity and thermal conductivity as well as many other material properties are strongly influenced by the vibrational density of states (VDOS). For this reason, a better understanding of the rules governing the vibrational properties of nanostructured materials is of high technological and must be given a high priority. The vibrational properties play a major role in structural stability [25,26,33].

The size-dependent properties of metallic clusters are currently of considerable interest, both experimentally [34] and theoretically [24,35]. Although the size effect on specific heat capacity has recently attracted much attention [36], many publications only focus on low temperature [37]. Some of the theoretical studies on the thermodynamic

*Corresponding author: Vishwanathan K, Physical and Theoretical Chemistry, University of Saarland, 66123 Saarbrucken, Germany
E-mail: vishwanathan7@yahoo.com

properties of clusters are based on molecular-dynamic simulations [11] from which the caloric curve, the heat capacity of clusters and the phase transitions can be determined. Nano clusters are interesting because their physical, optical and electronic characteristics are strongly size dependent. Often changing the size by only one atom can significantly alter the physical chemical properties of the system [38]. Many new periodic tables can thus be envisioned classifying differently-sized clusters of the same material as new elements. Potential applications are enormous, ranging from devices in nano-electronics and nano-optics [39] to applications in medicine and materials.

Reyes-Nava et al. [40] calculated the heat capacity for a few Na_N systems using the simple many-body Gupta potential, which approximates the atomic interactions with an analytical description that does not explicitly include electronic degrees of freedom. It was found that solid-liquid phase transitions occurring over a certain temperature range depend critically on the size N of the system. Lee et al. [41] carried out accurate molecular dynamics simulations. From a literature search [42], one can see that only a very few theoretical studies attempt to calculate the thermodynamically properties of the clusters directly through de-termination of the partition function Z. Doye and Calvo calculated the partition function for Lennard-Jones clusters with N ≤ 150 (Figure 1) [43].

In this study we combine numerical finite-difference approach and DFTB method. At T=0, the vibrational frequency of a re-optimized neutral gold cluster is obtained for optimized cluster of $Au_N=3-20$ which was calculated by Dong and Springborg [44]. The desired set of system Eigen frequencies (3N-6) is obtained by a diagonalization of the symmetric positive semi definite Hessian matrix. The effect of the range of interatomic forces has been studied. We found that the lower frequencies made an excellent contribution to the heat capacity (to be able to store significant heat energy) even at low temperatures (Figure 2). This novel and reliable methodology is constructed to explain the essence of the physical picture.

In our case, we are studying the finite-temperature behavior, which is sensitive to the size of the cluster [45] and the recent interest in planar gold clusters [46,47]. For example, according to Y. Dong and M. Springborg's [44] results, the gold clusters with up to N = 6 have a two-dimensional structure, whereas from N=7, the gold clusters form three-dimensional structures. However, in some of the research studies say that for the clusters with N=7-15, the structure can be either 2D or 3D or even both. However, experimental and theoretical studies have found that planar structures are stable up to around 15 atoms [48,49].

In the present work, we shall at first remember our earlier

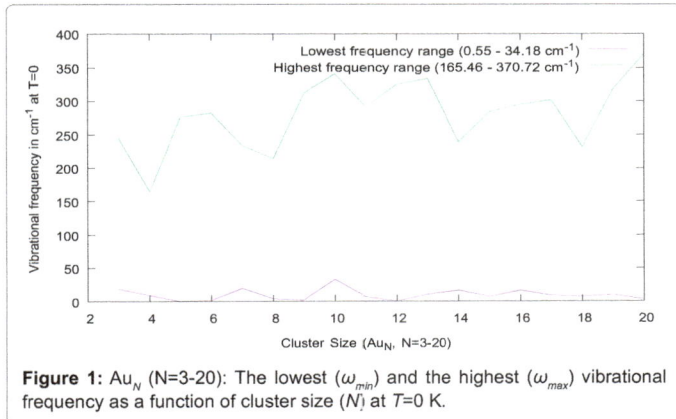

Figure 1: Au_N (N=3-20): The lowest (ω_{min}) and the highest (ω_{max}) vibrational frequency as a function of cluster size (N) at T=0 K.

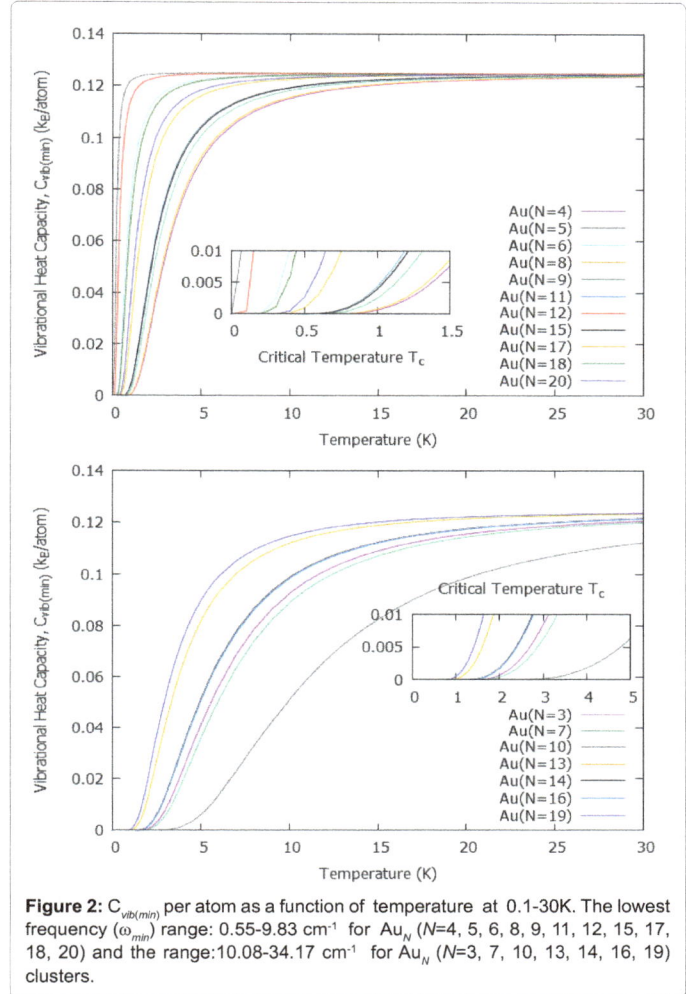

Figure 2: $C_{vib(min)}$ per atom as a function of temperature at 0.1-30K. The lowest frequency (ω_{min}) range: 0.55-9.83 cm^{-1} for Au_N (N=4, 5, 6, 8, 9, 11, 12, 15, 17, 18, 20) and the range:10.08-34.17 cm^{-1} for Au_N (N=3, 7, 10, 13, 14, 16, 19) clusters.

results on Vibrational Heat Capacity of Gold Cluster Au_N=14 at Low Temperatures [50]. As an extension of that work, we shall subsequently present new results devoted to the vibrational contributions to the thermodynamic low-temperature properties of the other clusters. Nevertheless, the purpose of the last part is to explore which kind of information can be obtained by studying the heat capacities of the Au_N=14 cluster and, in particular, to see whether the heat capacities can be correlated to structural and/or energetic properties of the clusters.

Theoretical and Computational Procedure

The DFTB [51-55] is based on the density functional theory of Hohenberg and Kohn in the formulation of Kohn and Sham. In addition, the Kohn-Sham orbitals $\psi_i(r)$ of the system of interest are expanded in terms of atom-centered basis functions $\{\phi_m(r)\}$,

$$\psi_i(r) = \sum_m c_{im}\phi_m(r), \ m = j \tag{1}$$

While so far the variational parameters have been the real-space grid representations of the pseudo wave functions, it will now be the set of coefficients c_{im}. Index m describes the atom, where ϕ_m is centered and it is angular as well as radially dependent. The ϕ_m is determined by self-consistent DFT calculations on isolated atoms using large Slater-type basis sets.

In calculating the orbital energies, we need the Hamilton matrix elements and the overlap matrix elements. The above formula gives the secular equations

$$\sum_m c_{im}(H_{mn} - \in_i S_{mn}) = 0 \tag{2}$$

Here, c_{im}'s are expansion coe cients, $\hat{I}_i$ is for the single-particle energies (or where $\hat{I}_i$ are the Kohn-Sham eigenvalues of the neutral), and the matrix elements of Hamiltonian H_{mn} and the overlap matrix elements S_{mn} are defined as

$$H_{mn} = \langle \phi_m | \hat{H} | \phi_n \rangle, S_{mn} = \langle \phi_m | \phi_n \rangle \tag{3}$$

They depend on the atomic positions and on a well-guessed density $\rho(r)$. By solving the Kohn-Sham equations in an effective one particle potential, the Hamil-tonian $\hat{H}$ is defined as

$$\hat{H}\psi_i(r) = \in_i \psi_i(r), \hat{H} = \hat{T} + V_{eff}(r) \tag{4}$$

To calculate the Hamiltonian matrix, the effective potential V_{eff} has to be approximated. Here, $\hat{T}$ being the kinetic-energy operator $\sum (\hat{T} = -\frac{1}{2}\nabla^2)$ $V_{eff}(r)$ being the effective Kohn-Sham potential, which is approximated as a simple superposition of the potentials of the neutral atoms,

$$V_{eff}(r) = \sum_j V_j^0(|r - R_j|) \tag{5}$$

V_j^0 is the Kohn-Sham potential of a neutral atom, $r_j = r-R_j$ is an atomic position, and R_j being the coordinates of the j-th atom. The short-range interactions can be approximated by simple pair potentials, and the total energy of the compound of interest relative to that of the isolated atoms is then written as,

$$E_{tot} \simeq \sum_i \in_i - \sum_j \sum_{m_j}^{occ} \in_{jm_j} + \frac{1}{2} \sum_{j \neq j'} U_{jj'}(|R_j - R_{j'}|), \in_B \equiv \sum_i^{occ} \in_i - \sum_j \sum_{m_j}^{occ} \in_{jm_j} \tag{6}$$

Here, the majority of the binding energy ($\hat{I}_B$) is contained in the difference between the single-particle energies $_i$ of the system of interest and the single-particle energies $\hat{I}_{jmj}$ of the isolated atoms (atom index j, orbital index m_j), $U_{jj'}(|R_j - R_{j'}|)$ is determined as the difference between $\hat{I}_B$ and $\in_B^{SCF}$ for di-atomic molecules (with $\in_B^{SCF}$ being the total energy from parameter-free density-functional calculations). In the present study, only the 5d and 6s electrons of the gold atoms are explicitly included, whereas the rest are treated within a frozen-core approximation [53,55,56].

Re-optimization and numerical force constants (FCs)

The vibrational frequencies of the gold clusters were calculated within the harmonic approximation by diagonalization of the Hessian matrix. The finite-difference method has been implemented within DFTB approach for our calculation (a finite-difference approximation to calculate the force constants). We found a total energy over those gradients were extended for a small displacement $ds=(\pm 0.01)$ a.u. within the equilibrium coordinates of a previously optimized structure (at $T=0$) by Dong and Springborg [44]. However, the DFTB method has some difficulties to extract the force constants which are most important for our spectrum calculations. Mainly, to get one set of hessian matrix it is necessary to compute two times for both positive and negative gradients. It is a reasonable value and allowed us to discriminate between the translational, rotational motion (Zero-eigenvalues) and the vibrational motion (Non-Zero-eigenvalues).

In our case, we have calculated the numerical first-order derivatives of the forces (F_{ia}; $F_{j\beta}$) instead of the numerical-second-order derivatives of the total energy (E_{tot}). In principle, there is no difference, but numerically the approach of using the forces is more accurate,

$$\frac{1}{M}\frac{\partial^2 E_{tot}}{\partial R_{i\alpha}\partial R_{j\beta}} = \frac{1}{M}\frac{1}{2ds}[\frac{\partial}{\partial R_{i\alpha}}(-F_{j\beta}) + \frac{\partial}{\partial R_{j\beta}}(-F_{i\alpha})] \tag{7}$$

Here, d_s is a differentiation step-size and M represents the atomic mass, for homo-nuclear case. The complete list of these force constants (FCs) is called the Hessian H, which is a $(3N \times 3N)$ matrix. Here, i is the component of $(x, y$ or $z)$ of the force on the j'th atom, so we get $3N$.

Calculation of vibrational heat capacity

Finally, from the calculated vibrational frequencies and with the use of Boltzmann statistics [57-59], we can get the formula to investigate size, structure and temperature effects on the vibrational heat capacity of clusters,

$$C_{vib} = \frac{1}{N}\sum_{i=1}^{3N-6}\frac{\alpha_i^2 e^{\alpha i}}{(e^{\alpha i} - 1)^2}, \text{with } \alpha_i = \frac{h\omega_i}{k_B T} \tag{8}$$

Here, N is the total number of atoms in a cluster, h is the reduced Planck's constant, ω_i is vibrational frequencies (low(min); high(max)), k_B is Boltzmann's constant and T is the absolute temperature. Naturally, zero frequencies are excluded from summation ($3N$-6) in eqn. (8). In this study, we focus only on the vibrational part of the heat capacity, i.e., on C_{vib}.

The Classical Treatment of $C_{trans} + C_{rot} + C_{elec}$: The heat capacity of a system with independent degrees of freedom can be ap-proximated as a sum of its individual contributions,

$$C_{tot} = C_{trans} + C_{rot} + C_{vib} + C_{elec} \tag{9}$$

Here, C_{trans}, C_{rot}, C_{vib} and C_{elec} are called as translation, rotational, vibrational and electronic heat capacity. We have proceeded to calculate the specific heat C_{vib} contribution due to the vibrational energy. In this study, we focus only on the vibrational part of the heat capacity, i.e., on C_{vib}. The other contributions are treated classically. Most often, the electronic excitation energies are much larger than kT and we neglected the electronic partition function.

Results and Discussion

In this article, we present an in-depth study on the behavior of vibrational frequency at $T=0$ K, the heat capacity (C_{vib}) of a re-optimized neutral gold cluster (Au$_N$, N=3-20) at $T=0.5$ to 300 K and the Boson peaks C_{vib}/T^3 at $T=0.5$-30 K. Our results are of special interest since the Boson peaks are thermally activated as well as the local atomic arrangements (within a cluster or group of strongly correlated atoms) but still are constrained by surrounding atoms [60]. This is an interesting point and may shed light on the enhancement of the heat conductance process mentioned in Physical Review Letters 103, 048301 (2009) [61] and PLOS ONE 8, e58770 (2013) [62]. As a function of temperature: at low temperatures the Boson peak corresponds to these local rearrangements with (mutual) reaction of one atom to another [63].

We applied the method outlined above for studying the properties of neutral gold atomic clusters with sizes Au$_N$=3-20 [44]. We analyzed vibrational spectrum and vibrational heat capacity which is being a striking feature and a classical size effect. Our investigation revealed that the C_{vib} curve shows a very strong influence of temperature, size and structure dependency. In addition to that we have also plotted the $C_{vib(min)}$ and $C_{vib(max)}$ curves separately for the lowest (ω_{min}) and the highest (ω_{max}) vibrational frequency modes and then studied their physical behaviors.

We have compared our theoretical results with the theoretical and experimental results calculated by Bishea and Morse [64], Gruene et al., [65] Mancera [66], Molina et al. [67], and with Nose [68], for the vibrational spectrum of $Au_{N=3,4,5,6,7,19,20}$ clusters. Interestingly, our results were in excellent agreement with their results.

The vibrational frequency (ω_l) and properties

Figure 1 shows the low (at the least) and the high (at the most) frequency ranges for cluster $Au_{N=3-20}$. The lowest and the highest frequency ranges are 0.55-34.18 cm^{-1} and 165.46-370.72 cm^{-1}, respectively.

The relative importance of high and low frequencies naturally depends on the size, structure and frequency spectra of the clusters. The size is super critical to the physiochemical properties of Au cluster. It is certainly affected by the ratios of dangling bond to overall bulk bonding number. Nanoparticles have a substantial fraction of their atoms on the surface. The surface energy is the (thermodynamically unfavorable) energy of making dangling bonds' at the surface. Atoms at the surface are under-coordinated, and because breaking bonds results in a loss of energy, surface atoms always have higher energy than atoms in the bulk. It is surely expected that such a vibrational spectrum depends on the material, size, and shape of clusters and nanoparticles [22]. Most importantly, the vibrational properties of atomic clusters are a fingerprint of their structures and can be used to investigate their thermodynamic behavior at low temperatures [23].

Of course, there is an increasing amount literature illustrating the conceptual and practical relevance of two dimensional (2D) systems with long range interactions [69-71]. The vibrational frequency of FCs contributions comes from both the symmetric and the asymmetric stretch and bending modes. We believe that bond-stretching (k_s) and bond-bending (k_b) force constants depend on the nearest neighbor distance obtained from lattice vibrations [72]. It is important to note that the stretching modes moves to lower frequency (approaching zero), and the bending mode b moves to higher frequency. The low frequencies are probably due to the influence of large collective motions of atoms and the high frequencies due to localized motions of atoms [73], which diminish bond length fluctuations. The interatomic interaction that is responsible for the frequency ranges and the variation of the FCs led to a shift in the mode frequencies [74].

The vibrational heat capacity $C_{vib(min)}$ for the lowest frequency ω_{min} at 0.1-30 K

Figure 2 has shown $C_{vib(min)}$ for all the clusters. The frequency of the lowest vibrational state of a nanoparticle is ω_{min}. The asymptotic behavior of heat capacity at low temperatures takes the form with respect to $\alpha_i=\xi_{min}=\omega_{min}=T$ of eqn. (8). The ω_{min} value is determined by the size, shape, and defect structure of the nanoparticle.

Interestingly, for all the clusters the $C_{vib(min)}$ starting at ranges (critical temperature, T_c) are within the temperature range of about $T=Tc=0.1-3$ K. As a result, the effect of the lowest frequency modes is dominant for all the clusters. This is very clear evidence that shape of $C_{vib(min)}$ curves differ with respect to the size and structure of the clusters.

The clusters Au_N (N=4, 5, 6, 8, 9, 11, 12, 15, 17, 18, 20) have extremely low ω_{min} frequencies between 0.55-9.83 cm^{-1}. Which is even below the range of Far Infrared FIR, IR-C 200-10 cm^{-1}? The starting at ranges are within the temperature range of about $T=T_c=0.1-0.7$ K. In the same manner, for the clusters Au_N (N=3, 7, 10, 13, 14, 16, 19) the frequencies are in between 10.08-34.17 cm^{-1}. Which is within the range of Far Infrared FIR, IR-C 200-10 cm^{-1}. Due to this the starting at ranges

are within the temperature range of about $T=T_c=0.7-3.0$ K.

In Figure 2, particularly the cluster Au_{14} (black line) and Au_{16} (blue line) both are very similar due to their ω_{min} values, 17.02 and 17.13 cm^{-1}. However, even though they have different sizes, they have the very same C_s symmetry.

Moreover, the $C_{vib(Min)}$ curve shows the asymptotic behavior, rising smoothly and finally reaching a linear flat at temperatures above $T=15$ K for all Au_N (N=3-20) clusters. This is scientifically significant with respect to the cluster size and shape. The minimum vibrational frequency ω_{min} plays an important role in determining the shape of the C_{vib} curve at low temperature.

Anomalous behavior of neutral gold clusters: In Figure 2, we have noticed some anomalous behavior in the shoulder of the heat capacity $C_{vib(min)}$ curves between the clusters Au_N (N=10-19). And the same behavior is exhibited between the clusters Au_N (N=4-5), which is due to the size and structure de-pendency of the clusters. The data reveal an anomalous contribution to the heat capacity at low temperatures. It was identified that this anomaly in heat capacity is caused by the effect of disorder in the cluster size and structure. Probably, gold nanostructures exhibit anomalous thermal behavior such as the shape transformation of Nano rods, generation of nanoparticles by heating a mesh etc., at extremely low temperature [75]. The results of our work within the numerical model caused us to consider the intrinsic relation between normal (anomalous) particle diffusion and normal (anomalous) heat conduction, along with size dependent thermal conductivity.

The vibrational heat capacity $C_{vib(max)}$ for the highest frequency ω_{max} at 10-1000 K

Figure 3 has shown $C_{vib(max)}$ for all the clusters. The frequency of the highest vibrational state of a nanoparticle is ω_{max}. The asymptotic behavior of heat capacity at low temperatures takes the form with respect to $\alpha_i=\xi_{max}=\omega_{max}/T$ of eqn. (8). The ω_{max} value is determined by the size, shape, and defect structure of the nanoparticle.

Here at $C_{vib(max)}$ curve, we did not see much of a difference on the shoulder (asymptotic curve) that means they are very much closer to each other. Of course, here we do not see anomalous behavior as we have observed in the case of low frequency (Figure 2). The reasons are, the high frequencies do contribute to the heat capacity when the temperature increases from 100 to 300 K ($C_{vib(max)}$ increases from 0.04 $k_B/$ atom to 0.11 $k_{B/}$atom). Most importantly, as this takes place, vibrational modes with lower frequencies give larger contribution to the heat capacity at ambient temperatures. However, the high frequencies do not generate large $C_{vib(max)}$ differences when comparing different cluster sizes (Figures 2 and 3). The increase in heat capacity depends on the particle size, shape and the clusters perfection.

In Figure 3 you can see $C_{vib(max)}$ for all the clusters, and, most importantly, you can see that $C_{vib(max)}$ starting at ranges are within the temperature range of about T=10-37 K, where the effect of the highest frequency modes are dominant for all the clusters. It is evident, that, the starting points varies with respect to the critical temperature T_c from which the $C_{vib(max)}$ curve raises either suddenly or smoothly. The overall temperature ranges are 10-1000 K. The maximum high frequency falls within the range of Mid Infrared MIR, IR-C 3330-200 cm^{-1} for the corresponding harmonic frequency ranges of 165.46-370.72 cm^{-1}, with the one exception of the frequency 165.46 cm^{-1} which is for the Au_N (N=4) cluster.

Comparison: The intensity of the enhancement of the specific heat

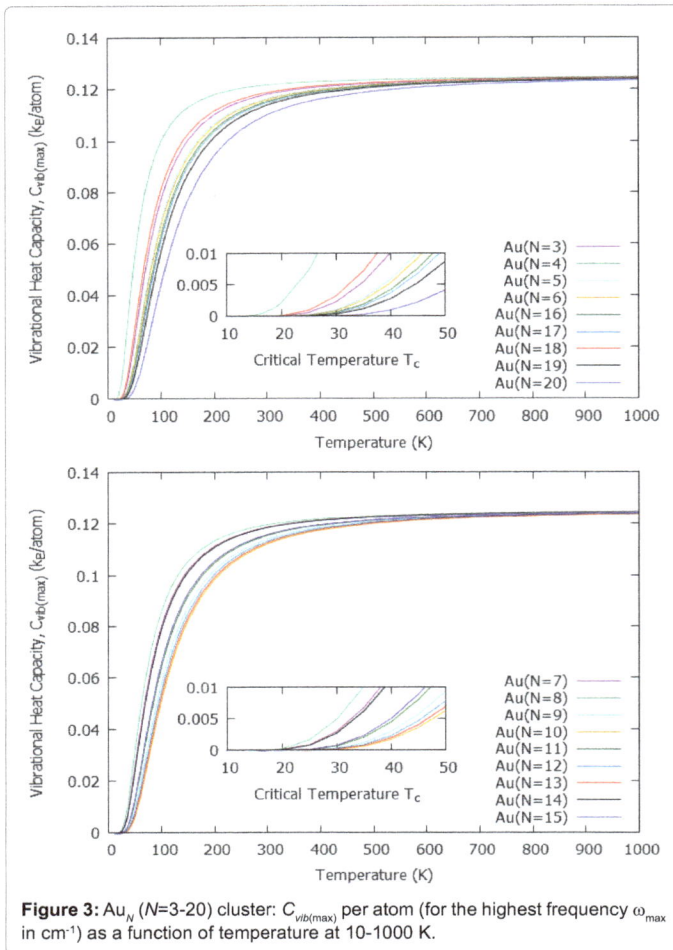

Figure 3: Au_N (N=3-20) cluster: $C_{vib(max)}$ per atom (for the highest frequency ω_{max} in cm^{-1}) as a function of temperature at 10-1000 K.

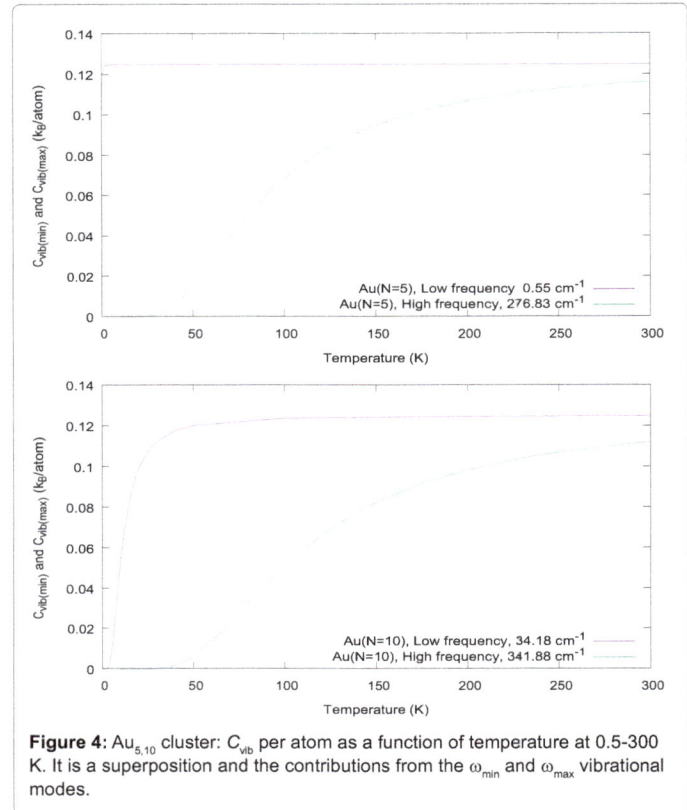

Figure 4: $Au_{5,10}$ cluster: C_{vib} per atom as a function of temperature at 0.5-300 K. It is a superposition and the contributions from the ω_{min} and ω_{max} vibrational modes.

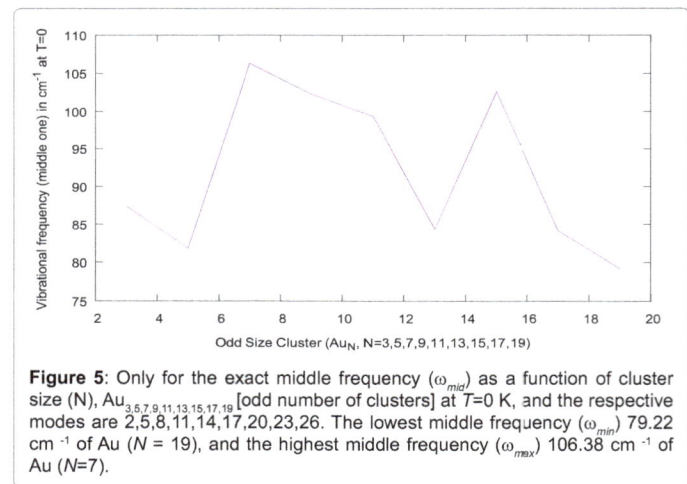

Figure 5: Only for the exact middle frequency (ω_{mid}) as a function of cluster size (N), $Au_{3,5,7,9,11,13,15,17,19}$ [odd number of clusters] at T=0 K, and the respective modes are 2,5,8,11,14,17,20,23,26. The lowest middle frequency (ω_{min}) 79.22 cm^{-1} of Au (N = 19), and the highest middle frequency (ω_{max}) 106.38 cm^{-1} of Au (N=7).

capacity for the lowest vibrational frequency comes in the temperature range of 0.1-15 K (Figure 2) but for the highest vibrational frequency it covers the temperature span 10-500 K (Figure 3). In both of these cases, above these maximum temperatures it gets saturated. Nevertheless, the vibrational heat capacity $C_{vib(min)}$ and $C_{vib(max)}$ is zero at T=0 but there are variations in the at ranges which is only because of the rst mode of the vibrational frequencies, and it nally asymptotes towards 0.125 k_B/atom at high temperatures.

Superposition of the lowest, the middle and the highest frequencies

The ESI † shows the frequency ranges for all the clusters, Au_N (N=3-20). Figures 4-7 shows how C_{vib} varies with cluster size, structure and temperature. These results show that $C_{vib(min)}$ curve for the lowest frequency, by that we mean very low frequency shoulder, varies for different clusters with respect to their size and structure. It is almost identical with the middle $C_{vib(mid)}$ and the high frequency $C_{vib(max)}$ curve. These results will be important to describe heat transfer at Nano scale as well as Casimi-Lifshitz forces between clusters due to thermal quantum fluctuations [76].

The lowest frequency: Figure 4 shows the very beginning of the low and the high frequencies, Au (N=5), 0.55 cm^{-1} and Au (N=10), 34.18 cm^{-1}. However, both the frequency values are only for the lowest frequency range. Here, one of the frequency ranges is not within the Far Region of IR (Far Infrared FIR, IR-C), 200-10 cm^{-1}.

A very interesting observation that Au (N=5) has very high heat capacities ($C_{vib(min)}$= ~ 0.125 k_B/atom at 5 K), whereas Au (N=10) has low heat capacities ($C_{vib(min)} \leq 0.01$ k_B/atom at 5 K). Comparisons of the relative low-frequency modes which were produced at close and distant locations for departures and arrivals at $C_{vib(min)}$. The effects of temperature on the low-frequency vibrational spectrum and local structural arrangements, and the low-frequency density of states distributions reveal that increasingly transverse atoms motions play a dominant role in controlling the band corresponding to the bending or transverse oscillations of the nano-particles at low temperatures.

The exact middle frequency (only possible for the odd sized clusters): Figure 5 shows that the exact middle frequencies for odd number of clusters, $Au_{3,5,7,9,11,13,15,17,19}$, out of which Au ($N=19$), has the lowest middle frequency, 79.22 cm^{-1} and Au ($N=7$), has the highest middle frequency, 106.38 cm^{-1} which is the frequency range and the rest of the odd number of clusters which falls within this range (Figures 5 and 6). Here, the frequency ranges are within the Far Region of IR (Far Infrared FIR, IR-C), 200-10 cm^{-1}.

The highest frequency: In the same manner, Figure 7 shows mainly only the very end of low and high frequencies, Au ($N=4$), 165.46 cm^{-1} and Au ($N=20$), 370.72 cm^{-1}. However, both the frequency values are only for the highest frequency range. Here, the frequency ranges are within the Far Region of IR (Far Infrared FIR, IR-C), 200-10 cm^{-1} and also within the Middle Region of IR (Mid Infrared MIR, IR-C), 3330-200 cm^{-1}. From the Figures 4-7, we certainly confirmed that the low frequency contributions are much higher than the middle and the high frequency contributions to the heat capacity, C_{vib}. Amazingly, C_{vib} curves differ significantly with respect to the low/middle/high frequency as a result of temperature as well as size and structure of the clusters.

The vibrational spectrum (ω_i) of $Au_{N=3}$ and the reliability of our model

Bishea and Morse [64] did research on the spectrum of Au_3. For example, they found that the totally symmetric breathing mode in the excited electronic state had a frequency of 182.9 cm^{-1}. We should expect the totally symmetric breathing mode in the ground state will have a somewhat higher frequency, perhaps around 200-250 cm^{-1}. The gold trimer, Au_3, has three normal modes, two of which may be degenerate, depending on the symmetry. Moreover, additional modes may be in other experiments. In our case, after the re-optimization the vibrational frequencies were found to be 19.21, 87.47 and, 246.21 cm^{-1}.

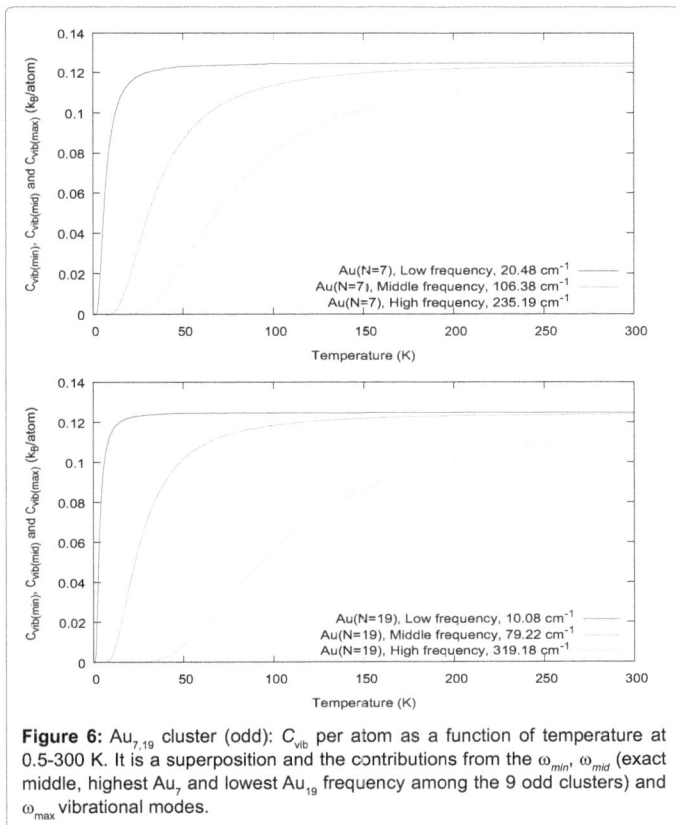

Figure 6: $Au_{7,19}$ cluster (odd): C_{vib} per atom as a function of temperature at 0.5-300 K. It is a superposition and the contributions from the ω_{min}, ω_{mid} (exact middle, highest Au_7 and lowest Au_{19} frequency among the 9 odd clusters) and ω_{max} vibrational modes.

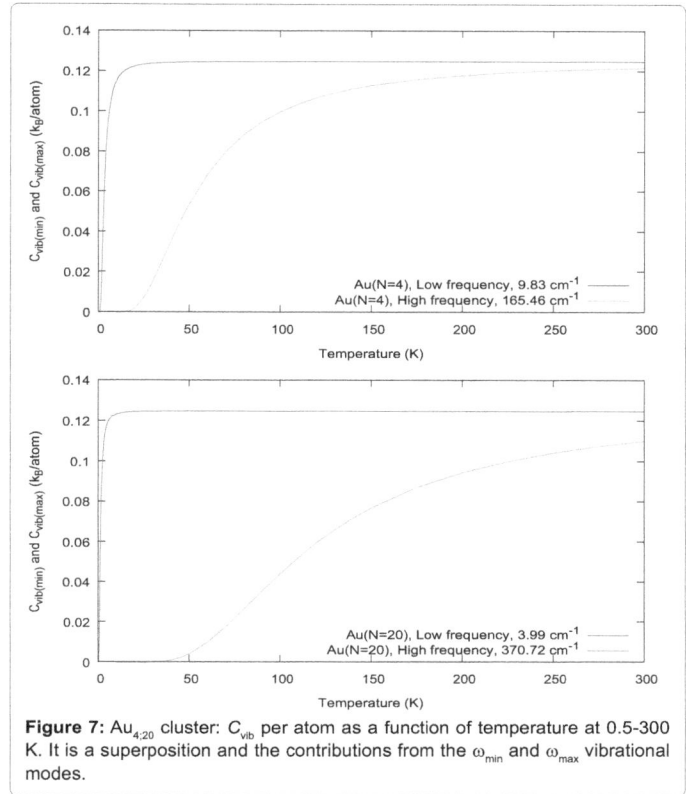

Figure 7: $Au_{4,20}$ cluster: C_{vib} per atom as a function of temperature at 0.5-300 K. It is a superposition and the contributions from the ω_{min} and ω_{max} vibrational modes.

Throughout our calculated vibrational frequencies, we assume that the ground state of Au3 must have either C_{2v} or C_s geometry. We consider the various possible ways that we can arrange the three atoms in Au_3, as the Jahn-Teller distortion will drive it away from the equilateral D_{3h} configuration [77-79]. For example, Au_3 without the spin-orbit coupling (SOC) effects exhibits a Jahn-Teller distortion towards the C_{2v} symmetry; however, with spin-orbit coupling it recovers the D_{3h} symmetry.

An important point: The very low frequency mode 19.21 cm^{-1} (Figure 1), corresponds to a motion that is not well-approximated as a harmonic motion, since motion on the potential energy surface along this coordinate converts the molecule from one equivalent isosceles triangle to another. As we move along this coordinate, the potential energy goes up (initially quadratically), then becomes anharmonic, then reaches a maximum, then descends into a different minimum where a different gold atom lies at the apex of the isosceles triangle. This will have implications for the heat capacity (C_{vib}) (Figure 8), which we have modeled as a pure harmonic oscillator.

With this confirmation of Au_3 vibrational modes, and with the help of Gabedit package [80]. Tolerance for principal axis classi cation: 0.00500 in angstrom (A) and Precision for atom position: 0.09399 in angstrom (A) we went back to the total energies, structures and then verified their symmetry of global structure optimization which were predicted by Dong and Springborg [44]. As a result, some of those Au_n clusters symmetries were differed (For example, Au_3 (D_2), Au_9 (D_{2v}/D_2), Au_{10} (D_2), Au_{11} (C_l), Au_{12} (C_l), Au_{13} (C_s), Au_{15} (C_l), Au_{17} (C_l), Au_{18} (C_2), Au_{19} (C_l), and Au_{20} (C_l)) (Figures 1-6 and 9-14).

The discrepancy at the frequency modes: For the global minimum energy structure, if the $\theta > 90^0$ then the modes are 24.7, 127.0, 183.1 cm^{-1} and if the $\theta > 60^0$ with acute structure the modes are 69.0, 89.3,

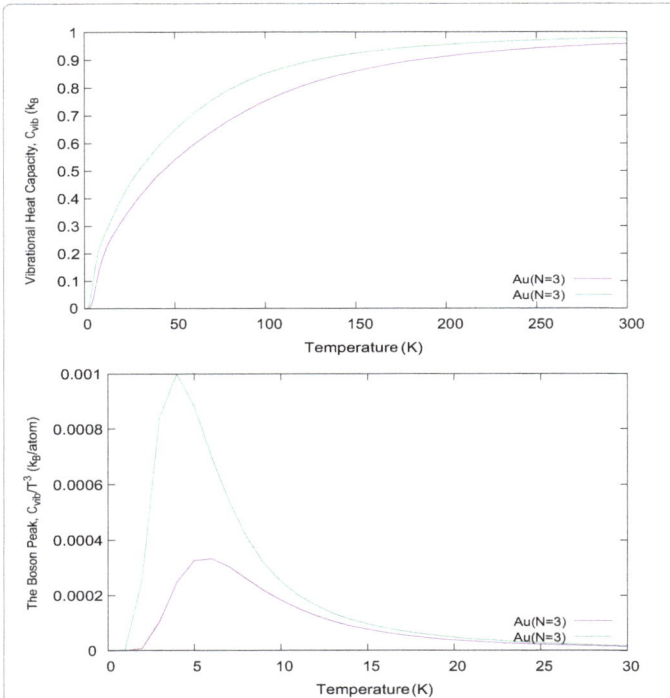

Figure 8: Cluster Au_3: The C_{vib} per atom as a function of temperature at 0.5-300 K. $\alpha_i = \omega/T$ [violet line] and $\alpha_i = \omega i/(0.6950356^* T)$ of eqn. (8) [green line (Aquamarine 4)] and the corresponding Boson peak C_{vib}/T^3 vs. T.

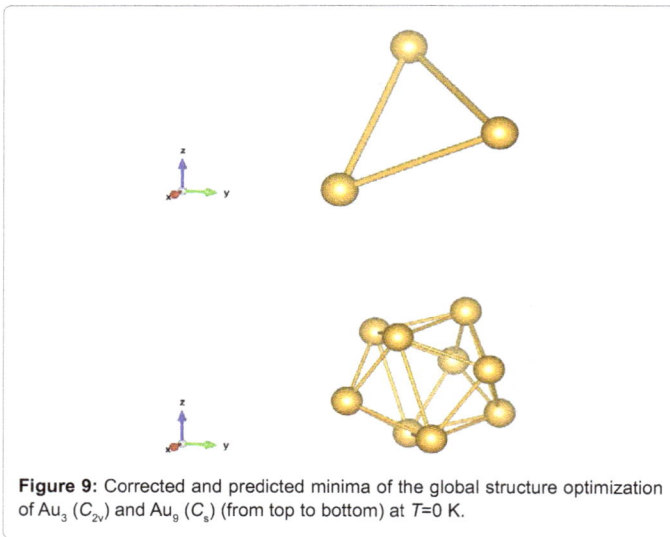

Figure 9: Corrected and predicted minima of the global structure optimization of Au_3 (C_{2v}) and Au_9 (C_s) (from top to bottom) at T=0 K.

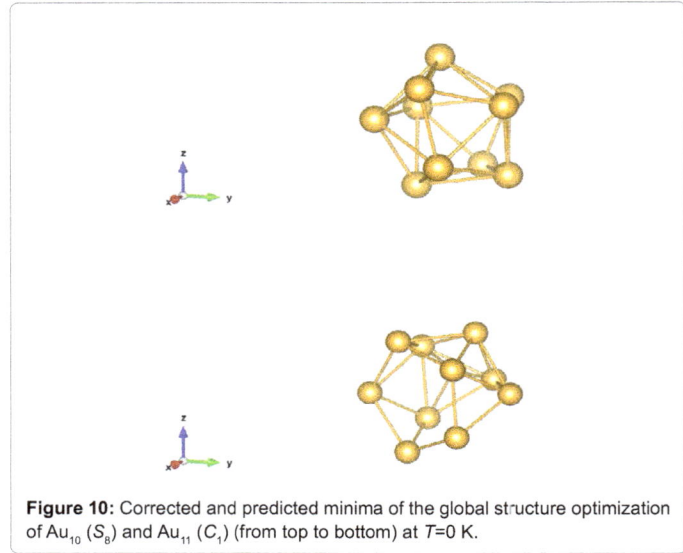

Figure 10: Corrected and predicted minima of the global structure optimization of Au_{10} (S_8) and Au_{11} (C_1) (from top to bottom) at T=0 K.

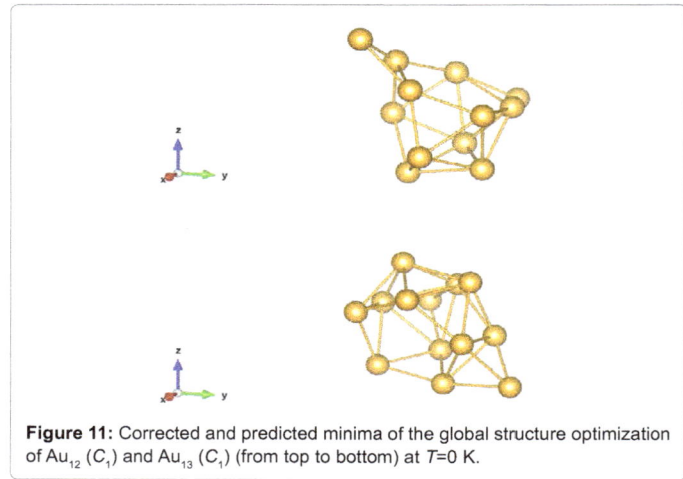

Figure 11: Corrected and predicted minima of the global structure optimization of Au_{12} (C_1) and Au_{13} (C_1) (from top to bottom) at T=0 K.

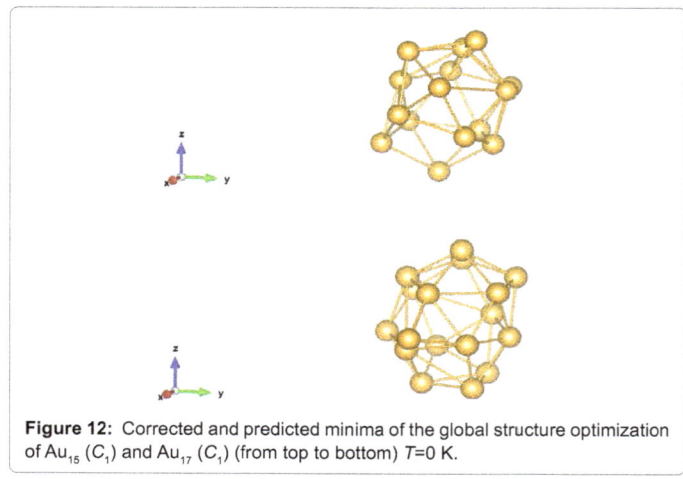

Figure 12: Corrected and predicted minima of the global structure optimization of Au_{15} (C_1) and Au_{17} (C_1) (from top to bottom) T=0 K.

167.1 cm^{-1} again, and if the $<60°$ with acute structure the modes are 5.3, 109.2, 168.6 cm^{-1}, which are Harmonic frequencies for various isomers of Au_3 calculated using PBE/VDB by Mancera and Benoit [66]. So this gives a confirmation that the vibrational modes varies with respect to the bond angle and isomers. Even for the experimental values [64] the modes are 61.9, no value (silent) and 179.9 cm^{-1} as well as, in another experimental calculation [81], no value (silent), 118 and 172 cm^{-1}.

The vibrational heat capacity C_{vib} of gold neutral clusters

Figures 15-17 shows C_{vib} as a function of temperature for di erent cluster sizes. Out of Au_N=3-20 cluster we have selected some of the

14 special clusters which display different interesting properties. The vibrational heat capacity C_{vib} has been plotted with respect to $\alpha_i = \omega_i/T$ and $\alpha_i = \omega_i/(6950356^* T)$ of eqn. (8) at temperature, T=0.5-300 K.

The C_{vib} curve is a sum of C_{vib} curves, one for each normal mode of

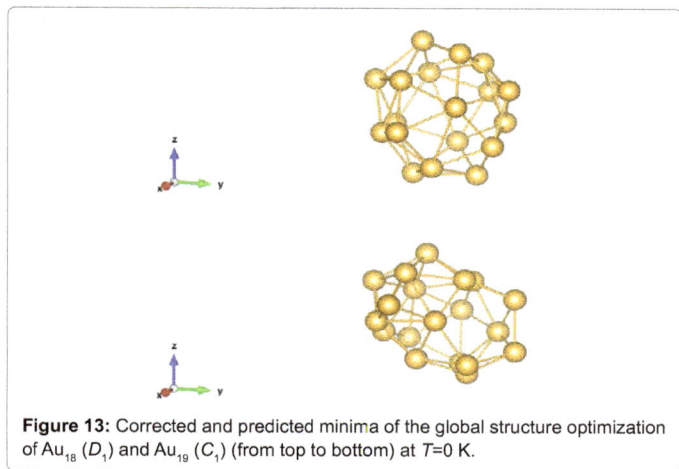

Figure 13: Corrected and predicted minima of the global structure optimization of Au_{18} (D_1) and Au_{19} (C_1) (from top to bottom) at T=0 K.

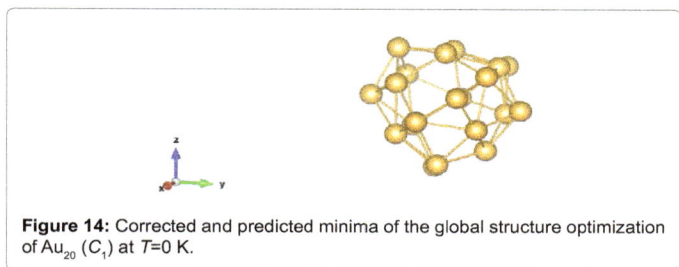

Figure 14: Corrected and predicted minima of the global structure optimization of Au_{20} (C_1) at T=0 K.

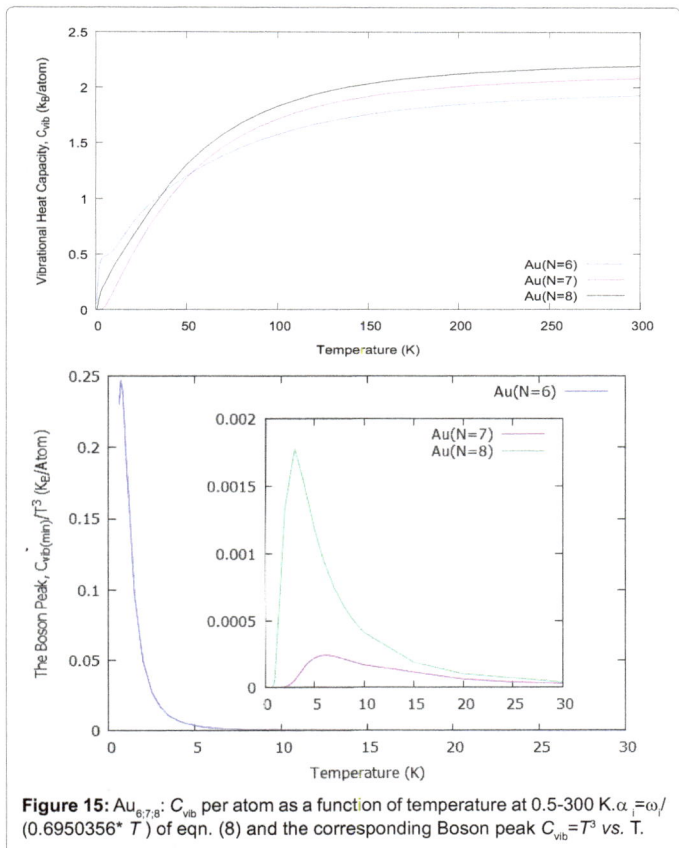

Figure 15: $Au_{6;7;8}$: C_{vib} per atom as a function of temperature at 0.5-300 K. $\alpha_i=\omega_i/$ (0.6950356* T) of eqn. (8) and the corresponding Boson peak C_{vib}=T^3 vs. T.

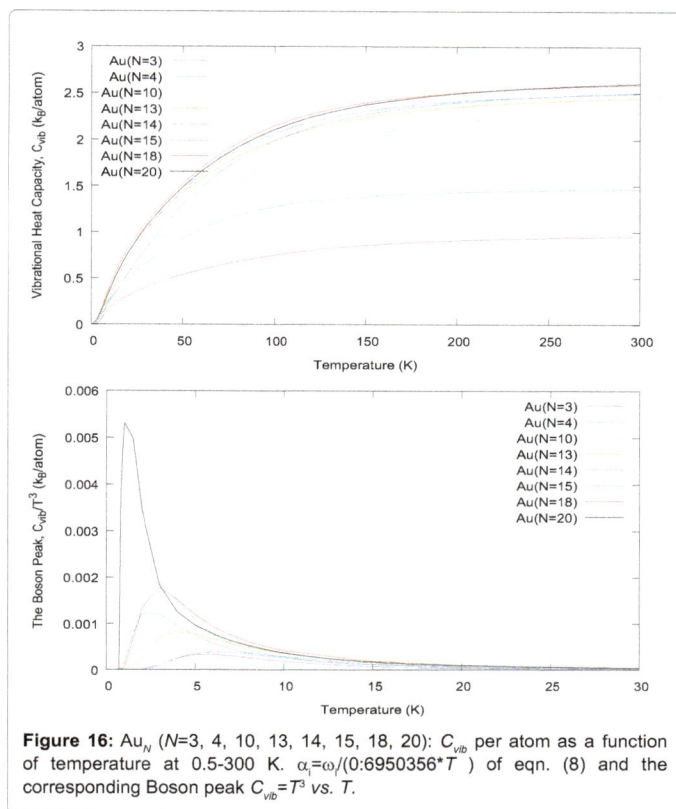

Figure 16: Au_N (N=3, 4, 10, 13, 14, 15, 18, 20): C_{vib} per atom as a function of temperature at 0.5-300 K. $\alpha_i=\omega_i/(0:6950356*T$) of eqn. (8) and the corresponding Boson peak C_{vib}=T^3 vs. T.

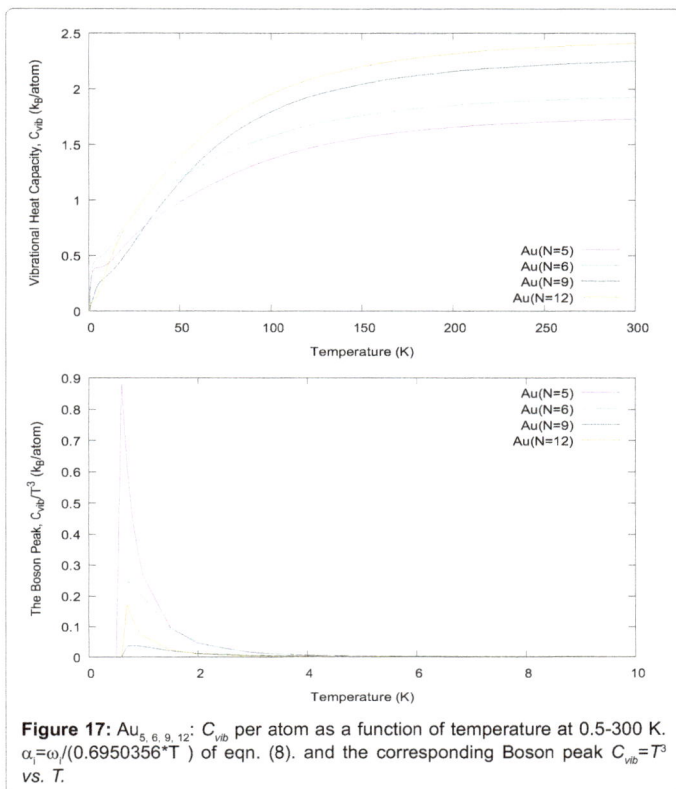

Figure 17: $Au_{5, 6, 9, 12}$: C_{vib} per atom as a function of temperature at 0.5-300 K. $\alpha_i=\omega_i/(0.6950356*T$) of eqn. (8). and the corresponding Boson peak C_{vib}=T^3 vs. T.

vibration. Each of these individual C_{vib} curves is S-shaped, but with the inflection points (where the curve changes from concave up to concave down) occurring at different temperatures. The temperature of the inflection point is proportional to the vibrational frequency for that particular normal mode. Since each individual C_{vib} curve is S-shaped, the sum must be S-shaped (to a degree) also. However, because our molecules/clusters have some normal modes of vibration with extremely low vibrational frequencies, some of the individual curves being summed to give the total C_{vib} curve will have their inflection points at extremely low temperatures. If we examined the C_{vib} curve at extremely low temperatures (perhaps lower than we have calculated, at 0.4 K → 0 K), we would see that they start out at with C_{vib}=0, then rise from C_{vib}=0 (concave up), then go through an inflection point at a low temperature. All of this follows logically from the fact that the overall C_{vib} curve for a single molecule/cluster is a sum of C_{vib} curves for each vibrational mode.

Au_3 (a planar-triangular C_{2v}): In fact, under careful observation, in the C_{vib} curve, there is a minute \Jerk/bending", in between 0.15 and 0.25 k_B/atom at the temperature range about 0.5 to 10 K (Figure 8). Thus, there seems to be a strong correlation between the atom located coordination numbers and the bond angle, which result in a huge variation, due to the fact that the normal mode energy intervals (19.21; 87.47 and; 246.21 cm^{-1}) are so large which has a great influence on the stability of the cluster.

Au_6 (a planar-triangular D_{3h}), Au_7 (a decahedron D_{5h}) and Au_8 (a tetrahedron T_d):

In Figure 15, the C_{vib} curve starts with the temperature at 0.6 K, and the motion of the atom at 10 K, and makes a hump-like shape which is an a fascinating phenomenon to observe. Most importantly, N=6 is a much more rapidly increasing function of T at low temperature than in the cases of the other cluster sizes (but still C_{vib} rises smoothly and reaches the size effect value). However, it has more energy stored at low temperature, but the C_{vib} curve crosses over at about 40 K of the Au (N=8) and 50 K of the Au (N=7) clusters which demonstrates its rapidness. Nevertheless, the main reason could be, the minimum frequency start with degenerate, followed by a double state degenerate, followed again double state degenerate and some single state degenerates and finally end with double state degeneracy, which also provides another confirmation that atoms are located in a periodic and zigzag arrangements with respect to the nearest neighboring atoms coordination. In the case of Au_7 and Au8 both C_{vib} starts at 0.5 K and rises smoothly with asymptotic behavior, and these are the only clusters which have more double state degeneracy and triplet state degeneracy, respectively. So the structural dependency of heat capacity is being confirmed with the triple, double and single state degeneracies.

Interatomic interactions in the clusters reflect a frequency distribution with a high degree of degeneracy due to the high symmetry of their lowest energy configurations (the potential energy with respect to the atomic coordinates). No degeneracy was obtained for the other gold clusters (except Au_N, (N=6; 7; 8)) due to the absence of symmetry in their minimum energy.

Au_4 (D_{2h}); Au_{10} (S_8); Au_{13} (C_1); Au_{15} (C_1); and Au_{20} (C_1): In Figure 16, it is shown that the C_{vib} curve heat capacity goes sharply up from zero, without the small region where the heat capacity is very close to zero. The very low frequency modes begin contributing to the heat capacity at lower temperatures. It is possible to have very low lying excited electronic states that can be thermally excited at low temperatures. If that is the case, there will be an additional contribution to the heat capacity due to electronic excitation. The heat capacity of the real substance would be higher than what we have calculated. However, no experimental data is available for comparison. Surprisingly, for each size C_{vib} there is a monotonously increasing function of T (which tends

asymptotically). The temperature dependence of the individual modes (Figures 4-7) led to the total vibrational heat capacity for all the clusters.

The calculated heat capacity curve remains near zero at the lowest temperatures, then begins monotonically rising near 1.75 K. Similar behavior is reflected in all of the clusters that were studied, although the precise location of the transition from near-zero values to the monotonic increase varies. The existence of this low-temperature, near-zero heat capacity regions arises because all of these clusters have a lowest vibrational frequency. The heat capacity only begins to deviate significantly from zero when the thermal energy, kT, has a significant likelihood of exciting the lowest frequency vibrational mode. This behavior is in effect a quantum size effect, resulting from the sparse vibrational density of states.

In addition at low temperatures it is the lowest-energy (lowest-frequency) vibrational states that mainly contribute to heat capacity. Indeed, if the frequency of the lowest vibrational state of a nanoparticle is at a minimum, the asymptotic behavior of heat capacity at low temperatures arises. This is a region T=0.5-1.75 K where the effect of the lowest frequency internal modes are dominant for all the clusters.

$Au_{14}(C_s)$: In our case, Au_N=14 has C_s symmetry structure and indeed of low symmetry [44]. The vibrational spectrum in structures spans the range from 17.02 up to 240.20 cm^{-1}. The Au_{14} spectrum includes several low-frequency modes. However, the frequencies are not necessarily all distinct due to degenerations state; i.e., some of the roots of the secular equation may occur more than once. In Figure 16, we show the heat capacity C_{vib} as a function of T, for N=14 cluster size. As shown by this figure, at very low temperatures (T→1.75-0.5 K) the C_{vib} curve falls off to zero. Surprisingly, at lower temperatures (1.75-0.5 K) the value of C_{vib} is very small and close to zero. This shows that some of the modes of vibration of the cluster are frequencies which cannot be accessed at low temperatures.

Figure 16 also shows that interatomic vibrations in heat capacity can be neglected at around (1.75-0.5 K) and the C_{vib} curve rises gradually with temperature, which is an exact signature of the vibrational changes. The smooth change, especially pronounced at low temperatures (1.75-25 K), is even more interesting.

It is due to the energy increase of the system. For a given size, the reduction of the heat capacity is more significant at lower temperatures. As the temperature is raised, the difference between the vibrational energies becomes progressively more conspicuous. There can never be any method which explains the difference as long as we retain a simple harmonic oscillator. However, the C_{vib} curve gives a confirmation that is temperature dependent. In the quantum limit, T (1.75- 0.5 K) →0, the C_{vib} curve approaches zero, while in the classical limit, T (300 K) →high temperature →size effect, the curve moves towards a value of C_{vib}=2.50 k_B/atom, which clearly indicates a temperature influenced size effect/dependency.

Au_{14} and Au_{14}: For Au clusters, Koskinen et al., [51] calculated above the ground states of N=11, 12, 13, and 14, respectively, corresponding roughly to T=1000 K. Particularly, they have shown Au_{14}^- as an example to demonstrate the features of a novel liquid-liquid coexistence (LLC), for Au_{14}^- the LLC ends at E_{tot}~2 eV followed by a 3D-liquid-3D-solid coexistence for E_{tot}<1.1 eV corresponding roughly to temperatures around 300 K (Figure 3c) for the partial caloric curves $T(E_{tot})$ of both dimensionalities [51]). The planar clusters form the hot, low-potential energy phase. The 3D liquid phase has the heat capacity of C_{vib}=4.05 kB and the 3D solid phase C_{vib}=3.06 k_B per atom, which are greater than our estimated value C_{vib}=2.50 kB=atom at 300 K*, for neutral cluster Au$^-_{14}$.

Moreover, the expected absolute value C_{vib} should be 2.57 k_B/atom. In our case, the difference is only 0.07 k_B/atom which are reasonable within the numerical accuracy. Nevertheless, C_{vib} has been achieved almost an accurate value, 2.56(389) k_3/atom when the temperature is high enough, 950 K*. But in the Koskinen et al., [51] case, the di erences were higher 1.48 and 0.49 k_B/atom than the expected absolute value for both 3D liquid and 3D solid phases. This discrepancy is due to the minimum energy difference of the anionic [51] and neutral clusters (which are more stable) [44]. So we conclude that the 2D planner structure is preferential over the 3D Structure.

Au$_{18}$(D$_1$): In our case, Au$_{N=18}$ has D$_1$ symmetric structure and indeed of low symmetry [44]. The vibrational spectra of eigenvalues were found in the region between 8.07 and 232.46 cm^{-1}. In a recent study, Bulusu et al., [82] found that hollow gold cages exist for Au with N=16-18 both according to experimental and theoretical results. In the study of Dong and Springborg [44], they found many AuN clusters with N=7-20 to have cage-like structures. We found that the C_{vib} curve of a neutral gold cluster increases smoothly towards the high temperatures and approaches a constant value of about C_{vib}=2.59, in k_B/atom at 300 K*, essentially becoming at.

Moreover, the expected absolute value (for size-dependent) should be C_{vib}=2.67 k_B/atom for neutral cluster Au$_{18}$. In our case, the difference is 0.08 k_B/atom from the expected value. Again, nevertheless, C_{vib} has been achieved almost an accurate value, 2.65(895) k_B/atom when the temperature is high enough, 950 K*. With this we can be sure that the structure can be D1 for Au$_{18}$. However, the C_{vib} curve shows that the heat capacity goes sharply up from zero, excluding the small region where the heat capacity is very close to zero, at T=0.5-1.25 K. The rest of the clusters size effect has been addressed well.

The acoustic vibrations are more important at low temperatures, because they dominate the heat capacity [20-26]. Low-frequency modes of harmonic systems can be related to the small amplitude of acoustic waves, which are experimentally observed in all elastic bodies. This implies that, at low temperatures, the specific heat is largely determined by the low frequency part of the vibrational spectrum and it is only at high temperatures that a substantial portion of the spectrum comes into play.

The vibrational (Phonon) density of states has been calculated by Sauceda and Garzon [25]. Eigenvectors of normal modes are associated with low and high frequency vibrations. This behavior is caused by the stiffening of the bonds (i.e., the frequency shifts to a higher value as the strength increases). Such a local stiffening of the diagonal terms of the FCs matrix is known to lead to the formation of localized oscillation modes [83]. The nature of the bond is most readily interpreted through the eigenvectors and eigenvalues of this FCs Hessian matrix. The most important contribution is a short-range order in the disordered state of the cluster. FCs (stretching and bending modes) matrices, independent of symmetry and magnitudes, are strongly correlated with the bond lengths. A small change in bond length upon disordering can have a large effect on FCs and vibrational entropy. The general increase in the bond angle in this series indicates an increasing repulsion between the bonds and this is consistent with the increasing bond order.

The contribution of the vibrational free energy is related to the disorder in the FCs. In general, gold clusters are not so strongly disordered, having only minor positional disorder. Whereas the zigzag structure (for nonlinear clusters) has lower energy, in contrast,

the zigzag structure within the clusters readily changes into 2D or 3D structures with its nearest neighbors towards better stabilization by multi-coordination [84]. We must remember that each normal mode acts like a simple harmonic oscillator, with a concerted motion of many atoms. The center of mass does not move. All atoms pass through their equilibrium positions simultaneously and normal modes are independent; they do not interact. This means that normal modes do not exchange energy. This is only true in the absence of harmonic terms, which is a theoretical approximation that is never achieved in reality.

For example, if the symmetric stretch is excited, the energy stays in the symmetric stretch. Asymmetry mainly affects the heat capacity at low temperatures; fragmentally both the internal energy and the entropy. Obviously, the C_{vib} curves are fairly different: the lower frequencies make a larger contribution to heat capacity. The absolute heat capacity value depends strongly not only on the size of cluster, but also on their shape and structural ordering.

Au$_5$(C$_{2v}$); Au$_6$(D$_{3h}$); Au$_9$(C$_s$) and Au$_{12}$(C$_1$): The energy gap of vibrational modes, the influence of the edge atoms and its effects (\ hump" shape) in Figure 17. The energy intervals are large in between the modes which vary with respect to the cluster size, structure, symmetry, molar mass and the arrangements of bond orders within the clusters. Particularly, at the end of the last two modes of these following clusters. Au$_5$(C$_{2v}$): Mode 8 (224.53 cm^{-1}) and Mode 9 (276.83 cm^{-1}), Au$_6$(D$_{3h}$): Mode 10 (178.30 cm^{-1}) and degeneracy Modes 11, 12 (282.99 cm^{-1}), Au$_9$(C$_s$): Mode 20 (220.93 cm+) and Mode 21 (313.24 cm^{-1}), Au$_{12}$(C$_1$): Mode 29 (264.76 cm^{-1}) and Mode 30 (325.89 cm^{-1}). In addition to that the degrees of degeneracy also played a major role, due to the fact of motion of the atoms at the temperature 0.5 to 7 K range, the C_{vib} curve shows a hump like shape which is an excellent phenomena on these clusters, Au$_5$, Au$_6$, Au$_9$ and Au$_{12}$ (Figure 17). Thus, we believe that more energy can be stored below 7 K. This also could be the cause of the edge atoms occupying the surface of the clusters.

Thus, in these Au$_N$ (N=3-20) cases, frequencies are low and the curve is squeezed so that this point on the curve falls at a low temperature. If the frequencies are much higher, the same point on the curve will fall at a much higher temperature. The heat capacity remains significantly above zero at lower temperatures. This is because the neutral gold cluster has much lower vibrational frequencies than most molecules have. Nevertheless, our investigation revealed, the vibrational heat capacity curve shows a very strong influence of size, temperature, structure and bond-length (stretching, bending) dependency (Figure 18).

The Boson peak (BP) C_{vib}/T³ vs. T of the neutral gold clusters: As shown

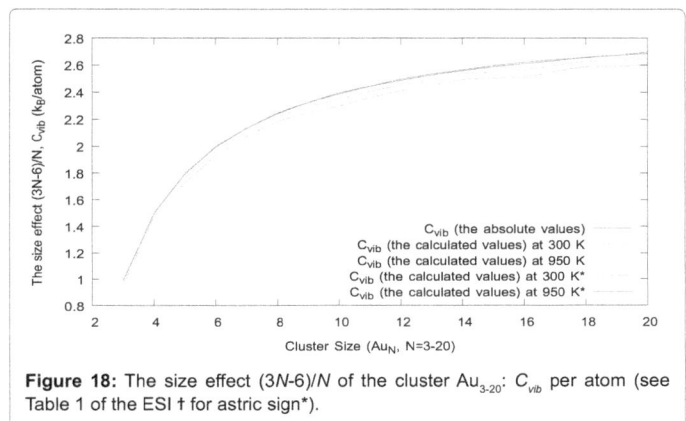

Figure 18: The size effect (3N-6)/N of the cluster Au$_{3-20}$: C_{vib} per atom (see Table 1 of the ESI † for astric sign*).

in Figures 8, 15-17, with respect to the eqn. (8) and the corresponding Boson peak C_{vib}/T^3 vs. T were plotted at temperature within the range T=0.5-30 K. Truly, we are surprised by our observation of a Boson peak in our nanoparticles. Hao Zhang and Jack F. Douglas [85,86] usually study the Boson peak from the velocity autocorrelation function, but they are both concerned about a \excess" contribution to the vibrational density of states. This feature has been observed in metal nanoparticles and zeolites and attributed to the coordinated harmonic motions of groups of atoms in the boundary region of the particle. There are similarities here to a glass because the surface of a nanoparticle has many features in common with this class of materials. In cases where the modes have been resolved, the Boson peak has corresponded to a ring of oscillating particles, the relatively low mass being related to the relatively high mass of these modes. Farrusseng and Tuel [87] also studied the perspectives on zeolite-encapsulated metal nanoparticles and their applications in catalysis.

One of the universal features of disordered glasses is the "Boson peak", which is observed in neutron and Raman scattering experiments. The Boson peak is typically ascribed to an excess density of vibrational states. Shintani and Tanaka [88] studied the nature of the boson peak, using numerical simulations of several glass-forming systems. They have discovered evidence suggestive of the equality of the Boson peak frequency to the Io e-Regel limit for "transverse" phonons, above which transverse phonons no longer propagate. Their results indicate a possibility that the origin of the Boson peak is transverse vibrational modes associated with defective soft structures in the disordered state. Furthermore, they suggest a possible link between slow structural relaxation and fast Boson peak dynamics in glass-forming systems.

However, Malinovsky and Sokolov [89] found that the form of a low-frequency Boson peak in Raman scattering is universal for glasses of varying chemical composition. They have shown that from the shape of the Boson peak one can determine the structural correlation function, i.e., the character of violations of ordered arrangement of atoms within several coordination spheres in non-crystalline solids. Most importantly, very recently Milkus and Zaccone found [90] that bond-orientation order is not so important for the boson peak. Whereas a much more important parameter is the local breaking of inversion symmetry.

Indeed, there is similar behavior of heat capacity in glasses and clusters at low temperatures. Glasses have some distribution of interatomic distances and modification of atomic coordination induced by disorder, compared to periodic crystals [91]. In clusters or nanoparticles, it is caused by the reduced atomic coordination of the surface atoms. Because the ratio of surface to volume is large, the number of atoms with reduced coordination is significant. The vibrations of surface atoms enhance the VDOS at low energies and the heat capacity increases.

The Boson peak is usually studied in glasses, where enhancement of heat capacity is induced by disorder. The modeling of clusters may be important for understanding the mechanism which lead to this effect. In glasses, there are different local con gurations of atoms that may be simulated by isolated clusters. In particular, it was found that the vibrational density of states (VDOS) exhibit an excessively low-frequency contribution. A corresponding low-temperature peak is observed in the temperature dependence of the specific heat if plotted as $C_{vib}(T)/T^3$.

$Au_3(C_{2v})$: Figure 8 shows that the increase in Boson peak curve $C_{vib}=T^3$ from 0-0.00035 k_B/atom [violet line] for $_i=\omega_i/T$ and 0-0.001 k_B/

atom [green line (Aqua-marine 4)] for $\alpha_i =_i /(0:6950356^*T)$ found for T→0→high temperature; 30K, and the maximum deviation at 6.00 K and 4.9 K, respectively, are due to a strong disorderly nature in the cluster. In addition to that conformed with the parameter dependency.

$Au_5(C_{2v})$; $Au_6(D_{3h})$; $Au_9(C_s)$ and $Au_{12}(C_1)$: Figure 17 shows that the calculated C_{vib}/T^3 vs. T for Au_N (N=5; 6; 9; 12) in the temperature 0.5-10 K range. The increase in Boson peak C_{vib}/T^3 from 0.88503 k_B/atom (the maximum deviation with higher amplitude at 0.6 K, Au_5. Nevertheless, with the lesser amplitude the deviation occurs at 0.6 at 0.6 K for the other clusters, Au_6, Au_9 and Au_{12}) found for T →0 to higher temperature, the rest of the clusters are in the interval between 0.5 to 4 K.

Howsoever, Figure 16 also shows the calculated C_{vib}/T^3 vs. T for Au_N (N = 10; 14; 18; 20) in the temperature 0.5-30 K range. The detailed confirmation is given below.

$Au_{10}(S_8)$: There is strong evidence that the maximum deviation of the Boson peak C_{vib}/T^3 is 6.88469×10^{-5} k_B/atom at 12 K, is lesser amplitude than the all other clusters, which is only because of their large number of high frequency modes.

$Au_{14}(C_s)$: Based on the experimental observations, the maximum in silica glass is placed at about 10 K and 8.5 K for neutral gold cluster (Figure 16 and Richet et al., [92,93]), and the maximum deviation C_{vib}/T^3 attains $2,4 \times 10^{-4}$ $J/(molK)$ for silica glass and 0.0004 k_B/atom for gold cluster, respectively. Please notice that the units are not the same.

$Au_{18}(D_1)$: The C_{vib}/T^3 deviation is 0.00178 k_B/atom found for T →3.7 K temperature. It shows the disorder nature of the cluster.

$Au_{20}(C_1)$: The C_{vib}/T^3 deviation with a higher amplitude and the intervals in between 5.31×10^{-3}-4.97×10^{-3} k_B/atom at 1-1.5 K, which has a large number of high frequencies modes.

To be noticed: Let us remember the Figure 2, and the vibrational heat capacity $C_{vib(min)}$ at above 30 K, that at high temperatures it gets saturated 0.12 k_B/atom (as a linear). Similarly, in the Boson peak, C_{vib}/T^3 also gets saturated (a linear) at above 30 K, but here it is towards zero values (0 k_B/atom). The lower frequencies certainly make a larger contribution to heat capacity. Nevertheless, the Boson peaks are highly visible, i.e., and the strength of the peaks strongly depends on the atomic coordination number. There are several origins, most of them related to the thermal fluctuations. If the atomic coordination is low, a single negative force constant renders the atomic arrangement much closer to an unstable situation than in the highly coordinated case [60].

The size effect of the gold neutral cluster Au_N=3-20

Interestingly, from the Figure 18, the clusters Au_{16} and Au_{20} possess the most significant and stable cluster of all other clusters which have C_s and C_1 symmetry (comparable with Au20 tetragonal cluster). In fact, the cluster Au16 has a greater number of single state degeneracy modes.

Au_{16}: For the size effect at about temperature 950/950* K, the C_{vib} curve reached 2.62/2.61 k_B/atom but for the absolute value it should be 2.63 k_B/atom. So there is a minute difference on both C_{vib} values which is 0.01/0.02 k_B/atom.

Au_{20}: For the size effect at about temperature 950/950* K, the C_{vib} curve reached 2.70/2.69 $k_B/atom$. However, here the absolute value should be 2.70 k_B/atom. However, only at $T=950^*K$, there is a minute difference 0.01 $k_B/atom$.

However, this minute differences are also clearly seen at temperature 300/300 K* for the rest of the clusters. The size effect on the heat capacity should be very sensitive to the accuracy because the absolute differences are small as shown in Figure 18.

Physical parameters influence at the C_{vib} curves and its shapes

If one carefully observes in Figures 2-8 and 15-17 they shows very clearly that the shape of the C_{vib} curves are not only dependent upon the low and high frequency ranges, but also with respect to the physical parameters (h, k_B). However, along with this we have noticed that the size dependency values of C_{vib} vary a little bit. Nevertheless, with high temperature at 950 K/950 K*, our calculated C_{vib} values are overlapping with the absolute values of C_{vib}.

This confirms the accuracy of the size dependency (Figure 18, violet, blue and red line curve). Nevertheless, near to the room temperature at about 300 K/300 K*, the C_{vib} values are still very close to the absolute values.

Most importantly, at both the above mentioned temperatures, smoothness and the asymptotic behavior of heat capacity C_{vib} for $\alpha_i = \omega_i/T$, looks like much better than $\alpha_i = \omega_i / (0.6950356*T)$ of eqn. (8) (Figures 2, 3 and 8). In addition to that, the changes at the shape of the C_{vib} curves are only due to existing some single state modes, but many double or triple state degree of degeneracies modes, and they are the contributing factors at C_{vib}, see for example, cluster sizes $Au_{6,7,8}$ at T=0 K (Figure 19). So with this in view, we can confidently say that our new methodology is a trustworthy and a successful model, with which one can take one further in making more accurate experimental calculations.

Comparison with the theoretical and experimental results

The spectral frequencies are the most recent studies covering neutral clusters are reported using far-infrared multiphoton-dissociation

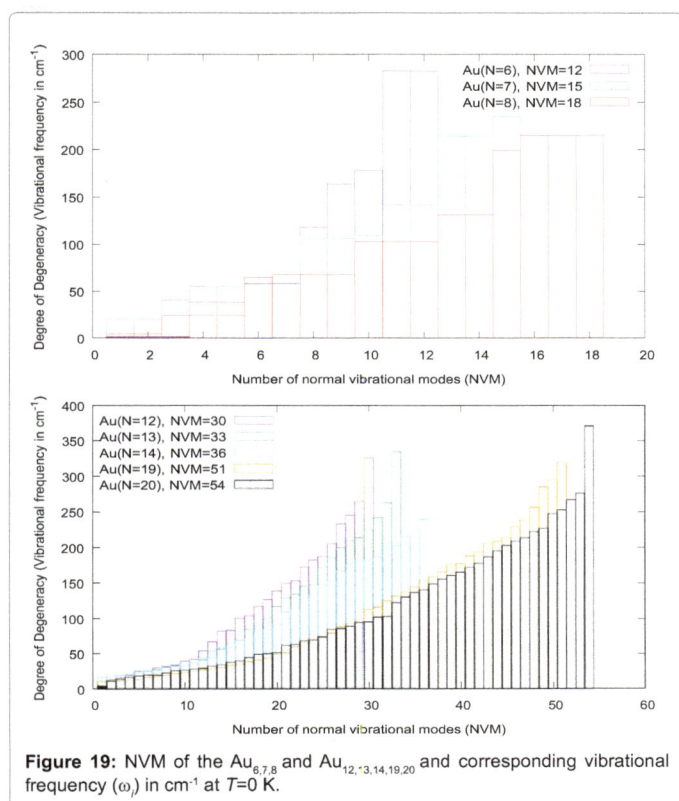

Figure 19: NVM of the $Au_{6,7,8}$ and $Au_{12,13,14,19,20}$ and corresponding vibrational frequency (ω_i) in cm⁻¹ at T=0 K.

spectroscopy (FIR-MPD) for $Au_{N=3,4}$ [94] and for $Au_{N=7,19\ 20}$ [65]. In closing we would like to comment on a trend in the experimentally observed low-frequency vibrational spectra which was realized recently by Gruene et al. [65]. This work experimentally investigated neutral gold clusters ($Au_{N=7}$, $Au_{N=19}$ and $Au_{N=20}$) in the gas phase by means of vibrational spectroscopy, which is inherently sensitive to structure. However, before they performed experiments, they used Gaussian package with DFT calculation.

The gold cluster, $Au_{N=7,19,20}$ has 15, 51, 54 normal vibrational modes (NVM) respectively, some of which may be degenerate, depending on the symmetry. More-over, some modes may appear in one or another experiment. We calculated the normal modes based on the structures which were predicted by Dong and Springborg [44]. However, after re-optimization the vibrational frequencies were found as mentioned in the Figures 2-19. Our calculated spectrum ranges are the peculiar values, which are in excellent agreement with the theoretical (DFT) and experimental results calculated for gold clusters by Gruene et al. [65]. Their visible modes at IR absorption coefficients, and the (ω_i) frequencies are Au_{20} (148 cm⁻¹), Au_{19} (149 and 166 cm⁻¹), and for Au_7 (165, 186 and 201 cm⁻¹). These values are in excellent agreement with our calculated value of Au_{20} (148.64 cm⁻¹) and for Au_{19} (within 144.15, 152.87 158.47, 165.28 cm⁻¹) (Figure 19).

However, they have only predicted vibrational frequencies in between 47 and 220 cm⁻¹ wavenumbers. We have calculated even lower (3.99 cm⁻¹) and higher (370.72 cm⁻¹) wavenumbers than those. This is because in infrared (IR) absorption spectroscopy the number of allowed transitions is restricted by selection rules, and thus directly reflects the symmetry of the particle.

Overall for cluster $Au_{N=3-20}$, certainly and in particularly, our lowest minimum frequency value of Au_8 cluster is 4.66 cm⁻¹ closer to the harmonic frequency of the Au8 cluster which was calculated by Mancera and Benoit [66]. Their harmonic frequency is 6.9 cm⁻¹, as a minimum. Additionally, our highest maximum frequency values are comparable with Nose S calculated (ω_i=221 cm⁻¹) as a highest frequency for Au20 Tetrahedron symmetry [68] using Molecular Dynamics Simulations.

Our calculated $Au_{20}(C_1)$ cluster's vibrational modes are comparable and in excellent agreement with the Au20 Tetrahedron (Td) symmetry calculated using density functional theory (DFT) with two different approximations: the hybrid-B3LYP and GGA-BP86 by Molina et al. [67]. Their ranges for BLYP and BP86 are 26.42-162.22 cm⁻¹ (BLYP) and 29.13-172.35 cm+ (BP86), respectively. Certainly, the most stable structure for Au20 was found through frequency calculations for the (T_d) Au_{20} structure by Jun Li, Xi Li, Hua-Jin Zhai, and Lai-Sheng Wang [95,96].

The harmonic frequencies (ω_i) for the global minimum energy structure of $Au_{4,5}$ calculated using PBE/VDB, in cm⁻¹ are being reported by Mancera and Benoit [66], which are in good agreement with some of our calculated normal modes of Au_4 (a planar rhombus D_{2h}): 165.46 (167.2) cm⁻¹; 147.12 (152.2) cm⁻¹; 98.00 (96.5) cm⁻¹ and 31.63 (34.5) cm⁻¹, as well as, Au_5 (a planar trapezoid C_{2v}): 132.74 (137.5) cm⁻¹; 47.70 (47.6) cm⁻¹ and 43.25 (47.6) cm⁻¹.

Moreover, our verified symmetry and calculated vibrational frequency modes of $Au_{N=3-6}$ clusters are almost in complete agreement with Mancera and Benoit [66].

$Au_{N=6-8}$: Double and Triple state degeneracy at T= 0

These clusters are unique among the other clusters, due to their nature. Very strong and visible evidence of the vibrational heat capacity

C_{vib} per atom as a function of temperature is shown in Figure 15.

Au$_{N=6}$ (D$_{3h}$): The harmonic frequencies for the Au$_6$ cluster are in the range of 2.44 cm^{-1} to 282.99 cm^{-1}, and there are 3 doubled and 1 tripled degenerate frequencies. 12 normal modes are represented as the following, one mode with degeneracy 3, three modes with degeneracy 2, and three single modes with non-degeneracy.

Au$_{N=7}$ (D$_{5h}$): The harmonic frequencies for the Au$_7$ cluster the range of 20.48 cm^{-1} to 235.19 cm^{-1}, and there are only 6 doubly degenerate frequencies, which is an exceptional case of all other clusters. Out of 15 normal modes, 12 of these modes are twofold-degenerate and the remaining 3 are non-degenerate.

Au$_N$=8 (T$_d$): The harmonic frequencies for the Au$_8$ cluster are in the range of 4.66 cm^{-1} to 215.08 cm^{-1}, there are 2 doubly and 4 triply degenerate frequencies. Out of 18 modes, 4 threefold, 2 double fold and 2 non-degenerate normal frequencies occurred.

Conclusions with Perspective

We have extracted vibrational frequencies at T=0 K, and investigated the vibrational heat capacities at a temperature range 0.5-300 K, of the re-optimized gold atomic clusters (Au$_N$=3-20). Especially, the vibrational modes with lower frequencies are given a significant contribution to the heat capacity at low temperatures. The C_{vib} is a monotonic function that tends asymptotically. The vibrational heat capacity is a strong function of cluster size and temperature, particularly in the low temperature regime.

Our first step towards the understanding of the so-called finite temperature effects was to take into account the vibrations of clusters. We focused our interest on the cluster size and temperature dependence of the heat capacities and the free energies of the clusters. Finally, we have studied vibrational and thermodynamic properties of the clusters in order to explore the interaction between stability, structure, and heat capacities of clusters. Our approach is worthy of further investigation and would pave a way in realizing numerical values which would allow for an experimental vibrational spectrum and heat capacity, which would prove crucial in development of Nano electronic devices. Nevertheless, our work gives a possible cause for the size, temperature, and structures effect of Au atomic clusters.

Supplementary Material

Details about the harmonic frequencies and the predicted minima of the global structure optimization of Au clusters can be found in the supplementary material (see the ESI†).

Acknowledgements

A part of this work was supported by the German Research Council (DFG) through project Sp 439/23-1. We gratefully acknowledge their very generous support.

References

1. RHM Smit et al. (2001) Physical Review Letters 87: 266102.

2. Rodrigues V, Bettini J, Silva PC, Ugarte D (2003) Physical Review Letters 91: 096801.

3. Young Cheol Choi, Han Myoung Lee, Woo Youn Kim, Kwon SK, Tashi Nautiyal, et al. (2007) How Can We Make Stable Linear Monoatomic Chains? Gold-Cesium Binary Subnanowires as an Example of a Charge-Transfer-Driven Approach to Alloying. Physical Review Letters 98: 076101.

4. Feynman RP (1991) There's plenty of room at the bottom. Science 254: 1300-1301.

5. Gajjar P, Pettee B, Britt DW, Huang W, Johnson WP, et al. (2009) Antimicrobial activities of commercial nanoparticles against an environmental soil microbe, Pseudomonas putida KT2440. Journal of Biological Engineering 3: 9-22.

6. Nel A, Xia T, Madler L, Li N (2006) Toxic potential of materials at the nanolevel. Science 311: 622-627.

7. Donaldson K, Stone V, Clouter A, Renwick L, MacNee W (2001) Ultra-fine particles. Occupational and Environmental Medicine 58: 211-216.

8. Oberdorster G, Oberdorster E, Oberdorster J (2005) Nanotoxicology: an emerging discipline evolving from studies of ultra ne particles. Environmental Health Perspectives 113: 823-839.

9. Service RF (2004) Nanotechnology grows up. Science 304: 1732-1734.

10. Borm PJA, Kreyling W (2004) Toxicological hazards of inhaled nanoparticles-potential implications for drug delivery. Journal of Nanoscience and Nanotechnology 4: 521-531.

11. Baletto F, Ferrando R (2005) Structural properties of nanoclusters: Energetic, thermodynamic and kinetic effects. Reviews of modern physics 77: 371-421.

12. Wales DJ (2003) Energy Landscapes with Applications to Clusters, Biomolecules and Glasses. Cambridge University, England.

13. Fehlner T, Halet JF, Saillard JY (2007) Molecular clusters: a bridge to solid-state chemistry. Cambridge University Press.

14. Pyykkö P (1997) Strong closed-shell interactions in norganic chemistry. Chemical reviews 97: 597-636.

15. Lemire C, Meyer R, Shaikhutdinov S, Freund HJ (2004) Surface chemistry of catalysis by gold. Gold Bulletin, 37(1-2), 72-124.

16. Mills G, Gordon MS, Metiu H (2003) Catalysis by Nanostructures: Methane, Ethylene Oxide, and Propylene Oxide Synthesis on Ag, Cu or Au nanoclusters. J. Chem. Phys. 118: 419.

17. Valden M, Lai X, Goodman DW (1998) Onset of catalytic activity of gold clusters on titania with the appearance of nonmetallic properties. Science 281: 1647-1650.

18. Daniel MC, Astruc D (2004) Gold nanoparticles: assembly, supramolecular chemistry, quantum-size-related properties, and applications toward biology, catalysis, and nanotechnology. Chemical reviews, 104: 293-346.

19. Garzón IL, Posada-Amarillas A (1996) Structural and vibrational analysis of amorphous Au 55 clusters. Physical Review B 54: 11796.

20. Bravo-Pérez G, Garzón IL, Novaro O (1999) Ab initio study of small gold clusters. Journal of Molecular Structure: THEOCHEM 493: 225-231.

21. Bravo-Pérez G, Garzón IL, Novaro O (1999) Non-additive effects in small gold clusters. Chemical physics letters 313: 655-664.

22. Sauceda HE, Mongin D, Maioli P, Crut A, Pellarin M, et al. (2012) Vibrational properties of metal nanoparticles: Atomistic simulation and comparison with time-resolved investigation. The Journal of Physical Chemistry C 116: 25147-25156.

23. Sauceda HE, Pelayo JJ, Salazar F, Pérez LA, Garzón IL (2013) Vibrational Spectrum, Caloric Curve, Low-Temperature Heat Capacity, and Debye Temperature of Sodium Clusters: The Na139+ Case. The Journal of Physical Chemistry C 117: 11393-11398.

24. Sauceda HE, Salazar F, Pérez LA, Garzón IL (2013) Size and shape dependence of the vibrational spectrum and low-temperature specific heat of Au nanoparticles. The Journal of Physical Chemistry C 117: 25160-25168.

25. Sauceda HE, Garzón IL (2014) Structural determination of metal nanoparticles from their vibrational (phonon) density of states. The Journal of Physical Chemistry C 119: 10876-10880.

26. Sauceda HE, Garzon IL (2015) Vibrational properties and speci c heat of core-shell Ag-Au icosahedral nanoparticles. Physical Chemistry Chemical Physics 17: 28054.

27. Wales DJ (2013) Surveying a complex potential energy landscape: Overcoming broken ergodicity using basin-sampling. Chemical Physics Letters 584: 1-9.

28. Ballard AJ, Martiniani S, Stevenson JD, Somani S, Wales DJ (2015) Wiley Interdisciplinary Reviews: Computational Molecular Science 5: 273-289.

29. Martiniani S, Stevenson JD, Wales DJ, Frenkel D (2014) Superposition enhanced nested sampling. Physical Review X 4: 031034.

30. Mandelshtam VA, Frantsuzov PA, Calvo F (2006) Structural transitions and melting in LJ74-78 Lennard-Jones clusters from adaptive exchange Monte Carlo simulations. The Journal of Physical Chemistry A 110: 5326-5332.

31. Sharapov VA, Mandelshtam VA (2007) Solid-Solid Structural Transformations in Lennard-Jones Clusters: Accurate Simulations versus the Harmonic Superposition Approximation J. Phys. Chem. A 111: 10284-10291.

32. Sharapov VA, Meluzzi D, Mandelshtam VA (2007) Low-temperature structural transitions: Circumventing the broken-ergodicity problem. Phys. Rev. Lett 98: 105-701.

33. Dugan NS, Erkoc (2008) Stability analysis of graphene nanoribbons by molecular dynamics simulations Phys. Stat. Sol. B 245: 695.

34. Heer de WA (1993) The physics of simple metal clusters: experimental aspects and simple models. Rev. Mod. Phys 65: 611.

35. Brack M (1993) The physics of simple metal clusters: self-consistent jellium model and semiclassical approaches. Rev. Mod. Phys 65: 677.

36. Wang BX, Zhou LP, Peng XF (2006) Surface and Size Effects on the Specific Heat Capacity of Nanoparticles. Int. J. Thermophys 27: 139.

37. Comsa GH, Heitkamp D, Rade HS (1977) Solid State Commun 24: 547-550.

38. Liangliang Wu, Fang W, Chen X (2016) Photoluminescence mechanism of ultra-small gold clusters Phys. Chem. Chem. Phys 18, 17320-17325.

39. Andres RP, Bein T, Dorogi M, Feng S, Henderson JI et al. (1996) Coulomb Staircase at Room Temperature in a Self-Assembled Molecular Nanostructure. Science 272: 1323-1325.

40. Reyes-Nava JA, Garzon IL, Michaelian K (2003) Negative heat capacity of sodium clusters. Phys. Rev. B 67: 165-401.

41. Lee MS, Chacko S, Kanhere DG (2005) First-principles investigation of nite-temperature behavior in small sodium clusters. J. Chem. Phys 123: 1643-1710.

42. Boscheto E, Souza M de, Lopez-Castillo A (2016) The Einstein speci c heat model for nite systems. Physica A 451: 592-600.

43. Doye JPK, Calvo F (2002) Entropic e ects on the structure of Lennard-Jones clusters. J. Chem. Phys, 116: 8307-8317.

44. Dong and Y, Springborg M(2007) Global structure optimization study on Au2 20. Eur. Phys. J. D 43: 15-18.

45. Breaux GA, Hillman DA, Neal CM, Benirschke RC, Jarrold MF (2004) Gallium Cluster "Magic Melters" J Am Chem. Soc 126: 8628.

46. Hakkinen H, YoonB, Landman U, Li X, Zhai HJ (2003) On the Electronic and Atomic Structures of Small AuN- (N = 4–14) Clusters: A Photoelectron Spectroscopy and Density-Functional Study. J. Phys. Chem. A 107: 6168–6175.

47. Bonaci –Koutecky V, Burda J, Mitrie R, Ge M, Zampella G (2002) J. Chem. Phys117: 3120.

48. S. Gilb, P. Weis, F. Furche, R. Ahlrichs and M. M. Kappes (2002) Structures of small gold cluster cations (Au+n, n<14): Ion mobility measurements versus density functional calculations. J. Chem. Phys 116: 4094.

49. Furche F, Ahlrichs R, Weis P, Jacob C, Gilb S et. al. (2002) Adiabatic time-dependent density functional methods for excited state properties J. Chem. Phys 117: 6982.

50. Vishwanathan K, Springborg M (2016) Vibrational Heat Capacity of Gold Cluster AuN=14 at Low Temperatures. J Phys Chem Biophys 6:232.

51. Koskinen P, Hakkinen H, Huber B (2007) Bernd von Issendor and Michael Moseler. Phys Rev Lett 98: 015701.

52. Koskinen P, Häkkinen H, Seifert G, Sanna S, Frauenheim Th, et al. (2006) Density-functional based tight-binding study of small gold clusters. New J Phys 8: 9.

53. Porezag D, Frauenheim Th, Kohler Th, Seifert G, Kaschner R (1995) Phys. Rev. B, 51, 12947.

54. Seifert G, Schmidt R (1992) New J Chem 16: 1145.

55. Seifert G, Porezag D, Frauenheim Th (1996) Calculations of molecules, clusters, and solids with a simplified LCAO-DFT-LDA scheme. Int J Quantum Chem 58: 185-192.

56. Seifert G (2007) Tight-Binding Density Functional Theory: An Approximate Kohn−Sham DFT Scheme. J. Phys Chem A 111: 5609-5613.

57. Mc Quarrie, Donald A (1975) Statistical mechanics. New York: Harper Row.

58. Landsberg PT (1990) Thermodynamics and statistical mechanics. Oxford University Press, New York: Dover.

59. Waldram JR (1985) The theory of thermodynamics, Cambridge: University Press.

60. Luo P, Li YZ, Bai HY, Wen P, Wang WH (2016) Memory E ect Mani-fested by a Boson Peak in Metallic Glass. Phys Rev Lett 116: 175901.

61. Perez-Madrid A, Lapas LC, Miguel Rub J (2009) Heat Exchange between Two Interacting Nanoparticles beyond the Fluctuation-Dissipation Regime. Phys Rev Lett 103: 048301.

62. Perez-Madrid A, Lapas LC, Miguel Rub J (2013) A Thermokinetic Approach to Radiative Heat Transfer at the Nanoscale. PLOS ONE 8: e58770.

63. McIntosh C, Toulouse J, Tick P (1977) The Boson peak in alkali silicate glasses. Journal of Non-Crystalline Solids 222: 335-341.

64. Gregory AB, Michael DM (1991) Resonant two-photon ionization spectroscopy of jet-cooled Au$_3$. J Chem Phys 95: 10.1063-1.461213.

65. Gruene P, Rayner DM, Redlich B, Alexander FG, van der Meer et al., (2008) Science., 321, 674.

66. Luis AM, David M (2016) Vibrational an harmonicity of small gold and silver clusters using the VSCF method. Phys Chem Chem Phys 18: 529-549.

67. Molina B, Soto JR, Calles A (2008) DFT normal modes of vibration of the Au$_{20}$ cluster. Rev Mex Fs 54(4), 314-318.

68. Nose S (1984) Mol Phys 52: 255.

69. Booth I, MacIsaac AB, Whitehead JP, De Bell K (1995) Phys. Rev. Lett. 75: 950-953.

70. Ifti M, Li Q, Soukolis CM, Velgakis MJ (2001) Modern Phys Lett B15: 895-903.

71. Ifti M, Li Q, Soukolis CM, Velgakis MJ (2001) A Study of 2d Ising Ferromagnets with Dipole Interactions. Modern Physics Letters B 15: 895-903.

72. Casartelli M, Dall'Asta L, Rastelli E, Regina S (2004) Metric features of a dipolar model. Journal of Physics A: Mathematical and General 37: 11731.

73. Verma AS (2008) Bond-stretching and bond-bending force constant of binary tetrahedral (A IIIB V and A IIB VI) semiconductors. Physics Letters A 372: 7196-7198.

74. Tama F (2003) Introduction to Normal Mode Analysis (NMA). Department of Molecular Biology, California 92037.

75. Loi MA, Cai Q, Chandrasekhar HR, Chandrasekhar M, Graupner W, et al. (2001) High pressure study of the intramolecular vibrational modes in sexithiophene single crystals. Synthetic metals 116: 321-326.

76. Nanda KK, Maisels A, Kruis FE, Rellinghaus B (2007) Anomalous thermal behavior of gold nanostructures. EPL (Europhysics Letters) 80: 56003.

77. Munday JN, Capasso F, Parsegian VA (2009) Measured long-range repulsive Casimir–Lifshitz forces. Nature 457: 170-173.

78. Jahn HA, Teller E (1937) Stability of polyatomic molecules in degenerate electronic states. I. Orbital degeneracy. In Proceedings of the Royal Society of London A: Mathematical, Physical and Engineering Sciences 161: 220-235. The Royal Society.

79. Senn P (1992) A simple quantum mechanical model that illustrates the Jahn-Teller effect. J. Chemical Educ 69: 819.

80. O'Brien MCM, Chancey CC (1993) The Jahn-Teller effect: An introduction and current review. Am. J. Physics 61: 688.

81. Allouche AR (2011) Gabedit-a graphical user interface for computational chemistry softwares. Journal of computational chemistry 32: 174-182.

82. Guo R, Balasubramanian K, Wang X, Andrews L (2002) Infrared vibronic absorption spectrum and spin-orbit calculations of the upper spin-orbit component of the Au 3 ground state. The Journal of chemical physics 117: 1614-1620.

83. Bulusu S, Li X, Wang LS, Zeng XC (2006) Evidence of hollow golden cages. Proceedings of the National Academy of Sciences 103: 8326-8330.

84. Harrison WA (1980) Solid state theory. Courier Corporation.

85. Geng W (2003) Physical Review Letters B 67: 233403.

86. Zhang H, Douglas JF (2013) Glassy interfacial dynamics of Ni nanoparticles: Part II Discrete breathers as an explanation of two-level energy fluctuations. Soft matter 9: 1266-1280.

87. Zhang H, Kalvapalle P, Douglas JF (2010) String-like collective atomic motion in the interfacial dynamics of nanoparticles. Soft Matter 6: 5944-5955.

88. Farrusseng D, Tuel A (2016) Perspectives on zeolite-encapsulated metal nanoparticles and their applications in catalysis. New Journal of Chemistry 40: 3933-3949.

89. Shintani H, Tanaka H (2008) Universal link between the boson peak and transverse phonons in glass. Nature materials 7: 870-877.

90. Malinovsky VK, Sokolov AP (1986) The nature of boson peak in Raman scattering in glasses. Solid state communications 57: 757-761.

91. Milkus R, Zaccone A (2016) Local inversion-symmetry breaking controls the boson peak in glasses and crystals. Physical Review B 93: 094204.

92. Chumakov AI, Monaco G, Monaco A, Crichton WA, Bosak A, et al. (2011) Physical Review Letters 106: 225501.

93. Richet NF (2009) Heat capacity and low-frequency vibrational density of states. Inferences for the boson peak of silica and alkali silicate glasses. Physica B: Condensed Matter 404: 3799-3806.

94. Richet NF, Kawaji H, Rouxel T (2010) The boson peak of silicate glasses: The role of Si-O, Al-O, and Si-N bonds. The Journal of chemical physics 133: 044510.

95. Ghiringhelli LM, Gruene P, Lyon JT, Rayner DM, Meijer G, et al. (2013) Not so loosely bound rare gas atoms: finite-temperature vibrational fingerprints of neutral gold-cluster complexes. New Journal of Physics 15: 083003.

96. Li J, Li X, Zhai HJ, Wang LS (2003) Au20: a tetrahedral cluster. Science 299: 864-867.

4

Investigation of Effect on Prefabricated Light Steel Structure Non-coating and Coating Using Transogard Zinc Chromate Paint after Exposed to Seawater

Shah MMK, Ismail A and Sarifudin J*

Faculty of Engineering, Universiti Malaysia Sabah, Sabah, Malaysia

Abstract

The main objective of this research is to determine corrosion rate and weight loss for the coated and non-coated with several layer of mild steel AISI 1020. This research discussed and focusing on prefabricated material coating and non- coating of light steel structure effect on seawater and study on the research have been done before by others researcher for better understanding on this research. Basically, most of the researches conducted before are only focusing on the type of prefabricated materials. However, in this research the coating will be used in other to improve the properties of materials.

Keywords: AISI 1020; Coating; Optical microscope; Zinc chromate; $ZnCrO_4$

Introduction

This research is conducted due to the problems faced by the marine area. Firstly, there is no specific coating can be used to coat the light steel for the application in sea water, the right selection for light steel material should be done properly for better outcome of experimental result without neglecting the cost [1-3].

The following objective need to achieve:

- To investigate the profile of prefabricated material light steel coated and non- coated structure after exposed to sea water.

- To determine weight loss of the coated and non- coated light steel material due to sea water.

The coatings based in chromium (VI) already used since a long time ago and now prohibited because of the toxic content especially chromium (VI) [4-6]. A solution came out, which is trivalent chromate coating, but this coat does not produce the yellow color like the hexavalent chrome coating does. But the function is quite same as the hexavalent chrome coating, so it is used to replace hexavalent chrome in coating, since trivalent chromate coating is more environmental friendly [7-10].

"Many factors in surface preparation affect the integrity of coating which includes residues of oil grease, rust on the surface and mill scale which can decrease adhesion or mechanical bonding of coating to the surface" [11]. Coating is mostly used in Steel piling and other maritime structures because coating constitute the most commonly used and most effective method of corrosion control of marine structures.

The coating used in this experiment will be transogard zinc chromate Paint. The chromate type used is trivalent chromate coating which is a non- toxic paint. The paint Chemical formula is $ZnCrO_4$ and the appearance of the coating is yellow- green crystals (Figure 1).

This is shows that before applying any kind of coatings the penetration playing an important roles in order to make sure the coating fully covered the specimen surface. Besides that, the surface finishing also important to make sure the coating fully used as the protective surface to minimize the corrosion occurs. In seawater, the reduction of dissolved oxygen molecules is the primary cathode reaction on the specimen surface, and such a steady current is due to the limitation of oxygen diffusion. Based on the result also shows, the corrosion rate of the coating depended on its porosity.

The weight loss is converted to a corrosion rate (CR) or a metal loss (ML), as follows:

$$\text{Corrosion (CR)} = \frac{\text{Weight loss(g)} \times K}{\text{Density(g / cm}^3) \times \text{Exposed Area(A)}}$$

The constant used, k coefficient is 3.45×10^6, the desired corrosion rate unit used is Mils/ year (mpy) and the area units used is in cm^2. The density used is 7.87 g/cc @ g/cm^3.

$$\text{Percentage of Weight loss(\%)} = \frac{\text{Weight loss(g)} \times 100\%}{\text{Initial Weight(g)}}$$

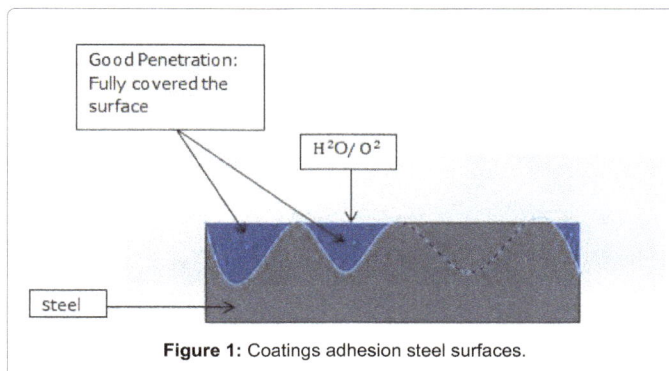

Figure 1: Coatings adhesion steel surfaces.

*Corresponding author: Sarifudin J, Faculty of Engineering, Universiti Malaysia Sabah, Jalan UMS, 88400 Kota Kinabalu, Sabah, Malaysia
E-mail: jumafisabilillah92@gmail.com

Methodology

The specimen used will be in rectangular design, before the experiment conducted, preparation of the specimen is the complex system (Figure 2).

Surface finishing methods vary across a broad range for better result. Type of steel used is AISI 1020 in the experiment. Firstly, cut the material into rectangular design with size 3 mm × 30 mm × 80 mm. Clean of specimens before weighing to remove any contaminants that could affect test results. The specimens clean by using grinding and polished using sand paper (Figure 3). Before coat the specimen, the samples will be degreased with ethanol and labeling to easier to identify the specimens (Figure 4).

After the cleaning process, 12 pieces of mild steel rectangular design will be coated with one layer of Zinc chromate, $ZnCrO_4$ paint with the thickness of 0.09692 mm. 12 pieces of mild steel rectangular design will be coated with two layer using the same paint which are Zinc chromate, $ZnCrO_4$ with the thickness of 0.14792 mm. the second layer applied when the first layer is completely dry. The other 12 pieces will be in non- coated condition.

The digital scale weigh will be used in order to weigh the specimen, before and after the experiment or test conducted (Figure 5).

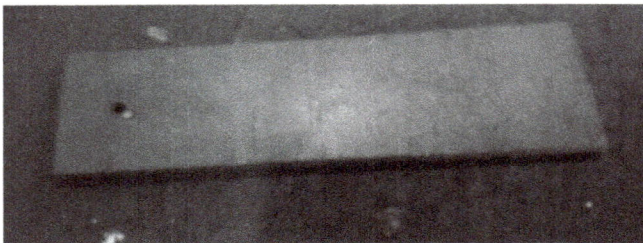

Figure 2: AISI 1020 mild steel.

Figure 3: Cleaning the specimens by grinding using sand paper.

Figure 4: Samples a after degreased with ethanol and labeling.

A closer picture of the corrosion products formed after progressive corrosion for long-term atmospheric exposure and compared to neutral salt fog and ASST accelerated methods is shown in Figure 5. The progressive corrosion product formation follows the same pattern as during initial corrosion, as described in the preceding paragraph for each corrosion exposure type [12]. From the figure above it is showed that how we can examine the corrosion happened effect of exposed to seawater (Figure 6).

Result

Optical microscopy without coating

The reddish colour covered most of the surface and the red participate more thick than 10th day, 20th day and 30th day. The based with silver and white colour show the area of the metal without any coating (Figure 7) [13-15].

Optical microscopy with one layer of coating

The size of water bubble on the paint layer is decreases compare to the 10th, 20th and 30th days. Besides that, the number of the bubble produce after submerged into the water is decreases covered the surrounding of the specimen surface but some of reddish colour is form on the surface (Figure 8).

Optical microscopy with two layer of coating

"Many factors in surface preparation affect the integrity of coating which includes residues of oil grease, rust on the surface and mill scale which can decrease adhesion or mechanical bonding of coating to the surface". This shows; the coating cannot be function fully as protective to the surface because of surface preparation (Figure 9). The mechanical bonding of coating to the surface also plays an important role, to avoid

Figure 5: Specimens weighed using digital scale weigh.

Figure 6: Pitting and corrosion in a crude oil tank (International Marine Coating, 2011).

the bubble to form between the coating paint and the mild steel. Hence, effect of the bonding between the coating paint and the mild steel is not strong to prevent the corrosion process of mild steel during exposed

to sea water [16,17]. Thus, the water bubble is form, the corrosive sea water inside the bubbles do the corrosion process (Tables 1 and 2).

The samples without coating showed increment in percentage of weight loss as the time increases. The percentage of weight loss increase directly proportional with time when the specimens exposed to sea water without any protective layer to reduced corrosion process (Figure 10). The specimens applied with one layer of coating, the percentage weight loss during the 10th, 20th and 30th day is negative. This is means when the percentage of weight loss is negative there is no weight loss occur, the specimens gain weight due to the layer of the coating produce bubble and stored water inside the specimens surface through bubbles. On the 40th day, the percentage of weight loss is 0.0006%. This is shows the corrosion start to take place at the 40th day. For the two layer of coating the percentage weight loss is negative along the period [18-21].

While the rate of corrosion, of the samples without coating showed decrement in corrosion rate as the time increases. The Corrosion rate decreases as the time increases, but start to increases on the 40th day. The specimens applied with one layer of coating, the percentage weight loss is negative on the 10th, 20th and 30th day. This is means when the percentage of weight loss is negative there is no weight loss occur, the specimens gain weight due to the layer of the coating produce bubble and stored water inside the specimens surface through bubbles [22,23]. On the 40th day, the corrosion rate is increases to 0.07mmy. This is shows the corrosion start to take place at the 40th day. The corrosion rate

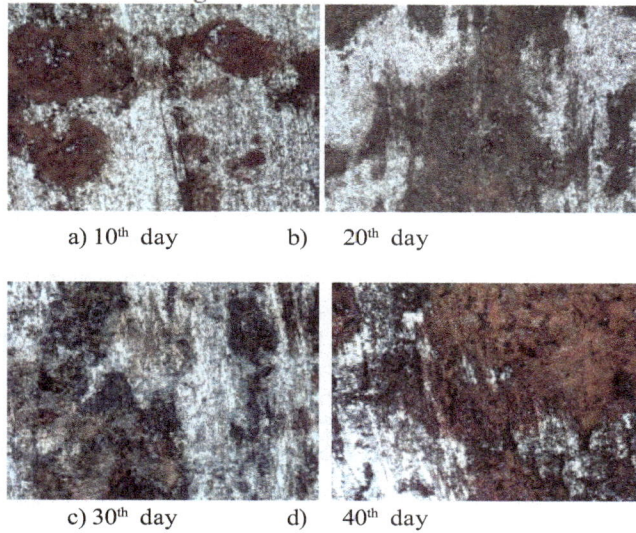

Figure 7: Showing optical microscopy without coating.

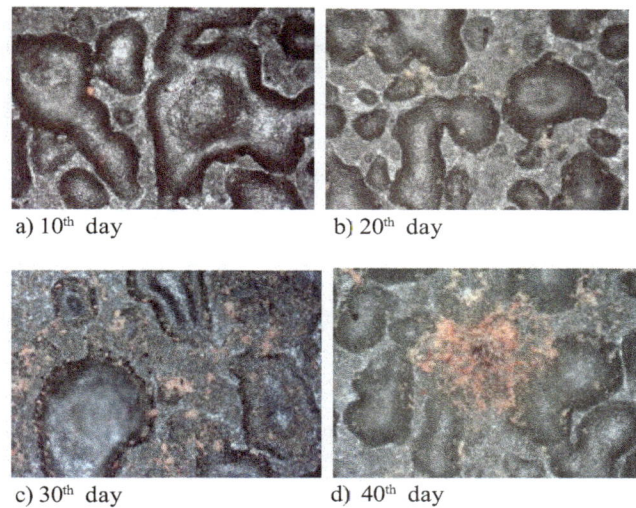

Figure 8: Showing optical microscopy with one layer of coating.

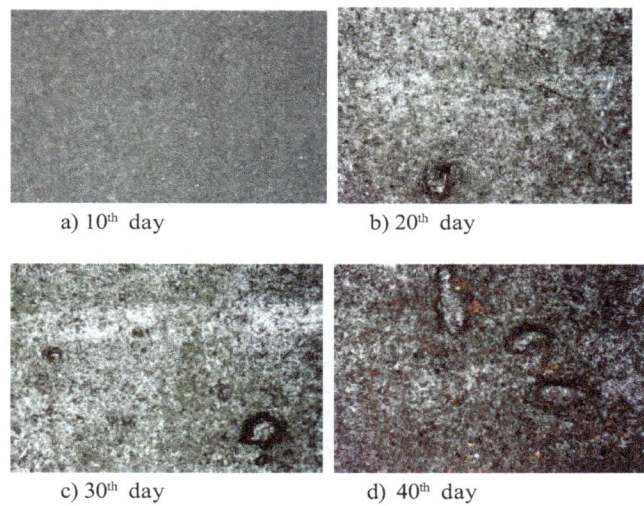

Figure 9: Showing optical microscopy with two layer of coating.

Days	Condition	Initial Weight (g)	Final Weight (g)	Weight loss (g)	Surface Area (cm²)	Corrosion rate (mmY)	Percentage of weight loss (%)
10	Without coating	135.0286	134.9.95	0.1191	5.418313	4.015	0.0882
	One -layer coating	134.9469	135.1761	-0.2292	5.43548	-77.02	-0.1698
	Two-layer coating	137.6914	137.7231	-0.0317	5.468987	-10.59	-0.023
20	Without coating	140.4065	140.2331	0.1734	5.46264	28.99	0.1235
	One -layer coating	136.0834	136.1232	-0.0397	5.44794	-6.655	-0.0292
	Two-layer coating	138.4652	138.4998	-0.0347	5.47076	-5.79	-0.0251
30	Without coating	139.5726	139.3480	0.2246	5.463780	25.03	0.1609
	One -layer coating	134.4546	134.4814	-0.0267	5.429353	-2.99	-0.0199
	Two-layer coating	139.9052	139.9421	-0.0370	5.479760	-4.11	-0.0264
40	Without coating	139.3340	139.0227	0.3112	5.461087	26.00	0.2233
	One -layer coating	143.9855	143.9846	0.0009	5.511080	0.07	0.0006
	Two-layer coating	145.3728	145.3904	-0.0176	5.528207	-1.45	-0.0121

Table 1: Corrosion rate.

Symbol	Quantity	Unit
CR	Corrosion Rate	1Mils/year (mpy)
ρ	Density	7.87 g/cc@g/cm³
A	Area	cm²
k	coefficient	3.45 × 10⁶

Table 2: Units properties.

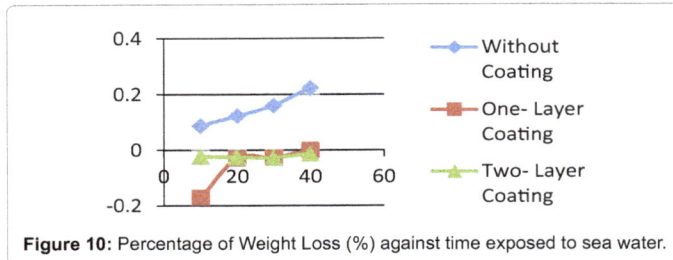

Figure 10: Percentage of Weight Loss (%) against time exposed to sea water.

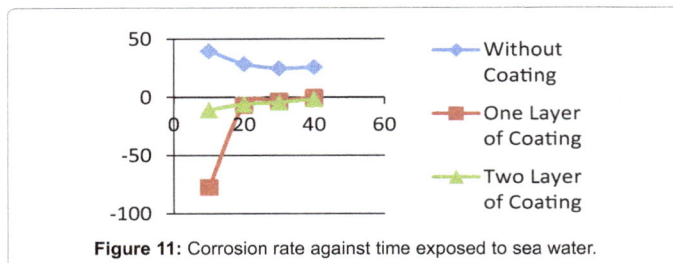

Figure 11: Corrosion rate against time exposed to sea water.

is slowly increases after undergoes the highest increment in corrosion rate due to the corrosion rate approaching the positive corrosion rate. For the two layer of coating, the corrosion rate is negative along the experiment period of time. The specimens start to undergo the corrosion rate after approaching the positive corrosion rate (Figure 11).

The corrosion take place after the water bubbles is formed. This is because the corrosion occurs between mild steel and sea water and the increases the weight loss of the samples. The surface finishing is also some of causes to the water bubbles to forms [24,25].

Conclusion

The data that was recorded show that there are different outcome from different condition of specimens surface. It is showed that with some protection of the mild steel surface can reduce the weight loss and also reduce corrosion rate. In addition, the water bubbles produce is due to the bonding attraction between the water and the mild still is high compare to the bonding attraction between the paint coatings and the mild steel. This is because the corrosion processes occur due to the presence of some ion presence on the mild steel and sea water. The coating only acts as to reduce the corrosion rate.

Acknowledgment

We would like to take this opportunity to express our deepest appreciation and gratitude to those people who had guided and assisted us doing this research. We thank our colleagues from University of Malaysia Sabah who provided insight and expertise that greatly assisted the research, although any errors are our own and should not tarnish the reputations of these esteemed University of Malaysia.

References

1. Ismail A, Loren NE, Abdul Latiff NF (2013) Effect of Anions in Seawater to Corrosion Attack on passive alloys. International Journal of Research in Engineering and Technology 2: 733-737.

2. Deen KM, Ahmad R (2009) Corrosion Protection Evaluation of Mild Steel Painted Surface by Electrochemical Impedance Spectroscopy. Journal of Quality and Technology Management.

3. Ahn WS (1995) Durability Testing of Reinforced Concrete Beams under Fatigue Loading in a Simulated Marine Environment. Simulated Marine Environment.

4. AISI 1020 Low Carbon/Low Tensile Steel.

5. Ajide MA (2011) Microstructural Analysis Of Selected Corroded Materials From Nigeria Oil and Gas Industry. American Journal of Materials Science 1: 108-112.

6. Anakin (2010) Cruise Shipping Wall and Cruise Ship.

7. Malik AU, Ahamd S (1999) Corrosion Behavior of Steels in Gulf. Research & Development Center Saline Water Conversion Corporation.

8. Anthoni DJ (2006) The Chemical Composition of Sea Water.

9. ASTM (1973) Coated Metal Specimens At 100 Percent Relative Humidity. Standard Method for Testing.

10. Bergeler E (2012) Why Can Heavy Steel Ships Float? Science Niblets.

11. Calle HM (2002) Steelwork Corrosion Control (2ndedn) London and New York: Spon Press.

12. Montgomery EL, Calle LM, Curran JP, Kolody MR (2011) Timescale Correlation Between Marine Atmospheric Exposure And Accelerated Corrosion testing.

13. Ferry MM (2013) Study of Corrosion Performance of Zinc Coated Steel In Seawater. Journal of Naval Architecture.

14. Abdulaziz MK (2013) The Effect Of Inhibition On Corrosion Resistance.

15. International Marine Coating (2011) What Is Corrosion? Coatings Technology.

16. Kawakita J, Fukushima T, Kuroda S, Kodama T (2001) Corrosion Behaviour of Hvof Sprayed Sus316l Stainless Steel in Seawater. Corrosion Science 44.

17. Harry LA (2005) Cadmium Replacement for Propellant Actuated Devices (Pads). Journal Cadmium Replacement for Propellant Actuated Devices (Pads).

18. Li Y (2001) Corrosion Behaviour of Hot Dip Zinc And Zinc–Aluminium coatings on steel in seawater. Bulletin of Materials Science 24: 355-360.

19. Michele (2001) Salt Water Corrosion of Boarding Area on Offshore Platform.

20. Norsok Standard (1999) Surface Preparation And Protective Coating 2-26.

21. Ajide OO, Agara KW (2012) Comparative Assessment Of Corrosion Behaviour of MCS And KS7 SS in Saline and Carbonate Environments. Journal of Minerals and Materials Characterization and Engineering 11: 836-840.

22. Pittsburgh (2005) Use of Coatings To Control Corrosion of maritime structures. The Society For Protective Coatings.

23. Razaqpur AIO (2009) Prediction of Reinforcement Corrosion In Concrete Structures.

24. Saleh A, Malik AU (2008) Effect of Seawater Level On Corrosion Behaviour Of Different Alloys. Desalination.

25. Watts M (2015) Application of Fiber-Reinforced Composite in Steel Material. Journal Application of Fiber-Reinforced Composite in Steel Material.

5

Large-Scale Bose-Einstein Condensation in a Vapor of Cesium Atoms at Normal Temperature (T=353K)

You PL*

Institute of Quantum Electronics, Guangdong Ocean University, Zhanjiang, China

Abstract

Large-scale BEC of cesium at T=353 K was first observed. Until now, scientists have applied magnetic fields and lasers, but never applied electric fields, and atoms are oriented at random, so observation of BEC is very difficult. Our innovation lies in the application of electric fields. We theoretically proved that alkali atom (include Cs) may be polar atom doesn't conflict with quantum mechanics. Variation of the capacitance with temperature offers a means of separating the polar and non-polar atom. Cs vapor was filled in cylindrical capacitor. Our experiment shows that Cs is polar atom because its capacitance is related to temperature. In the past, to realize the phase transition, ultralow temperature is necessary. But now we don't require ultralow temperature, because we use the critical voltage V_c to achieve the phase transition. From the entropy $S=Nk \ln 2\pi e /a=0$, $a=dV/kTH =2\pi e$, $V_c \approx 63$volts. When $V < V_c$, $S > 0$; when $V > V_c$, $S < 0$, phase transition occurred. When V=350 volts, the capacitance decreased from $C=1.97C_0$ to $C \approx C_0$ (C_0 is the vacuum capacitance), this result implies that almost all Cs atoms (more than 98.9%), like as dipoles, are aligned with the field. We create BEC with 1.928×10^{17} atoms, these atoms have the same momentum. Cs material with purity 99.95% was supplied by Strem Chemicals Co., USA. Both BEC and superconductivity are condensed in the momentum space, therefore these two kinds of condensation can't be observed with the naked eye. When superconductivity occurs, the resistance $R \approx 0$, a simple and direct method to observe superconductivity is to measure the resistance by voltammetry. Similarly, when BEC occurs, the electric susceptibility $\chi_e = C/C_0 - 1 \approx 0$, a simple and direct method to observe BEC is to measure the capacitance by cylindrical capacitor. BEC is also a quasi-superconducting state.

Keywords: PACS numbers; Bose-Einstein condensation; Entropy; Polar atom; Cylindrical capacitor; Permanent dipole moment; Quasi-superconducting state

Introduction

BEC experiment is quite intriguing and a challenging task for the experimentalists. In 1925 Einstein pointed out that I maintain that, in this case, a number of molecules steadily growing with increasing density goes over in the first quantum state (which has zero kinetic energy) while the remaining molecules distribute themselves according to the parameter value λ=1. "A separation is effected; one part condenses, the rest remains a 'saturated ideal gas' (A=0, λ=1)" [1]. The famous physicist Abraham Pais emphasized that this is now called Bose-Einstein condensation. Now we estimate how many atoms in this part of Bose gas. In order to achieve BEC, the suitable density of Bose gas is $10^{13} - 10^{15}$ cm^{-3}, and its volume is usually greater than 10 cm^3, so according to Einstein's prediction, the number of condensed atoms at least greater than $10^{13} - 10^{15}$ (take 1/10 as "one part"). But what is about the actual situation? A typical example of BEC is provided by the dilute gases of alkali-metal atoms that can be prepared inside magnetic ion traps [2-6]. All stable alkali species---Li (2), Na (3), K (4), Rb (5), and Cs (6) ---have been condensed. The atoms, usually only $10^4 - 10^6$ of them, can be trapped and cooled [2-6]. And therefore the previous condensate fraction is usually less than 10^{-9}. This result shows that in the one billion alkali atoms which participating in the experiment, equivalent to only one atom was condensed. Strictly speaking, the results of previous experiments did not conform to Einstein's predictions.

In the past, there are two ways to achieve phase transition: given atomic density, lower the temperature of Bose gas, making

$$T < T_c = \frac{2\pi\hbar^2}{km}\left(\frac{n}{2.612}\right)^{2/3}$$; or given temperature, increase the density of

Bose gas, making n > n$_c$ (the critical density $n_c = 2.612\left(\frac{mkT}{2\pi\hbar^2}\right)^{1/3}$). But the

latter method is also not successful, because the density of Bose gas can't

be increased indefinitely, and therefore the laser cooling is now widely used. What is the definition of BEC? "Bose-Einstein condensation is a macroscopic occupation of the ground state" [7]. Until now, scientists have applied magnetic field (used to trap atoms) and laser (used to cool atoms), but never applied electric field, despite the use of many advanced technology, condensed atomic number is still very small. This fact shows that according to the current method, the "macroscopic occupation" is very difficult to achieve.

Can we find a new way to get out of the current predicament? This article provides a new idea for the implementation of BEC. In BEC experiments, scientists never applied an electric field because they think that alkali atoms are non-polar atoms. If alkali atom is non-polar atom, has been verified by experiment? Answer: no! No such experimental results have been reported in the history of physics! In fact, this traditional concept is an untested hypothesis, and it has misled physicists all over the world. Variation of the capacitance with temperature offers a means of separating the polar and non-polar atom. If alkali atom is a non-polar atom, its capacitance should be independent of the temperature due to the nucleus located at the center of the electron cloud. We measured the capacitance at different temperatures. Our experiment proved that Cs atom is polar atom, and it becomes a dipole. When an electric field is applied, the dipoles tend

*Corresponding author: You PL, Institute of Quantum Electronics, Guangdong Ocean University, Zhanjiang, 524025 China, E-mail: youpeli@163.com

to align with the field because of the torque it experiences. So a large-scale BEC at normal temperature can be observed.

Our previous research found that rubidium atom has a non-zero permanent dipole moment (PDM), $d_{Rb} \geq 8.6 \times 10^{-9}$ e.cm, the saturation polarization of rubidium vapor has been observed [8,9]. In particular, recent report showed that sodium is polar atom, BEC of Na atoms at normal temperature has been observed [10]. This article has been rigorously peer reviewed for up to two months. This reviewer has published three theoretical articles about BEC in Physical Review, and these interesting theoretical results may be useful for the design of BEC experiment [11-13]. This fact indicates that this peer review is authoritative and persuasive. Reviewer comments pointed out that "The author presented a good idea (using the critical voltage) to observe the BEC. Moreover, the author shows that the ultra-low temperature is not a necessary condition to verify the existence of BEC. This paper is interesting and certainly deserves publication in the journal". This objective evaluation is a great inspiration to us, and encourages us to report the results of new BEC experiment of cesium atoms.

Let us take the iron filings as an example for further explanation. In the absence of magnetic field, iron filings are oriented at random, although use laser cooling technology, but the condensation of these iron filings is still very difficult. When a magnetic field is applied, however, the iron filings immediately arranged along tangent to the magnetic field lines like as little compass needles, and the condensation of these iron filings is easily observed. This example shows the previous research is caught in a misunderstanding because they never applied an electric field, and therefore scientists have missed out on this significant discovery.

Theoretical Breakthrough

Normally atoms do not have PDM because of their spherical symmetry; however our article proved that alkali atom forms an exception. ns and np states of alkali atoms are not degenerate, and therefore the expectation value of PDM is zero: $<\psi_E|-er|\psi_E>=0$($d=-er$ is the dipole moment operator). However, many physicists strictly proved that the state function $|\psi_E>$ doesn't describe an individual particle but an ensemble of particles with the same energy [14-16]. "The expectation value is the average of repeated measurements on an ensemble of identically prepared systems, not the average of repeated measurements on one and the same system" [17]. So $<\psi_E|-er|\psi_E>=0$ only means that the average PDM of large number of alkali atoms is zero, but doesn't mean that the PDM of individual alkali atom is zero.

The hydrogen atom is a typical example. "The shift in the energy levels of an atom in an electric field is known as the Stark effect" [14]. Normally the effect is quadratic in the field intensity, which corresponds to an induced electric dipole moment (EDM) [14]. The quadratic Stark effect occurs in general in all states. But LD Landau once stated that "the hydrogen atom forms an exception; here the Stark effect is linear in the field". "The energy levels of the hydrogen atom, unlike those of other atoms, undergo a splitting proportional to the field (the *linear Stark effect*)" [18]. Evidently, the linear effect corresponds to a PDM. This effect showed that hydrogen atom (n=2) has a PDM of magnitude $-3ea_0=1.59 \times 10^{-8}$ e.cm (a_0 is the Bohr radius) [14,18]. LD Landau once stated that "The presence of the linear effect means that, in the unperturbed state ψ_{2lm}, the hydrogen atom has a dipole moment whose mean value is $d=-3ea_0$." [18]. LI Schiff also stated that "It is also possible, as in the case of the hydrogen atom, that unperturbed degenerate states of opposite parities can give rise to a

permanent electric dipole moment" [19]. That is, $d(\psi_{200}) \neq 0$, $d(\psi_{210}) \neq 0$, $d(\psi_{211}) \neq 0$ and $d(\psi_{21-1}) \neq 0$.

However, quantum mechanical calculations indicated that although ψ_{2lm} is four-fold degenerate, but the expectation value of PDM is zero:$<\psi_{2lm}|-er|\psi_{2lm}>=0$! That is, $d(\psi_{200})=d(\psi_{210})=d(\psi_{211})=d(\psi_{21-1})=0$. Evidently, the zero result is inconsistent with above conclusion. Up to now, no quantum mechanical textbook explains this contradictory result. This fact shows that individual hydrogen atom (n=2) has a non-zero PDM, but quantum mechanics can't obtain this non-zero result in any way. It convincingly proved that $<\psi_{2lm}|-er|\psi_{2lm}>=0$ only means that the average PDM is zero, but an individual hydrogen (n=2) may have a non-zero PDM.

Recall that alkali atoms having only one valence electron in the outermost shell can be described as hydrogen-like atoms [20]. So similar to the first excited state of hydrogen, the ground state of alkali atom may be polar atom doesn't conflict with quantum mechanics. A neutral alkali atom (include Cs) is or is not polar atom must be determined by experiments.

A new formula of atomic PDM is obtained for the first time

Experiments to search for PDM of atoms began half a century ago. In all experiments, they measured the spin resonance frequency v of individual atom by $hv=2\mu B \pm 2dE$, where h is Planck's constant, μ and d is magnetic and electric dipole moments [21,22]. But "experimental searches for PDM have so far yielded null results" [21]. This fact shows that this formula is not successful, because it measured the microscopic quantity d by using another microscopic quantity v.

However, measuring the average kinetic energy of a gas molecule using the temperature is easy: $E_k=3kT/2$. Similarly, we measure the PDM of an atom using the change of the capacitance is easy: $d=(C - C_0)V/L(a)$ nS. This formula is easy to verify. The magnitude of the PDM is $d=e$ r. $L(a)$ equals the percentage of Cs atoms lined up along an electric field, n is its density, S is the plate area. When the electric field is applied, the change of the charge of the capacitor is $\Delta Q=(C-C_0)V$. On the other hand, its volume is SH, the total number of oriented atoms of the capacitor is SH n $L(a)$. The number of layers of oriented atoms is H/r. Because inside the Cs vapor the positive and negative charges cancel out each other, the polarization only gives rise to a net positive charge on one side of the capacitor, and a net negative charge on the opposite side. Therefore $\Delta Q=$SH n $L(a)e/(H/r)=$n S $L(a)e$ r=n S $L(a)$ d=(C-C_0)V, so $d=(C-C_0)V/$ n $L(a)$ S.

Our Innovations

According to W. Ketterle's standard, our experiment is a truly ideal BEC

Wolfgang Ketterle, in the Nobel Prize winning paper, proposed the objective standard of an ideal BEC: "An ideal Bose condensate shows a macroscopic population of the ground state of the trapping potential. This picture is modified for a weakly interacting Bose gas" [3]. This standard emphasizes that the ground state is the ground state of the trapping potential, which is very important correction. Because this description doesn't involve temperature, and therefore this standard is also suitable for the evaluation of our experiments. Cesium has a non-zero PDM, and it becomes an electric dipole. When an electric field is applied, the dipoles tend to orient in the direction of the field because of the torque it experiences. The magnitude of the torque is $\tau=d$ E $sin\theta$, and θ is the angle between d and E. When $\theta=0$, the dipole is in equilibrium. In our experiment, the trapping potential is the electric potential energy of Cs atoms: $\varepsilon = -dE\cos\theta$. When $\theta=0$, the potential energy is a minimum, which indicates that the dipole is oriented parallel to the filed. But atomic collisions tend to disarrange the dipoles. When $V < V_c$, many atoms are in random directions, this state has high entropy $S > 0$; when $V > V_c$, the atoms become aligned with the field, this state has low entropy $S < 0$, phase transition occurred. When $V \gg V_c$, $C \approx C_0$, this result implies that the alignment would be perfect, these atoms have the same momentum, this is condensation in momentum space. In effect, BEC is the perfect alignment of bosons. So BEC at normal temperature can be observed. When $V \gg V_c$, Bose condensate contained up to 1.928×10^{17} atoms really achieved "macroscopic population". The atoms condensed into the ground state of trapping potential, because their electric potential energy is a minimum along the field. Therefore, Ketterle's standard proved that although we didn't use laser cooling atoms, but our experiment is an ideal BEC. There are few groups that use electric fields to cool atoms. This article will not discuss this situation.

Variation of the capacitance with temperature offers a means of separating the polar and non-polar atom experimentally. The classical electrodynamics textbook plotted the relationship between χ_e and $1/T$ [23].

For the polar atom $\chi_e = A + B/T$, for non-polar atom $\chi_e = A$ $\quad$ (1)

Where, A and B is constant [23]. If Cs is polar atom, the form $\chi_e = A + B/T$ should be expected. As a contrast, the capacitance of Hg has been measured, but its capacitance is independent of temperature, Hg is non-polar atom.

Our experimental type is quantitative

The condensate fraction is a very important physical quantity in BEC, but previous experiments didn't provide a suitable formula, and they are qualitative. We strictly proved that the Langevin function L(a) equals the condensate fraction, and it can be expressed as $L(a) = a\,(C - C_0)/\eta$, where η is the capacitance constant, and $(C - C_0)$ is the change of the capacitance. For example, when the voltage V=350 volts, $\eta=192$ pF, $C - C_0 = 2$ pF and the coefficient $a=95$, we obtain $L(a) = 0.9896$. These facts indicate that our experiments not only don't conflict with their experiments, but also this experimental type is quantitative.

The most striking characteristic of BEC is that BEC is condensed in momentum space

Einstein first noticed that Bose gas would condense to the lowest energy state. However, the result of the condensation doesn't produce crystals, this fact shows that BEC doesn't occur in the position space but occurs in the momentum space. "The term 'condensation' often implies a condensation in space, as when liquid water condenses on a cold window in a steamy bathroom. However, for Bose-Einstein condensation it is a condensation in k-space, with a macroscopic occupation of the lowest energy state" [7]. Note that the wave vector k equals the momentum p divided by Planck constant $\hbar$, and therefore BEC is a condensation in momentum space. Superconductivity is also the condensation in the momentum space because the Cooper pairs act like bosons, therefore these two kinds of condensation can't be observed with the naked eye. When superconductivity occurs, the resistance $R \approx 0$ (the resistance of vacuum is zero). Although BCS theory is difficult, however, a relatively simple and direct observation method is to measure the resistance by voltammetry. Similarly, when BEC occurs, the electric susceptibility $\chi_e = C/C_0 - 1 \approx 0$ or $C \approx C_0$, a simple and direct observation method is to measure the capacitance. BEC is quasi superconducting state. Note that the resistance and capacitance are two

macroscopic electrical quantities, and their changes are the decisive evidence of these two kinds of condensation. When they(R or C) are reduced to close to the vacuum value, these two kinds of condensation occur. This fact fully reflects the harmony and unity of nature.

The entropy of a system is a measure of the disorder of molecular or atomic motion. No doubt, it is the most important concept in BEC. Consider a system composed of N cesium atoms which are placed in an electric field E, θ is the angle between d and E. Note that the collision between Cs atoms is always through their mass centers, and therefore the nucleus has no contribution to the rotational energy of the atom. When orientation polarization occurs, its rotational energy can be neglected. The potential energy of Cs atom in the field can be expressed as $\varepsilon = -dE\cos\theta$. Unlike the orientation quantization of magnetic moment, the orientation of Cs atoms can be changed continuously in the field. Note that $\beta = 1/kT$ and the chemical potential $\mu \approx 0$[7], the partition function is given by

$$Z = \int_0^{2\pi} d\phi \int_0^{\pi} e^{-\beta\varepsilon} \sin\theta\, d\theta = \int_0^{2\pi} d\phi \int_0^{\pi} e^{dE\cos\theta/kT} \sin\theta\, d\theta = 2\pi kT(e^{dE/kT} - e^{-dE/kT})/dE \quad (2)$$

The entropy is given by $S = Nk(\ln Z + T\frac{\partial}{\partial T}\ln Z)$ [7]. Let the coefficient $a = dE/kT = dV/kTH$, we obtain

$$S = Nk\,[\,\ln 2\pi e\,(e^a - e^{-a})\,/a - a\coth a\,] \quad (3)$$

When $a = dE/kT \gg 1$ or $V \geq 37$ volts $(a \geq 10)$, $e^{-a} \approx 0$ and $\coth a \approx 1$, we obtain a simplified formulas

$$S = Nk\ln 2\pi e\,/a \quad (4)$$

The critical coefficient is $a_c = 2\pi e \approx 17.08$. The formula contains two fundamental constants in nature ($\pi \approx 3.14159$ and $e \approx 2.71828$), it reflects the objective laws of BEC.

Experiment and Interpretation

The preparatory experiment: we measured the density of Cs vapor. The longitudinal section of the apparatus is shown in Figure 1. This closed glass container resembles a Dewar flask in shape. Its internal and external diameters are $D_1 = 5.6$ cm and $D_2 = 8.12$ cm respectively. The external and internal surfaces was paste with aluminum foil, also can be plated with silver, they form the outer and inner electrode (Figure 1) [10]. Their length is L=33.4 cm. This capacitor is equivalently connected in series by two capacitors. One is called C', and contains the Cs vapor of thickness $H_0 = 9.60$ mm; the other is called C'', and contains the glass medium of thickness h=1.5 mm. The total capacitance is $C = C'C''/(C'+C'')$, where C'' and C can be directly measured. The magnitude of capacitance was measured by a digital capacitance meter. The model of the meter is DM6031A and was made in Shenzhen, China. The accuracy of the meter was 0.5%, the measuring voltage was $V_0 = 1.2$ volts, the measuring frequency was 800Hz, and the definition was 0.1

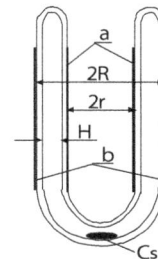

Figure 1: The longitudinal section of a cylindrical capacitor, which filled with Cs vapor and surplus liquid cesium, but in the behind two experiment, it keeps a fixed density without liquid cesium.

pF ($C \leq 200$ pF), 1 pF (200 pF$\leq C \leq$ 2 nF) or 10 pF (2 nF$\leq C \leq$20 nF). In order to remove impurities such as oxygen, when the capacitor was empty, it was pumped to vacuum pressure $P \leq 10^{-9}$ torr for 20 hours. The vacuum capacitance is $C'_0 = (54.0 \pm 0.1)$ pF. Next, a small amount of Cs material (about 10 grams) was put into the container, and it is again pumped to $P \leq 10^{-9}$ torr, and then it is sealed. We obtain a cylindrical glass capacitor filled with Cs vapor and surplus liquid cesium (Figure 1) [10]. We put the capacitor into a temperature-control stove, raise its temperature slowly, and keep it at $T_0 = 473$ K for 6 hours. It means that these results are obtained under the saturated vapor pressure. We measured the capacitance is $C'_t = (5140 \pm 10)$ pF. Note that $P = 10^{6.949 -3833.7/T}$ psi (473 K$\leq T \leq$623 K, 1 psi=6894.8 Pa) is the saturated vapor pressure of Cs atoms [24]. We obtained $P = 481.3$ Pa at $T_0 = 473$ K. From the ideal gas law, the density of Cs vapor was $n_0 = P/kT_0 = 7.370 \times 10^{16}$ cm^{-3}, where $k = 1.3807 \times 10^{-23}$ J/K. The statistical error is $(\Delta n_j/n)^2 = (\Delta P/P)^2 + (\Delta T/T)^2$, due to $\Delta T = 0.5$K and $\Delta T/T \leq 0.001$, $\Delta P = P(T_0 + \Delta T) - P(T_0) = 9.6$ Pa and $\Delta P/P \leq 0.02$, so $\Delta n_j/n \leq 0.03$. Considering all systematic error, we have that $\Delta n_j/n \leq 0.03$, and the density of Cs vapor is $n_0 = [7.37 \pm 0.22\text{(stat)} \pm 0.22\text{ (syst)}] \times 10^{16}$ cm^{-3}.

The first experiment: we measured capacitances of Cs and Hg vapor at different temperatures. A glass cylindrical capacitors fill with cesium at a fixed density n_1 without liquid cesium, and $n_1 << n_0$. Another cylindrical capacitors fill with Hg vapor at a fixed density n_3.

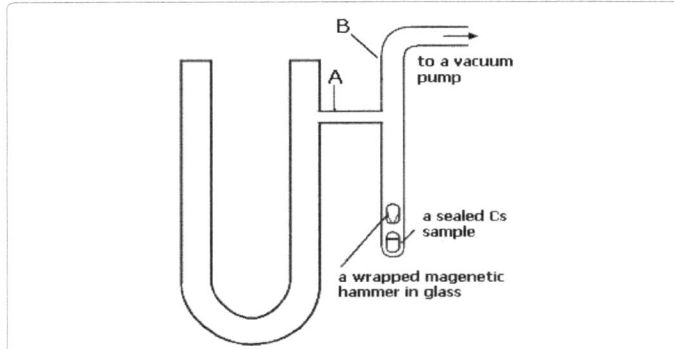

Figure 2: This diagram shows the perfusion method. We let the left capacitor to connect a vacuum pump through a glass tube. In this glass tube, a small magnetic hammer and a sealed cesium sample are put together. When the pump works, the magnetic hammer is raised by a magnet outside the tube. We suddenly release the magnetic hammer, which breaks the bottle sealed cesium sample, and then the tube is sealed at B point. Next step, we put the two containers into the heating furnace, and keep at $T_1 = 453$K for 6 hours, the glass tube was sealed once again at A point. Thus we obtain a capacitor filled with Cs vapor at a fixed density (i.e. ensure the number of Cs atoms remained constant).

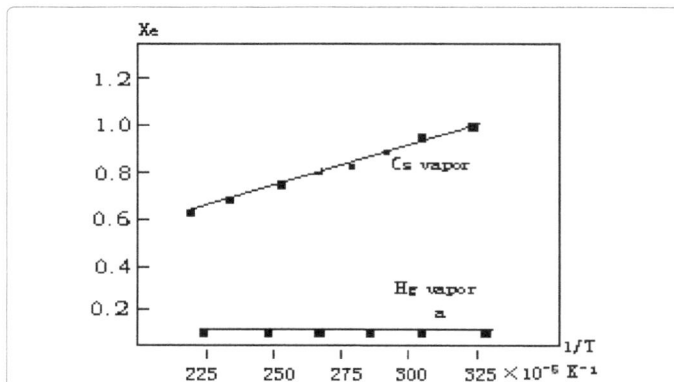

Figure 3: The diagram shows χ_e versus T^{-1}. For Cs vapor $\chi_e = 0.007 + 282.3/T$, but $\chi_e = 0.003$ for Hg vapor. So Cs is polar atom but Hg is non-polar.

T(K)	308	326	345	357	377	392	425	448
1/T(×10⁻³)	3.2468	3.0675	2.8985	2.8011	2.6525	2.5510	2.3529	2.2321
C(pF)	112.7	109.6	106.9	105.4	102.9	101.3	98.0	95.8
χ_e	0.9232	0.8703	0.8242	0.7986	0.7560	0.7287	0.6724	0.6348

Table 1: The electric susceptibility χ_e of cesium vapor at different temperature T.

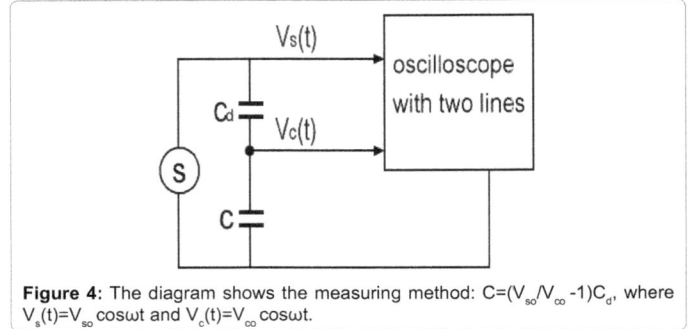

Figure 4: The diagram shows the measuring method: $C = (V_{so}/V_{co} - 1)C_d$, where $V_s(t) = V_{so}\cos\omega t$ and $V_c(t) = V_{co}\cos\omega t$.

Their capacitances were still measured by the digital meter, the vacuum capacitance is $C_{10} = 58.6$ pF (for Cs) and $C'_{10} = 63.9$ pF (for Hg). Next, a small amount of Cs or Hg material was put into the two capacitors. In order to ensure the number of atoms remained constant, we adopted an ingenious technique (Figure 2) [10]. After fill with Cs or Hg vapor, the capacitances are $C_1 = 112.9$ pF (for Cs) and $C'_1 = 64.1$ pF (for Hg) respectively. We measured their capacitances at different temperatures. As was expected, the capacitance of Hg vapor, $C'_1 = 64.1$ pF, remains constant at different temperatures (i.e., B=0) and $\chi_e = A \approx 0.003$. But the capacitance of Cs vapor decreases gradually from 112.7 pF to less than 96 pF, and $\chi_e = A + B/T$!

The Figure 3 shows the two experimental results. Table 1 gives the electric susceptibility of cesium vapor. By least-square method we obtain $B = (282.3 \pm 1.2)$ K and $A = 0.007$, because $(\Delta\chi_e/\chi_e)^2 = (\Delta B/B)^2 + (\Delta T/T)^2 = (\Delta C/C)^2 + (\Delta C_0/C_0)^2$, so $(\Delta B/B)^2 < (\Delta C/C)^2 + (\Delta C_0/C_0)^2$, and $\Delta B/B < 0.004$. They can be expressed as follows

For cesium vapor $\chi_e = 0.007 + 282.3/T$, for mercury vapor $\chi_e = 0.003$ (5)

The two results formed a sharp contrast because Cs atom is polar atom but Hg atom is non-polar.

The second experiment: we measured the capacitance of Cs vapor at different voltages under a fixed density n_2 and $T_2 = 353$ K. The vacuum capacitance of the apparatus is $C_{20} = (66.0 \pm 0.1)$ pF, where $H_2 = 6.8$ mm. The Figure 4 shows the experimental method [10]. C is the measured capacitor, which filled with cesium vapor. C_d is a reference capacitor. Two signals $V_c(t) = V_{co}\cos\omega t$ and $V_s(t) = V_{so}\cos\omega t$ were measured by a two channel digital oscilloscope (Tektronix TDS 210 USA). From Figure 4, we have $(V_s - V_c)/V_c = C/C_d$ and $C = (V_{so}/V_{co} - 1)C_d$. The voltages V_{so} could be adjusted from zero to 800 V. The frequency could be adjusted from one to 10^6 Hz. The measurement was started in 0.01 volt. When $V_1 = V_{co} \leq 0.3$ volts, $C_1 = 130.0$ pF is approximately constant. When $V < V_C$, the peak difference of the two signals waveforms is large, this image means that $C >> C_0$ (Figure 5). When $V_2 = V_{co} = 350$ volts $>> V_C$, $C_2 = 68.0$pF $\approx C_0$. This oscilloscope shows that the peak of the two signals is very close and they almost overlap, this precious image means that large-scale BEC at normal temperature has been observed (Figure 6). If almost all the dipoles in a gas were to line up with an external electric field, this effect is called the saturation polarization. The experimental C–V curve shows that the saturation polarization of Cs vapor is easily observed when $V_2 >> V_C$ (Figure 7) [10]. Table 2 gives the measured values of Cs vapor at different voltages.

RP Feynman, he was awarded the 1965 Nobel Prize, once stated that "*The electric field tends to line up the individual dipoles to produce a net moment per unit volume. If all the dipoles in a gas were to line up, there would be a very large polarization, but that does not happen*" [25]. Our experiment confirmed his prediction, and it tells us that the saturation polarization of Cs vapor is an unexpected discovery. When the peak of the two signal waveforms is very close, and the two waveforms are almost overlapped, this image means the BEC occurs. The capacitance $C \approx C_0$, it implies that Cs vapor entered a quasi-vacuum state! Figure 8 shows that the condensate fraction versus the external voltages, a large-

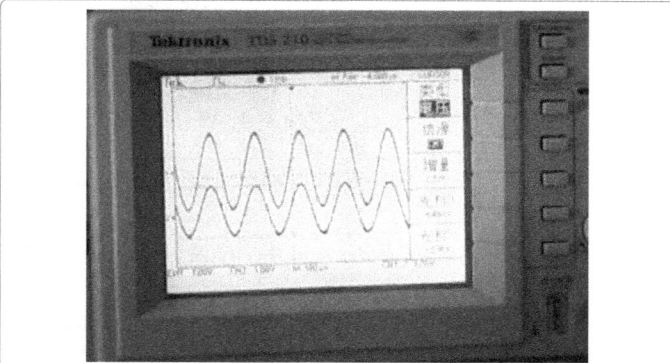

Figure 5: This diagram shows that when $V_{co} < V_c$, the peak difference of the two signals is large, this image means that $C >> C_0$ (C_0 is the vacuum capacitance). The calculated result shows that this state is far from BEC.

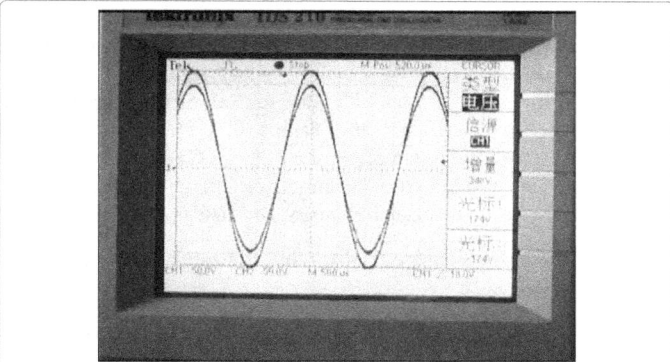

Figure 6: This diagram shows that when $V_{co} = 350V >> V_c$, the two signals almost overlap, and $C \approx C_0$, this precious image means that large scale BEC at normal temperature has been observed. Note that this oscilloscope uses an attenuator, its attenuation rate is 2.

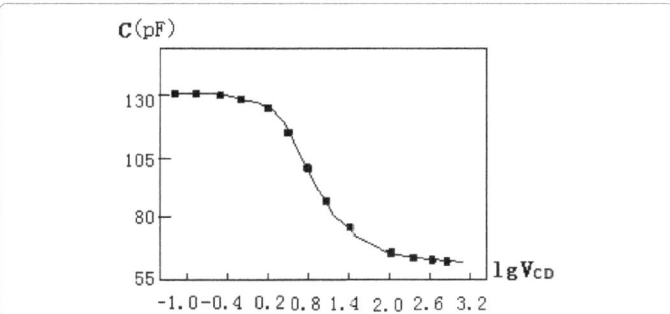

Figure 7: The C – V curve shows that when $V >> V_c$, the capacitance decreased to $C \approx C_0$, it implies that almost all the Cs atoms were to line up along the field, and BEC occurs.

V(volt)	0.01	0.3	7.1	63	173	232	350	∞
C(p F)	130	130	118	76.6	70	69	68	66
X_e	0.9697	0.9697	0.7879	0.1606	0.0606	0.0454	0.0303	0
P(e. cm⁻²)	7.9×10^3	2.4×10^5	4.6×10^6	8.2×10^6	8.5×10^6	8.58×10^6	8.63×10^6	8.72×10^6

Table 2: The measured values of Cs vapor at different voltages T=353K.

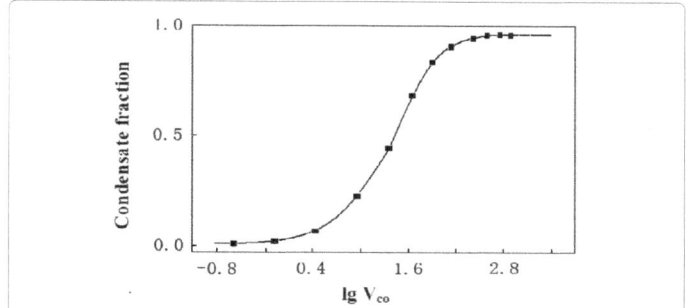

Figure 8: Condensate fraction versus voltages: when $V_{co} \geq 350$volts, condensates contained up to 1928×10^{17} atoms. A large-scale BEC of cesium at T=353K has been observed.

scale BEC has been observed when $V_2 \geq 350$ volts [10]. When $V >> V_c$, the capacitance $C \approx C_0$, $\chi_e \approx 0$, and the polarization increase to $P = P_{max}$, the changes of these electrical quantities show that a large-scale BEC at normal temperature has been observed.

The condensate fraction

The local field acting on a molecule or atom in a gas is almost the same as the external field E [23]. For polar molecules or atoms, $\chi_e = n\alpha + n d_0 L(a) / \varepsilon_0 E$, where $a = d_0 V/kTH$. Langevin function $L(a) = [(e^a + e^{-a}) / (e^a - e^{-a})] - 1/a = \coth a - 1/a$. $L(a)$ equals the average value of $\cos\theta$ [26]:

$$L(a) = < \cos\theta > = f \int_0^\pi \cos\theta \exp(d_0 E \cos\theta / kT)\sin\theta d\theta,$$
$$f = [\int_0^\pi \exp(d_0 E \cos\theta / kT)\sin\theta d\theta]^{-1} \qquad (6)$$

Where, f is normalized constant, θ is the angle between d_0 and E [25]. The electric polarizability of cesium is $\alpha = 59.6 \times 10^{-30} m^3$ [27], the density $n < 8.0 \times 10^{22} m^{-3}$, and induced susceptibility $\chi_e = n\alpha < 4.8 \times 10^{-6}$ can be neglected. We obtain $\chi_e = nd L(a)/\varepsilon_0 E$, where d is PDM of Cs atom. When $a << 1$, $L(a) \approx a/3$, and $\chi_e = n d_0^2 /3kT\varepsilon_0$, this is the familiar Langevin formula. The polarization P, the average dipole moment per unit volume, is defined as

$$P = \chi_e \varepsilon_0 E = n d L(a) \qquad (7)$$

Formula (6) and (7) are clearly indicated that $L(a)$ equals the condensing fraction of BEC. When $a >> 1$, $L(a) \approx 1$ and $P_{max} \approx nd$, it implies the saturation polarization. Note that E=V/H and the formula of the parallel-plate capacitor $\varepsilon_0 = C_0 H/S$, from $\chi_e = n d L(a)/\varepsilon_0 E$, we obtain

Condensate fraction: $L(a) = a(C - C_0)/\eta$ or $C = \eta L(a)/a + C_0$ $\qquad (8)$

Where, $\eta = Snd^2/kTH$ is the capacitance constant. When $V_1 = 0.3$ V, $a_1 << 1$, $L(a) = a/3$ and $\eta = 3(C_1 - C_{20}) = 192$ pF. When $V_2 = 350$ V, $C_2 - C_{20} = 2.0$ pF, $a_2 = \eta L(a_2)/C_2 - C_{20} = 95$, $L(95) = 0.9896$. The critical voltage is $V_c = V_2 2\pi e /a_2 = 62.9$ V.

Notice that we deduced equation (8) from the formula of the

V(volt)	0.01	0.30	7.1	63	173	232	350	∞
C(p F)	130	130	118	76.6	70.0	69.0	68.0	66.0
a	2.7×10^{-3}	0.0814	1.93	17.1	47.0	63.0	95.0	∞
L(a)	9×10^{-4}	0.02714	0.5249	0.9415	0.9787	0.9841	0.9896	1.0
N_c (10^{17})	0.0018	0.0529	1.023	1.834	1.906	1.917	1.928	1.948
S (Nk)	2.531	2.530	2.076	-1.2×10^{-4}	-1.012	-1.305	-1.716	$-\ln\infty$

Table 3: a, L(a), N_c and S of Cs vapor at different voltages (V) T=353K.

Electric susceptibility of alkali atoms	χ_e =A	χ_e =A+ B/T
PDM of alkali atom	d=0	d= (C – C_0)V/L(a)Sn
Entropy of the system		S=Nk ln 2π e /a
The energy density of the system		U=–n d L(a) E
Condensate fraction formula		L(a)=a (C – C_0)/η(η is constant)
The density of alkali gas	$10^{13} – 10^{15}$ cm^{-3}	5.65×10^{14} cm^{-3}
Critical condition of BEC	Tc ≤ 10 μK	Vc= 63 volts
Number of condensed atoms	10^4-10^6	1.928×10^{17}
Most atoms are in BEC	T << Tc is difficult	V >> Vc is easily
Macroscopic occupation of ground state	No	Yes
A large-scale BEC	No	Yes

Table 4: A sharp contrast between previous and our BEC experiment.

parallel-plate capacitor ε_0 =C_0 H/S, so the cylindrical capacitor must be regarded as an equivalent parallel-plate capacitor with the plate area S=C_0H/ε_0. In the three experiment, the equivalent area is S_0'=C_0' H_0/ε_0=5.855 × 10^{-2} m^2 and S_2'=$C_{20}$$H_2$/$\varepsilon_0$=5.069 × 10^{-2} m^2. Since the density n_2 is unknown, from equation (8) we obtain n_2=(C_2 – C_{20}) V_2 L(a) S_0' n_0/(C_t' – C_0')V_0 L(a_2) S_2'=5.652 × 10^{14} cm^{-3}, where a=a_2 $V_0$$T_2$$H_2$/$T_0$ $H_0$$V_2$=0.1722 and L($a$) =0.05729. According to the root mean square rule, the statistical error of the measured value is Δn_j/n ≤ 0.05, and the systematic error is Δn_j/n ≤ 0.04, so n_2= [5.65 ± 0.28(stat) ±0.23(syst)] × 10^{14} cm^{-3}. The volume of C_{20} is $S_2'$$H_2$=3.447 × 10^2 cm^3, the total number of Cs atoms is N=1.948 × 10^{17} and the number of condensing atoms is N_c=N L(a). Table 3 provides a complete analysis of BEC of Cs atoms.

From equation (8) we can deduce the formula of the atomic PDM

$$d=(C – C_0)V/L(a)n S \qquad (9)$$

For example, by second experimental data, we obtain d_{Cs} =2.469×10^{-29} C. m=1.543 × 10^{-8} e. cm, where (C – C_0)=2pF, V=350 volts, L(a) =0.9896, S=S_2'=5.069 × 10^{-2} m^2 and n=n_2=5.652 × 10^{20} m^{-3}. The statistical error of the measured value is Δd_j/d ≤ 0.08 due to $\Delta C_t'$/C_t' ≤ 0.005, $\Delta C_0'$/C_0' ≤0.003, ΔV_0/V_0≤ 0.03, Δn_0/n_0 ≤ 0.06, $\Delta S_0'$/S_0' ≤ 0.04. Considering all systematic error, Δd_2/d ≤ 0.06, so d_{Cs}=[1.54 ± 0.12(stat) ± 0.09 (syst)] × 10^{-8} e. cm.

In our experiment, the maximum field intensity is E_{max}=V_2/H_2=5.15 × 10^4 V/m, and therefore the maximum induced dipole moment is d_{ind} ≤ 2.72 × 10^{-35} C.m =1.70 × 10^{-14} e.cm, and it can be neglected.

Discussion

If cesium atom has a non-zero PDM, why it does not violate the time reversal and parity symmetry?

According to quantum theory, an atom in its ground state at most has an extremely small PDM, d ≤ e × 10^{-20} cm, it points along the nuclear spin axis and arise mainly from the nuclear spin [19,21,22]. Under time reversal the direction of the spin changes, while the direction of the PDM does not change. Therefore, the PDM of an atom would violate time reversal symmetry. By the CPT theorem it also implies a violation of CP symmetry [21]. A representative result as follows: d(Hg)=[0.49 ± 1.29(stat) ± 0.76 (syst)]×10^{-29} e.cm [22]. In short, if the PDM of atom violates time reversal symmetry, it must have two characteristics: the

first is very small, $d \ll 10^{-20}$ e. cm, and the second is arises from the nuclear spin.

However, the linear Stark effect of hydrogen atom provides a new example. This effect showed that the hydrogen atom (n=2) has a PDM of magnitude -3ea_0=1.59×10^{-8} e.cm. Since the nuclear spin was completely irrelevant to the calculation of the PDM, and Bohr radius (a_0=0.53 × 10^{-8} cm) is far greater than the nuclear radius (r_0 ≈10^{-11} cm), so this PDM has nothing to do with the nuclear spin, and only arises from the asymmetrical charge distribution of hydrogen atom. LD Landau once stated that the hydrogen atom has a dipole moment whose mean value is d=-3ea_0, "This is in accordance with the fact, in a state determined by parabolic quantum numbers, the distribution of the charges in the atom is not symmetrical about the plane z=0" [18]. So similar to the first excited state of hydrogen atom, the PDM of Cs atom doesn't arise from the nuclear spin but from asymmetrical charge distribution, and it doesn't violate time reversal and parity symmetry [21,22].

If cesium atom has a large PDM, why its linear Stark effect has not been observed?

This is a challenging question. As two concrete examples, first let us deal with the fine structure and the linear Stark shifts of the hydrogen (n=2). The wavenumber of the fine structure of the hydrogen (n=2) is only 0.33 cm^{-1} for the Hα lines of the Balmer series, where λ=656.3 nm. The splitting is only Δ λ=0.33 × $(656.3 \times 10^{-7})^2$=0.014 nm, therefore the fine structure is difficult to observe [28]. The linear Stark shift of the hydrogen (n=2) is proportional to the field intensity: ΔW=d_HE=1.59 × 10^{-8} E e.cm. When E=10^5 V/cm, ΔW=1.59 × 10^{-3} eV, this corresponds to a wavenumber of 12.8 cm^{-1}. So the linear Stark shifts is Δλ=ΔW λ^2/hc =12.8 × $(656.3 \times 10^{-7})^2$=0.55 nm, and the linear Stark shift of the hydrogen (n=2) is easily observed [28]. However, when V =350 V, the most field intensity is E_{max}=V/H=515 V/cm. When the external electric field increases to E=E_{max}, almost all the Cs atoms (more than 98.9%) were to line up along the field, Cs vapor no longer absorb energy, this Stark effect will not occur. if the PDM of cesium is d_{Cs}=1.54 × 10^{-8} e.cm, and the most splitting of the energy levels is ΔW$_{max}$=$d_{Cs}$$E_{max}$=7.93 × 10^{-6} eV. This corresponds to a wavenumber is ΔW$_{max}$/hc ≤ 0.064 cm^{-1}. On the other hand, observed values for a line pair of the first primary series of Cs atom (Z=55, n=6) are λ_1=894.3 nm and λ_2=852.1 nm [24,28]. So the most linear Stark shift of Cs atoms is only Δ λ=ΔW $(\lambda_1 + \lambda_2)^2$/4hc=0.0048 nm. It is so small, in fact, that a direct observation of the linear Stark shifts of Cs atom is not possible by conventional spectroscopy!

A sharp contrast between previous and our BEC experiment

In order to illustrate the original innovation of our experiments, Table 4 gives a detailed comparison between our and previous BEC experiments [2-6]. From the formula of PDM, if $(C - C_0) \neq 0$, and is bound to get: $d \neq 0$. But in the past few decades, scientists have never measured the capacitance of the alkali atoms, so they missed this significant discovery.

Rigorous mathematical proof that BEC of cesium vapor is bound to occur

From equation (8) $C = \eta L(a)/a + C_0$, we construct a new function

$$f(a) = L(a)/a = [(e^a + e^{-a}) / a (e^a - e^{-a})] - 1/a^2 \qquad (10)$$

Now we find the inflection point of the function $f(a)$. From the second order derivative of this function is zero, we obtain $f''(a) = [(2e^{-3a} -2e^{-a} -8ae^{-a} +8a^2e^{-a} +2e^{3a} -2e^a +8ae^a +8a^2e^a)/ a^3(e^a - e^{-a})^3] - 6/a^4 = 0$. From $f''(1.9296812) = -10^{-9} < 0$ and $f''(1.9296814) = 2 \times 10^{-9} > 0$, we obtain its inflection point is $a_p = 1.9296813 \approx 1.93$. The voltages of the inflection point is $V_p = 7.1V$ and $lgV_p = 0.85$. By contrast with Fig.4, it is clear that our polarization equation Eq.(8) is correct. This result shows that when $lgV_{CD} < 0.85$, the C–V curve is upper convex; when $lgV_{CD} > 0.85$, the curve is down convex. This result shows that when the voltage increases to thousands of volts, C will inevitably approach C_0.

A discussion of many interesting questions, such as the definition of Boltzmann constant, can be found in the reference 10.

Conclusions

In theory, $<\psi_E|-er|\psi_E> = 0$ only means that the average PDM of large number of alkali atoms is zero, but doesn't mean that the PDM of individual alkali atom is zero. Despite 6s and 6p states of cesium are not degenerate, but Cs may be polar atom doesn't conflict with quantum mechanics because it is hydrogen-like atom.

All kinds of alkali atoms are non-polar atom, which is an untested hypothesis, and we must test it by experiments. We measured the capacitance at different temperatures, our experiment proved that Cs atom is polar atom, and it becomes a dipole. But in the past, scientists have never measured the capacitance of alkali atoms, so they missed this significant discovery.

BEC has three main features: BEC is a macroscopic occupation of the ground state; BEC is a condensation in momentum space; Bose gas would undergo a phase transition. Our experiments are fully in line with these three main features, so although we have not used laser cooling techniques, our experiments are an ideal BEC.

Ultra-low temperature is in order to make Bose gas phase transition, and we use the critical voltage V_c to describe phase transition, and therefore ultra-low temperature is not necessary. Because there are only a handful of laboratories in the world that can achieve ultra-low temperatures, it limits the majority of scientists involved in the study. This new technology will completely change the current situation, and it will be used and praised by the vast majority of scientists.

The entropy of a system is a measure of the disorder of molecular or atomic motion. No doubt, it is the most important concept in BEC. This formula, $S = Nkln2\pi e/a$, contains two fundamental constants in nature

($\pi \approx 3.14159$ and $e \approx 2.71828$), it reflects the objective laws of BEC.

The presence of the inflection point proved that the saturation polarization of Cs vapor is inevitable, and therefore BEC of Cs vapor is also a certainty. Including the introduction of entropy, this is the two successful examples of applying mathematics to explain abstruse physical phenomena.

Both BEC and superconductivity are condensed in the momentum space. When the resistance R or the capacitance C was reduced to close to the vacuum value, these two kinds of condensation occur. So BEC is also a quasi-superconducting state. This fact fully reflects the harmony and unity of nature.

Our experiments are easily repeated in other laboratories, because the details have been described in this paper. Once scientists completed these measurements, they will obtain the same results as our experiments. They will discover that a large-scale BEC of cesium atoms is easily observed when an electric field is applied!

Acknowledgements

This research was supported by the NSF of Guangdong Province (Grant No. 021377). The author thanks Prof. Xiang-You Huang (Peking University), Dr. Yu-Sheng Zhang, Director Xun Chen, Engineer Yi-Quan Zhan (Peking University), Engineer Jia You and our colleagues Rui-Hua Zhou, Ming-Jun Zheng, Xue-Ming Yi, Zhao Tang and Xin Huang for their help with this work.

References

1. Pais A (1982) Subtle is the Lord: The Science and the Life of Albert Einstein. Oxford University Press, USA. p: 430.

2. Bradley C, Sackett C, Tollett J, Hulet R (1995) Evidence of Bose-Einstein Condensation in an Atomic Gas with Attractive Interactions. Phys Rev Lett 75: 1687.

3. Davis K, Mewes MO, Andrews MR, van Druten NJ, Durfee DS, et al. (1995) Bose-Einstein Condensation in a Gas of Sodium Atoms. Phys Rev Lett 75: 3969.

4. Modugno G, Ferrari G, Roati G, Brecha RJ, Simoni A, et al. (2001) Bose-Einstein Condensation of Potassium Atoms by Sympathetic Cooling. Science 294: 1320-1322.

5. Anderson MH, Ensher J, Matthews M, Wieman C, Cornell E (1995) Observation of Bose-Einstein Condensationin a dilute Atomic Vapour. Science 269: 198.

6. Weber T, Herbig J, Mark M, Nägerl H, Grimm R (2003) Bose-Einstein Condensation of Cesium. Science 299: 232-235.

7. Blundell SJ, Blundell KM (2009) Concepts in Thermal Physics. (2nd edn) Oxford University Press, USA. p: 369.

8. Huang XY, You PL (2002) Permanent Electric Dipole Moment of an Rb Atom. Chin Phys Lett 19: 1038.

9. Huang XY, You PL, Du WM (2004) The Experiment on the Saturation Polarization of Rb Vapour. Chin Phys 13: 11.

10. You PL (2016) Bose-Einstein Condensation in a Vapor of Sodium Atoms in an Electric Field. Physica B 401: 84-92.

11. Yan Z, Konotop VV (2009) Exact Solutions to Three-Dimensional Generalized Nonlinear Schrödinger Equations with Varying Potential and Nonlinearities. Phys Rev E 80: 036607.

12. Yan Z, Konotop VV, Yulin AV, Liu WM (2012) Two-dimensional Superfluid Flows in Inhomogeneous Bose-Einstein Condensates. Phys Rev E 85: 016601.

13. Yan Z, Konotop V, Akhmediev N (2010) Three-dimensional Rogue Waves in Nonstationary Parabolic Potentials. Phys Rev E 82: 036610.

14. Ballentine LE (1998) Quantum Mechanics a Modern Development. World Scientific Publishing Co. Pte. p: 286.

15. Ballentine LE (1970) The Statistical Interpretation of Quantum Mechanics. Rev Mod Phys 42: 358.

16. Ballentine LE, Yang Y, Zibin JP (1994) Inadequacy of Ehrenfest's Theorem to Characterize the Classical Regime. Phys Rev A 50: 2854.

17. Griffiths DJ (2005) Introduction to Quantum Mechanics. (2nd edn) Pearson Education. p: 15.

18. Landau LD, Lifshitz EM (1999) Quantum Mechanics. Beijing World Publishing Corporation. p: 285-290.

19. Schiff LI (1968) Quantum Mechanics. McGraw-Hill, New York. p: 252.

20. Greiner W (1994) Quantum Mechanics an Introduction. Springer-Verlag Berlin/ Heidelberg. p: 213.

21. Fortson N, Sandars P, Barr S (2003) The Search for a Permanent Electric Dipole Moment Physics Today. pp: 33-39.

22. Griffith WC, Swallows MD, Loftus TH, Romalis MV, Heckel BR, et al. (2009) Improved Limit on the Permanent Electric Dipole Moment of Hg199. Phys Rev Lett 102: 101601.

23. Jackson JD (1999) Classical Electrodynamics. (3rd edn) John Wiley and Sons, Inc, USA. p: 162-173.

24. Dean JA (1998) Lange's Handbook of Chemistry. McGraw-Hill, Inc, New York.

25. Feynman RP, Leighton RB (1964) The Feynman Lectures on Physics: Volume 2. Addison-Wesley Publishing Co, Boston. p: 11-3.

26. Bottcher CJF (1973) Theory of Electric Polarization. Elsevier, Amsterdam. p: 161.

27. Lide DR (1998) Handbook of Chemistry and Physics. CRC Press, Boca Raton New York.

28. Haken H, Wolf HC (2000) The Physics of Atoms and Quanta. Springer-Verlag Berlin Heidelberg. p: 172.

6

Development of Graphite-DNA Polymer Composites as Electrode for Methanol Fuel Cells

Chtaini A[1]*, Touzara S[1], Cheikh Ould S'Id E[2,4], Chamekh M[2], Mabrouki M[3] and Kheribech A[4]

[1]Molecular Electrochemistry and Inorganic Materials Team, Sultan Moulay Slimane University, Faculty of Science and Technology Béni Mellal, Morocco
[2]Faculty of Science and Technology, University of Science, Technology and Medicine, B.P. 5026, Nouakchott, Mauritania
[3]Laboratory of Industrial Engineering, Sultan Moulay Slimane University, Faculty of Science and Technology Béni Mellal, Morocco
[4]Faculty of Sciences, Laboratory of water and environment (team Biomaterials and Electrochemistry), University Chouaib Doukkali, El Jadida, Morocco

Abstract

DNA aggregates were electro less deposited onto carbon paste electrode, firstly and electrode surface was coated by polymer film to protect the DNA film. Prepared electrode has shown great activity towards the oxidation of methanol, and no effect of empoisoning is observed. The effect of various parameters such as scan rate and methanol concentration, on the electro catalytical oxidation of methanol has also been investigated. The morphological study of electrode surface was investigated by Atomic Force Microscopy (AFM) and optical microscopy.

Keywords: Fuel cells; Methanol; Cyclic voltammetry; AFM

Introduction

Fuel cells are high energy density, energy conversion devices, which have application for portable power [1]. Electric current is generated in the fuel cell by the direct electrochemical oxidation of either hydrogen (proton exchange membrane fuel cell, PEM) or methanol (Direct Methanol Fuel Cell, DMFC). The electrochemical processes that yield energy are essentially pollution free. Water formed during the operation of the device is beneficial in space travel and submarines. Applications of fuel cells are diverse ranging from stationary (individual homes or district schemes) or mobile (transportation as cars, buses, etc.), mobile phones and lap top computers [2,3]. Hydrogen is currently the only practical fuel for use in the present generation of fuel cells. The main reason for this is this is its high electrochemical reactivity compared with that of the more common fuels from which it is derived, such as hydrocarbons, alcohols, or coal. Also, its reaction mechanisms are now rather well understood [4,5] and are characterized by the relative simplicity of its reaction steps, which lead to no side products. Pure hydrogen is attractive as a fuel, because of its high theoretical energy density, its innocuous combustion product (water), and its unlimited availability so long as a suitable source of energy is available to decompose water.

One of the disadvantages of pure hydrogen is that it is a low density gas under normal conditions, so that storage is difficult and requires considerable excess weight compared with liquid fuels. Methanol offers several advantages as a fuel. It is inexpensive but has a relatively high energy density and can be easily transported and stored. It can be supplied to the fuel cell unit from a liquid reservoir which can be kept topped up, or in cartridges which can be quickly changed out when spent.

To obtain high current densities, the used electrode must be a thin, porous and high surface area body activated by the presence of a suitable catalyst, which must make electronic contact with the remainder of the electronically conducting structure. The electrode must provide a large number of suitably active reaction sites where both the reactant and the electrolyte can come into contact. It must also maintain a stable interface between the electrolyte and the active species.

The carbon paste electrodes (CPEs) are cheaper and are suitable for preparing the electrode material with desired composition and pre-determined properties [6,7]. The electrochemical response of CPE mainly depends on the properties of the modifying species. The modification of the carbon paste electrode can be done by different ways like grinding in an agate mortar [8,9] electro polymerization [10,11] and immobilization method [12]. The DMFC using superior membrane materials have been subject for extensive research [13-17].

In This study, the morphology of the Polyacrylic Acid membrane coating DNA deposit film onto CPE has been characterized via atomic force microscopy (AFM) and optical microscopy. The catalytic oxidation of methanol was studied at the prepared electrode, using cyclic voltammetry (CV) and square wave voltammetry (SWV). This study which uses graphite-DNA polymer composites as oxidation catalysts, indicating the DNA aggregates can be an anodic electro catalyst candidate for fuel cells.

Experimental

Apparatus

Electrochemical experiments were performed using a Volta lab potentiostat (model PGSTAT 100, Eco Chemie B.V., Utrecht, The Netherlands) driven by the general purpose electrochemical systems data processing software (Volta lab master 4 software). A conventional three-electrode systems consisting of the HAP-modified carbon paste working, platinum counter and SCE reference electrodes were used.

Reagents

All solutions used in this work, were prepared by dissolving the initial product, without further purification step. Carbon paste was supplied from (Carbone, Lorraine, ref 9900, French). All other regents used were of analytical grade. Bid stilled deionized water (BDW) was

*Corresponding author: Abdelilah C, Molecular Electrochemistry and Inorganic Materials Team, Faculty of Science and Technology, Sultan Moulay Slimane University, Béni Mellal, Morocco, E-mail: a.chtaini@usms.ma

used throughout the work. The DNA used in this work is taken from quail blood, according to the protocol below:

- 5 l of blood taken from the axillary vein is poured into an Eppendorf containing 500 µl of danazol

- The solution is stirred for 5 min by vortexing and centrifuged at 6000 rpm/5 min

- The supernatant is removed and then 400 µl of isopropanol is added to the residue remaining in the Eppendorf

- The mixture was vortexed and centrifuged at 6000 rpm/5 min

- The supernatant was removed and 500 µl of pure water was added to suspend the DNA.

Preparation of the CPE

The carbon paste electrode (CPE) was prepared by thoroughly hand-mixing of graphite powder (CP). The obtained paste was dried at room temperature then a portion of the resulting paste was grounded and packed firmly into homemade PTFE cylindrical tube (geometric area 0.1256 cm^2) electrode. Electrical contact was established with a bar of carbon. DNA-CPE's were prepared by immobilizing the DNA system by soaking the preformed carbon paste electrode in a solution containing the DNA solution. And thereafter the surface of the prepared electrode is covered by the polymer, glued on the edges with Araldite.

Results and Discussion

Morphological study

Surface morphologies were characterized with AFM (Figure 1). The surface of the polymer shows a structure with reliefs, the size of the aggregates is not uniform over the entire surface. This structure shows the presence of pores of different sizes.

The electrochemical properties of carbon paste electrode coated or not by polymer in 0.1 M NaCl solution, were characterized with cyclic voltammetry. Figure 2 shows a typical cyclic voltammograms of the CPE and Polymer-CPE. We can see that the presence of the polymer on the surface of the CPE (Figure 2b), leads to a slight decrease in the current density, but the shape of CV is not reached.

Figure 3 shows the representative cyclic voltammograms recorded

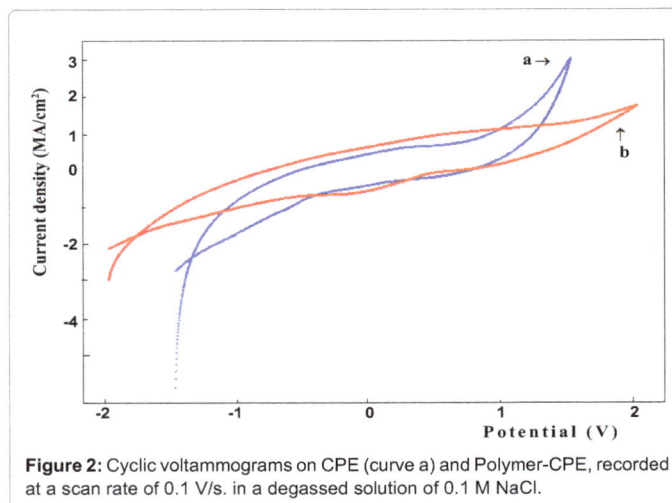

Figure 2: Cyclic voltammograms on CPE (curve a) and Polymer-CPE, recorded at a scan rate of 0.1 V/s. in a degassed solution of 0.1 M NaCl.

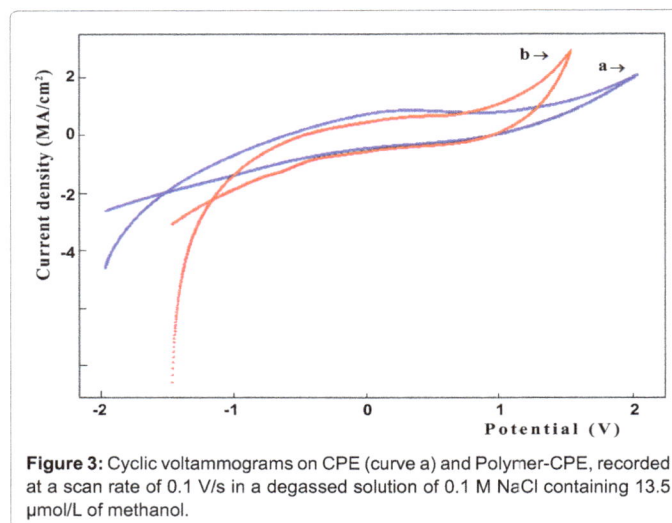

Figure 3: Cyclic voltammograms on CPE (curve a) and Polymer-CPE, recorded at a scan rate of 0.1 V/s in a degassed solution of 0.1 M NaCl containing 13.5 µmol/L of methanol.

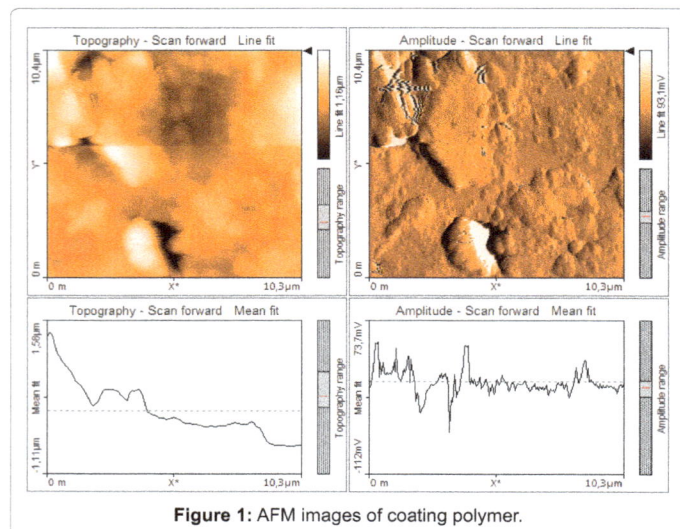

Figure 1: AFM images of coating polymer.

respectively, at CPE (curve a) and Polymer-CPE (curve b), in methanol fuel solution in 0.1 M NaCl. CPE is used control. The current densities are relatively high on the Polymer-CPE compared to the CPE. The methanol oxidation starts at 0.8 V, and characterized by a sudden rise in the current, which corresponds to the onset potential.

The electrochemical characterization of DNA-CPE and Polymer-DNA-CPE in a blank 0.1 M NaCl solution is given by Figure 4. The electroless deposition of DNA on the CPE surface is manifested by the appearance of two redox peaks on cyclic voltammograms, the first in the anodic scan direction at about 0.7 V and the second one on the of cathodic scan at -1.1 V.

By coating the surface of the DNA-CPE with the polymer, the VC retains its shape, whereas the current densities decrease, due to the non-conductive nature of the polymer (Figure 4b).

The electrochemical behavior of methanol was studied at the surface of different electrodes (DNA-CPE and Polymer-DNA-CPE) by CV experiments in 0.1 M NaCl. The results are shown in Figure 5. The addition of 13.6 µmol/L methanol to the electrolytic solution causes an increase in the current density, in particular in the presence of the Polymer-DNA-CPE. The methanol oxidation starts at -1 V, this potential value corresponds to the onset point.

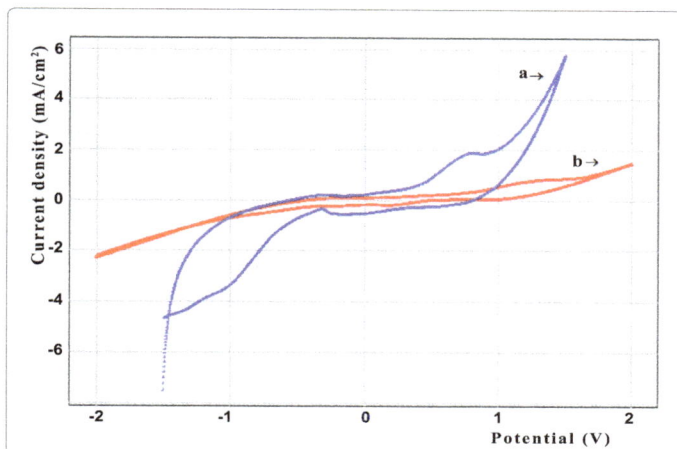

Figure 4: Cyclic voltammograms on DNA-CPE (curve a) and Polymer-DNA-CPE, recorded at a scan rate of 0.1 V/s in a degassed solution of 0.1 M NaCl.

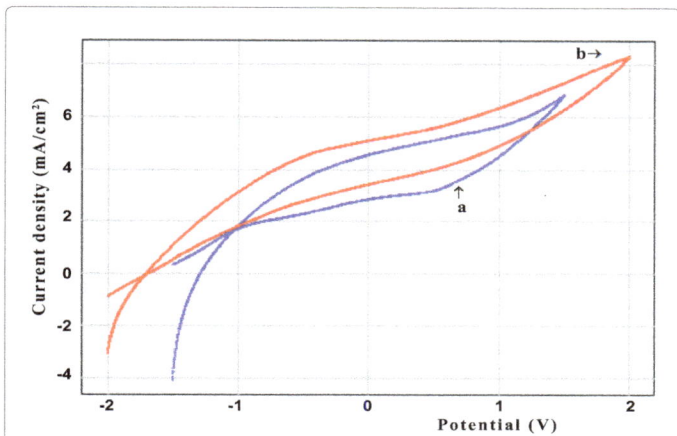

Figure 5: Cyclic voltammograms on DNA-CPE (curve a) and Polymer-DNA-CPE, recorded at a scan rate of 0.1 V/s in a degassed solution of 0.1 M NaCl containing 13.5 µmol/L of methanol.

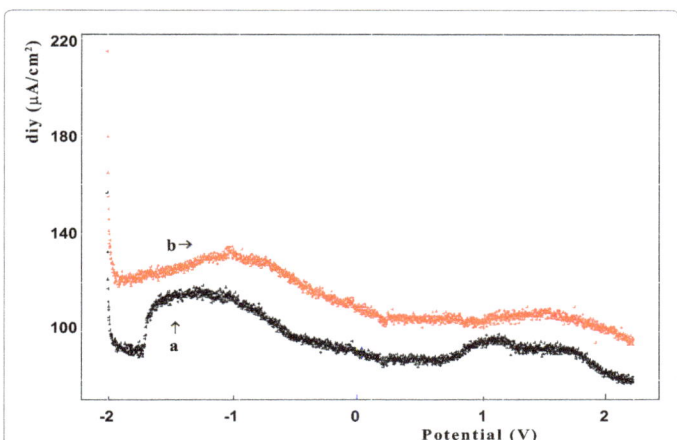

Figure 6: Square wave voltammograms on DNA-CPE (curve a) and Polymer-DNA-CPE, recorded in a degassed solution of 0.1 M NaCl containing 13.5 µmol/L of methanol.

The methanol oxidation is manifested in square wave volatmmograms by the appearance of two peaks, the first towards

about -1 V and the second at 1 V. The current densities are remarkably higher in the case of the Polymer-DNA-CPE (Figure 6).

CVs for methanol electro oxidation on Polymer-NDA-CPE at different scan rates are illustrated in Figure 7. The methanol oxidation current density increase with increasing scan rate ranging from 10 mV/s to 100 mV/s in the presence of 13.6 µmol/L methanol. This linearity indicates that the methanol electrocatalytical oxidation is a diffusion controlled process.

The Figure 8 shows the cyclic voltammograms for different concentration of methanol from 3.4 µmol/L to 17 µmol/L in 0.1 M NaCl with a scan rate of 100 mV/s at Polymer-NDA-CPE. The methanol oxidation current density increase with increasing the concentration of methanol.

The optical microscopy images of the Polymer-DNA-CPE, Polymer-CPE, CPE, DNA-CPE and Polymer-CPE after methanol electro oxidation are presented in Figure 9. The images show adhesion of the polymer to the electrode surface, and protect the DNA film deposited on the carbon paste surface.

Conclusion

In this work, we have successfully prepared a novel electrode. The

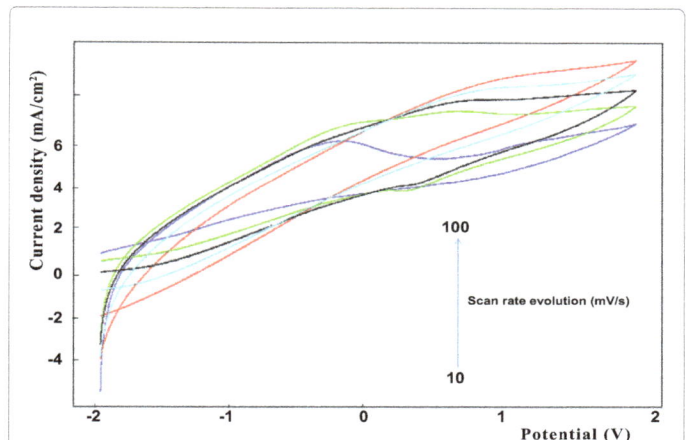

Figure 7: Cyclic voltammograms on Polymer-DNA-CPE, recorded at different scan rates in a degassed solution of 0.1 M NaCl containing 13.5 µmol/L of methanol.

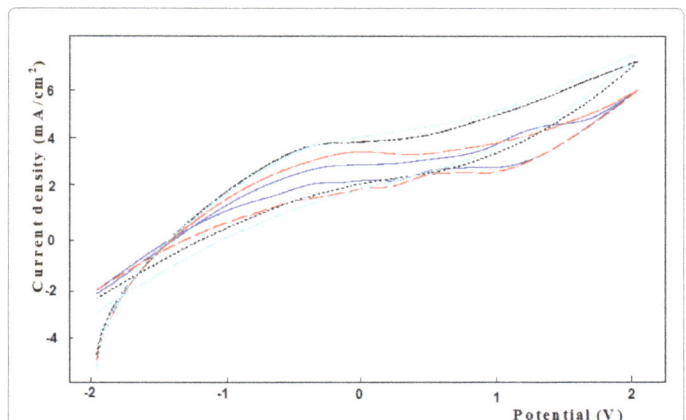

Figure 8: Cyclic voltammograms on Polymer-DNA-CPE, recorded at a scan rates of 0.1 mV/s in a degassed solution of 0.1 M NaCl containing different concentrations of methanol.

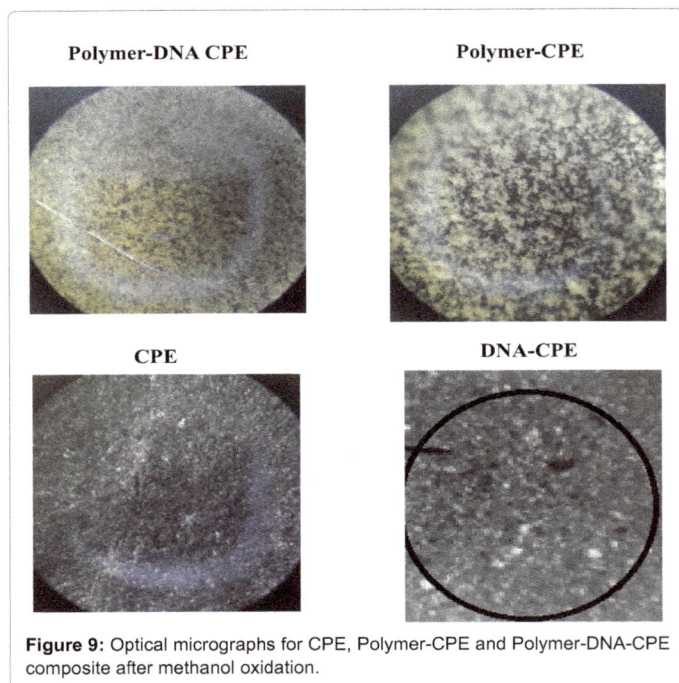

Figure 9: Optical micrographs for CPE, Polymer-CPE and Polymer-DNA-CPE composite after methanol oxidation.

Polyacrylic Acid membrane coating DNA film electro less deposited at carbon paste electrode. We have shown that the new electrode has excellent activity in the presence of methanol. The polymer prevents deterioration of the DNA film, and therefore improves the life of the electrode. The prepared electrode (Polymer-DNA-CPE) is very simple and easy to prepare, and has the properties necessary for the methanol electro catalysis. No empoisoning effect is observed. The Volta metric techniques showed that Polymer-NDA-CPE increased the oxidation current of methanol with increasing methanol oxidation.

References

1. Zhao TS (2009) London WC1X 8RR. UK.

2. Neel PJ (2008) MVC Sastry Hall. NCCR.

3. Carabineiro SAC, David Thompson T (2007) Catalytic applications of gold nanotechnology in nanoscience and technology, nanocatalysis. Springer, Berlin, p: 463.

4. Appleby AJ (1988) Fuel cell handbook. Van Nostrand Reinhold, New York.

5. Chtaini A (1993) Thesis, Poitiers (French).

6. Svancara I, Vytras K, Barek J, Zima J (2001) Carbon paste electrodes in modern electroanalysis. Critical Reviews in Analytical Chemistry 31: 311-345.

7. Rice ME, Galus Z, Adams RN (1983) Graphite paste electrodes: Effects of paste composition and surface states on electron-transfer rates. J Electroanal Chem Interfacial Electrochem 143: 89-102.

8. Reddy S, Swamy BK, Chandra U, Sherigara BS, Jayadevappa H (2010) Synthesis of CdO nanoparticles and their modified carbon paste electrode for determination of dopamine and ascorbic acid by using cyclic voltammetry technique. Int J Electrochem Sci 5: 10-17.

9. Tanuja SB, Swamy BEK, Vasantakumar KP (2016) Electrochemical Response of Dopamine in Presence of Uric Acid at Pregabalin Modified Carbon Paste Electrode: A Cyclic Voltammetric Study. J Anal Bioanl Tech 7: 1.

10. Gilbert O, Chandra U, Swamy BK, Char MP, Nagaraj C, et al. (2008). Poly (alanine) modified carbon paste electrode for simultaneous detection of dopamine and ascorbic acid. Int J Electrochem Sci 3: 1186-1195.

11. Gilbert O, Swamy BK, Chandra U, Sherigara BS (2009) Electrocatalytic oxidation of dopamine and ascorbic acid at poly (Eriochrome Black-T) modified carbon paste electrode. Int J Electrochem Sci 4: 582-591.

12. Shankar SS, Swamy BK, Ch U, Manjunatha JG, Sherigara BS (2009) Simultaneous determination of dopamine, uric acid and ascorbic acid with CTAB modified carbon paste electrode. Int J Electrochem Sci 4: 592-601.

13. Pu C, Huang W, Ley KL, Smotkin ES (1995) A methanol impermeable proton conducting composite electrolyte system. J Electrochem Soc 142: L119-L120.

14. Morse JD, Jankowski AF, Graff RT, Hayes JP (2000) J Vac Sci Technol A 18:2003-2005.

15. Hobson LJ, Nakano Y, Ozu H, Hayase S (2002) Targeting improved DMFC performance. J Power Sources 104: 79-84.

16. Peled E, Duvdevani T, Aharon A, Melman A (2000) A direct methanol fuel cell based on a novel low-cost nanoporous proton-conducting membrane. Electrochemical and Solid-State Letters 3: 525-528.

17. Jörissen L, Gogel V, Kerres J, Garche J (2002) New membranes for direct methanol fuel cells. J Power Sources 105: 267-273.

7

Magnetic Mirtazapine Loaded Poly(propylene glycol) bis(2aminopropylether) (PPG-NH2, MW_2000) Nanocarriers for Controlled Drug Release

Jibowu T*, Rohani S and Ragab D
University of Western Ontario, London

Abstract

Mirtazapine is an antidepressant that was introduced in 1996 for the treatment of moderate and severe depression. Mirtazapine is the only tetracyclic antidepressant that is approved by the Food and Drug Administration to treat depression. Mirtazapine is devoid of most side effects but has antihistamine side effects of drowsiness and weight gain. Its bioavailability is only fifty percent. The low bioavailability and side effects can be improved by altering the pharmokinetic profile of the drug by controlling the release of the drug. The slow release of the drug will reduce the harmful affect it has on the cells decreasing the side effects, as well as the loading of the drug in the nanocarrier will allow for a longer residence time in the body before it is removed by the gastrointestinal tract.

In this research paper the pharmokinetic profile of Mirtazapine will be altered by surrounding the drug with a biodegradable polymer called poly(propylene glycol) bis(2-aminopropylether) (PPG-NH$_2$, MW _ 2000) chains. This profile will be done at different polymer concentrations, drug concentrations and solubilizer concentration to see how this will affect the release of the drug.

In this research project it was found that using a lower concentration of poly(propylene glycol) bis (2-aminopropylether) (PPG-NH$_2$, MW_2000) chains of 0.5 g/mL led to a slower release in comparison to the other polymer concentrations with an encapsulation of 10 mg of Mirtazapine. When the drug weight was increased but the polymer concentration stayed the same (0.95 g/mL) the release rate increased with drug concentration. Also when the stabilizer concentration was increased, but the polymer concentration and drug concentration remained the same (0.95 g/mL and 10 mg respectively) the release rate increased. Therefore in order to allow for a slower release rate one should use the lower polymer concentration of 0.95 g/mL, with the lower concentration of stabilizer. This will allow for a slower release of the drug Mirtazapine which will lower the side effects and increase the bioavailability percentage.

Keywords: Nanomedicine; Nanotechnology; Pharmaceutical; Engineering; Controlled release

Nomenclature

PBS: Phosphate Buffer Solution

PPG-NH$_2$: Poly(propylene glycol) bis (2-aminopropylether)

BCD: Beta Cyclodextrin

MRZ: Mirtazapine

Objective

The objective of this report is to improve the bioavailability and decrease the side effects of MTZ in the human body. This will be done by incorporating MTZ in a PPG-NH$_2$ nanocarrier which will allow for controlled drug release. Controlled drug release allows for a certain amount of drug to be released from the nanocarrier reducing the amount of drug interacting with cells. Also since the drug will be incorporated in a biodegradable nanocarrier this will allow a reduction in uptake of the nanocarriers by the gastrointestinal tract.

Introduction

Mirtazapine (MTZ) is an antidepressant introduced by Organon International in 1996 which was used to treat moderate to severe depression. MTZ has a tetracyclic chemical structure and is an antidepressant that has been approved by the Food and Drug Administration to treat depression. MTZ is unique because it is virtually devoid of anticholinergic effects, serotonin-related side effects and adrenergic side effects [1]. However side effects such as drowsiness and weight gain are prominent [2]. Mirtazapine has a bioavailability of fifty percent [3].

This bioavailability can be enhanced by controlling the release of the drug from a formulation. Therefore biodegradable nanocarriers were created to prolong the release of the drug in systemic circulation. The antihistaminic side effects can also decreased by decreasing the amount of dose that can be used for treatment [4]. This can also be achieved by using nanocarriers. Biodegradable and injectable microspheres have been studied in the past thirty years, they can significantly prolong the duration of the drug which are metabolized in the gastrointestinal tract. The total dose of the drug and some of the adverse reactions can be reduced because it allows for steady plasma concentrations [4]. Biodegradable polymers are therefore becoming more important for the development of sustained release drug delivery systems and implantable biomaterials [3]. The common biodegradable polymers used are polylactide co-glycolide. However in this project magnetic drug loaded nanoparticles were created using poly(propylene glycol) bis(2-aminopropylether) (PPG-NH$_2$) as the biodegradable polymer and Iron oxide magnetic nanoparticles. Magnetic nanoparticles allow for the drug encapsulated nanocarriers to be transferred to the place of interest using a magnet, this leads to greater therapeutic efficacy. In this report there

*Corresponding author: Jibowu T, University of Western Ontario, London
E-mail: ToyinJibowu@gmail.com

will be a comparison of the cumulative release of Mirtazapine (MTZ) from polymers of different concentrations, MTZ weight and Betacyclodextrin (BCD) concentration. This information will then be used to determine which polymer concentration and BCD concentration leads to the optimal drug. The encapsulation efficiency and drug loading of the nanocarrier can be calculated using these equations below:

Entrapment Efficiency: $\dfrac{\textit{Actual weight of drug Loaded in Nanoparticle} * 100\%}{\textit{Theoretical weight of drug loaded in nanoparticle}}$

The cumulative release of the drug was calculated using this equation:

$$\frac{\textit{Concentration of drug released into PBS solution} * 100\%}{\textit{Concentration of total drug loaded}}$$

Materials

The materials used in this lab are Mirtazapine, Beta Cyclodextrin and ferrous sulphate heptahydrate, PPG-NH$_2$.

▶ PPG-NH$_2$ is a biodegradable polymer that is starting to be used as a nanocarrier in drug delivery, allowing for a slow release of the drug.

▶ Mirtazapine, as described before is a drug that treats depression. It has a low bioavailability of around 50% and side effects such as weight gain.

▶ BetaCyclodextrin, is a stabilizer. It has a hydrophilic exterior and a hydrophobic core. It forms a complex around hydrophobic drugs providing stability. It also allows for an increase in the bioavailability of the drug. This occurs because betacyclodextrin and drug are released together, and the complex does not allow it to be recognized as something to be removed by the body along for greater residence time in the body.

▶ Ferrous sulphate heptahydrate creates Iron oxide nanoparticles when Iron ferrous sulphate heptahydrate in water had Ammonium oxide dripped into it while stirring occurred creating iron oxide. Incorporating this technique into the creating of MRZ loaded PPG-NH$_2$ magnetic nanoparticles allowed for the creation of controlled release magnetic particles.

Methodologies

Preparation of magnetic MTZ nanospheres

The magnetic MTZ nanoparticles were created by combining PPG-NH$_2$, water, ferrous sulphate heptahydrate and stirring this mixture at fifty degrees Celsius. Betacyclodextrin with ammonium hydroxide was added drop wise to this mixture while increasing the temperature to one hundred degrees Celsius. The magnetic nanoparticles were then washed with ethanol/water. After nanoparticles were mixed with mirtazapine and water, forming mirtazapine magnetic nanocarriers. The supernatant was then removed and entrapment efficiency was calculated.

Characterization of nanocarriers: The supernatant was removed and the drug content was determined spectrophotometric ally at 290 nm.

Equipment and Software Used to Analyze Data

Zeta Particle Sizer: The Zetasizer 3000 HS$_A$ (Malvern Instruments) laser diffraction particle size analyser delivers rapid and accurate particle sizes for wet and dry dispersions. It measures over the nanometer to millimeter particle size range. The Zetasizer was used to find the mean diameters of the individual particles Nanoparticle suspensions of 1.0 mg/L were prepared and diluted by a factor of 45

to obtain homogenous suspensions of nanoparticles. Three runs were carried out per sample in order to get a standard deviation.

UV-Vis Spectroscopy: The UV Spectroscopy was used to determine the concentration of the release of Mirtazapine using UV-visible spectroscopy based on the absorbance at 290 nm. This was completed when 1mg of mirtazapine loaded particles were dispersed in 10 mL of phosphate buffered saline solution. At predetermined intervals the particles were separated using a magnet and the liquid was collected and replaced with fresh buffer solution. The concentration of the mirtazapine in the liquid was measured using the UV-visible spectroscopy.

In vitro release: For *in vitro* release, weighed microspheres containing a specific amount of MTZ were suspended in Phosphate buffer solution pH 7.4. The drug release was assessed intermittently, and 2 ml of the media were removed, filtered and the amount of MTZ released in the buffer solution was quantified by UV spectrophotometer at 290 nm.

Results

Table 1 shows entrapment efficiency, particle size, PDI.

[1] 0.95 mg/mL + 1.5 mmol BCD + 10 mg MRZ

[2] 0.95 mg/mL + 0.5 mmol BCD + 10 mg MRZ

[3] 2.8 mg/mL+ 1.5 mmol BCD + 10 mg MRZ

[4] 1.9 mg/mL+ 1 mmolBCD + 10 mg MRZ

[5] 0.95 mg/mL+ 1 mmolBCD +10 mg MRZ

[6] 0.95 mg/mL + 1 mmol BCD +20 mg MRZ

[7] 0.95 mg/mL + 1 mmol BCD + 15 mg MRZ

Figure 1 shows Mirtazapine Standard Curve.

Figure 2 shows the effect of polymer concentration on Cumulative drug release.

Figure 3 shows the effect of the weight of drug on Cumulative drug release.

Figure 4 shows the effect of the amount of BCD on Cumulative drug release.

Discussion

In this lab one created magnetic nanoparticles with MRZ loaded drugs. From the data in Table 1 and the data from Appendix 1.1 one can see that the entrapment efficiencies obtained were similar regardless of the concentration of polymer, amount of drug or concentration of BCD. The entrapment efficiency was around 97 percent.

Through *in vitro* studies one measured the cumulative concentration vs. time of the drug in phosphate buffer solution (PBS).

S.no	Entrapment Efficiency (%)	Particle Size(nm)	PDI
1	97.2348 ± 0.091	13.403	0.356
2	97.8192 ± 0	13.143	0.312
3	98.0205 ± 0.499	13.992	0.295
4	97.8249	11.097	0.125
5	98.7190 ± 0.0554	9.173	0.213
6	98.7395 ± 0.00501	15.7	0.321
7	98.8543	16.5	0.238

Table 1: Entrapment efficiency, particle size, PDI.

Absorbance vs. Concentration for Mirtzapine

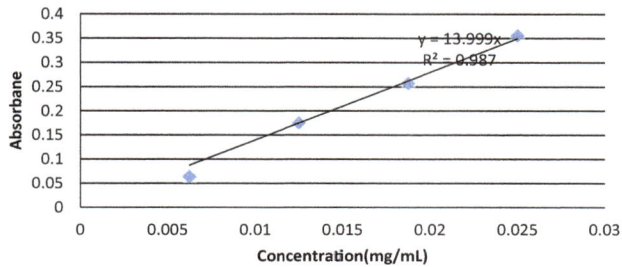

Figure 1: Mirtazapine standard curve.

The effect of polymer concentration on Cumulative drug Release

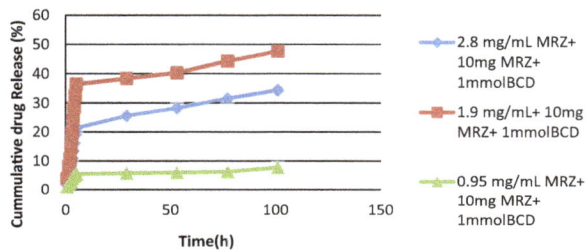

Figure 2: The effect of polymer concentration on cumulative drug release.

The effect of the weight of drug on Cumulative drug Release

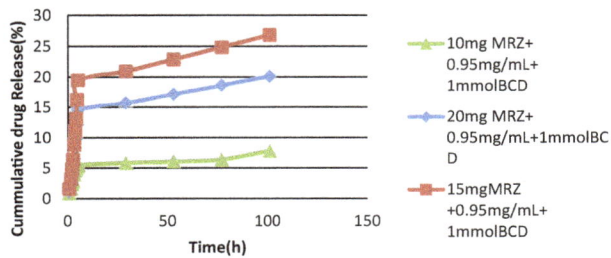

Figure 3: The effect of the weight of drug on cumulative drug release.

The effect of the amount of BCD on Cumulative drug Release

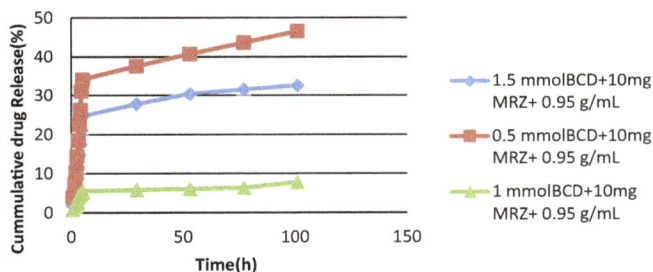

Figure 4: The effect of the amount of BCD on cumulative drug release.

Figure 1 and the data in Appendix 1.2 were used to determine the concentration of each drug variation at different time intervals. The cumulative concentration vs. time graph was completed by finding the absorbance for each variation of drug at different time intervals. Then the concentration of each drug was determined using Figure 1 and the results can be seen in Appendix 1.3. The cumulative concentration was then calculated and the values can be seen in Appendix 1.4, Appendix 1.5 and Appendix 1.6. The resulting data was graphed in Figures 2-4.

As one can see from figure two, as the polymer concentration increased the rate of drug dissolution increased. However when one compares a concentration of 1.94 mmol/mL to 2.84 mmol/mL the dissolution rate is higher than 0.95 mmol/mL but the 1.94 mmol/mL concentration has a higher dissolution rate than the 2.84 mmol/mL concentration. This could be due to the amount of BCD in the polymer. BCD allows for a faster dissolution of the drug out of the nanocarrier [1], however it has also been known to cause a slower dissolution rate [1] because it takes a longer time for BCD to diffuse through the matrix of the polymer depending on the concentration of the polymer. As the concentration, of the polymer increases the ability for BCD to become entrapped in the matrix increases.

In Figure 3 it is shown that as the beta cyclodextrin concentration increased from 1 mmol to 1.5 mmol the rate of dissolution increased. However when the amount of beta cyclodextrin decreased to 0.5 mmol the rate of dissolution was higher than 1 mmol but lower than 1.5 mmol. The results similar to what is discussed here has been found in journals. As the BCD amount increases dissolution increases due to drug stability [1] however the higher the amount of BCD the more bulky the compound which means it is harder for the compounds to diffuse through the polymer membrane.

Figure 4 shows as the drug concentration increases from ten milligrams to twenty milligrams the rate of dissolution increases. However the nanoparticles with 15 mg of drug have a higher dissolution rate than those with twenty milligrams of drug. Information similar to this could not be found in journals however one can assume that as the amount of drug increases the BCD cannot incorporate the entire drug which can lead to a lower dissolution rate in comparison to other releases based on concentration.

Conclusion

In conclusion the entrapment efficiency and drug loading percentage using the double emulsion technique was around 98 percent regardless of the drug concentration, polymer concentration or beta cyclodextrin concentration.

The cumulative release vs time graph showed different dissolution rates depending on what concentration of polymer, concentration of beta cyclodextrin and concentration of drug. As the drug weight increased from 10 mg the dissolution rate increased. However the nanocarrier with 20 mg of drug had a slower dissolution rate then the nanocarrier with fifteen mg of drug. The cumulative release vs. time graph also differed regarding the beta cyclodextrin concentration as the beta cyclodextrin concentration increased from 1mmol to 1.5 mmol there was an increase in drug dissolution. However the beta cyclodextrin concentration of 0.5mmol had a higher dissolution rate than the one mmol nanocarrier. Also when one compares the polymer concentration one can see as the polymer concentration increased the rate of dissolution increased in comparison to 0.95 mmol/mL, however the 2.8 mmol/mL concentrations had a slower dissolution rate then the 1.94 mg/mL concentration.

Recommendations

Author would recommend that one would use a MRTZ loaded nanocarrier, for the treatment of depression. This will lead to greater bioavailability from BCD as well as the nanocarrier allows for the drug not to be removed from the bodies system. The nanocarrier will also allow for a controlled release of the drug as one can see from Figures 2-4. Due to its controlled release this will reduce the side effects normally experienced by this drug because a controlled concentration of it will be released. If one wants to reduce the amount of drug affecting healthy cells one can add ligands to this nanocarrier which will allow it to specifically target certain cells.

There are several options to creating a controlled release drug; one can vary the polymer concentration and solubizer concentration to get a better controlled release. If one wants a slow release it is best to use the nanocarrier with the concentration of 0.95195 mmol/mL with a BCD weight of 1 mmol because even with a higher concentration of drug it allows for a slow release of the drug over time in comparison the other polymer concentrations.

Future Work

In the future in order to improve this product one should crosslink specific ligands onto the nanocarriers in order to allow for targeted delivery of the drug. This will improve therapeutic efficacy as well as reduce side effects since the drug will target only specific cells. Also one should find a way to control the size of the nanoparticles.

References

1. Carrier RL, Miller LA, Ahmed I (2007) The utility of cyclodextrins for enhancing oral bioavailability. Journal of Controlled Release 123: 78-99.

2. Palamoor M, Jablonskia MM (2013) Synthesis, characterization and in vitro studies of celecoxib-loaded poly(ortho ester) nanoparticles targeted for intraocular drug delivery. Colloids and Surfaces B: Biointerfaces 112: 474-482.

3. Ragab DM, Rohani S, Consta S (2012) Controlled release of 5-flourouracil and progesterone from magnetic nanoaggregates. Int J Nanomedicine 7: 3167-3189.

4. Ranjan OP, Shavi GV, Nayak UY, Arumugam K, Averineni RK, et al. (2011) Controlled release chitosan microspheres of mirtazapine: in vitro and in vivo evaluation. Arch Phar Res 34: 1919-1929.

8

Iron Oxide Nanoparticles Coated with Polymer Derived from Epoxidized Oleic Acid and Cis-1,2-Cyclohexanedicarboxylic Anhydride: Synthesis and Characterization

Pereira da Silva S, Costa de Moraes D and Samios D*

Laboratory of Instrumentation and Molecular Dynamics, Department of Physical Chemistry, Chemistry Institute, Federal University of Rio Grande do Sul Av. Bento Gonçalves 9500, Porto Alegre, Brazil

Abstract

This study investigated the use of polymer derived from oleic acid for coating iron oxide nanoparticles. The purpose of this study was to provide the magnetic nanoparticles an appropriate surface for stabilization in organic solution. The magnetic nanoparticles coated were produced by mixing of the polymer solution with the ferromagnetic fluid by mechanical stir, followed by magnetic separation. These nanoparticles generated a core-shell behavior, in which the core provides the magnetic properties and the external layer formed by the polymer. The interaction between iron nanoparticles and oleic acid polymer occurred by the affinity of carboxylic group. This interaction makes the nanoparticles hydrophobic, moving to the organic media. The carbon content of the coated nanoparticles was approximately 14%, when analyzed by scattering electron microscopy (SEM–EDX), and 12%, when analyzed by Elemental Analysis of carbon, hydrogen and nitrogen. This percentage confirms the presence of the polymer on the surface of magnetic nanoparticles. The average diameter of the coated and uncoated nanoparticles obtained by transmission electron microscopy was around 13 nm and 11 nm and the average diameter of crystallite by X-ray diffraction was around 8 nm and 12 nm respectively. Averaging all this values we obtain 11 ± 2 nm. The thermogravimetric analysis showed the degradation temperatures starting from 200°C to 500°C, attributed to the polymer, and another one degradation temperature between 650-750ºC, relative to the polymer-nanoparticles interaction. Furthermore, the vibrating sample magnetometer indicated that coated nanoparticles remain magnetic, with increasing saturation magnetization value, when a magnetic field was applied.

Keywords: Magnetic nanoparticles; Iron oxide; Polymers; Oleic acid and cis-1,2-cyclohexanedicarboxylic anhydride

Introduction

Magnetic materials in nano-scale have been studied because these materials have a diversity of biomedical applications, such as drug delivery in hyperthermia, contrast agents for magnetic resonance imaging (MRI) and bio-separation [1,2]. These magnetic nanoparticles have dimensions of 1 to 100 nm. The particle size depends of the synthesis applied to it and can be controlled by the use of stabilizing agents which act on the surface of the nanoparticles [3,4].

Some preparation methods of magnetic nanoparticles include thermal decomposition [5], microemulsion [6] or co-precipitation [7,8]. The co-precipitation method is the simplest and most efficient for obtaining magnetic particles. By this method, the iron oxides (Fe_3O_4 and $\gamma-Fe_2O_3$) are prepared from the stoichiometric mixture of aqueous solutions of ferrous and ferric salts in alkaline medium [9]. The chemical reaction of synthesis of magnetite (Fe_3O_4) can be described according to Equation 1. However, magnetite is not very stable and can be oxidized when exposed to atmospheric oxygen turning into maghemite ($\gamma-Fe_2O_3$), according to Equation 2.

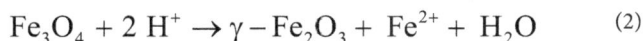

$$Fe^{2+} + Fe^{+} + 8\ OH^{-} \rightarrow Fe_3O_4 + 4\ H_2O \qquad (1)$$

$$Fe_3O_4 + 2\ H^{+} \rightarrow \gamma - Fe_2O_3 + Fe^{2+} + H_2O \qquad (2)$$

In this stage, according to literature [10,11] the oleic acid, in this case the polymer, is used to coat the magnetic nanoparticles due to its surfactant action that generates high affinity to the nano-magnetite. The action of the stabilizer on the surface of the nanoparticles eliminates, in other words cut out the aggregation process caused by magnetic force attraction and Van der Waals forces. Furthermore, this facilitates the stabilization and dispersion of the particles in organic solutions [2,12,13].

Other factors affecting the formation and shape of the nanoparticles as the salts, the Fe^{3+}/Fe^{2+} ratio, the reaction temperature and the pH [14]. These particles are low cost, low toxicity and eco-friendliness. Also, have excellent physical, chemical and magnetic properties as supermagnetism, high surface area, easy separation under an external magnetic field and strong adsorption capacity [15].

Oleic acid has been one of the fatty acids most commonly used for coating of magnetite. The oleic acid is composed by the carboxylic group and the long chain hydrocarbon, $(CH_3(CH_2)_7CH=CH(CH_2)_7COOH)$ [4]. The carboxylic group interacts with the hydrophilic surface of the iron oxide and makes it hydrophobic by the presence of the long chain fatty acid. The same action is expected by the polymers produced by using oleic acid and different dicarboxylic anhydrides. Polymers that have functional groups -COOH, -NH$_2$, -OH, are also compatible with these particles, increasing electrostatic attraction between them [16]. Among the polymers used for this purpose are the gum arabic, dextran, chitosan, poly(ethylenimine) (PEI) and poly(ethylene glycol) (PEG) [15,17-19]. Recently, different groups work with the polymerization of fatty acids products like oleic acid, linoleic acid and linolenic acid and

***Corresponding author:** Samios D, Laboratory of Instrumentation and Molecular Dynamics, Department of Physical Chemistry, Chemistry Institute, Federal University of Rio Grande do Sul Av. Bento Gonçalves 9500, Porto Alegre, Brazil E-mail: dsamios@ufrgs.br

others [20,21]. These polymers are synthesized by epoxidition of the double bond of the fatty acids, and subsequent opening of the epoxide ring with use of triethylamine (initiator) and different kind of anhydride, as phtalic, maleic, succinic or cis-1,2-cyclohexanedicarboxylic.

The aim of this study was to prepare magnetic nanoparticles coated with polymer, produced by the reaction between epoxidized oleic acid, cis-1,2-cyclohexanedicarboxylic anhydride and triethylamine, for stabilizing in organic media. This polymer presents a hydrophilic part characterized by the structure of carboxylic group that provide high chemical affinity to the surface of the magnetic nanoparticles. Moreover, the hydrophobic chain is exhibiting a shell behavior and causes the movement of the nanoparticles to organic medium.

Experimental Part

Material

The reagents were used as supplied. Ferric chloride (FeCl2•4H2O, 99%), and cloridric acid (99.5%) were procured from Sigma-Aldrich, (Darmstadt, Germany). Ferric chloride (FeCl3•6H2O, 97%), ammonium hydroxide (NH4OH, minimum assay: 24-26%), sodium bisulfate (99%), anhydrous sodium sulfate (58,5%) and triethylamine P.S. (99%) were obtained from Vetec (Rio de Janeiro, Brazil). Oleic acid P.A, formic acid (85%), toluene (99.5%) and hydrogen peroxide (30% w/w) were purchased from Synth (São Paulo, Brazil). Cis-1,2-cyclohexanedicarboxylic anhydride (99%) was purchased from Acros Organics, (Geel, Belgium).

Synthesis of iron oxide nanoparticles

The magnetic nanoparticles were produced using the co-precipitation method. The synthesis was carried out by mixing a solution of iron (III) chloride hexahydrate and a solution of iron (II) chloride tetrahydrate by alkaline hydrolysis. The molar ratio of Fe (III) and Fe (II) was 2:1. The solution of FeCl3. 6 H2O was prepared by dissolving the reagent in 40 ml of solution of HCl (2 mol L-1), while FeCl2. 4H2O was dissolved in 10 ml of solution of HCl (2 mol L-1). After, aqueous solution of ammonia to 0.7 mol L-1was added. The products were stirred mechanically for 30 min. The black precipitate formed was washed with distilled water and excess water was dropped by magnetic sedimentation [22]. The sample was heated at 70°C for 30 min to evaporation of traces of ammonia [4]. The black precipitate formed was kept in water (ferrofluid) for later use.

Epoxidation of oleic acid

The epoxy rings of the oleic acid were produced by the substitution of the double bonds of the oleic acid. Thus, oleic acid was stirred with toluene and formic acid. After that, the hydrogen peroxide was added dropwise to the reaction for 1 h. The molar ratio of hydrogen peroxide/formic acid/unsaturation was 20/2/1.The reaction continued for 3 hours at 80°C [23]. Finally, a solution of sodium bisulfite 10% was added under stirring for 10 min at room temperature. The aqueous phase was discarded and the organic phase was separated and washed until the pH 6.0-7.0. Anhydrous sodium sulfate was added to remove moisture. The system remained at rest overnight. The epoxide was filtered off and the solvent (toluene) was evaporated in rotary evaporator at 80°C [24,25]. The obtained product was named Epoxidized Oleic Acid (EOA).

Polimerization reaction

The polymerization process occurred by mixing 20 g of epoxidized oleic acid, 10 g of cis-1,2-cyclohexanedicarboxylic anhydride, 160 μl of triethylamine (TEA) as initiator. The reactants were stirred for 3h

30 min at 150°C. The polymer obtained was presented as a yellowish and translucent product [23,26]. This material was named as oleic acid polymer (OAP).

Interaction between magnetic nanoparticles and oleic acid polymer

The magnetic fluid was mixed in the toluene polymer solution. The suspension was mechanically stirred for 1h at room temperature. The coated magnetic nanoparticles were separated by magnetic sedimentation. The samples were dried at 70°C to constant weight. The resulting product was named Iron oxide nanoparticles–Oleic acid polymer (INP-OAP), as indicated in Figure 1.

Characterization

The morphological images of the particles were obtained by scanning electron microscopy (SEM) on a JEOL, model JSM 5800, operating at 20kVand coupled with energy dispersive X-ray spectroscopy (EDX), model Norman SistemSix. The preparations of samples were carried out by deposition of the nanoparticles, with and without coating, in aluminum holders and subsequent metallization with gold. The average diameter of the particles and nano-scale images were obtained by transmission electron microscopy (TEM) in a Libra microscope, operating at 120kV. The interaction between the magnetic nanoparticles and the polymer was investigated by Fourier Transform Infrared spectroscopy (FTIR) and Thermogravimetric Analysis (TGA). The equipment used for the analysis of infrared spectroscopy was carried out on Shimadzu FTIR-8300. The spectra were obtained by measurements of transmittance in the range 4000-500 cm-1and averaged over 32 scans. The samples were prepared in KBr pellets. The equipment used for thermogravimetric analysis was a Universal analyzer V2.6D (TA Instruments). Samples (10 mg) were heated from room temperature to 900°C at a heating rate of 20°C/min under N2atmosphere. The equipment used for the analysis of X-ray diffraction was a X-ray diffractometer of Rigaku brand, model DMAX 2200, operating at 40 kV, 17.5 mA, 200 V, equipped with copper tube (λ=1.54178 Å). The angular range was set at 20° to 75°. The magnetic properties were measured from a vibrating sample magnetometer (VSM), Microsensebrand, model EZ9, at room temperature.

Results and Discussion

The behavior of the iron nanoparticles in water or toluene can be observed in Figure 2. The pure magnetic nanoparticles remain suspended in aqueous medium due the presence of the hydrophilic

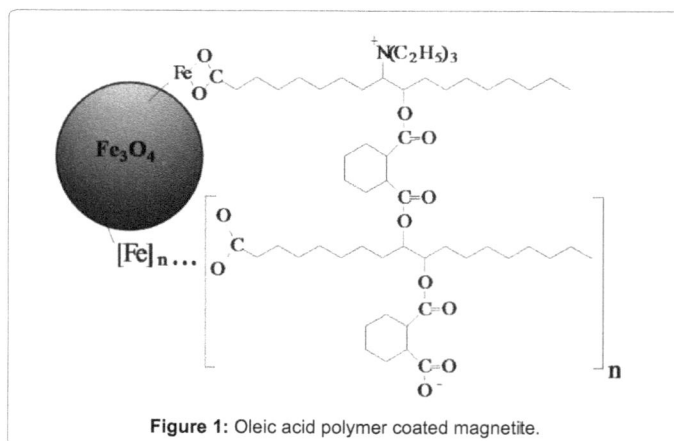

Figure 1: Oleic acid polymer coated magnetite.

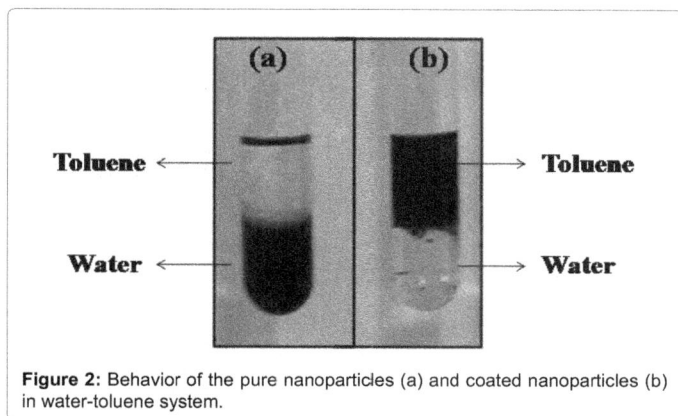

Figure 2: Behavior of the pure nanoparticles (a) and coated nanoparticles (b) in water-toluene system.

surface of the material, as shown in Figure 2a. The Figure 2b demonstrates the suspension of iron nanoparticles coated by oleic acid polymer in toluene. As we can see, the coating with polymer makes them to leave from the water in to organic solvent. This stabilization of the modified nanoparticles in toluene confirmed the chemical interaction between the polymer and iron oxide nanoparticles [27].

The Figure 3a shows the SEM images of pure magnetic nanoparticles, Figure 3b shows morphology of the polymer and Figure 3c shows nanoparticles coated by the polymer. The corresponding EDX results are presented parallel to these images. The results of percentage of carbon element for nanoparticles uncoated, pure polymer and nanoparticles coated presented values of 1.3, 83.4 and 14.1, respectively. In the case of the iron element the values were 98.7, 16.6 and 85.9, respectively. These results confirm the coating of the nanoparticles by the polymer, because the percentages of carbon, present in chain of the polymer, increased approximately 10% after coating.

The carbon percent on the surface of the nanoparticles coated was also determined from the elementary analysis of carbon, hydrogen and nitrogen (CHN). This technique showed a percentage of 12.9% of carbon, 2.4% of hydrogen and 0.1% of nitrogen. The carbon content in the sample is derived from oleic acid polymer, confirming the interaction of the nanoparticles with the polymer. The CHN analysis showed an organic mass content of 19.0% from the chemical formula of the compound.

The TEM micrographs of the pure nanoparticles and the coated nanoparticles are shown in Figure 4a and 4b respectively, and the respective histograms. The image show that the particles have a reasonably spherical shape, where in the obtained average diameter of the particles was approximately 11 nm for the uncoated and 14 nm for the coated, setting them as nanoparticles. However, it is worth to observe that the distribution of the particles is not the same. The coating makes the distribution wider in comparison to the uncoated nanoparticles.

The analysis of X-ray diffraction confirmed the existence of iron oxide particles as shown in Figure 5a. The XRD pattern showed values of 2θ at 30.3°, 35.7°, 43.1°, 53.6°, 57.4° and 62.6° concerning plans reflection of the crystal structure of magnetite (220) (311), (400), (440), (531) and (533), respectively [28-30]. The same can be observed for the magnetic nanoparticles coated with polymer (Figure 5b). The broadening of width of half maximum of the peak in the INP-OAP can be seen in all the peaks in the Figure 5b, as well as the amplified signal relative to the crystallographic plane 311 (Figure 5c).

The average diameter of the crystallites of pure and coated nanoparticles were obtained from the half height of the peaks 311 XRD diffraction and calculated according to the Scherrer equation. The values found were 12 nm (INP) and 8 nm (INP-OAP). This difference of the magnitude can be explained by the method used in the preparation of the coated nanoparticles. The coated nanoparticles are submitted in a new heating treatment, which is related to a selective process, selecting the smaller nanoparticles to be coated. However, the process needs more studies. The polymer possesses a lower optical density compared to the pure magnetite, thus, the size of the crystalline refer only to the magnetic core.

The FTIR spectra of Figure 6 shows a comparison of the pure polymer (OAP), the pure nanoparticles (INP) and the iron oxide coated by polymer (INP-OAP). The asymmetric and symmetric stretching of CH groups present in the polymer chain can be checked at 2925 and 2856 cm^{-1}, respectively [31]. The band at 1724 cm^{-1}is related to the stretching of the carbonyl group of the carboxylic anhydride and generates an overlapping of peaks (C=O) [32,33]. The band in 1170 cm^{-1}is a characteristic one for the group C-O-C, present in the ester of the polymer [23]. The nanoparticles of iron oxide (INP) showed bands in the region of 620 and 571 cm^{-1} characteristic of the Fe-O group. The bands at 3400 and 1630 cm^{-1} assigned to the vibration of OH group coated on the surface of iron oxide [34]. A shoulder at 1540 cm^{-1} can be related to the bond of the carboxylate with the structure of magnetite [35,36]. These results confirm the interaction between the oleic acid polymer and the nanoparticles of the Iron oxide [32].

Thermogravimetric analysis was employed to determine the loss of mass of the polymer used in the coating of the nanoparticles (Figure 7). The degradation curve of pure nanoparticles demonstrated a loss percentage of approximately 1%, under the temperature of 100°C, which can be related to the water adsorbed on the surface of magnetite. In relation to the mass loss of the pure polymer, it was observed two significant losses. The first loss is between 150-300°C and the second loss is between 300-400°C [32]. This two-step degradation is a characteristic of the polymers produced by plant oils. Observing the mass loss curve of the iron nanoparticles coated with the polymer, we see three degradation stages: the two stages characteristic of the polymer and the third one that is related to the interaction between them, which can be observed at a temperature between 600- 800ºC [30,37].

According to the Figures 8 and 9, it is possible to make a comparison of the degradation temperature of the pure polymer and coated nanoparticles with it. The two main degradation of the polymer are around 260 and 370°C (Figure 8). The Figure 9 indicates in the region between 200°C and 500°C more than five degradation superposed processes. However, what differentiates the Figures 8 and 9 is the mass loss at temperature 713°C and percentage of the loss of 7.55%. This is an additional argument which proves the interaction between iron nanoparticles and polymer obtained using oleic acid.

The presence of magnetic property in the nanoparticles coated was tested by the application of an external magnetic field [38]. The magnetic attraction of these nanoparticles has shown that the magnetic properties are maintained even after coating. This can be seen in Figure 10a where the nanoparticles coated by polymer are suspended in organic phase (toluene), without external magnetic field and Figure 10b shows the same composition is suspended under an external magnetic field. It can be observed the attraction of the nanoparticles under the magnetic field.

The magnetic behavior of the Iron oxide nanoparticles and the same nanoparticles coated with the polymer can also be observed

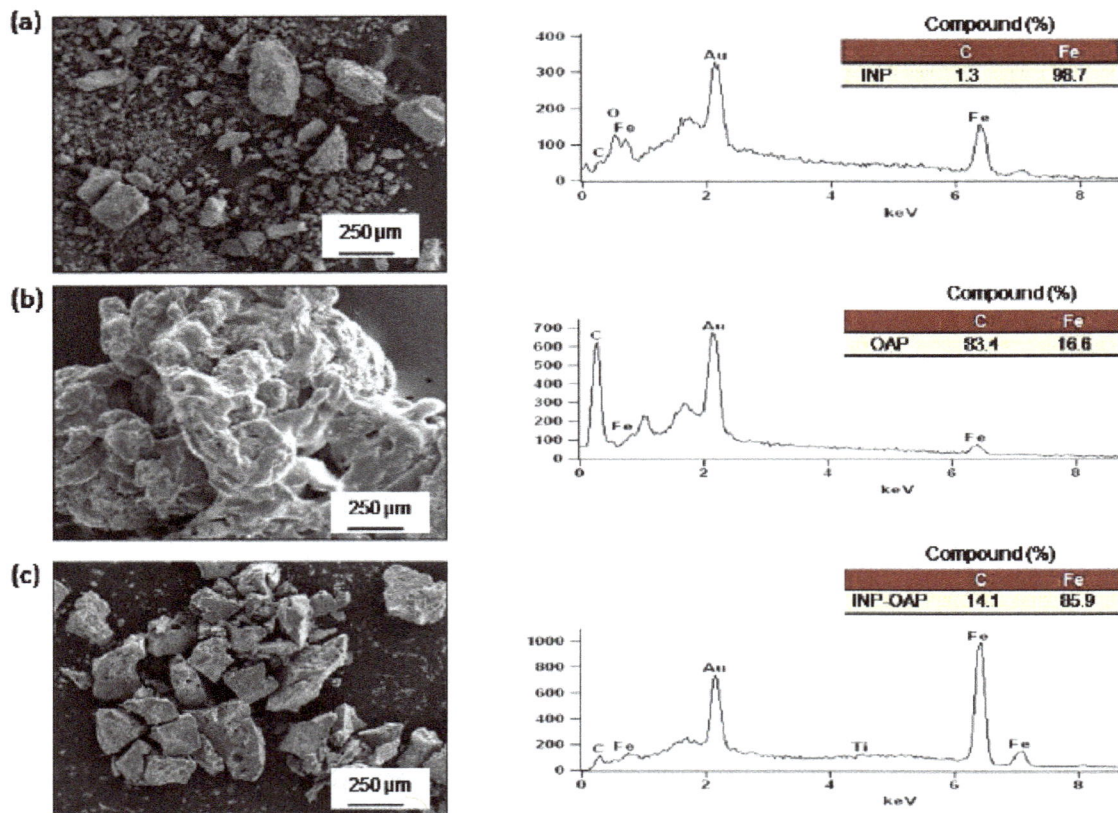

Figure 3: SEM-EDX images: nanoparticles INP (3a), the polymer OAP (3b) and the coated nanoparticles INP-OAP (3c).

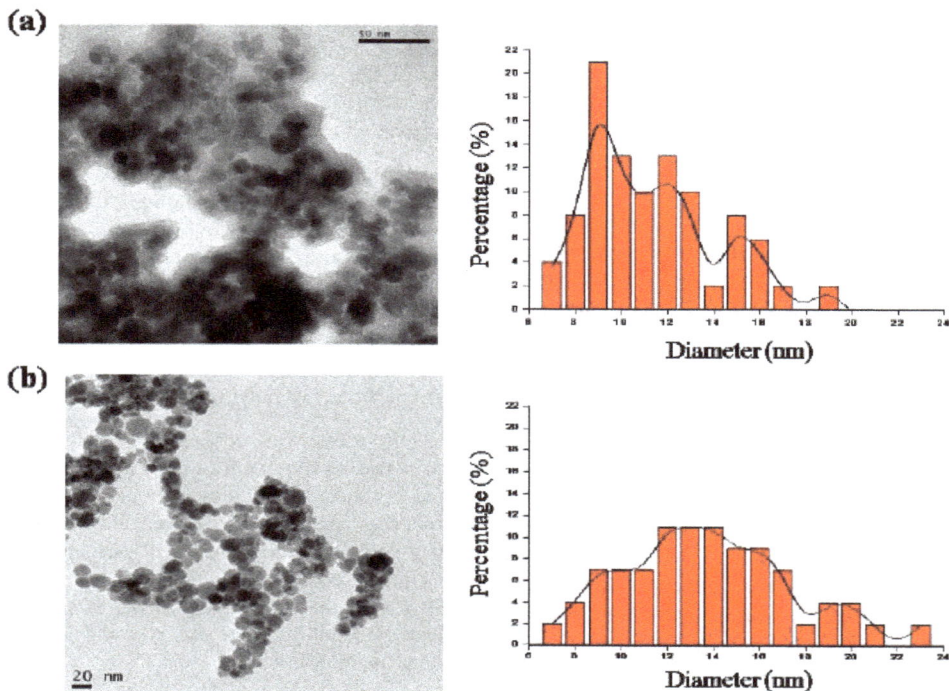

Figure 4: TEM of INP (a), INP-OAP (b) and the histogram of average particle diameters, respectively.

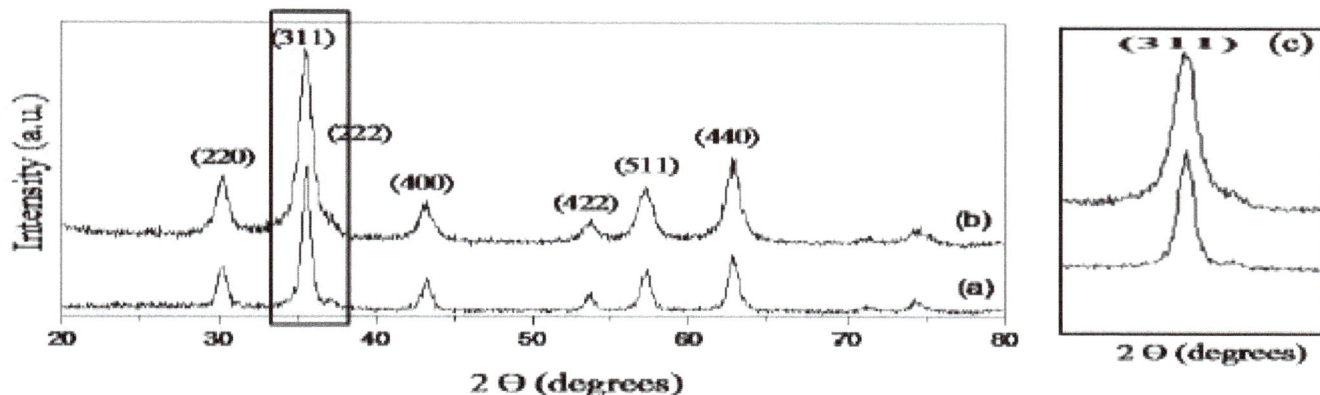

Figure 5: The XRD pattern of iron oxide nanoparticles pure (a), iron oxide coated with oleic acid polymer (b) and amplification of the signal 311 (c).

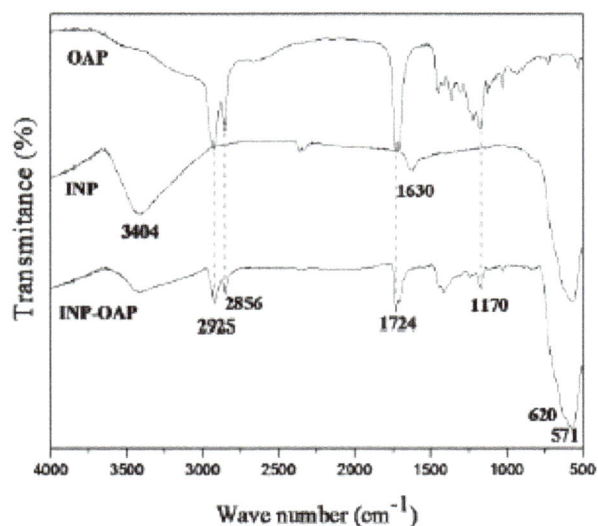

Figure 6: FTIR spectra of iron nanoparticles pure, oleic acid polymer and magnetite coated by OAP.

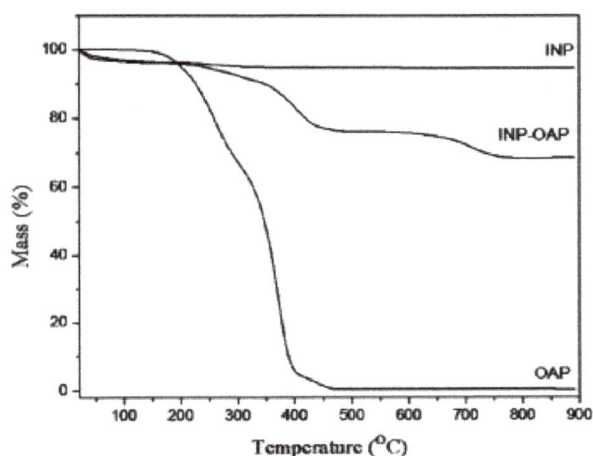

Figure 7: TGA thermograms of Iron nanoparticles (INP), oleic acid polymer (OAP) and the nanoparticles coated by the OAP.

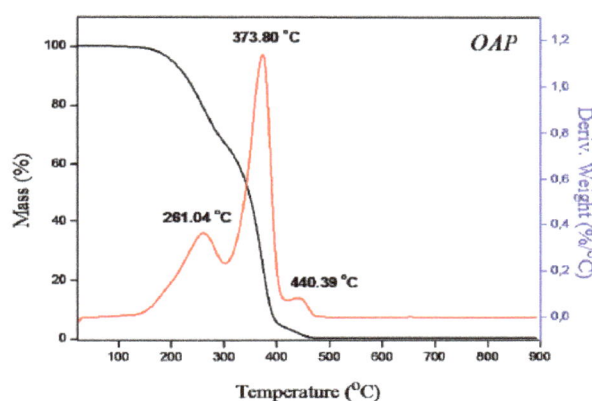

Figure 8: TG-DTA curve of OAP.

from the measurements of the magnetization at room temperature, as shown in Figure 11. According to the magnetization curves could be observed that both of the samples possess typical superparamagnetic behavior. The saturation magnetization of the uncoated particles was 17.7emu/g and the corresponding value, for the nanoparticles coated with polymer, was 21.1 emu/g. This can be explained by the size and the broader distribution of the coated the nanoparticles, as well as, the fact that the nanoparticles uncoated when exposed to the environment may undergo oxidation and consequently losing magnetism. The uncoated nanoparticles show smaller diameter and, consequently, smaller protection against oxidation [39].

Conclusion

Nanoparticles of Iron oxide (Fe_3O_4) were prepared by using the method of co-precipitation. The prepared nanoparticles were by the coating with a polymer produced by Epoxidized Oleic Acid and cis-1,2-cyclohexanedicarboxylic anhydride using triethylamine as initiator.

The core-shell behavior of the coated particles was proved by the stabilization of the nanoparticles in organic medium. By the TEM was possible to obtain particles with diameter average around 11 nm for the uncoated and around 13 nm for the coated nanoparticles however with a broader distribution for the second. XRD confirmed the structure of magnetite due to the presence of their crystallographic planes and the same XRD pattern was obtained for the INP-OAP nanoparticles. The

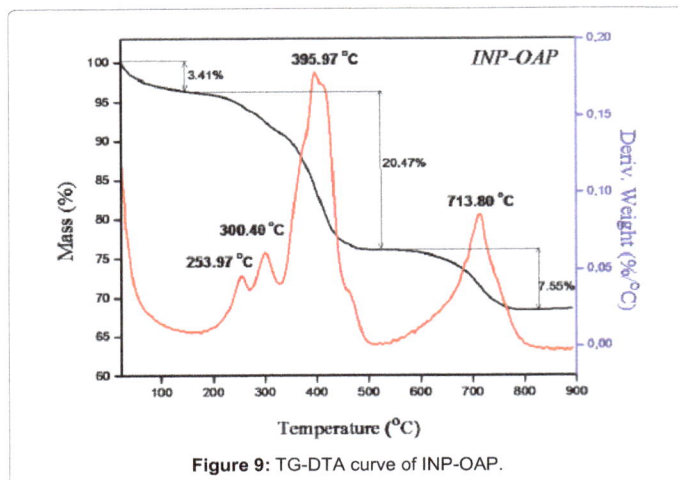

Figure 9: TG-DTA curve of INP-OAP.

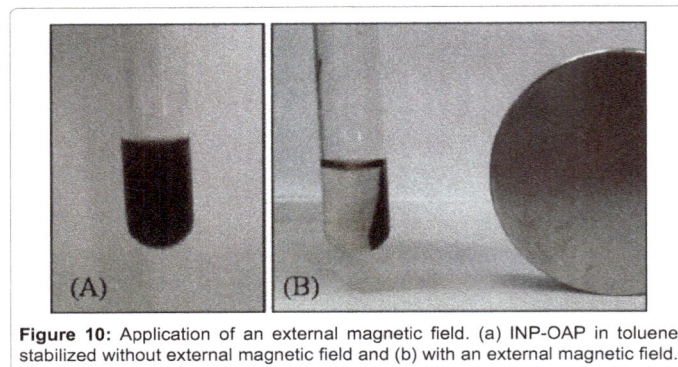

Figure 10: Application of an external magnetic field. (a) INP-OAP in toluene stabilized without external magnetic field and (b) with an external magnetic field.

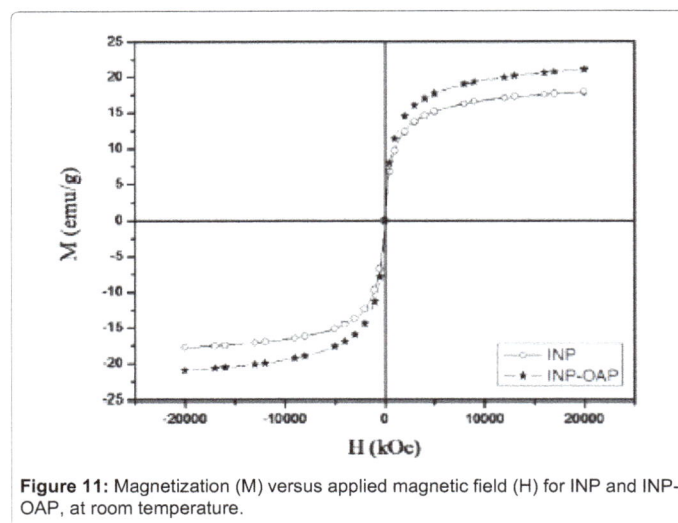

Figure 11: Magnetization (M) versus applied magnetic field (H) for INP and INP-OAP, at room temperature.

broadening of the XRD peaks, compared to the magnetite, indicates the interaction between magnetite and polymer. The FTIR of the product polymer-particle showed characteristic bands of separate samples, such as C=O and CH, referring to the polymer chain, as well as the Fe-O, present in the iron oxide. Both materials, uncoated and coated, indicate magnetic properties, however, the magnetization curves showed the presence of superparamagnetic behavior. The coated nanoparticles showed the magnetization value of 21.1 emu/g and the uncoated the value of 17.7 emu/g.

Acknowledgements

The authors thank the CNPq (Projects 551116/2010-2 and 405011/2013-0) for financial support, the CNANO (Nanoscience and nanotechnology center) and CME (center of electron microscopy), Federal University of Rio Grande do Sul, Porto Alegre, Brazil, by SEM and TEM images, and the Laboratory of Magnetism, Federal University of Rio Grande do Sul, Porto Alegre, Brazil, for the analyzes of magnetism.

References

1. Wu Y, Song M, Xin Z, Zhang X, Zhang Y, et al. (2011) Ultra-small particles of iron oxide as peroxidase for immune histochemical detection. Nanotechnology 22: 225703.

2. Jadhav NV, Prasad AI, Kumar A, Mishra R, Dhara S, et al. (2013) Synthesis of oleic acid functionalized Fe_3O_4 magnetic nanoparticles and studying their interaction with tumor cells for potential hyperthermia applications. Colloids Surfaces B 108: 158-168.

3. Yang TI, Brown RNC, Kempel LC, Kofinas P (2011) Controlled synthesis of core-shell iron-silica nanoparticles and their magneto-dielectric properties in polymer composites. Nanotechnology 22: 105601.

4. Darwish MSA, Peuker U, Kunz U, Turek T (2011) Bi-layered polymer-magnetite core/shell particles: synthesis and characterization. J Mat Sci 46: 2123-2134.

5. Teja AS, Koh PY (2009) Synthesis, properties, and applications of magnetic iron oxide nanoparticles. Prog Cryst Growth Ch 55: 22-45.

6. Wongwailikhit K, Horwongsakul S (2011) The preparation of iron (III) oxide nanoparticles using W/O microemulsion. Mat Lett 65: 2820-2822.

7. Mahdavi M, Ahmad MB, Haron MJ, Namvar F, Nadi B, et al. (2013) Synthesis, Surface Modification and Characterisation of Biocompatible Magnetic Iron Oxide nanoparticles for Biomedical Applications. Molecules 18: 7533-7548.

8. Willard MA, Kurihara LK, Carpenter EE, Calvin S, Harris VG (2004) Chemically preparedmagnetic nanoparticles. Inter Mat Rev 49: 125-170.

9. Kikuchi T, Kasuya R, Endo S, Nakamura A, Takai T, et al. (2011) Preparation of magnetite aqueous dispersion for magnetic fluid hyperthermia. J Magn Magn Mater 323: 1216-1222.

10. Rodríguez C, Bañobre-López M, Kolen'ko Y, Rodríguez B, Freitas P, et al.(2012) Magnetization Drop at High Temperature in Oleic Acid-Coated Magnetite Nanoparticles. IEEE Transactions on Magnetics 48: 3307-3310.

11. Cano M, Sbargoud K, Allard E, Larpent C (2012) Magnetic separation of fatty acids with iron oxide nanoparticles and application to extractive deacidification of vegetable oils. Green Chem 14: 1786-1795.

12. Gyergyek S, Makovec D, Drofenik M (2011) Colloidal stability of oleic- and ricinoleic- acid- coated magnetic nanoparticles in organic solvents. J Colloid Interf Sci 354: 498-505.

13. Cohen H, Gedanken A, Zhong ZY (2008) One-Step Synthesis and Characterization of Ultrastable and Amorphous Fe_3O_4 Colloids Capped with Cysteine Molecules. J Phys Chem C 112: 15429-15438.

14. Omer M, Haider S, Park SY (2011) A novel route for the preparation of thermally sensitive core-shell magnetic nanoparticles. Polymer 52: 91-97.

15. Laurent S, Forge D, Port M, Roch A, Robic C, et al. (2008) Magnetic Iron Oxide Nanoparticles: Synthesis, Stabilization, Vectorization, Physicochemical Characterizations, and Biological Applications. Chem Rev 108: 2064-2110.

16. Kalska-Szostko B, Wykowska U, Piekut K, Satuła D (2014) Stability of Fe_3O_4 nanoparticles in various model solutions. Colloid Surface A 450: 15-24.

17. Kadar E, Batalha IL, Fisher A, Roque ACA (2014) The interaction of polymer-coated magnetic nanoparticles with seawater. Science of the Total Environment 487: 771-777.

18. Mahdavi M, Ahmad MB, Haron MJ, Namvar F, Nadi B, et al. (2013) Synthesis of Naphthylpyridines from Unsymmetrical Naphthylheptadiynes and the Configurational Stability of the Biaryl Axis. Molecules 18: 7533-7548.

19. Fathi M, Entezami AA (2014) Stable aqueous dispersion of magnetic iron oxide core–shell nanoparticles prepared by biocompatible maleate polymers. Surf Interface Anal 46: 145-151.

20. Santos EF, Oliveira RVB, Reiznautt QB, Samios D, Nachtigall SMB (2014) Sunflower-oil biodiesel-oligoesters/polytactide blends: Plasticizing effect and ageing. Polym Test 39: 23-29.

Iron Oxide Nanoparticles Coated with Polymer Derived from Epoxidized Oleic Acid...

53

21. Nicolau A, Samios D, Piatnick CMS, Reiznautt QB, Martini DD, et al. (2012) On the polymerisation of the epoxidized biodiesel: The importance of the epoxy rings position, the process and the products. Eur Polym J 48: 1266-1278.

22. Bruce IJ, Taylor J, Todd M, Davies MJ, Borioni E, et al. (2004) Synthesis, characterization and application of silica-magnetite nanocomposites. J Magn Mater 284: 145-160.

23. Nicolau A, Mariath RM, Martini EA, Martini DS, Samios D (2010) The polymerization products of epoxidized oleic acid and epoxidized methyl oleate with cis-1,2-cyclohexanedicarboxylic anhydride and triethylamine as the initiator: Chemical structures, thermal and electrical properties. J Magn Magn Mater 30: 951-962.

24. Martini DS, Braga BA, Samios D (2009) On the curing of linseed oil epoxidized methyl esters with different cyclic dicarboxylic anhydrides. Polymer 50: 2919-2925.

25. Reiznautt QB, Garcia ITS, Samios D (2009) Oligoesters and polyesters produced by the curing of sunflower oil epoxidized biodiesel with cis-cyclohexane dicarboxylic anhydride: Synthesis and characterization. Mater Sci Eng 29: 2302-2311.

26. Nicolau A, Mariath RM, Samios D (2009) Study of the properties of polymers obtained from vegetable oil derivatives by lightscattering techniques. Mater Sci Eng 29: 452-457.

27. Lee SY, Harris MT (2006) Surface modification of magnetic nanoparticles capped by oleic acids: Characterization and colloidal stability in polar solvents. J Colloid Interf Sci 293: 401-408.

28. Harris LA, Goff JD, Carmichael AY, Riffle JS, Harburn JJ, et al. (2003) Magnetite Nanoparticle Dispersions Stabilized with Triblock Copolymers. Chem Mater 15: 1367-1377.

29. Habibi N (2014) Preparation of biocompatible magnetite-carboxymethyl cellulose nanocomposite: Characterization of nanocomposite by FTIR, XRD, FESEM and TEM. Spectrochim Acta A: Molecular and Biomolecular Spectroscopy 131: 55-58.

30. Yang G, Zhang B, Wang J, Xie S, Li X (2015) Preparation of polylysine-modified superparamagnetic iron oxide nanoparticles. J Magn Magn Mater 374: 205-208.

31. Bloemen M, Brullot W, Luong TT, Geukens N, Gils A, et al. (2012) Improved functionalization of oleic acid-coated iron oxide nanoparticles for medical application. J Nanopart Res 14: 1100.

32. Zhang L, He R, Gu HC (2006) Oleic acid coating on the monodisperse magnetite nanoparticles. Appl Surf Sci 253: 2611-2617.

33. Lan Q, Liu C, Yang F, Liu S, Xu J, et al. (2007) Synthesis of bilayer oleic acid-coated Fe_3O_4 nanoparticles and their application in pH-responsive Pickering emulsions. J Colloid Interf Sci 310: 260-269.

34. Yuen-Jian C, Juan T, Fei X, Jia-Bi Z, Nin G, et al. (2010) Oleic acid were chemisorbed onto the Fe_3O_4 nanoparticles as a carboxylate. Drug Dev Ind Pharm 36: 1235-1244.

35. Soares PIP, Alves AMR, Pereira LCJ, Coutinho JT, Ferreira IMM, et al. (2014) Effects of surfactants on the magnetic properties of iron oxide colloids. J Colloid Interf Sci 419: 46-51.

36. Meiorin C, Muraca D, Pirota KR, Aranguren MI, Mosiewicki MA (2014) Nanocomposites with superparamagnetic behavior based on a vegetable oil and magnetite nanoparticles. Eur Polym J 53: 90-99.

37. Zhou J, Fa H, Yin W, Zhang J, Hou C, et al. (2014) Synthesis of superparamagnetic iron oxide nanoparticles coated with a DDNP-carboxyl derivative for in vitro magnetic resonance imaging of Alzheimer's disease. Mater Sci Eng 37: 348-355.

38. Ying XY, Du YZ, Hong LH, Yuan H, Hu FQ (2011) Magnetic lipid nanoparticles loading doxorubicin for intracellular delivery: Preparation and characteristics. J Magn Magn Mater 323: 1088-1093.

39. Daou TJ, Pourroy G, Begin-Colin S, Greneche JM, Ulhaq-Bouillet C, et al. (2006) Hydrothermal synthesis of monodisperse magnetite nanoparticles. Chem Mater 18: 4399-4404.

9

Mechanical Behaviour of Skin

Kalra A*, Lowe A and Al-Jumaily AM

Institute of Biomedical Technologies, Auckland University of Technology, New Zealand

Abstract

Objective: The mechanical behaviour or the Young's Modulus of the skin is measured as a ratio of the stress applied to the skin *in vitro* or *in vivo* over the skin deformation. The Young's Modulus of skin is an important factor to estimate the characteristics of skin, to determine the course of a disease or to follow a cosmetic application.

Methods: The mechanical behaviour of the skin is measured by changing the shape of skin by employing tensile, indentation, and suction and torsion tests.

Results: Out of all the skin's mechanical testing methods, suction tests are a common choice for skin testing, as they are easy to apply *in vivo* and consider both in-plane and normal loading conditions. Skin is found to be highly anisotropic and viscoelastic, with a range of Young's Modulus between 5 kPa and 140 MPa.

Conclusion: This paper reviews *in vivo* and *in vitro* reported values for Young's Modulus of human skin for tensile, indentation, suction and torsion mechanical testing methods.

Keywords: Young's modulus; Skin structure; Skin barrier

Introduction

Skin is composed of three layers: Epidermis, Dermis, and Hypodermis [1]. The outermost layer epidermis acts as a skin barrier. Pereira [2] considered skin to be viscoelastic, where there is a dynamic alteration in the stress-strain relationship, until a stable state is attained [3].

The stress-strain behaviour of the skin is typically explained in three phases: When a strain of up to 0.3% is applied, the elastin fibres offer low resistance to the applied strain [4]. The skin exhibits isotropic behaviour and collagen fibres remain tangled and intertwined and do not contribute to the stiffness as seen in Figure 1. Phase 1 offers a linear stress-strain relationship and a low Young's Modulus (0.1-2MPa) [5].

In Phase 2, the collagen fibres offer some resistance to the deformation [6] and the crimped collagen fibres begin to stretch, thus introducing non-linearity into the stress strain relationship. In the final Phase 3, for applied strain above 0.6%, the crimps begin to disappear and a linear stress-strain relationship can be observed. The collagen fibres break after the application of an ultimate tensile strain of 0.7% [5].

Young's Modulus measurements differ with many factors, including the type of test performed (*in vivo* or *in vitro*), method of testing (tensile or indentation), test velocities (in tensile testing) or depth (in indentation techniques). This paper summarises reports of the range of Young's Modulus of the human skin, considering all of the above mentioned factors. The structure of this paper can be summarised in Figure 2.

Significance of Skin's Young's Modulus

Young's Modulus of the skin is a vital parameter to estimate the characteristics of skin. One of the striking features of a healthy skin is its ability to get back to normal after being pulled. Cosmetic surgeons use a variety of topical and invasive methods to maintain the skin's elasticity to prevent ageing [7]. The mechanical testing of skin can be useful to determine the mechanical behaviour of skin in the field of dermatology, to determine the course of a disease (Scleroderma, morphea, radio dermatitis etc.) or to follow a cosmetic application. It can be used in detection of diseases in connective tissues such as mid-

dermis elastolysis [8]. The UV radiation has been found to induce skin contractions causing photo ageing which can be analysed using Young's Modulus through the stress-strain relationship [9]. Quantification of hardness, elasticity and viscosity of the skin can help estimate the skin's thickness which is a significant index for diagnosing patients with systemic sclerosis [10]. The paper summarises the many different techniques for measuring skin stiffness as a guide to interpreting results obtained in clinical practice. It also assists in the choice of techniques to be used for measuring the skin's elasticity. Knowing the Young's Modulus of skin can help in calibrating the elasticity of bio-sensors to measure skin-stretch induced motion artifacts.

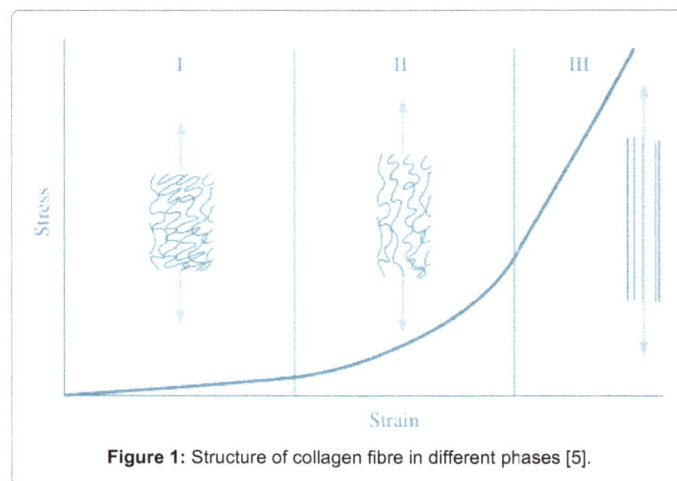

Figure 1: Structure of collagen fibre in different phases [5].

*Corresponding author: Kalra A, Institute of Biomedical Technologies, Auckland University of Technology, New Zealand
E-mail: anubha.kalra@aut.ac.nz

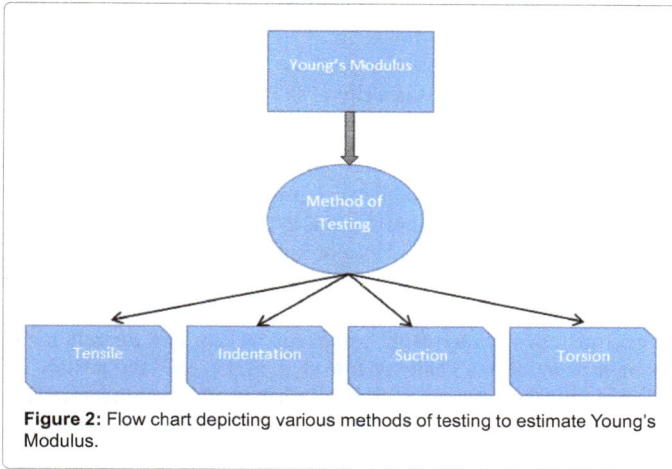

Figure 2: Flow chart depicting various methods of testing to estimate Young's Modulus.

In previous reviews, Hendriiks [11] discussed several innovative techniques to determine the mechanical and structural properties of the skin such as Ultrasound, Confocal Microscopy, Optical Coherence Tomography and Nuclear Magnetic Resonance. The use of the above methods is however restricted to the measurement of skin's thickness and tomography.

This paper provides a comparative study of various mechanical testing methods used *in vivo* and *in vitro* and reviews the works of various authors, thereby covering a broad range of factors affecting the Young's Modulus of skin.

Methods: Mechanical Testing of Skin

The mechanical behaviour of the skin is measured by changing the shape of skin by employing different techniques such as stretching (tensile test), applying normal load on the skin (indentation test), elevating the skin in an aperture (suction test) and rotating the epidermis to different degrees (torsion test). All these tests have been discussed in detail in this section.

The mechanical testing of skin can be further classified into *in vivo* and *in vitro* tests. *In vitro* tests provide a simple and easy to model Stress-Strain relationship under controlled conditions with fewer confounding factors. *In vitro* tests can also be used to calculate the ultimate tensile stress and strain when the skin ruptures. However, it can be difficult to clamp samples without applying an axial load and structural integrity of the excised skin is altered particularly at the edges of the sample as it is no longer attached to the body [12]. In comparison, *in vivo* tensile measures are able to include anatomical and physiological effects on skin properties. For example, skin ageing provides a negative impact on skin's ability to perform functions like body temperature regulation and water loss prevention. Longitudinal studies of Young's Modulus values of skin must therefore be done *in vivo*.

Tensile test

Tensile testing is the most common type of test performed *ex vivo* under controlled conditions [1]. In tensile tests, the skin is stretched parallel to the plane of the skin. The load can either be uniaxial or biaxial. In early work, Manschot and Brakkee [13] performed uniaxial strain measurements on human skin (calf) and observed a non-linear relationship between stress and applied strain. The maximum and minimum values of the Young's Modulus across the tibial axis were found to be 0.32 and 4 MPa respectively and 0.3 and 20 MPa, respectively, along it. Meijer et al. [14] performed uniaxial tensile

measurements on the forearm and found the stiffness value (K_c) to be 25 MPa. The work proposed a combined numerical-experimental method based on Lanir's Skin model [15] which considers the strain-energy function to be the sum of individual strain-energy values of the tissues.

Several investigations relating to tensile testing of the skin at dynamic [16-19] and quasistatic (low level) speeds [19-21] have been reported and a summary of results is given in Table 1. Ottenio [20] performed tensile testing by clamping an *ex vivo* sample from two sides while stretching at a speed of 10 mm/min and at a maximum strain of 20%. The values of Young's Modulus were found to be dependent on the orientation of the Langer's lines.

Annaidh et al. [12] carried out uniaxial tensile tests on a human skin excised from the back at a strain rate of 0.012 s^{-1} using a Universal Tensile Test machine. The test was carried out on 7 subjects in the age group of 81-97 years and strain was evaluated using Digital Image Correlation. The mean Young's Modulus was found to be 83.33 ± 4.9 MPa.

A customised tensile device was used to measure the ultimate stress along with the longitudinal, transverse and shear strain field in an I-shaped tissue sample (taken from an 85-year old male) using Image Correlation Method [17]. The machine had been divided into an upper chamber and a lower chamber to clamp the tissue from both ends. Young's Modulus was calculated for longitudinal, transverse and shear strains by pulling down the lower chamber at a velocity of 3 ms^{-1}.

Dynamic tensile stress tests were performed by Gallagher [18] using an Instron type 8802 testing machine at different stretch velocities (1-1.5 ms^{-1}). The results obtained through this study indicated maximum and minimum strain energies when the sample was placed perpendicular and at 45 degrees with respect to the Langer's lines respectively. Young's Modulus were obtained for 3 patients (aged 85, 77 and 82) for human skin excised from their backs, with stretch velocities of 1 ms^{-1}, 1.5 ms^{-1} and 2 ms^{-1}.

From Table 1, it can be inferred that the Young's Modulus measured at quasistatic speeds (0.1-0.9 mm s^{-1}) varies from 4–15 MPa while for dynamic speeds (2–30 ms^{-1}), it varies from 14-100 MPa. Significant fluctuations in these values have been found with different orientations like transverse and shear, however, the overall Young's Modulus increased monotonically with speed.

Indentation test

Indentation is one of the most widely used and accepted means of measurement of skin's bio-mechanical properties *in vivo*. It employs the use of an indenter which comes in to contact with and applies a perpendicular force on a small area of skin. This method characterizes

References	Skin Source	Speed	~Young's Modulus
Ankerson et al. [19]	Abdomen	Quasistatic (0.83 mms^{-1})	14.96 MPa
Annaidh et al. [16]	Not Mentioned	Dynamic (29 ms^{-1})	100 MPa
Jacquemoud et al. [17]	Forehead	Dynamic (3 ms^{-1})	14 MPa, 140 MPa and 35 MPa (for longitudinal, transverse and shear strain)
Gallagher et al. [18]	Back	Dynamic (2 ms^{-1})	83.3 MPa
Ottenio et al. [20]	Abdomen	Quasistatic Speed (0.16 mms^{-1})	4.02 ± 3.81 MPa

Table 1: Values of Young's Modulus at quasistatic and dynamic speeds using tensile testing.

skin as a monolayer by restricting the indentation amplitude to microns. However, the accurate prediction of Young's Modulus can be done only by considering the effects of underlying layers. Delalleau et al. [22] proposed a combined numerical-experimental work to estimate the skin elasticity. The skin was assumed to be a linearly elastic semi-infinite layer. Pailler-Mattie et al. [23] investigated different mechanical models to determine the effects of the underlying tissue layers and developed a two layer elastic model for mechanical analysis. The indentation method delivers Young's Modulus in the perpendicular direction without any skin pre-stressing [24,25]. The obtained values for skin's Young's Modulus vary from 4.5-8 kPa.

The value of Poisson's ratio also contributes to the obtained Young's Modulus calculations using indentation. Choi [26] performed experiments on bovine patellar articular cartilage and estimated the Young's Modulus to be 1.33-2.21 MPa for a Poisson's ratio ranging from 0.45-0.47 using single indentation test. Jia [27] in his research identified the variation of Young's Modulus with indentation depth using finite element analysis. The dynamic analysis was performed on two gel samples with different Young's moduli between 0-500 Hz using Tissue Resonator Indenter Device (TRID).

Some of the works relating to quasistatic and dynamic speeds are summarized in Table 2. Zheng and Mak [28] proposed an Ultrasound Indenter system to obtain quasistatic indentation responses of softer tissues in the lower limb. Young's Modulus was found to depend on the area, posture, gender and subject. Khaothong [29] aimed at determining the biomechanical properties of skin and muscle using an inverse finite element method combined with indentation test and found that the non-linear properties were best suited by Jamus-Green-Simpson strain energy function.

Boyer et al. [30] developed a non-invasive dynamic indentation device using very small amplitude strain (1-10 µm) and indenter penetration (100-500 µm). These small amplitudes were obtained using a piezoelectric translation stage for moving the indenter. In 2009 [31], the same authors performed tests on elastic inert materials to validate the device. In [32], an advanced device called Tonoderm® has been used to measure the Young's Modulus on human forearm. The device exerts pressure on the skin using an air compressor. The distance/depth of indentation has been measured using a laser beam passing through a Laser Displacement Sensor.

The efficiency of simple indentation measurements in thin films can be compromised by ignoring the combined contributions of the film and indenter to measured properties, as has been analysed in [33-37] . As a correction, there must be some consideration of a 'reduced Young's Modulus' which constitutes the effect of the film and the indenter. Pailler-Mattie et al. [23] analysed the effect of changing indenter penetration to a reduced Young's Modulus (E*) of skin defined by:

$$E^* = \frac{\pi}{4} \frac{k_z}{\delta} \tan\left(\frac{\pi}{2} - \alpha\right) \tag{1}$$

where,

$k_z = (dF_N/d\delta)|_{FN=FN\,max}$ [38],

δ is penetration depth and α is measure of difference in 'plane strain modulus' [39].

They found the modulus to increase with increasing indenter depths. The test was carried out on different layers of tissues underlying skin (including hypodermis and dermis) and considered skin to be as a thin film over a rigid substrate (muscle).

Jia [27] measured tissue mechanical properties in terms of static stiffness and dynamic stiffness as a function of various indenter depths and found an increasing trend for both. Groves [1] conducted experiments to determine elasticity of skin at various indenter depths for spherical and cylindrical indenters, as summarized in Figure 3.

He observed that the cylindrical indenter measured a higher average value of Young's Modulus than the spherical indenter at higher indentation depths. Kuilenburg [40] also investigated the necessity of considering the geometry and size of indenters while considering the measurement of skin's elasticity. A comparative analysis of different works showed a decrease in Young's Modulus for indenter depth in microns and an increasing behaviour of elasticity for millimetre indenter penetrations. Pailler-Mattie et al. [23] carried out a study for different models accounting for skin's thickness (e) and indenter-skin contact radius (a). The apparent Young's Modulus decreased with an increasing penetration depth for a/e < 0.5. For the same load, the contact area of a spherical indenter is more than a cylindrical indenter; therefore, the spherical indenter exhibited a lower average value of Young's Modulus than the cylindrical indenter as observed from Figure 3.

Suction test

The mechanical properties of thin elastic membranes of materials like rubber can be determined using Diaphragm tests, where the membrane is clamped at two ends and inflated in the form of a dome (Figure 4) while the pressure of suction is controlled by a pressure controller.

Early work of Grahame [41], Alexander and Cook [42] adopted a

References	Skin Source	Speed	Young's Modulus
Khaothong [28]	Inner-forearm	Quasistatic (1 mms⁻¹)	0.1-2.4 MPa
Zheng and Mak [29]	Tibia/Fibula	Quasistatic (0.5 – 1 mms⁻¹)	10.4-89.4 kPa
Boyer et al. [30]	Forearm (Right)	Dynamic (1-10 µm for 10-60 Hz)	5.1-13.3 kPa
Boyer et al. [31]	Forearm (Right)	Dynamic (1-10 µm / 100-500 µm for 10-60 Hz)	13.2-33.4 kPa
Boyer et al. (Laser Displacement Method) [32]	Forearm	Dynamic (2-100 lₙ min⁻¹)	4.75-17.99 kPa

Table 2: Values of Young's Modulus at quasistatic and dynamic speeds using indentation technique.

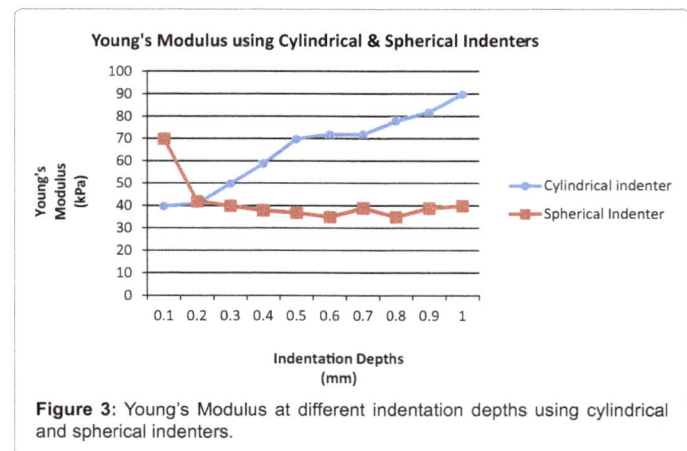

Figure 3: Young's Modulus at different indentation depths using cylindrical and spherical indenters.

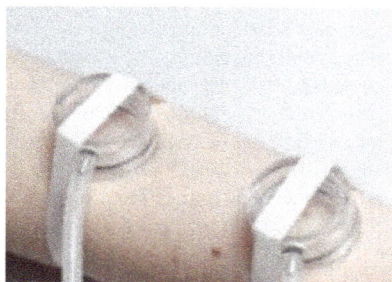

Figure 4: Circular aperture of the dome used to elevate the skin in suction test.

method of suction to stratum corneum considering skin to be isotropic. Following these works, the suction method to investigate anisotropy of skin has evolved to become a common procedure for skin mechanical testing. Generally, it employs the measurement of skin elevation in a circular aperture caused due to vacuum conditions (< 500 mBar) [43] using optical systems like Dermaflex and Cutometer.

Dermaflex is a device with an aperture size of 10 mm, the cup being adhered to the skin to prevent creep. It has been used to measure skin distensibility [44] and to account for mechanical properties of dermis in [45] by measuring elasticities as a percentage of skin retraction after the stretch. The Cutometer is a suction device employing probe apertures between 2-8 mm with the application of negative pressure through a vacuum pump [46]. Barel et al. [47] determined stress-strain and strain-time curves using a Cutometer at 2 mm aperture and found a linear response within 150-500 mBar. Skin elevations of 0.1-0.6 mm were observed yielding Young's Modulus values between 130-260 kPa at different skin sites. Diridollou et al. [48] developed a suction system with ultrasound scanning-an echo rheometer capable of measuring thickness of epidermis and dermis. It operated in 3 modes at a frequency of 20 MHz and provided an axial resolution of 0.07 mm. Table 3 represents different values of Young's Modulus obtained by the suction method, measuring deformation with different aperture sizes.

Several assumptions are typically made in applying suction measurements. Hendriks ignored the mechanical contribution of epidermis in his model, instead considering that the fat layer is a major contributor for elasticity as proposed by Diridollou [49,50]. Moreover, the values of skin thickness have an effect along with the aperture size and the magnitude of negative suction pressure. Khatyr et al. [51] accounted for this aspect and compared the suction results based on three geometrical considerations of skin: thin plate, Timoshenko's geometry [52] and finite element modelling as discussed below:

Thin plate geometrical model (based on analysis of Siqueira):

$$\frac{E}{1-\mu} = \frac{pa}{2e[Arc\sin\left(2au/a^2+u^2\right) - 2au/a^2+u^2]} \quad (2)$$

Where,

a is radius of probe,

e is skin thickness,

p is negative pressure applied,

u is the elevation of dome,

E is Young's Modulus of material, and

μ is Poisson's ratio [53].

Timoschenko's model: It is defined by following three equations:

$$Case\ I \rightarrow \frac{u_o}{e} + A\left(\frac{u_o}{e}\right)^3 = B\frac{p}{E}\left(\frac{a}{e}\right)^4 \quad (3)$$

$$Case\ II \rightarrow \sigma_r^t = \alpha_r E\frac{u_o^2}{a^2} \quad (4)$$

$$Case\ III \rightarrow \sigma_r^f = \beta_r E\frac{u_o e}{a^2} \quad (5)$$

Where,

u_o is dome elevation,

σ_r^t is stress in median plane,

σ_r^f is flexion stress,

e is plate thickness,

a is radius of plate,

E is Young's Modulus,

p is pressure exerted, and

A, B, α_r and β_r are limiting parameters.

The work used Timoschenko's model, where the coefficients were optimised to fit FE modelling. The work illustrated a model for isotropic and orthotropic materials using specific initial conditions. According to the models proposed by Siqueira [53] and Timoschenko [52], the Young's Modulus exhibits an exponential increase with the increase in aperture size.

Torsion tests

Torsion measurements are carried out by applying a constant torque through a guard ring and an intermediary disc and measuring the resultant rotation of skin as seen in Figure 5.

The method is supposed to reduce the skin anisotropic effects since the underlying layers do not contribute to the readings as postulated by Escoffier et al. [54]. As the torque is applied, an immediate elastic deformation occurs followed by the occurrence of creeping viscoelastic deformation which is time dependent. The release of torque leads to immediate recovery followed by a slow recovery process which is usually not completed [55]. In torsion, the elongation is replaced by rotation and hence the measurement of elasticity becomes more complex.

Early work includes that of Sanders [56] who performed an *in vivo* analysis to determine the extensibility of skin subjected to torsion. A twist of 0.8 mN-m was applied to a disc of diameter 8.7 mm. Young's Modulus was calculated using the formula [7]

$$YM = \frac{2M(1+\mu)}{4eR^2\theta} \quad (6)$$

References	Skin Source	Deformation measurement/ Aperture Size	Young's Modulus
Diridollou [49]	Forearm	100 mBar suction/ 6 mm	130 kPa
Hendriks [11]	Forearm	350 mBar suction, Ultrasound detection/ 6 mm	56 kPa
Barel [47]	Cheek	150-500 mBar suction/ 2 mm	130-260 kPa
Liang [50]	Palm, Forearm	450 mBar/ 2 mm	25 kPa, 100 kPa

Table 3: Measurement of Young's Modulus using suction at different deformations and aperture sizes.

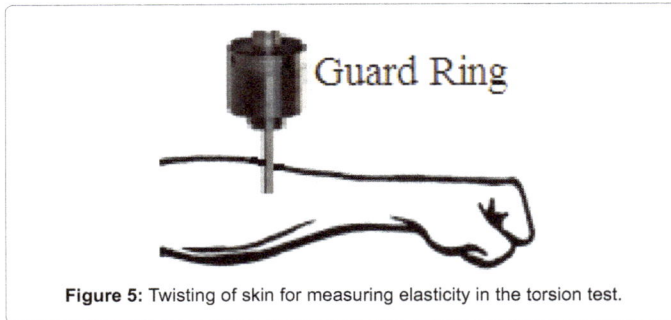

Figure 5: Twisting of skin for measuring elasticity in the torsion test.

where,

M is the applied torque,

e is the skin thickness,

μ is Poisson's ratio,

R is disc radius and θ is the rotation.

Agache et al. [57,58] studied the skin ageing through their experiment to determine skin stiffness through torsion. A torque of 28.6×10^{-3} N-m was applied through a disc and guard ring of 25 and 35 mm diameters respectively and rotations of 2-6° were obtained. The Young's Modulus was calculated through:

$$YM = \frac{M}{0.8\pi \, e \, r_1 \, r_2 \, \theta} \qquad (7)$$

where,

M is the applied torque,

e is the skin thickness,

r_1 is the disc radius,

r_2 inner radius of guard ring and θ is the rotation in radians.

The values of Young's Modulus obtained by using torsion techniques are shown in Table 4.

Other significant works include the study of Grebenyuk and Uten'kin [59] who worked on different anatomical sites on children resulting in rotations of 7 -10° at an application of a constant torque.

Discussion

Considering indentation and tensile testing, the relevance of the chosen technique mainly depends on the application. McKee et al. [60] reviewed the various indentation and tensile tests and provided a comparative insight. Tensile testing was described as a more 'direct and economical' approach. He also suggested that the value of Young's Modulus is dependent on model specific constraints such as type of test performed, controlled conditions etc. Tensile tests at a dynamic speed are generally conducted to investigate skin failure, while quasistatic speed is used to carry out conventional tests to measure the skin stiffness. In one of the works [61], dynamic speed can be used to estimate the injury levels in human skin, owing to the accurate representation of human ligament from the properties obtained at high strain rate, while quasi-static speed exhibit a linear stress-strain relationship. The values of Young's Modulus measured at quasistatic speeds are lower than those at dynamic speeds.

In general, Young's moduli found by indentation are significantly lower than those found by tensile tests, indicating that skin is highly anisotropic when thickness and in-plane directions are considered. A contributing factor may be that Young's Modulus values are dependent on contact dimensions and range of fit. The indentation contact is very small whereas the tensile tests are macroscopic. Furthermore, in an indentation test, the Young's Modulus depends on the depth of the indenter in contact with the underlying tissues. Therefore when the depth of the indenter is small; the skin poses lower resistance from the collective effect of the underlying tissues/fibres or matrices. Conversely, these structures play a significant role in resisting tensile deformations.

In many applications *in vivo* testing would provide more relevant information than over *in vitro* testing. However, *in vivo* tests are subject to more confounding factors that are difficult to control. Conversely, *in vitro* tests typically require excised skin samples, which may be difficult to obtain. Mostly, tensile tests are performed *in vitro* by taking excised skin samples or skin patches from animals and in some cases by manufacturing skin-mimicking materials. However, some works [13,62,63] also experimented with tensile testing *in vivo* under different conditions of loading. Indentation tests are mostly performed *in vivo*, and are relatively easily applied. However the dimensions of the test site significantly affect the results. Furthermore, in applications requiring in-plane measurements, indentation is not suitable as it computes a thickness-mode response.

Torsion measurements are an accepted and reproducible means of in-plane skin elasticity analysis. However, they assume an isotropic behaviour of skin layers and a uniform deformation for the entire skin thickness. However, this consequently assumes that the applied force gradient reaches uniformly to the deeper layers of the skin. Also, since the measure of torsion is the rotational angle, it obtains, essentially, the shear modulus of the skin, which is theoretically related to the Young's Modulus.

Suction tests are a common choice for skin testing, as they are easy to apply *in vivo* and also allow for additional deformation detection through, for example, imaging ultrasound. However, this technique involves the skin undergoing both in-plane and normal loading and depends on theoretical models to determine elastic properties.

Conclusion

Skin is a highly anisotropic material. Young's Modulus in the thickness-direction typically measures between 5 to 100 kPa by indentation tests. However, measured values can depend on indenter geometry and whether quasistatic or dynamic testing is being performed. Values of between 25 kPa and 140 MPa are typical for both tensile and torsion tests. Tensile tests indicate higher Young's Modulus at higher strain rates, indicating that skin is viscoelastic. Young's Modulus measured by suction tests span 25 kPa to 260 kPa, which is between the ranges found from indentation (thickness-mode) and tensile/torsion (in-plane mode). This may be as suction tests involve both in-plane and perpendicular deformations.

Acknowledgements

The work was supported by the Institute of Biomedical Technologies and School of Engineering, Computing and Mathematical Sciences, Auckland University of Technology, New Zealand.

References	Skin Source	Torque/ Disc diameter/ Guard ring diameter	Young's Modulus
Sanders [56]	Forearm	0.8 mN-m/ 8.7 mm/	0.02-0.1 MPa
Agache et al. [57]	Forearm	28.6 mN-m /25 mm/ 35 mm	0.42-0.85 MPa
Escoffier et al. [54]	Forearm	2.3-10.4 mN-m /18 mm/ 24 mm	1.12 MPa

Table 4: Young's Modulus obtained using torsion using different parameters.

References

1. Groves R (2012) Quantifying the mechanical properties of skin in vivo and ex vivo to optimise microneedle device design. Cardiff University.

2. Pereira BP, Lucas PW, Swee-Hin T (1997) Ranking the fracture toughness of thin mammalian soft tissues using the scissors cutting test. J Biomech 30: 91-94.

3. Matsumura H, Yoshizawa N, Watanabe K, Vedder NB (2001) Preconditioning of the distal portion of a rat random-pattern skin flap. Br J Plast Surg 54: 58-61.

4. Oxlund H, Manschot J, Viidik A (1988) The role of elastin in the mechanical properties of skin. J Biomech 21: 213-218.

5. Holzapfel GA (2000) Biomechanics of soft tissue. Handbook of material behavior nonlinear models and properties, France.

6. Silver FH, Freeman JW, DeVore D (2001) Viscoelastic properties of human skin and processed dermis. Skin Res Technol 7: 18-23.

7. Vlasblom DC (1967) Skin elasticity. Bronder-Offset.

8. Agache P, Humbert P (2004) Measuring the Skin. Springer Science & Business Media.

9. Oba A, Edwards C (2006) Relationships between changes in mechanical properties of the skin, wrinkling, and destruction of dermal collagen fiber bundles caused by photoaging. Skin Res Technol 12: 283-288.

10. Kuwahara Y, Shima Y, Shirayama D, Kawai M, Hagihara K, et al. (2008) Quantification of hardness, elasticity and viscosity of the skin of patients with systemic sclerosis using a novel sensing device (Vesmeter): a proposal for a new outcome measurement procedure. Rheumatology 47: 1018-1024.

11. Hendriks FM, Brokken D, van Eemeren JTWM, Oomens CWJ, Baaijens FPT, et al. (2003) A numerical-experimental method to characterize the non-linear mechanical behaviour of human skin. Skin Res Technol 9: 274-283.

12. Annaidh AN, Bruyère K, Destrade M, Gilchrist MD, Otténio M (2012) Characterization of the anisotropic mechanical properties of excised human skin. J Mech Behav Biomed Mater 5: 139-148.

13. Manschot JF, Brakkee AJ (1986) The measurement and modelling of the mechanical properties of human skin in vivo--I. The measurement. J Biomech 19: 511-515.

14. Meijer R, Douven LFA, Oomens CWJ (1999) Characterisation of anisotropic and non-linear behaviour of human skin in vivo. Comput Methods Biomech Biomed Engin 2: 13-27.

15. Lanir Y (1983) Constitutive equations for fibrous connective tissues. J Biomech 16: 1-12.

16. Annaidh AN, Destrade M, Ottenio M, Bruyere K, Gilchrist MD (2014) Strain rate effects on the failure characteristics of excised human skin. 9th International Conference on the Mechanics of Time Dependent Materials, Ireland.

17. Jacquemoud C, Bruyere-Garnier K, Coret M (2007) Methodology to determine failure characteristics of planar soft tissues using a dynamic tensile test. J Biomech 40: 468-475.

18. Gallagher AJ, Anniadh AN, Bruyere K, Otténio M, Xie H, et al. (2012) Dynamic tensile properties of human skin. International Research Council on the Biomechanics of Injury.

19. Ankersen J, Birkbeck AE, Thomson RD, Vanezis P (1999) Puncture resistance and tensile strength of skin simulants. Proc Inst Mech Eng 213: 493-501.

20. Ottenio M, Tran D, Annaidh AN, Gilchrist MD, Bruyère K (2015) Strain rate and anisotropy effects on the tensile failure characteristics of human skin. J Mech Behav Biomed Mater 41: 241-250.

21. http://studylib.net/doc/7654896/instron-tensile-testing--material-properties-of-sutured-c...

22. Delalleau A, Josse G, Lagarde JM, Zahouani H, Bergheau JM (2006) Characterization of the mechanical properties of skin by inverse analysis combined with the indentation test. J Biomech 39: 1603-1610.

23. Pailler-Mattei C, Bec S, Zahouani H (2008) In vivo measurements of the elastic mechanical properties of human skin by indentation tests. Med Eng Phys 30: 599-606.

24. Bader DL, Bowker P (1983) Mechanical characteristics of skin and underlying tissues in vivo. Biomaterials 4: 305-308.

25. Falanga V, Bucalo B (1993) Use of a durometer to assess skin hardness. J Am Acad Dermatol 29: 47-51.

26. Choi APC, Zheng YP (2005) Estimation of Young's modulus and Poisson's ratio of soft tissue from indentation using two different-sized indentors: Finite element analysis of the finite deformation effect. Medical and Biological Engineering and Computing 43: 258-264.

27. Jia M (2009) Nonlinear finite element analysis of static and dynamic tissue indentation. University of Toronto.

28. Zheng Y, Mak AFT (1999) Effective elastic properties for lower limb soft tissues from manual indentation experiment. IEEE Eng Med Biol Soc 7: 257-267.

29. Khaothong K (2010) In vivo measurements of the mechanical properties of human skin and muscle by inverse finite element method combined with the indentation Test. 6th World Congress of Biomechanics (WCB 2010). August 1-6, 2010 Singapore 31, pp: 1467-1470.

30. Boyer G, Zahouani H, Le Bot A, Laquieze L (2007) In vivo characterization of viscoelastic properties of human skin using dynamic micro-indentation. Conf Proc IEEE Eng Med Biol Soc 2007: 4584-4587.

31. Boyer G, Laquièze L, Le Bot A, Laquièze S, Zahouani H (2009) Dynamic indentation on human skin in vivo: ageing effects. Skin Res Technol 15: 55-67.

32. Boyer G, Pailler Mattei C, Molimard J, Pericoi M, Laquieze S, et al. (2012) Non contact method for in vivo assessment of skin mechanical properties for assessing effect of ageing. Med Eng Phys 34: 172-178.

33. Burnett PJ, Rickerby DS (1987) The mechanical properties of wear-resistant coatings: II: Experimental studies and interpretation of hardness. Thin Solid Films 148: 51-65.

34. Joslin DL, Oliver WC (1990) A new method for analyzing data from continuous depth-sensing microindentation tests. J Mater Res 5: 123-126.

35. Korsunsky AM, McGurk MR, Bull SJ, Page TF (1998) On the hardness of coated systems. Surf Coat Technol 99: 171-183.

36. Menčík J, Munz D, Quandt E, Weppelmann ER, Swain MV (1997) Determination of elastic modulus of thin layers using nanoindentation. J Mater Res 12: 2475-2484.

37. Tsui TY, Pharr GM (1999) Substrate effects on nanoindentation mechanical property measurement of soft films on hard substrates. J Mater Res 14: 292-301.

38. Loubet JL, Georges JM, Marchesini O, Meille G (1984) Vickers indentation curves of magnesium oxide (MgO). J Tribol 106: 43-48.

39. Johnson KL (1987) Contact mechanics. Cambridge university press, England.

40. Van Kuilenburg J, Masen MA, Van der Heide E (2013) Contact modelling of human skin: What value to use for the modulus of elasticity? J Eng Tribol 227: 349-361.

41. Grahame R (1970) A method for measuring human skin elasticity in vivo with observations on the effects of age, sex and pregnancy. Clin Sci 39: 223-229.

42. Alexander H, Cook TH (1977) Accounting for natural tension in the mechanical testing of human skin. Journal of Investigative Dermatology 69: 310-314.

43. Hendriks F (1969) Mechanical behaviour of human skin in vivo. Biomed Eng pp: 322-327.

44. Jemec GB, Gniadecka M, Jemec B (1996) Measurement of skin mechanics. Skin Research and Technology 2: 164-166.

45. Gniadecka M, Serup J (2006) Suction chamber method for measuring skin mechanical properties: the dermaflex.

46. Li Y, Dai DX (2006) Biomechanical engineering of textiles and clothing. Woodhead Publishing, USA.

47. Barel AO, Courage W, Clarys P (2006) Suction chamber method for measurement of skin mechanics: the new digital version of the cutometer (2nd edn.) Handbook of non-invasive methods and the skin, CRC Press, Boca Raton.

48. Diridollou S, Berson M, Vabre V, Black D, Karlsson B, et al. (1998) An in vivo method for measuring the mechanical properties of the skin using ultrasound. Ultrasound in medicine & biology 24: 215-224.

49. Diridollou S, Patat F, Gens F, Vaillant L, Black D, et al. (2000) In vivo model of the mechanical properties of the human skin under suction. Skin Research and technology 6: 214-221.

50. Liang X, Boppart SA (2010) Biomechanical properties of in vivo human skin from dynamic optical coherence elastography. IEEE Transactions on Biomedical Engineering 57: 953-959.

51. Khatyr F, Imberdis C, Varchon D, Lagarde JM, Josse G (2006) Measurement of the mechanical properties of the skin using the suction test. Skin research and technology 12: 24-31.

52. Timoshenko SP, Woinowsky-Krieger S (1961) Theory of plates and shells.

53. De Mesquita Siqueira CJ (1993) Development of a biaxial test on composite sheets and use for modeling the behavior of a material SMC (Doctoral dissertation).

54. Escoffier C, de Rigal J, Rochefort A, Vasselet R, Lévêque JL, et al. (1989) Age-related mechanical properties of human skin: an in vivo study. Journal of Investigative Dermatology 93: 353-357.

55. Serup J, Jemec GBE, Grove GL (2006) Handbook of non-invasive methods and the skin (2nd edn.) CRC Press, USA.

56. Sanders R (1973) Torsional elasticity of human skin in vivo. Pflüg Arch 342: 255-260.

57. Agache PG, Monneur C, Leveque JL, Rigal JD (1980) Mechanical properties and Young's modulus of human skin in vivo. Arch Dermatol Res 269: 221-232.

58. Leveque JL, De Rigal J, Agache PG, Monneur C (1980) Influence of ageing on the in vivo extensibility of human skin at a low stress. Archives of Dermatological Research 269: 127-135.

59. Grebenyuk LA, Uten'kin AA (1994) Mechanical properties of the human skin. Communication I. Human Physiology 20: 149.

60. McKee CT, Last JA, Russell P, Murphy CJ (2011) Indentation versus tensile measurements of Young's modulus for soft biological tissues. Tissue Engineering Part B: Reviews 17: 155-164.

61. Warhatkar H, Chawla A, Mukherjee S, Malhotra R (2008) Experimental study of variation between quasi-static and dynamic load deformation properties of bovine medial collateral ligaments.

62. Evans JH, Siesennop W (1967) Controlled quasi-static testing of human skin in vivo. In Digest of the 7th International Conference on Medical Electronics and Biological Engineering.

63. Abas WW, Barbenel JC (1982) Uniaxial tension test of human skin in vivo. Journal of biomedical engineering 4: 65-71.

10

Enhanced Interlayer Coupling of CuO$_2$ Planes Promote the Superconducting Properties of Cu$_{0.5}$Tl$_{0.5}$Ba$_{2-Y}$Sr$_Y$Ca$_2$Cu$_3$O$_{10-\delta}$ (Y=0, 0.15, 0.25) Samples

Muzaffar MU[1]*, Khan NA[2], Rehman UU[2] and Ali SA[3]

[1]PMAS Arid Agriculture University Attock Campus, Pakistan
[2]Materials Science Laboratory, Quaid-i-Azam University, Islamabad, Pakistan
[3]Department of Physics, Government College, University Lahore, Pakistan

Abstract

The CuTl-1223 superconducting samples, doping the Sr atom at Ba site, have been synthesized at 860°C pressure. The charge reservoir layer (CRL) of Cu$_{0.5}$Tl$_{0.5}$-1223 superconductor is modified by doping Sr atom. The decrease in c-axis length which is most probably due to smaller size of Sr atom as compared to Ba. The substitution of Sr atom at Ba is confirmed by the Fourier Transform Infrared Spectroscopy (FTIR). The critical temperatures i.e., Tc(R=0). Tc^{onset} are increased with the Sr content which shows that superconducting magnitude enhanced. The excess conductivity analysis has been done using Aslamazov-Larkin and Lawrance-Donaich models. The crossover temperatures i.e., TCR-3D=TG, T3D-2D and T2D-SWF and c-axis coherence length ξc(0) are slanted to lower values. Moreover, the inter-plane coupling (J) increases due to decrease in c-axis length. From fluctuations induced conductivity, it is found that there is an inverse relationship between critical temperatures and coherence length.

Keywords: Cu$_{0.5}$Tl$_{0.5}$-1223; Synthesis; Sr substitution; X-ray diffraction; Excess conductivity

Introduction

The cup-rate intrinsic superconducting parameters are structure dependent. There are two parts i.e., i) MBa$_2$O [M=Cu, Tl, Bi, Hg, C] a charge reservoir layer (CRL) and ii) conducting copper oxide planes nCuO$_2$ [1] of general unit cell in HTSC. In unit cell, the CRL provides the carriers (cooper pairs) to the cooper oxide planes and due to these carriers superconductivity exit [2-4].

Hence, the modification in CRL has a vital role in superconducting properties. In order to enhance the magnitude of superconductivity, numerous scientists have tried to modification in CRL [Co, Fe, Al] [5-7] and in CuO$_2$ planes [Zn, Ni] [8-10]. The increase in anisotropy and reduce in inter-plane coupling may be possible due to thicker CRL. Although, in periodic table, both atoms i.e., Sr and Ba lie in the same group but Sr atom is smaller in ionic radius (1.12 Å) as compared to Ba (1.35 Å). It is expected that dopant Sr atom would help to squeeze the CRL for enhanced interlayer coupling and hence improved the efficiency of CRL to the conducting CuO$_2$ planes. In contrast to the fixed Cu valence (~2+) in the Tl-bilayer cuprate superconductors, the average formal valence of Cu in the Tl-monolayer compounds TlBa$_2$Ca$_{n-1}$Cu$_n$O$_{2n+3}$ varies as (2+n−1)+. This characteristic is reflected in linear augmented-plane-wave band-structure results for the simplest n=1 member of this Tl-monolayer homologous series, TlBa$_2$CuO$_5$, where the filling (~0.16) of the planar Cu(3d)-O(2p) σ* band is reduced well below one-half. It is shown that the 50-50 Ba-La alloy is an appropriate "parent" compound for this n=1 phase since the half-filled-band condition is restored. For any member of this Tl-monolayer series, the optimal doping for high-temperature superconductivity should involve a combination of structural and chemical contributions.

Experimental Details

Synthesis

The superconducting samples Cu$_{0.5}$Tl$_{0.5}$Ba$_{2-Y}$Sr$_Y$Ca$_2$Cu$_3$O$_{10-\delta}$ (y=0, 0.15, 0.25) were synthesized by using the solid state reaction method. In first step, we prepared the Cu$_{0.5}$Ba$_{2-Y}$Sr$_Y$Ca$_2$Cu$_3$O$_{10-\delta}$ (y=0, 0.15, 0.25) by mixing the Ca(NO$_3$)^2.4H$_2$O, Cu(CN)^2, SrCO^3 and Ba(NO$_3$)^2 as starting compounds. These compounds were thoroughly mixed for almost 2 hours in mortal and pestle. The chamber furnace at 860°C is used for heat treatment. After 24 hours continuously firing, the furnace was put off. Repeat the process under the same atmosphere.

In second step, well calculated amount of thallium oxide (Tl$_2$O$_3$) was added in precursor material and thoroughly mixed. The material was pelletized and these pellets were sintered for nearly 10 min to get finally Cu$_{0.5}$Tl$_{0.5}$Ba$_{2-Y}$Sr$_Y$Ca$_2$Cu$_3$O$_{10-\delta}$ (y=0, 0.15, 0.25) samples.

Characterizations

To measure the resistivity, we used four-probe method was used. In this method, four uniformly spaced silver paste contacts were applied. The crystal structure of the material was determined by X-ray diffraction (XRD) measurements using Bruker diffractometer at X-ray wavelength of 1.5418 Å.

Results and Discussion

The crystal structure of Cu$_{0.5}$Tl$_{0.5}$(Ba$_{2-y}$Sr$_y$)Ca$_2$Cu$_3$O$_{10-\delta}$ where; y=0, 0.15 and 0.25 superconducting samples have been determined from the x-ray diffraction data, shown in Figure 1. The dimensions of unit cell were calculated from check cell. All the samples have orthorhombic structure with PMMM space group. It can be seen from Figure 1 that there is a slight shift in peak positions to higher 2 theta values with the increase of Sr content, which is most probably due to the decrease in the c-axis length of the unit cell. The overall contraction of the unit cell can be seen from the decrease of both a and c axis with the Sr content

---

*Corresponding author: PMAS Arid Agriculture University Attock Campus, Pakistan, E-mail: m.usman@uaar.edu.pk

(Figures 2a and 2b). Due to smaller in size, Sr atom prompt such modifications in the unit cell.

The resistivity versus temperature measurements of $Cu_{0.5}Tl_{0.5}(Ba_{2-Y}Sr_Y)Ca_2Cu_3O_{10-\delta}$ (y=0, 0.15, 0.25) superconducting samples are in Figure 3a and inset is shown the variation of ρ290 K(Ω-cm) versus Sr content. These samples have shown metallic behaviour from room temperature down to onset of superconductivity. From resistivity analysis, it is observed that these samples have shown Tc(onset) around 114.5, 116.3 and 116.3 K whereas Tc(R=0) at 90.26, 95.47 and 96.5 K respectively, shown in Figure 3b. The systematic increase of Tc (R=0) and with increasing Sr content is mostly suggested that the dopant Sr atom at Ba site in Tc(onset) $Cu_{0.5}Tl_{0.5}Ba_2O_{10-\delta}$ (CRL) promotes efficient transfer of the carriers to the conducting CuO_2 planes and hence as a result, the aforementioned superconducting parameters enhanced. The room temperature resistivity ρ 290 K(Ω-cm) is systematically decreased with the Sr content which shows that dopant atom boosts an extra metallic trend in the final compound.

FTIR absorption measurements of $Cu_{0.5}Tl_{0.5}Ba_{2-Y}Sr_YCa_2Cu_3O_{10-\delta}$ (y=0, 0.15, 0.25) samples are shown in Figure 4. Three vibrational modes are witnessed around 613.4, 483.7 and 421.2 cm⁻¹ in un-doped sample. However, with Sr doping in CRL, the former two modes are related to apical oxygen atoms of nature Tl-O_A-$Cu_{(2)}$, $Cu_{(1)}$ -O_A-$Cu_{(2)}$ are softened from 483.7 to 482.0 cm⁻¹ and 421.2 to 415.1 cm⁻¹ whereas the third mode related to the planar oxygen atoms of nature Tl-O_p-$Cu_{(2)}$, are hardened from 613.4 to 623.4 cm⁻¹. The softening/hardening of the oxygen modes are most possibly related with the relaxation and compression of apical and planar bond lengths due to stresses and strains created after the doping of Sr atom at Ba site.

Excess conductivity analysis (FIC) $Cu_{0.5}Tl_{0.5}Ba_{2-Y}Sr_YCa_2Cu_3O_{10-\delta}$ (y=0, 0.15 and 0.25) samples

By fitting the experimental data of resistivity in theoretical models i.e., Aslamazov-Larkin and Lawrance-Donaich, the excess conductivity analyses is done in the temperature regime around Tc and beyond. According to Aslamsov Larkin (AL) model, the conductivity Δσ(T) is given by:

$$\Delta\sigma AL = C^-$$ (1)

$$\Delta[\sigma AL]^{3D} = C^{3D} \; -0.5$$ (2)

$$\Delta[\sigma AL]^{2D} = C^{2D} \; -1.0$$ (3)

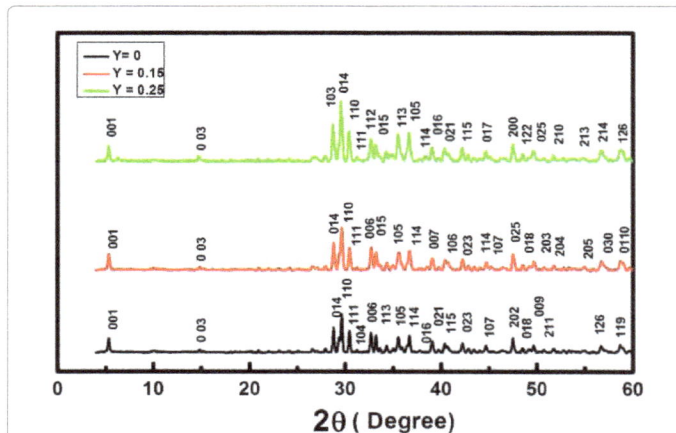

Figure 1: X-ray diffraction spectra of $Cu_{0.5}Tl_{0.5}$ $(Ba_{2-y}Sr_y)Ca_2Cu_3O_{10-\delta}$ superconductor with Y=0, 0.15, 0.25.

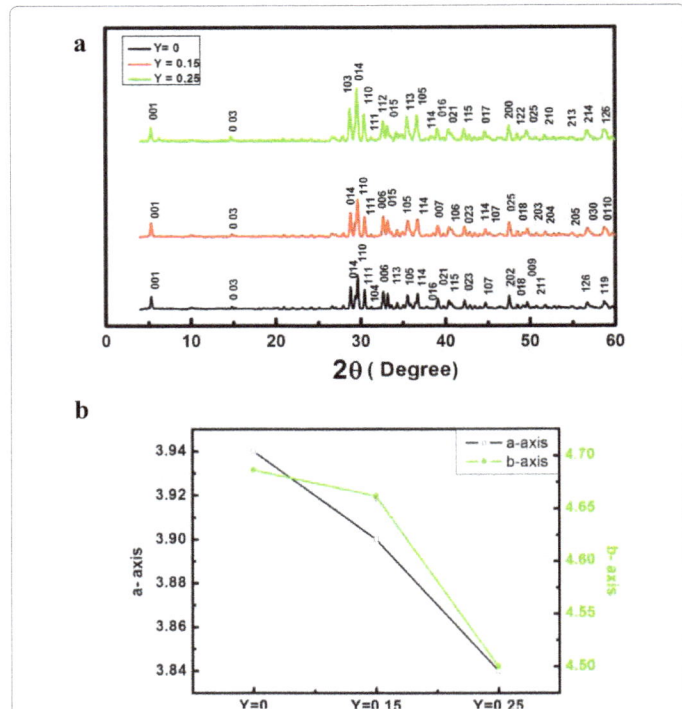

Figure 2: (a) Variation of c-axis length and volume of unit cell with Sr content; (b) Variation of a-axis and b-axis length with Sr content.

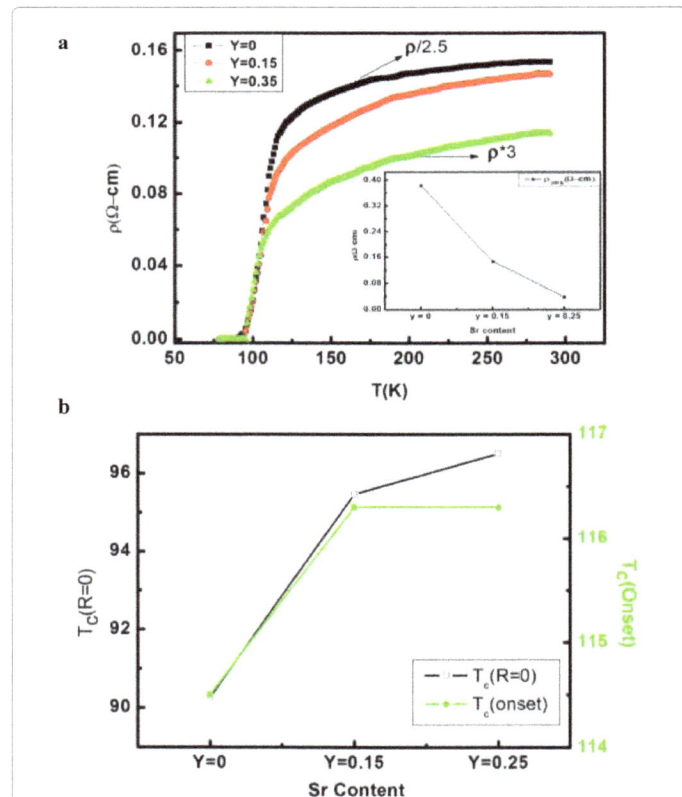

Figure 3: (a) Resistivity measurements of $Cu_{0.5}Tl_{0.5}(Ba_{2-y}Sr_y)Ca_2Cu_3O_{10-\delta}$ superconductor with y=0, 0.15, 0.25 as a function of temperature; (b): Variation of critical temperatures Vs Sr content.

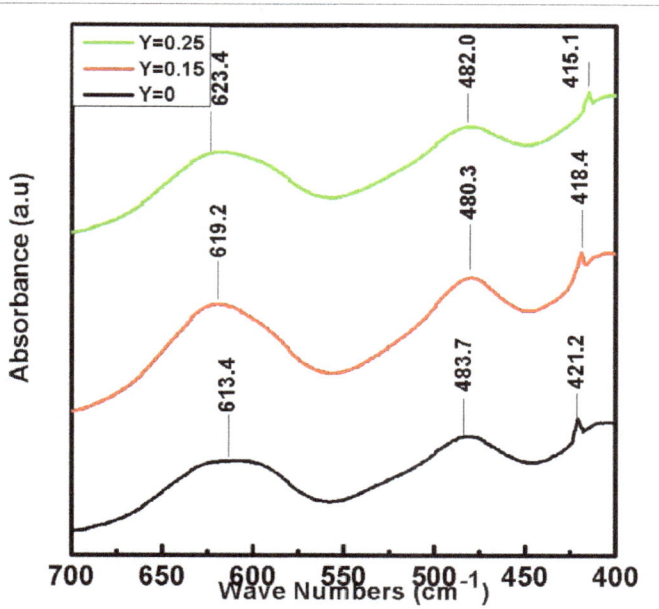

Figure 4: FTIR spectra of $Cu_{0.5}Tl_{0.5}(Ba_{2-y}Sr_y)Ca_2Cu_3O_{10-\delta}$ y=0, 0.15, 0.25 samples.

Where, C is dimensional exponent and its values vary with dimension of fluctuations i.e., 0.5, 1, 2 for 3D, 2D and 0D fluctuations respectively [11-13].

Moreover, $\frac{T_c T_{mf}}{T_c^{mf}}$ is the reduced temperature, Tc is usually referred as mean field critical temperature [14,15] and C is the fluctuation amplitude.

$$C^{3D} = \frac{e2}{32h\xi c(0)} \quad (3)$$

$$C^{2D} = \frac{e2}{16hd} \quad (4)$$

Where, e, d and $\xi c(0)$ are electronic charge, inter-layer thickness and $\xi c(0)$ coherence length respectively. Since Lowerence Donich is a modified form of Aslamsov- Larkin theory and it explained the fluctuations from 2D to 3D regimes. In the light of Lowerence Donich (LD model), the excess conductivity and cross over temperature is given below,

$$\Delta\sigma LD = A^x \varepsilon^{-1}\{1+[2\xi(0)/d]^2\} \quad (5)$$

$$T_{3D-2D} = T_{cmf}\{1+[2\xi(0)/d]^2 \quad (6)$$

The other parameters are given below, can be calculated by using TG and NG equations and GL theory.

$$N_G \left| \frac{T_G T_c^{mf}}{T_c^{mf}} \right| = 0.5[K_B T_c / B_{c(0)}^2 Y^2 \xi_{c(0)}^3]^2 \quad (7)$$

$$B_{c1} = \frac{Bc}{\sqrt{2}} \ln \quad (8)$$

$$B_{c2} = \sqrt{2}B_c \quad (9)$$

$$\frac{3 \times p.d(0)}{8k_B T_0} \quad (10)$$

$$\frac{3 \times p.d(0)}{8k_B T_0} \quad (11)$$

$$V_F = \frac{5k_B T_{cc(0)}}{2K} \quad (12)$$

$$E = \frac{h}{(1.6\times10^{19})(eV)} \quad (13)$$

The 2nd term, J i.e., Inter layer coupling is related with $J=\varepsilon/4$ [16-18] here ε is reduced temperature. Excess conductivity Δσ is determined by the expression:

$$\Delta\sigma=\rho n-\rho/\rho n^*\rho \quad (14)$$

Where, ρ is the resistivity measured experimentally and ρn is the extra-plotted normal state resistivity. The fluctuation induced conductivity analysis (FIC) has been done employing the above cited models (Aslamazov-Larkin and Lawrance-Donaich). The extracted superconducting parameters are given in Tables 1 and 2. The graphs are plotted between $\ln(\Delta\sigma)$ and $\ln(\varepsilon)$ of $Cu_{0.5}Tl_{0.5}Ba_{2-Y}Sr_YCa_2Cu_3O_{10-\delta}$ (y=0, 0.15, 0.25) samples are shown in Figures 5a-5c. It can be seen from Table 1 that all the crossover temperatures i.e., T_{CR-3D}, T_{3D-2D}, T_{2D-SW} have been suppressed with the Sr doping which shows that there is an inverse correlation between the crossover temperatures and superconductivity transition temperatures (Tc,0 and Tc^onset). The zero temperature coherence length, interlayer coupling strength, electron-phonon coupling parameter and critical magnetic field calculated from the excess conductivity analysis are given in Table 2. It is observed that the interlayer coupling strength has been increased with the increase of Sr in the charge reservoir layer (CRL).

The values of parameters such as $B_{c0}(T)$, $B_{c1}(T)$, and $J_{c(0)}$ are increased with the Sr content. These parameters appreciably dependent on thermodynamic critical magnetic field B_c and it is related to the free energy difference at the interface of normal and superconducting electrons. So in our case, the dopant atom seems to support in the difference of free energy and as a result, these parameters increases. The coherence length along the c-axis $\xi c(0)$ and the Fermi velocity vF of superconducting carriers are decreased with Sr content in $Cu_{0.5}Tl_{0.5}(Ba_{2-y}Sr_y)Ca_2Cu_3O_{10-}$ unit cell. Since $K_F=(32N/V)1/3$; [n=N/V], $\xi c=\hbar2K_F/2m\Delta$ and V_F. These parameters are dependent on density of carriers and doping of Sr atom, increases the density of carriers which suppresses the order parameter's values. It confirms that Sr atom promote the efficiency of transfer the charge carriers to the CuO_2 planes.

Conclusion

The present work has been resulted in the following conclusions as stated by the above study. Using solid state reaction method, the Sr-doped $Cu_{0.5}Tl_{0.5}$1223 (y=0, 0.15, 0.25) samples were synthesized at ambient pressure. The XRD analysis shows that samples have orthorhombic crystal structure. The fluctuation induced conductivity analysis (FIC) has been done employing the above cited models (Aslamazov-Larkin and Lawrance-Donaich). Three vibrational modes are witnessed around 613.4, 483.7 and 421.2 cm^{-1} in un-doped sample. However, with Sr doping in CRL, the former two modes are related to apical oxygen atoms of nature $Tl-O_A-Cu_{(2)}$, $Cu_{(1)}-O_A-Cu_{(2)}$ are softened from 483.7 to 482.0 cm^{-1} and 421.2 to 415.1 cm^{-1} whereas the third mode related to the planar oxygen atoms of nature $Tl-O_P-Cu_{(2)}$, are hardened from 613.4 to 623.4 cm^{-1}. The substitution of Sr at Ba site in the charge reservoir layer decreases the lattice parameters including c-axis, as a result, promotes the enhanced interlayer coupling. The magnitude of the superconductivity is notably increased with the inclusion of Sr which shows that dopant Sr atom at Ba site in $Cu_{0.5}Tl_{0.5}Ba_2O_{10-}$ (CRL) promotes efficiency of transfer the carriers to the conducting CuO_2 planes.

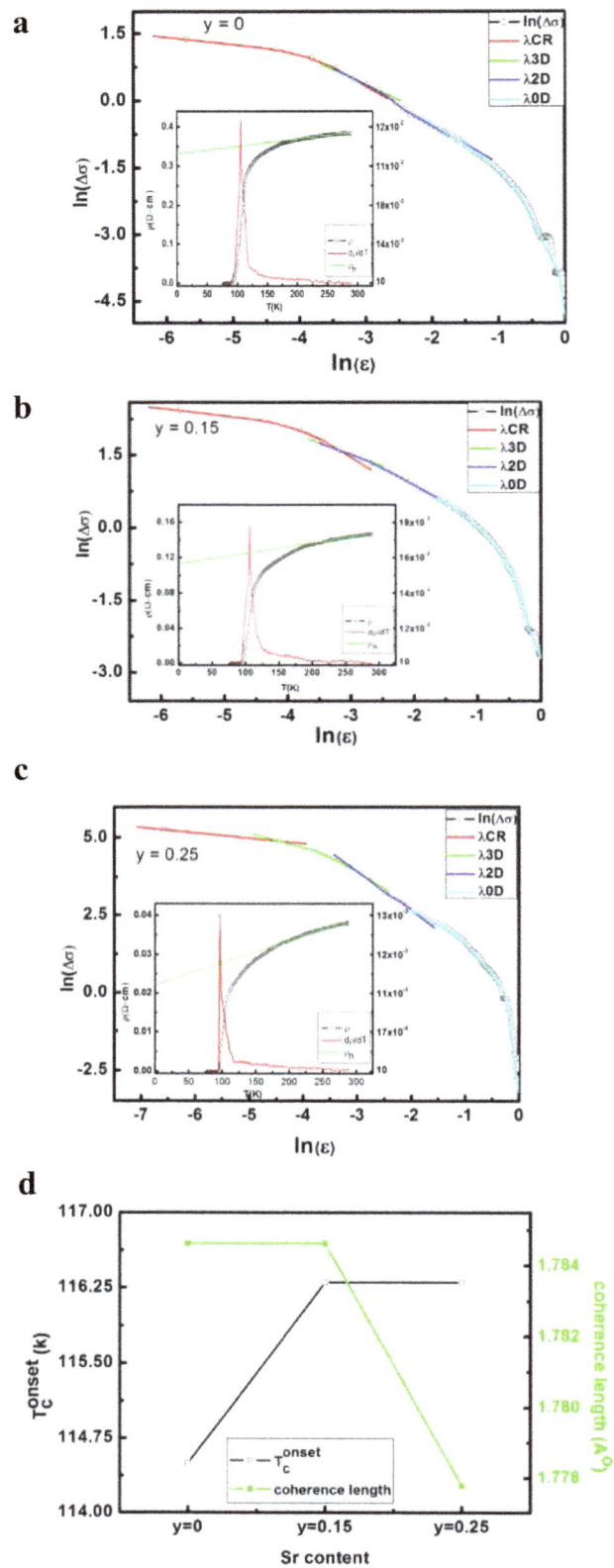

Figure 5: Representative figure of the excess conductivity analysis of $Cu_{0.5}Tl_{0.5}(Ba_{2-y}Sr_y)Ca_2Cu_3O_{10-\delta}$ (y=0, 0.15, 0.25) superconductors.

Sample	λCR	λ3D	λ2D	λSW	TCR-3D =TG (K)	T3D-2D (K)	T2D-SW (K)	Tcmf (K)
Y=0	0.32	0.66	0.97	2.09	110.38	111.38	126.0	106.1
Y=0.15	0.30	0.49	0.96	2.1	110.38	111.38	125.0	106.1
Y=0.25	0.29	0.64	1.1	2.01	98.80	102.0	109.8	97.2

Table 1: Widths of critical, 3D, 2D and 0D fluctuation regions observed from fitting of the experimental data of $Cu_{0.5}Tl_{0.5}Ba_{2-y}Sr_yCa_2Cu_3O_{10-}$ (y=0, 0.15, 0.25).

Sample	ξc(0) (Å)	J	NG	λp.d (Å)	Bc(0) (T)	Bc1 (T)	Bc2 (T)	K	Jc(0)*10³ (A/cm²)	VF*10⁷ (m/s)
Y=0	1.78	0.049	0.22	1020.6	1.426	0.06	128.7	63.79	0.76	1.47
Y=0.15	1.78	0.049	0.15	907.97	1.603	0.08	128.7	56.74	0.96	1.45
Y=0.25	1.77	0.051	0.02	561.22	2.594	0.18	128.7	35.07	2.51	1.37

Table 2: Superconductivity parameters calculated from FIC analysis of $Cu_{0.5}Tl_{0.5}Ba_{2-Y}Sr_YCa_2Cu_3O_{10-}$ (y=0, 0.15, 0.25).

From FTIR analysis, it has been observed that the planar and apical phonon modes are hardening and softening respectively. The library of FTIR spectra available in this laboratory includes more than 9400 spectra of organic, polymeric, and inorganic materials. These spectra are compared to the unknown sample spectra using computer software to identify the "best match". The changing in position of these phonon modes provided evidence of Sr atom in the unit cell. The FIC analysis shows that there is an inverse correlation between the superconductivity critical temperature Tc^{onset} and the crossover related to the fluctuations in the order parameter.

References

1. Park C, Snyder RL (1995) Structures of high-temperature cuprate superconductors. J Am Ceram Soc 78: 3171-3194.

2. Mattheiss LF (1990) Electronic structure and crystal chemistry of $TlBa_2CuO_5$ and related cuprate superconductors. Phys Rev B 42: 10108-10112.

3. Maarit K, Yamauchi H (1999) The doping routes and distribution of holes in layered cuprates: a novel bond-valence approach. Philos Mag B 79: 343-366.

4. Karppinen M, Yamauchi H, Morita Y, Kitabatake M, Motohashi T, et al. (2004) Hole concentration in the three-CuO_2-plane copper-oxide superconductor Cu-1223. J Solid State Chem 177: 1037-1043.

5. Gang X, Xiong P, Cieplak MZ (1992) Universal hall effect in $La_{1.85}Sr_{0.15}Cu_{1-x}A_xO_4$ systems (A= Fe, Co, Ni, Zn, Ga). Phys Rev B 46: 8687.

6. Gang X (1990) Magnetic pair-breaking effects: moment formation and critical doping level in superconducting $La_{1.85}S_{0.15}Cu_{1-x}A_xO_4$ systems (A=Fe, Co, Ni, Zn, Ga, Al). Phys Rev B 42: 8752.

7. Mary TA, Kumar NRS, Varadaraju UV (1993) Influence of Cu-site substitution on the structure and superconducting properties of the NdBa2Cu3−xMxO7+δ (M= Fe, Co) and NdBa2Cu3−xMxO7−δ (M= Ni, Zn) systems. Phys Rev B 48: 16727.

8. Awad R, Aly NS, Ibrahim IH, Abou-Aly AI, Saad AI (2000) Replacement study of Thallium by Zn and Ni in Tl-1223 superconductor phase. Physica C: Superconductivity 341: 685-686.

9. Iyo A, Tanaka Y, Hirai M, Tokiwa K, Watanabe T (2003) Zn and Ni doping effect on anomalous suppression of T_c in an over doped region of $TlBa_2Ca_2Cu_3O_{9-\delta}$. J Low Temp Phys 131: 643-646.

10. Kuo YK, Schneider CW, Skove MJ, Nevitt MV, Tessema GX, et al. (1997) Effect of magnetic and nonmagnetic impurities (Ni, Zn) substitution for Cu in $Bi_2(SrCa)_{2+n}(Cu_{1-x}M_x)_{1+n}O_y$ whiskers. Phys Rev B 56: 6201.

11. Ibrahim EMM, Saleh SA (2007) Influence of sintering temperature on excess conductivity in Bi-2223 superconductors. Supercond Sci Technol 20: 672.

12. Han SH, Axnäs J, Zhao BR, Rapp O (2004) Fluctuation conductivity at high temperatures in polycrystalline Hg, Tl-1223-Is there 1D fluctuation behavior? Physica C: Superconductivity 408: 679-680.

13. Mumtaz M, Hasnain SM, AA Khurram, Nawazish A Khan (2011) Fluctuation induced conductivity in $(Cu_{0.5}Tl_{0.5-x}K_x)Ba_2Ca_3Cu_4O_{12-\delta}$ superconductor. J Appl Phys 109: 023906.

14. Shun-Hui H, Lundqvist P, Rapp O (1997) Excess conductivity in $Y_{1-2x}Ca_xTb_xBa_2Cu_3O_{7-\delta}$. Physica C: Superconductivity 282: 1571-1572.

15. Sato T, Nakane H, Yamazaki S, Mori N, Hirano S, et al. (2003) Analysis of fluctuation conductivity in melt-textured DyBa 2 Cu 3 O y superconductors. Physica C: Superconductivity 392: 643-647.

16. Felix V, Veira JA, Maza J, Garcia-Alvarado F, Moran E, et al. (1988) Excess electrical conductivity above Tc in high-temperature superconductors, and thermal fluctuations. Journal of Physics C: Solid State Physics 21: L599.

17. Han SH, Eltsev Y, Rapp O (2000) Three-dimensional XY critical fluctuations of the dc electrical conductivity in $Bi_2Sr_2CaCu_2O_{8+\delta}$ single crystals. Phys Rev B 61: 11776.

18. Ben Azzouz F, Zouaoui M, Annabi M, Ben Salem M (2006) Fluctuation conductivity analysis on the Bi-based superconductors processed under same conditions. Physica Status Solidi 3: 3048-3051.

11

Assessing Biaxial Stress and Strain in 3C-SiC/Si (001) by Raman Scattering Spectroscopy

Talwar DN[1]*, Wan L[2], Tin CC[3] and Feng ZC[2]

[1]Department of Physics, Indiana University of Pennsylvania, USA
[2]Laboratory of Optoelectronic Materials and Detection Technology, Guangxi Key Laboratory for Relativistic Astrophysics, School of Physical Science & Technology, Guangxi University, China
[3]Department of Physics, Auburn University, Auburn, USA

Abstract

Highly strained 3C-SiC/Si (001) epilayers of different thicknesses (0.1 μm-12.4 μm) prepared in a vertical reactor configuration by chemical vapor deposition (V-CVD) method were examined using Raman scattering spectroscopy (RSS). In the near backscattering geometry, our RSS results for "as-grown" epilayers revealed TO- and LO-phonon bands shifting towards lower frequencies by approximately ~2 cm^{-1} with respect to the "free-standing" films. Raman scattering data of optical phonons are carefully analyzed by using an elastic deformation theory with inputs of hydrostatic-stress coefficients from a realistic lattice dynamical approach that helped assess biaxial stress, inplane tensile- and normal compressive-strain, respectively. In each sample, the estimated value of strain is found at least two order of magnitude smaller than the one expected from lattice mismatch between the epilayer and substrate. This result has provided a strong corroboration to our recent average-t-matrix Green's function theory of impurity vibrational modes – indicating that the high density of intrinsic defects at the 3C-SiC/Si interface are possily responsible for releasing the misfit stresses and strains. Unlike others, our RSS study in "as-grown" 3C-SiC/Si (001) has reiterated the fact that for ultrathin epilayers (d<0.4 μm) the optical modes of 3C-SiC are markedly indistinctive. The mechanism responsible for this behavior is identified and discussed. PACS: 78.20.-e 63.20.Pw 63.20.D.

Keywords: 3C-SiC/Si (001); Raman scattering; Stress and strain; Elastic theory

Introduction

Silicon carbide (SiC) is one of the very few IV-IV compound semiconductors – exhibiting exceptional mechanical, electrical and chemical properties [1-14]. The novel characteristics of SiC have made it suitable for fabrication of many important modern [1-10] devices for microelectronic, optoelectronic and sensor application needs. The scientific interest in SiC is stimulated by a strong chemical bond between Si and C atoms which provides the material a wider-bandgap, extreme hardness, high thermal stability, chemical inertness, higher thermal conductivity, high melting temperature, large bulk modulus, high critical (breakdown) electric field strength, and low dielectric constant. Among other wide-bandgap semiconductors, SiC is rather distinctive for controlling both *n*- and *p*-type dopants [11-18] across the broader concentration range (~10^{14}-10^{19} cm^{-3}). The ability of SiC to form native silicon dioxide (SiO_2) is an advantage [19-21] leading to its use in device fabrications. SiC has also been considered as a substitute for Si to make Schottky diodes and metal–oxide–semiconductor field-effect transistors (MOSFETs) for high-power, high-temperature, and high-frequency applications. Both crystalline and polycrystalline SiC have become attractive in recent years to design micro- and nano-electro-mechanical systems (MEMS, NEMS) [1-10]. While silicon carbide is currently being used to fabricate green, blue and ultraviolet light-emitting diodes (LEDs) – the emerging market [22,23] of utilizing heteroepitaxy/homoepitaxy SiC films is in high-power switches and microwave devices.

SiC occurs in more than 200 different crystalline structures [9-14] called polytypes. While every polytype is perceived by its own stacking sequence of Si-C bilayers – each structure displays its explicit set of distinct electrical and vibrational properties. The customary polytypes that are being developed for commercial needs include the cubic (3C-SiC), hexagonal (4H-SiC, 6H-SiC), and rhombohedral (15R-SiC, 21R-SiC) structures. The original work by Nishino et al [24] proposed a

multistep chemical vapor deposition (CVD) method to prepare 3C-SiC on Si. Many attempts have been made in recent years to improve the growth [25-30] mechanisms of 3C-SiC/Si (001) epifilms. The prospect of attaining large area 3C-SiC/Si (001) epilayers by CVD and molecular beam epitaxy (MBE) appears to be very encouraging. Despite the successful growth of 3C-SiC/Si (001) the quality of epilayers is still lacking thus impeding their use in the fabrication of electronic devices. On the other hand, a positive trade-off is its low cost advantage and there is a greater prospect of scalability in the fabrication of devices using Si as compared to 4H-SiC that sustained the current interest in the growth of 3C-SiC/Si for future applications.

There exists a considerable difference in the lattice constants (19.8%) and thermal expansion coefficients (8%) between 3C-SiC and Si. This leads to biaxial strain in 3C-SiC/Si (001) epilayers which might instigate modifying their physical and chemical properties. Whether one will be able to make practical use of such highly strained structures in electronic devices is still an open question. It is quite possible, however, that a large lattice mismatch in 3C-SiC/Si (001) leads to breaking of atomic bonds in epilayers which generates high density of dislocations or intrinsic defects [31]. It is likely that these defects stimulate releasing interfacial strains in epilayers. Therefore, it is imperative to evaluate biaxial stress in 3C-SiC/Si (001) epilayers for further progress in device

*Corresponding author: Talwar DN, Department of Physics, Indiana University of Pennsylvania, 975 Oakland Avenue, 56 Weyandt Hall, Indiana, Pennsylvania 15705-1087, USA, E-mail: talwar@iup.edu

engineering. Apart from the x-ray diffraction study (which is insensitive to ultrathin layers) [32,33] the most frequently used technique for stress estimation is the Raman scattering spectroscopy (RSS) [34-46]. Earlier, the method has been used successfully to appraise microstrains in bulk semiconductors under hydrostatic and uniaxial stress [36,37]. As the RSS approach is precise, sensitive, convenient and non-destructive it can be employed for studying the biaxial strain in thin epilayers grown on thick substrates including 3C-SiC/Si (001) [41-45].

The purpose of this work is to explore both theoretically and experimentally the problem of assessing residual stress and strain in 3C-SiC/Si (001) epilayers with large lattice mismatch. By using RSS, we will study the optical phonon shifts in a number of 3C-SiC/Si (001) samples grown by CVD method in the vertical reactor configuration (V-CVD) [30]. A T64000 Jobin Yvon triple advanced research Raman spectrometer, equipped with an electrically cooled charge coupled device (CCD) detector, is employed to measure the optical phonon frequencies in the near backscattering geometry. The observed phonon shifts will be assimilated in a conventional elastic deformation [40-46] theory to appraise the stresses and strains inside the 3C-SiC films. Our calculated biaxial stress in the V-CVD grown 3C-SiC/Si (001) epifilms of varied thickness fall well within the range of 0.45-0.94 GPa, i.e., an order of $\sim 10^9$ dyn/cm^2. Theoretical results are compared and discussed with the existing experimental and other simulated data. Despite a considerable difference ($\sim 19.8\%$) in the lattice constants [30] between the bulk Si and 3C-SiC materials – the study has offered significantly lower (two-order of magnitudes) values of inplane strains (i.e., ~ 0.1-0.2%) as well as normal (i.e., ~ -0.07 to -0.14%) strains within the 3C-SiC films grown on Si substrate. While the simulated results are quite surprising – the outcome has undoubtedly offered support to our earlier speculations of high density dislocations and/or intrinsic defects near the 3C-SiC/Si interface [31] which are likely to be responsible for releasing misfit stresses and strains in 3C-SiC films.

Experimental: V-CVD Growth of 3C-SiC/Si (001) Epifilms

3C-SiC epifilms used in the present RSS study are grown on (001) Si substrates under normal atmospheric pressure environment using CVD method in a vertical reactor configuration. Although, we employed 1 in diameter Si wafers as substrates – our reactor is capable of scaling up to handle 3 in diameter Si substrates for growing 3C-SiC epifilms. The V-CVD system utilizes a rotating SiC-coated susceptor heated by a radio-frequency (RF) induction power supply and is capable of operating at atmospheric and low pressure modes. The vertical configuration has several advantages including substrate rotation to give a large-area thickness uniformity, convenient in-situ monitoring of substrate parameters, and easy implementation of various growth enhancement procedures. The method used here to grow 3C-SiC/Si (001) epilayers consisted of three main steps described in details [30] elsewhere. It is to be noted that 3C-SiC/Si (001) epilayers are prepared at 1 atm and 1360°C with source ratio of Si/C (of ~ 0.33) using growth time τ between 2 min and 4 h at a rate of 3.2 ± 0.1 µm h^{-1} achieving the film thicknesses d, between 0.1 µm and 12.8 µm (Table 1). The set of single-crystalline 3C-SiC films used in the RSS study show uniformly smooth and mirror-like surfaces without macro-cracks – even for the thinnest film.

Raman Scattering Spectroscopy

RSS is a powerful and non-destructive technique to provide valuable information on the vibrational characteristics of materials

Sample #	Growth time	Epifilm thickness d µm
125 D	2 min	0.1
125 C	15 min	0.8
125 B	30 min	1.6
125 A	45 min	2.4
119 A	1 hr	3.2
119 B	3 hr	9.6
113	4 hr	12.8

Table 1: Properties of V-CVD grown 3C-SiC/Si (001) samples at 1 atm and 1360°C. The source ratio of Si/C was set at approximately $\cong 0.33$ with different growth times of 2 min, 15 min, 30 min, 45 min, and 1 hr, 3 hr, and 4 hr, respectively.

for assessing the epilayer thickness, strain, disorder, and site selectivity of defects [46,47]. The method is particularly suited for probing local atomic- and/or nanoscale structural changes in SiC materials while making careful analysis of its subtle spectral variations. Since RSS efficiency depends upon the polarizability of electron cloud – the process is quite sensitive to light elements involved in producing covalent bonds including SiC. The strong Si-C bonding in 3C-SiC with large bandgap stimulates higher Raman efficiency – requiring incident laser light of visible spectral range with reduced intensity to prevent significant heating of the material samples. We have performed room temperature RSS measurements in the near backscattering $x(y',y')\bar{x}$ geometry on several 3C-SiC epifilms of diverse thickness (~ 0.1 and 12.8 µm) grown on thicker (~ 200 and 400 µm) Si-substrates. A T64000 Jobin Yvon triple advanced research Raman spectrometer equipped with electrically cooled CCD detector is employed with an Ar$^+$ 488 nm line as an excitation source, while keeping the power level adjusted to 200 mW.

Assessing the lattice phonons in an ideal backscattering geometry for perfect diamond/zb materials requires strict wavevector conservation [47,48] and polarization selection rules. In the first-order RSS, this constraint limits the phonon wavevector to $\bar{q} = 0$ for observing the lattice modes. Thus, for Si crystal a triply degenerate phonon $\omega_{o(\Gamma)}$ at the center of the Brillouin zone (i.e., Γ-point) is allowed while in 3C-SiC material a doubly degenerate TO mode ($\omega_{TO(\Gamma)}$) is forbidden and a non-degenerate LO phonon ($\omega_{LO(\Gamma)}$) is permitted. By applying the hydrostatic pressure X up to 22.5 GPa in bulk 3C-SiC crystals, Raman scattering spectroscopy [37] was used earlier to measure the changes in the long wavelength optical phonon frequencies that helped evaluate the mode Grüneisen parameters γ_H^{TO} and γ_H^{TO} of the optical modes.

Near backscattering Raman spectra to assess stress and strain in 3C-SiC/Si (001)

One must note that the V-CVD grown 3C-SiC/Si (001) epilayers are perceived with biaxial stress in 3C-SiC films due to differences in Si and 3C-SiC lattice constants and thermal expansion coefficients [30]. An elastic deformation theory developed here can be applied to this system for assessing stressess and strains. We used RSS method to study the optical phonon shifts in (i) "as grown" 3C-SiC epilayers prepared on Si (001) with no processing done after V-CVD growth, and (ii) self-supported "free-standing" 3C-SiC films in which the Si substrate is removed with KOH etching solution. As an example, we have displayed (Figure 1a and 1b) our results of the RSS measurements recorded in the near back-scattering geometry for a 12.8 µm thick (a) "as-grown" and (b) "free-standing" film. Clearly, in the "as-grown" sample, the observed $\omega_{LO(\Gamma)}$, $\omega_{TO(\Gamma)}$ modes near ~ 972 cm^{-1}, ~ 796 cm^{-1} are seen shifting towards lower frequencies by approximately ~ 2 cm^{-1} when Si substrate is etched away. In Figure 1a we have also noticed a weak phonon feature appearing between ~ 938–950 cm^{-1} almost ~ 20-35

cm^{-1} lower than the $\omega_{LO(\Gamma)}$ phonon line. No such trait emerged, however, in the "free standing" film. By examining several "as-grown" 3C-SiC/Si (001) samples we have realized the weak feature in only a few cases and certainly not in the "free-standing" films. We strongly believe attributing this characteristic to interface states between 3C-SiC and Si. Again, the weak trait has neither affected the $\omega_{LO(\Gamma)}$, $\omega_{TO(\Gamma)}$ phonon lines nor it changed the stress calculations. By using a conventional elastic deformation theory we have evaluated the bi-axial strains in 3C-SiC epilayers by incorporating the observed Raman optical phonon shifts in the "as-grown" and "free-standing" films.

Thickness dependent Raman spectra of 3C-SiC/Si (001)

In Figure 2 we have reported RSS measurements in the near back-scattering geometry for seven of the V-CVD grown 3C-SiC/Si (001) samples (Table 1) in which the film thickness d is varied between 100 Å to 12.8 μm. Except for the 100 Å thick sample, we observed in all other "as-grown" materials (Figure 2) the $\omega_{o(\Gamma)}$ Si phonon line near ~520 cm^{-1} and $\omega_{LO(\Gamma)}$, $\omega_{TO(\Gamma)}$ modes of 3C-SiC near ~972 cm^{-1}, ~796 cm^{-1}, respectively. The relative peak intensities and lineshapes of $\omega_{o(\Gamma)}$ mode for Si substrate and the $\omega_{LO(\Gamma)}$, $\omega_{TO(\Gamma)}$ phonons of 3C-SiC epifilms showed variations with growth time τ or film thickness d. In thicker "as-grown"

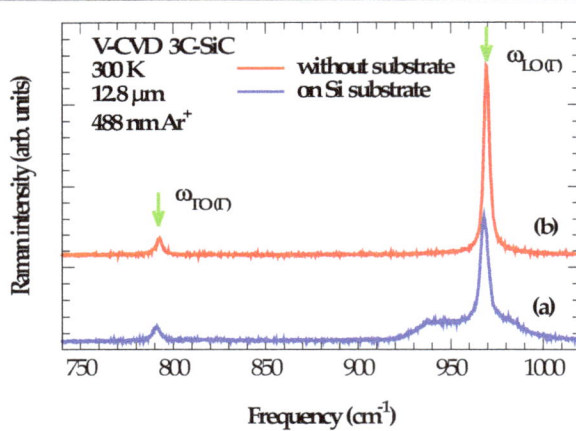

Figure 1: Raman spectra in the near back-scattering geometry of 3C-SiC/Si (sample #113 of film thickness 12.8 μm). An Ar$^+$ 488 nm line is used as the excitation source while keeping the laser power level adjusted to 200 mW (a) blue line on Si substrate and (b) redline without Si substrate.

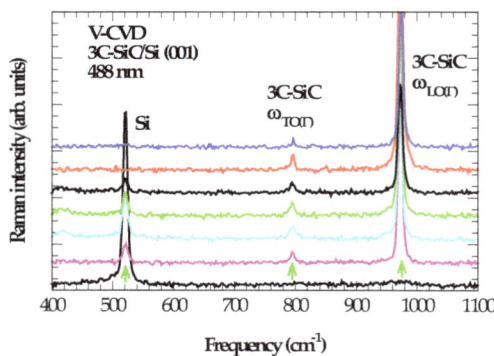

Figure 2: Thickness dependent Raman spectra in the near back-scattering geometry for seven V-CVD "as-grown" 3C-SiC/Si (001) samples, where d is varied between 0.1 μm to 12.8 μm. Except for 0.1 μm thick epifilm, in all other samples we observed $\omega_{o(\Gamma)}$ Si phonon line near ~520 cm^{-1} and $\omega_{LO(\Gamma)}$, $\omega_{TO(\Gamma)}$ modes of 3C-SiC near ~972 cm^{-1}, ~796 cm^{-1}, respectively.

samples, one expects less penetration of the laser light through 3C-SiC into Si substrate which inflicts a decrease in the intensity of the $\omega_{o(\Gamma)}$ phonon line and a rise of the $\omega_{LO(\Gamma)}$ mode intensity (Figure 2) with respect to $\omega_{TO(\Gamma)}$ phonon. This observation clearly indicates improvement in the crystalline quality of the thicker epilayers prepared with increased growth time τ. Unlike others [44] our RSS study has reiterated (Figure 2) the fact that in "as-grown" samples with thin epilayers ($d<0.4$ μm) the 3C-SiC optical modes are markedly indistinctive. We will incorporate the observed optical phonon shifts in "as-grown" and "free-standing" films to our elastic deformation theory to estimate both "in-plane" and "normal" strains in epilayers of different thickness.

Theoretical Background: Universal Axial Stress and Strain

To comprehend the "pressure-dependent" vibrational properties in semiconductors one can: (i) apply the "hydrostatic pressure" in bulk samples using diamond anvil cell, (ii) perform "uniaxial-stress" on large size specimens, and (iii) examine "biaxial-stress" in thin films prepared on mis-matched substrates. In each case the "pressure" is considerd as a perturbation. By adopting a conventional elastic deformation theory, one can derive expressions for the three cases in terms of Raman-stress coefficients and optical mode frequencies to empathize experimental data.

By using an elasticity theory [49,50] for a continuous media – the strain (ε) and stress (σ) tensors can be linked to the elastic-stiffness $\boldsymbol{C}$ and elastic-compliance $\boldsymbol{S}$ tensors via:

$$\sigma_{ij} = C_{ijkl}\varepsilon_{kl}, \tag{1a}$$

$$\text{and} \quad \varepsilon_{ij} = S_{ijkl}\sigma_{kl}. \tag{1b}$$

For uniaxial stress, the off-diagonal elements of stress $\sigma_{ij}(i \neq j) = 0$ and strain $\varepsilon_{ij}(i \neq j) = 0$ are zero. In a generalized form, the axial stress and strain tensors are symmetric about the z axis [41]:

$$\sigma = \begin{pmatrix} X & 0 & 0 \\ 0 & X & 0 \\ 0 & 0 & Z \end{pmatrix}, \quad \varepsilon = \begin{pmatrix} \varepsilon_{xx} & 0 & 0 \\ 0 & \varepsilon_{yy} & 0 \\ 0 & 0 & \varepsilon_{zz} \end{pmatrix} \tag{2}$$

In Eq. (2), if X or $Z<0$ the stress is compressive and it is tensile when X or $Z>0$. The elements of strain tensor can be rewritten as $\varepsilon_{xx} = \varepsilon_{yy}(=\varepsilon_\parallel)$ and $\varepsilon_{zz}(\equiv \varepsilon_\perp)$. Again, one can separate the axial stress into a hydrostatic term X and a uniaxial term P along the z-axis:

$$\begin{pmatrix} X & 0 & 0 \\ 0 & X & 0 \\ 0 & 0 & Z \end{pmatrix} = \begin{pmatrix} X & 0 & 0 \\ 0 & X & 0 \\ 0 & 0 & X \end{pmatrix} + \begin{pmatrix} 0 & 0 & 0 \\ 0 & 0 & 0 \\ 0 & 0 & P \end{pmatrix} \tag{3}$$

with $Z=X + P$. By using Eqs. (1 and 2) it is straight forward to link the strain and stress elements to the elastic compliance (S_{ij}) and elastic stiffness constants (C_{ij}), respectively as:

$$\varepsilon_\parallel = (S_{11} + 2S_{12})X + S_{12}P \tag{4a}$$

$$\varepsilon_\perp = (S_{11} + 2S_{12})X + S_{11}P \tag{4b}$$

$$X = (C_{11} + C_{12})\varepsilon_\parallel + C_{12}\varepsilon_\perp \tag{4c}$$

$$\text{and} \quad Z = 2C_{12}\varepsilon_\parallel + C_{11}\varepsilon_\perp \tag{4d}$$

Both compliance and stiffness tensors are related [51] through $\boldsymbol{S}=\boldsymbol{C}^{-1}$ for evaluating compliance coefficients (S_{11}, S_{12}, S_{44}) from the known elastic (C_{11}, C_{12}, C_{44}) constants. Again, the negative ratio of strains perpendicular and parallel to the stress axis is known as the

Poisson ratio ν. For the hydrostatic pressure, $\nu_H = -1$ and for uniaxial stress along the z-axis $\nu_s = -\varepsilon_\parallel / \varepsilon_\perp$.

If the zb crystal is deformed either internally (residual stress) or externally (applied stress) – the three optical phonon frequencies can be obtained by solving the secular equation [40]:

$$\begin{pmatrix} p\varepsilon_{xx} + q(\varepsilon_{yy}+\varepsilon_{zz}) - \Delta\omega_i^2 & 2r\varepsilon_{xy} & 2r\varepsilon_{xz} \\ 2r\varepsilon_{xy} & p\varepsilon_{yy} + q(\varepsilon_{zz}+\varepsilon_{xx}) - \Delta\omega_i^2 & 2r\varepsilon_{yz} \\ 2r\varepsilon_{xz} & 2r\varepsilon_{yz} & p\varepsilon_{zz} + q(\varepsilon_{xx}+\varepsilon_{yy}) - \Delta\omega_i^2 \end{pmatrix}\begin{pmatrix} u_1 \\ u_2 \\ u_3 \end{pmatrix} = 0 \quad (5)$$

where p, q, r are the symmetry allowed anharmonic parameters known as phonon deformation potentials. Here, the term $\Delta\omega_i^2 (\equiv \omega_\varepsilon^2 - \omega_o^2)$ $\left(\text{or } \omega_\varepsilon \approx \omega_o + \dfrac{\Delta\omega_i^2}{2\omega_o}\right)$ represents the shift of perturbed (strained: ω_ε^2) optical phonons from the unstrained $\vec{q}=0$ modes ω_o^2, and u_i are the components of eigen vectors. For axial stress with strains (Eq. 2) the non-trivial solutions of Eq. (5):

$$\begin{vmatrix} p\varepsilon_{xx} + q(\varepsilon_{yy}+\varepsilon_{zz}) - \Delta\omega_i^2 & 2r\varepsilon_{xy} & 2r\varepsilon_{xz} \\ 2r\varepsilon_{xy} & p\varepsilon_{yy} + q(\varepsilon_{zz}+\varepsilon_{xx}) - \Delta\omega_i^2 & 2r\varepsilon_{yz} \\ 2r\varepsilon_{xz} & 2r\varepsilon_{yz} & p\varepsilon_{zz} + q(\varepsilon_{xx}+\varepsilon_{yy}) - \Delta\omega_i^2 \end{vmatrix} = 0 \quad (6)$$

provide two phonon modes given by:

$$[p\varepsilon_\parallel + q(\varepsilon_\parallel + \varepsilon_\perp) - \Delta\omega_i^2]^2 [p\varepsilon_\perp + 2q\varepsilon_\parallel - \Delta\omega_i^2] = 0 \quad (7)$$

one having a doublet

$$\Delta\omega_{di}^2 = p\varepsilon_\parallel + q(\varepsilon_\parallel + \varepsilon_\perp)$$

or $\omega_{d\varepsilon} = \omega_o + [(p+q)\varepsilon_\parallel + q\varepsilon_\perp] / 2\omega_o$ (8a)

and the other a singlet

$$\Delta\omega_{si}^2 = p\varepsilon_\perp + 2q\varepsilon_\parallel$$

or $\omega_{s\varepsilon} = \omega_o + [p\varepsilon_\perp + 2q\varepsilon_\parallel] / 2\omega_o$ (8b)

Stress induced modes: Hydrostatic case

In a hydrostatic case ($P=0$), with a_o changing to a:

$\varepsilon_{xx} = \varepsilon_{yy} = \varepsilon_{zz} = \varepsilon = \dfrac{a - a_o}{a_o}$ and $\varepsilon_{ij}(i \neq j) = 0$,

$$\begin{pmatrix} X \\ X \\ X \end{pmatrix} = \begin{pmatrix} C_{11} & C_{12} & C_{12} \\ C_{12} & C_{11} & C_{12} \\ C_{12} & C_{12} & C_{11} \end{pmatrix}\begin{pmatrix} \varepsilon \\ \varepsilon \\ \varepsilon \end{pmatrix} \quad (9a)$$

or $X = (C_{11} + 2C_{12})\varepsilon = \dfrac{C_{11}+2C_{12}}{3}\dfrac{\Delta V}{V_o} = B_o \dfrac{\Delta V}{V_o}$ (9b)

where $\dfrac{\Delta V}{V_o} = (1+\varepsilon)^3 - 1 \approx 3\varepsilon$ and $\dfrac{C_{11}+2C_{12}}{3} = B_o$ is the bulk modulus.

The Grüneisen constant (hydrostatic stress) γ_o is:

$$\gamma_o = -\frac{\partial \ln \omega}{\partial \ln V} = \frac{B_o}{\omega_o}\frac{\partial \omega}{\partial X} \quad (10)$$

where ω is the mode frequency and V_o crystal volume. By using Eqs. [8(a-b) and 10] one gets:

$$\frac{\Delta\omega}{\omega_o} = -\gamma_o \frac{\Delta V}{V_o} = -3\varepsilon\gamma_o = (p+2q)\varepsilon / 2\omega_o^2 \quad (11)$$

where $\gamma_o = -(p+2q)/6\omega_o^2$.

If a cubic cell of lattice constant a_o is distorted to a tetragonal cell

of lattice constants a and c, respectively, then $\varepsilon_{xx} = \varepsilon_{yy} = \varepsilon_\parallel = \dfrac{a - a_o}{a_o}$; $\varepsilon_{zz} = \varepsilon_\perp = \dfrac{c - a_o}{a_o}$ and $\varepsilon_{ij}(i \neq j) = 0$. Consequently, one can rewrite Eqs. [8(a)-(b)] as:

$$\frac{\Delta\omega_{di}^2}{\omega_o^2} = [p\varepsilon_\parallel + q(\varepsilon_\parallel + \varepsilon_\perp)] / \omega_o^2 = \frac{(p+2q)}{3\omega_o^2}(2\varepsilon_\parallel + \varepsilon_\perp) + \frac{(p-q)}{3\omega_o^2}(\varepsilon_\parallel - \varepsilon_\perp),$$

or $= -2\gamma_o(2\varepsilon_\parallel + \varepsilon_\perp) + \dfrac{2}{3}\gamma_s(\varepsilon_\parallel - \varepsilon_\perp)$,   (12)

and $\dfrac{\Delta\omega_{si}^2}{\omega_o^2} = [p\varepsilon_\perp + 2q\varepsilon_\parallel] / \omega_o^2 = \dfrac{(p+2q)}{3\omega_o^2}(2\varepsilon_\parallel + \varepsilon_\perp) - 2\dfrac{(p-q)}{3\omega_o^2}(\varepsilon_\parallel - \varepsilon_\perp)$

or $= -2\gamma_o(2\varepsilon_\parallel + \varepsilon_\perp) - \dfrac{4}{3}\gamma_s(\varepsilon_\parallel - \varepsilon_\perp)$,   (13)

where $\gamma_s = \dfrac{(p-q)}{2\omega_o^2}$ is the shear-deformation parameter.

Stress induced modes: Uniaxial case

If a uniaxial stress is applied in the z-direction:

$$\begin{pmatrix} 0 \\ 0 \\ P \end{pmatrix} = \begin{pmatrix} C_{11} & C_{12} & C_{12} \\ C_{12} & C_{11} & C_{12} \\ C_{12} & C_{12} & C_{11} \end{pmatrix}\begin{pmatrix} \varepsilon_\parallel \\ \varepsilon_\parallel \\ \varepsilon_\perp \end{pmatrix} \quad (14)$$

one gets:

$(C_{11} + C_{12})\varepsilon_\parallel + C_{12}\varepsilon_\perp = 0$ (15a)

and $2C_{12}\varepsilon_\parallel + C_{11}\varepsilon_\perp = P$ (15b)

Solving the above two Eqs. [15(a)-(b)], it can be shown that:

$2\varepsilon_\parallel + \varepsilon_\perp = \dfrac{P}{C_{11}+2C_{12}}$ (15c)

and $\varepsilon_\parallel - \varepsilon_\perp = \dfrac{P}{C_{11}-C_{12}}$ (15d)

Substituting Eqs. [15(c-d)] into Eqs. [(12) and (13)] one can obtain

$\dfrac{\omega_{d\varepsilon}^2 - \omega_o^2}{\omega_o^2} = -2\gamma_o \dfrac{P}{C_{11}+2C_{12}} + \dfrac{2}{3}\gamma_s \dfrac{P}{C_{11}-C_{12}}$,   (16a)

and $\dfrac{\omega_{s\varepsilon}^2 - \omega_o^2}{\omega_o^2} = -2\gamma_o \dfrac{P}{C_{11}+2C_{12}} - \dfrac{4}{3}\gamma_s \dfrac{P}{C_{11}-C_{12}}$,   (16b)

Subtracting Eqs. [16(a)-(b)], we will have

$\dfrac{\omega_{s\varepsilon}^2 - \omega_{d\varepsilon}^2}{\omega_o^2} = -2\gamma_s \dfrac{P}{C_{11}-C_{12}}$ (17a)

$\dfrac{\omega_{s\varepsilon}^2 + 2\omega_{d\varepsilon}^2 - 3\omega_o^2}{\omega_o^2} = -6\gamma_o \dfrac{P}{C_{11}+2C_{12}}$ (17b)

or $\dfrac{3\omega_h^2 - 3\omega_0^2}{\omega_o^2} = -6\gamma_o \dfrac{P}{C_{11}+2C_{12}}$

$\dfrac{\omega_h^2 - \omega_o^2}{\omega_o^2} = -2\gamma_o \dfrac{P}{C_{11}+2C_{12}} = -\dfrac{2\gamma_o}{3B_o}P$ (17c)

where we defined $\omega_{s\varepsilon}^2 + 2\omega_{d\varepsilon}^2 = 3\omega_h^2$

Stress induced modes: Biaxial case

The 3C-SiC/Si(001) system is regarded as one with biaxial stress (i.e., $P = -X$: cf. Eq. 3) in the film due to difference in the substrate lattice constants and thermal expansion coefficients. With Eq. [4 (d)], one can rewrite Eqs. [8(a)-(b)] as

$$\omega_{d\varepsilon} = \omega_o + [(p+q)\varepsilon_{\parallel} + q\varepsilon_{\perp}]/2\omega_o = \omega_o + \left[\left(\frac{p+q}{2\omega_o}\right) - \left(\frac{qC_{12}}{\omega_o C_{11}}\right)\right]\varepsilon_{\parallel}$$

$$= \omega_o + \omega_o\left[\frac{1}{3}\gamma_s\left(1 + \frac{2C_{12}}{C_{11}}\right) + 2\gamma_o\left(\frac{C_{12}}{C_{11}} - 1\right)\right]\varepsilon_{\parallel}$$

$$= \omega_o + \omega_o\left[\frac{1}{3}\gamma_s(S_{11} - S_{12}) - 2\gamma_o(S_{11} + 2S_{12})\right]X \qquad (18a)$$

$$= \omega_o + \left[\frac{1}{3}\zeta_s - 2\zeta_H\right]X$$

$$\omega_{s\varepsilon} = \omega_o + [p\varepsilon_{\perp} + 2q\varepsilon_{\parallel}]/2\omega_o = \omega_o + \left[\left(\frac{q}{\omega_o}\right) - \left(\frac{pC_{12}}{\omega_o C_{11}}\right)\right]\varepsilon_{\parallel}$$

and
$$= \omega_o + \omega_o\left[-\frac{2}{3}\gamma_s\left(1 + \frac{2C_{12}}{C_{11}}\right) + 2\gamma_o\left(\frac{C_{12}}{C_{11}} - 1\right)\right]\varepsilon_{\parallel}$$

$$= \omega_o + \omega_o\left[-\frac{2}{3}\gamma_s(S_{11} - S_{12}) - 2\gamma_o(S_{11} + 2S_{12})\right]X$$

or
$$= \omega_o + \left[-\frac{2}{3}\zeta_S - 2\zeta_H\right]X \qquad (18b)$$

where the two stress coefficients

$$\zeta_S = \omega_o\gamma_s(S_{11} - S_{12}) \text{ and } \zeta_H = \omega_o\gamma_o(S_{11} + 2S_{12}) \qquad (19)$$

can be determined by RS experiments. Again it is customary to add superscripts TO, LO on ω, ω_o for classifying the doublet and singlet (cf. Eqs. 18(a-b)) modes i.e.,

$$\omega^{TO} = \omega_o^{TO} + \left[\frac{1}{3}\zeta_S - 2\zeta_H\right]X \qquad (20a)$$

$$\omega^{LO} = \omega_o^{LO} + \left[--\zeta_S - 2\zeta_H\right]X \qquad (20b)$$

Subtracting Eq. (20 a) from Eq. (20 b) one can obtain $\zeta_S X$:

$$\zeta_S X = (\omega_o^{LO} - \omega^{LO}) - (\omega_o^{TO} - \omega^{TO}), \qquad (21)$$

from the observed Raman mode frequencies ($\omega^{LO}, \omega^{TO}, \omega_o^{LO}$ and ω_o^{TO}).

Using Eq. (21) with hydrostatic stress coefficients $\zeta_H^{LO}, \zeta_H^{TO}$ known from the pressure dependent RS experiments, we can get two values of X from Eqs. [20(a-b)]:

$$X = [(\omega_o^{TO} - \omega^{TO}) + \frac{1}{3}\zeta_S X]/(2\zeta_H^{TO}) \qquad (22)$$

and
$$X = [(\omega_o^{LO} - \omega^{LO}) - \frac{2}{3}\zeta_S X]/(2\zeta_H^{TO}) \qquad (23)$$

If we take an average of the two X values and assign error bar, we can appraise both in-plane and normal strains in 3C-SiC films by using Eqs. [4(a)-(b)] and setting $P = -X$. It is to be noted that the classical elastic deformation theory assumes a linear relationship between stress and strain (Eqs. (10 and 15-17)) – which may not be very realistic at higher $X > 10$ GPa. In our analyses from RSS data of optical phonons the estimated values of hydrostatic pressure components are seen to fall well within $X < 1$ GPa in all V-CVD grown 3C-SiC/Si (001) samples. It is, therefore, strongly believed that in the elastic deformation theory one would anticipate nearly <5% error in assessing strains.

Results: Data Analysis

Elastic constants

Accurate knowledge of elastic constants for 3C-SiC is crucial for engineering MEMS and/or NEMS devices and evaluating stress and strains in 3C-SiC/Si (001) epilayers (Eqs. (4a-b)). The published data on elastic constants by exploiting diverse experimental and theoretical methods is, however, rather conflicting [52-66]. The complete result on lattice dynamics of 3C-SiC has been reported earlier[67] by inelastic x-ray scattering (IXS) method. The pressure dependent optical phonon shifts in bulk 3C-SiC crystals are investigated with applied hydrostatic pressure (X) up to 22.5 GPa [37]. Recently, we have adopted a realistic rigid-ion model (RIM) to simulate phonon dispersions of 3C-SiC at 1 atm [31] and 22.5 GPa [67]. In the RIM scheme, the short- and long-range Coulomb interactions are optimized by least-square fitting procedures using lattice constants, critical point phonon energies as input while elastic constants C_{ij} and their pressure derivatives are deliberated as constraints to match the IXS [67] and pressure-dependent phonon data. For 3C-SiC, the best fit values of phonon dispersions find recent experimental [54] and local density functional data of elastic constants more reliable than the results available from earlier measurements. In Table 2 we have listed the derived values of elastic compliances (S_{11}, S_{12}, S_{44}), $S_{11}+2S_{12}$, $\tilde{S}$ and Poisson ratio ν_s, from elastic constants (C_{11}, C_{12}, C_{44}) reported [52-66] by various research groups.

Grüneisen parameter

Earlier, Olego et al. [36,37] described the hydrostatic pressure dependent optical [$\omega_{LO(\Gamma)}, \omega_{TO(\Gamma)}$] phonons in bulk 3C-SiC crystals by using RSS with applied X up to 22.5 GPa:

$$\omega^{TO} = (796.2 \pm 0.3) + (3.88 \pm 0.08)X - (2.2 \pm 0.4) \times 10^{-2}X^2 \qquad (24a)$$

$$\omega^{LO} = (972.7 \pm 0.3) + (4.75 \pm 0.09)X - (2.5 \pm 0.4) \times 10^{-2}X^2 \qquad (24b)$$

where the mode frequencies ω^{TO}, ω^{LO} are expressed in cm^{-1} while X is in GPa (10^{10} dyn/cm^2). For 3C-SiC, the optical phonon shifts at low temperature (6 K) have also been reported with X up to 15 GPa [42]. In Figure 3, a linear fit to the phonon data by dotted lines has revealed that for $X < 10$ GPa the X^2 term in the right hand side of Eqs. (24a-b) is trivial and can be neglected. Therefore, one can determine the hydrostatic Grüneisen parameters $\gamma_H^{TO}, \gamma_H^{LO}$ from the linear relationship (Eq. (10))

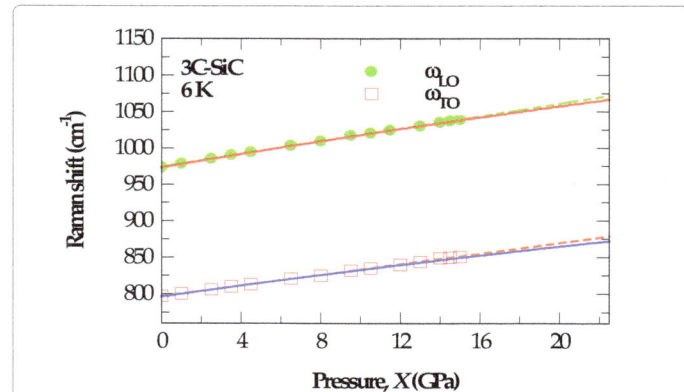

Figure 3: The long wavelength $\omega_{LO(\Gamma)}$, $\omega_{TO(\Gamma)}$ phonon mode frequencies of 3C-SiC vs. hydrostatic pressure (GPa). The symbols (●, □) represent experimental data [42] whereas the solid lines correspond to the data fits and dashed lines represent the linear fit to the experimental results.

				3C-SiC					
c_{11}	c_{12}	c_{44}	s_{11}	s_{12}	s_{44}	$s_{11} + 2s_{12}$	$\hat{s}$	v_s	Ref.
(x*10^12 dyn/cm²)			(x*10^-13 cm²/dyn)			(x*10^-13 cm²/dyn)			
5.4	1.8	2.5	2.222	-0.556	4	1.111	2.5	0.25	[52] Exp.
3.9	1.42	2.56	3.183	-0.85	3.906	1.484	2.718	0.267	[53] Exp.
3.95	1.36	2.36	3.074	-0.787	4.237	1.499	2.575	0.256	[54] Exp.
3.523	1.404	2.329	3.673	-1.047	4.294	1.58	2.988	0.285	[55] Cal.
3.489	1.384	2.082	3.7	-1.051	4.803	1.598	2.972	0.284	[56] Cal.
3.71	1.69	1.76	3.77	-1.18	5.682	1.41	3.51	0.313	[57] Cal.
2.89	2.34	0.554	12.56	-5.62	18.05	1.321	13.764	0.447	[58] Cal.
3.9	1.426	1.911	3.188	-0.854	5.233	1.481	2.729	0.268	[59] Cal.
4.2	1.26	2.87	2.764	-0.638	3.484	1.488	2.286	0.231	[60] Cal.
3.9	1.34	2.53	3.111	-0.795	3.953	1.52	2.57	0.256	[61] Cal.
4.05	1.35	2.54	2.963	-0.741	3.937	1.481	2.5	0.25	[62] Cal.
4.2	1.2	2.6	2.727	-0.606	3.846	1.515	2.2	0.222	[63] Cal.
3.84	1.32	2.41	3.16	-0.808	4.149	1.543	2.571	0.256	[64] Cal.
4.151	1.319	2.654	2.845	-0.686	3.768	1.473	2.397	0.241	[65] Cal.
3.63	1.54	1.49	3.687	-1.098	6.711	1.49	3.211	0.298	[66] Cal.

Table 2: Calculated elastic compliance and other parameters of 3C-SiC obtained from the existing sets of elastic constants available in the literature.

	ω_0 (cm⁻¹)	dω/dX	γ_0	B (GPa)	B
3C-SiC[a]	TO(Γ) 797.7	3.88	1.56,b 1.10[c]	227	4.1
	LO(Γ) 973.6	4.59	1.55,b 1.09[c]		
	LO-TO 175.9	0.654			
C (dia)[a]	1330	2.9	0.96	442	4.09
Si	523.9	5.1	0.96	99	4.24
Ge	304.6	4.02	1	75.8	4.55

[a]Ref. [42].
[b]Ref. [36-37].
[c]Our

Table 3: Zone-center optical phonon frequencies [TO(Γ), LO(Γ)] and their hydrostatic pressure derivatives dω/dX for 3C-SiC. The phonon frequencies ω_0 are in cm⁻¹ and the hydrostatic pressure X is in GPa. For comparison we have listed the related data for zone-center modes [ω_0 (cm⁻¹)] of diamond, Si, and Ge crystals. The mode Grüneisen parameters γ_0, the bulk moduli B (in GPa) and their pressure derivatives B' employed to calculate $\frac{\Delta V}{V_o}$ or $\frac{\Delta a}{a_o}$ for 3C-SiC are also listed.

Sample #	d_{SiC} μm	ω_o^{TO} cm⁻¹	ω_o^{LO} cm⁻¹	ω^{TO} cm⁻¹	ω^{LO} cm⁻¹	X(GPa)	$\varepsilon_\parallel$ (%)	$\varepsilon_\perp$ (%)	Ref.
125 A	2.4	795.5	972.8	794.2	971.6	0.446 ± 0.045	0.102	-0.072	[our]
119 A	3.2	795.7	972.9	793.9	971.8	0.552 ± 0.056	0.126	-0.087	[our]
119 B	9.6	795.8	973.2	793.5	971.5	0.739 ± 0.075	0.169	-0.116	[our]
113	12.8	796	973.4	793.3	971.4	0.868 ± 0.088	0.199	-0.137	[our]
487*	4	795.5	972.4	794.3	969.9	0.575 ± 0.058	0.132	-0.091	[41]
475 B	4.5	795.1	972.4	793.1	971.1	0.622 ± 0.063	0.142	-0.098	[41]
462	6	796.3	973.6	793.1	970.3	1.138 ± 0.115	0.26	-0.179	[41]
475	7	796.3	972.8	794.3	969.9	0.810 ± 0.082	0.185	-0.127	[41]

Table 4: Comparison of the Raman scattering data on optical phonons (cm⁻¹) used for assessing the stresses and strains in V-CVD grown 3C-SiC/Si (001). We used elastic compliance values of S_{11}=3.074 × 10⁻¹³ cm²/dyn and S_{12}=-0.787 × 10⁻¹³ cm²/dyn (Table 3) from the experimental data of elastic constants [54].

involving $\frac{\partial \omega}{\partial X}$. For 3C-SiC, the values [$\gamma_H^{TO} = 1.56$, $\gamma_H^{LO} = 1.55$] estimated by Olego et al. used an average bulk modulus data B_o (=321.9 GPa) of Si and C (Table 3) [36,37]. A correction was made to the γ_H^{TO}, γ_H^{LO} values by exploiting the experimental bulk modulus B_o (=227 GPa) of 3C-SiC. While the experimental results of hydrostatic mode Grüneisen parameters for 3C-SiC fall within <3% – our RIM calculations [67] of $\gamma_H^{TO} = 1.1$, $\gamma_H^{LO} = 1.09$ provided strong corroborations to the amended values. While Feng et al. [41] employed $\gamma_H^{TO} = 1.56$, $\gamma_H^{LO} = 1.55$ for assessing strains in 3C-SiC epifilms – one would expect improvement in the strain values if accurate results of γ_H^{TO}, γ_H^{LO} are adopted.

Stress and strain in V-CVD 3C-SiC/Si (001) films

In Table 4 we have displayed the RSS results of optical phonons

ω^{LO}, ω^{TO} and ω_o^{LO}, ω_o^{TO} in the near-backscattering geometry for several "as-grown" 3C-SiC/Si (001) and "free-standing" films. By using Eqs. [(21) and (22-23)] along with the Raman data of optical phonons and hydrostatic stress coefficients ζ_H^{TO}, ζ_H^{TO} we obtained two values of X – took their average result and assigned an error bar (Table 4). From Table 4, the uncertainty of the calculated stress X is about 10% which agrees very well with the estimated value from the experimental error of 10% to 20% from the pressure dependent RSS mode frequencies. The results of in-plane and normal strains are evaluated for 3C-SiC films by exploiting Eqs. [4(a)-(b): setting P=-X] along with the elastic compliance data of S_{11}=3.074x10⁻¹³ cm²/dyn and S_{12}=-0.787x10⁻¹³ cm²/dyn from Table 2 [54]. The calculated values reported in Table 4 of biaxial stress (X~0.45-0.87 GPa) and inplane ($\varepsilon_\parallel$ ~0.1-0.2%) strains as well as normal ($\varepsilon_\parallel$~-0.07 to -0.14%) strains for several V-CVD grown samples of different thickness are found in good agreement with the results reported in the literature. The positive and negative signs of $\varepsilon_\parallel$ and $\varepsilon_\perp$ indicate that in-plane strains are tensile while the normal strains are compressive respectively for all 3C-SiC/Si (001) samples. In order to justify the use of elastic deformation theory for 3C-SiC films, we have studied the volume dependent shifts of optical phonons by using a well known Murnaghan's equation of state with B_o ($\equiv 227$ GPa) and B_o' ($\equiv 4.1$):

$$\frac{\Delta V}{V_o} = \left[1 + X \frac{B_o'}{B_o}\right]^{-1/B_o'} - 1 \qquad . \tag{25}$$

The results displayed in Figure 4 for small X clearly reveal a linear dependence of the relative change for $\frac{\Delta\omega^{LO}}{\omega^{LO}}\left\{\approx \frac{\Delta\omega^{TO}}{\omega^{TO}}\right\}$ versus $\frac{\Delta V}{V_o}$ as compared to non-linear behavior of the hydrostatic pressure $X > 10$ GPa dependent shifts for optical modes ω^{LO}, ω^{LO} (Figure 3). This means that for $X > 10$ GPa, the deviation from a linear behavior of pressure dependent mode frequencies ω^{LO}, ω^{TO} (Figure 3) is caused by the non-linear pressure – volume relationship. In other words if X^2 terms from the right-hand side of Eqs. (24a-b)) are neglected – one would expect nearly ~5% error in the optical mode frequencies for $X > 10$ GPa and less than 0.5% for $X < 1$ GPa. This result is quite significant as the evaluated hydrostatic pressure component X in all V-CVD grown 3C-SiC/Si (001) samples are found to be < 1 GPa – validateing our elastic deformation theory of assessing strains in epifilms grown on mismatched substrates.

Summary and Conclusion

In summary, we have carried out extensive RSS measurements on several highly mismatched thin epitaxially grown 3C-SiC films on thick Si (001) substrates using V-CVD method. The 3C-SiC/Si (001) material system is perceived as one with a biaxial stress due to differences in the lattice constants and thermal expansion coefficients. In the V-CVD approach, while keeping Si/C ratio at ~0.33 we have prepared 3C-SiC films on Si by varying the growth time between 2 min to 4 h. A conventional elastic deformation theory is used to derive the necessary expressions involving stress coefficients – correlating them with Raman phonon shifts and hydrostatic, uniaxial and biaxial stresses. For bulk 3C-SiC, we used the existing pressure dependent phonon [37,42] measurements to estimate the hydrostatic-stress coefficients. The experimental data of long wavelength optical phonons [LO(Γ) and TO(Γ)] in the near-backscattering geometry is compared for several "as-grown" 3C-SiC/Si (001) and "free-standing" samples having film

thickness ranging between 2.4 μm to 12.8 μm. The analysis of Raman scattering phonon data has not only helped us appraise the crystalline quality of films but also facilitated assessing the stresses and strains in several V-CVD grown 3C-SiC/Si (001) samples of different thickness. Our estimated results of biaxial stresses (X~0.45-0.87 GPa) as well as inplane ($\varepsilon_\parallel$ ~0.1–0.2%) and normal ($\varepsilon_\perp$ ~-0.07 to -0.14%) strains using elastic deformation theory with elastic constants from Philippe Djemia provided values in good agreement with data reported in the literature by different research groups [41,44]. In 3C-SiC/Si epifilms, while the appraised average value of the biaxial stress (~0.651 GPa) is an order of magnitude smaller – the strain estimates are found two-order of magnitudes smaller than the lattice misfits between the bulk 3C-SiC and Si. Although this result is quite interesting – it provides strong corroboration to our recent study of impurity vibrational modes based on average-t-matrix Green's function theory implying that there exists a high density of intrinsic defects at the 3C-SiC/Si interface which is possibly responsible for releasing misfit stresses and strains.

Acknowledgements

The author (DNT) wishes to thank Dr. Deanne Snavely, Dean College of Natural Science and Mathematics at Indiana University of Pennsylvania for the travel support and the Innovation Grant that he received from the School of Graduate Studies making this collaborative research possible. The work at Guangxi University is supported by National Natural Science Foundation of China (NO. 61367004) and Guangxi Key Laboratory for the Relativistic Astrophysics-Guangxi Natural Science Creative Team funding (No. 2013GXNSFFA019001).

References

1. Zhuang H, NYang N, Zhang L, Fuchs R, Jiang X (2015) Low-Temperature Non-Ohmic Galvanomagnetic Effects in Degenerate n-Type InAs. ACS Appl Mater Interfaces 7: 10886-10895.

2. Frewin CL, Reyes M, Register J, Thomas SW, Saddow SE (2014) 3C-SiC on Si: A Versatile Material for Electronic, Biomedical and Clean Energy Applications. MRS Proceedings 1693: dd05-01.

3. Choi K, Choi DK, Lee DY, Shim J, Ko S (2012) Nanostructured thermoelectric cobalt oxide by exfoliation/restacking route. Appl Phys 108: 161.

4. Rajasekhara S, Neuner BH, Zorman CA, Jegenyes N, Ferro G, et al. (2011) Mixed-Mode Excitons in the Photoluminescence of Zinc Oxide-Reabsorption and Exciton Diffusion. Appl Phys Lett 98: 191904.

5. Yang N, Zhuang H, Hoffmann R, Smirov W, Hess J, et al. (2011) Growth and Characterization of 3C-SiC Films for Micro Electro Mechanical Systems (MEMS) Applications. Anal Chem 83: 5827.

6. Bosi M, Watts BE, Attolini G, Ferrari C, Frigeri C, et al. (2009) Growth and Characterization of 3C-SiC Films for Micro Electro Mechanical Systems (MEMS) Applications. Cryst Growth Design 9: 4852.

7. Reddy JD, Volinsky AA, Frewin CL, Locke C, Saddow SE (2008) Quantum Theory of a Basic Light-Matter Interaction. Mater Res Soc Symp Proc 1049: AA03-06.

8. Young DJ, Du J, Zorman CA, Ko WH (2004) Effects of Uniaxial Stress on Excitons in CuCl. IEEE Sensors 4: 464.

9. Baliga BJ (2005) Silicon Carbide – Power Devices. World Scientific, Singapore.

10. Neudeck PG (2003) VLSI Technology (Ed. Wai-Kai Chen). CRC Press, USA.

11. Choyke WJ, Matsunami H, Pensl G (2004) Silcon carbide: Recent Major Advances (Springer-Verlag, Berlin, USA).

12. Saddow SE, Agarwal A (2004) Advances in Silicon carbide - Processing and Applications. Artech House, Inc. Boston.

13. Feng ZC (2004) SiC Power Materials-Devices and Applications. Springer, Berlin.

14. Feng ZC, Zhao JH (2003) Silicon Carbide: Materials, Processing and Devices. Taylor & Francis Books Inc, New York.

15. Phan HP, Dinh T, Kozeki T, Qamar A, Namazu T, et al. (2016) Triply Differential Cross Sections for the Ionization of Helium by Fast Electrons. Scientific Reports 6: 28499.

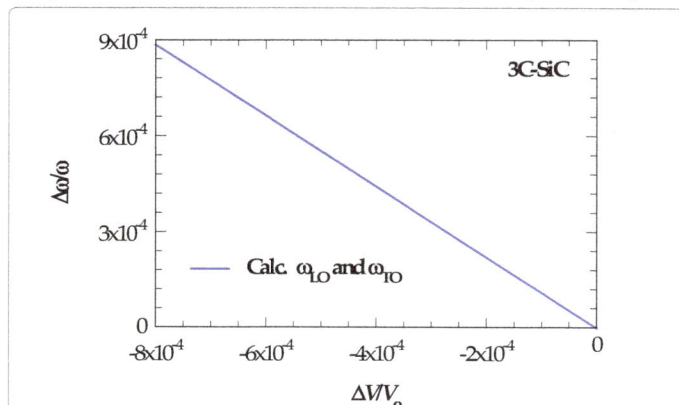

Figure 4: Plot of $\frac{\Delta\omega}{\omega}$ versus $\frac{\Delta V}{V_o}$ for the strained 3C-SiC grown on Si (001) substrate.

16. Wijesundara MBJ, Azevedo R (2011) Silicon Carbide Microsystems for Harsh Environments. Springer-Verlag New York.

17. Scaburri R (2011) The incomplete ionization of substitutional dopants in silicon carbide, Dottorato di Ricerca in Ingegneria dei Materiali, Universitá di Bologna, Italy.

18. Rurali R, Godignon P, Rebollo J,Hernández E, Ordejón P (2003) Spectrum of Small-Scale Density Fluctuations in Tokamaks. Appl Phys Lett 82: 4298.

19. Liu P (2014) Atomic Structure of the Vicinal Interface between Silicon Carbide and Silicon Dioxide. The University of Tennessee, Knoxville.

20. Gupta SK, Akhtar J (2011) Thermal Oxidation of Silicon Carbide (SiC) – Experimentally Observed Facts. Intech 1-26.

21. Hornos T (2008) Theoretical study of defects in silicon carbide and at the silicon dioxide Interface. Budapest University of Technology and Economics, USA.

22. Fu W (2015) Design and Comparison of Si-based and SiC-based Three-Phase PV Inverters. The University of Wisconsin-Milwaukee, USA.

23. Nawaz M (2015) Active and Passive Electronic Components 651527.

24. Shibahara K, Kuroda N, Nishino S, Matsunami H (1987) Magnetic Double Transition in Au-Fe near the Percolation Threshold. Jpn J Appl Phys 26: L1815.

25. Kong HS, Glass JT, Davis RF (1988) Direct Determination of Sizes of Excitations from Optical Measurements on Ion-Implanted GaAs. J Appl Phys 64: 2672.

26. Powell JA, Larkin DJ, Matus LG, Choyke WJ, Bradshaw JL, et al. (1990) Localized Modes and Cell-Model Limit in the Crystal Impurity Problem. Appl Phys Lett 56: 1442.

27. Nishino K, Kimoto T, Matsunami H (1995) Thermally Stimulated Exoelectron Emission. Jpn J Appl Phys 34: L1110.

28. Kimoto T, Yamashita A, Itoh A, Matsunami H (1993) Domain walls and ferroelectric reversal in corundum derivatives. Jpn J Appl Phys 32: 1045.

29. Steckl AJ, Roth MD, Powell JA, Larkin DJ (1993) Numerical analytic continuation: Answers to well-posed questions. Appl Phys Lett 62: 2545.

30. Tin CC, Hu R, Williams J, Feng ZC, Yue KT (1994) Origin of the second peak in the mechanical loss function of amorphous silica. Mat Res Soc Symp 339: 411.

31. Talwar DN (2015) Non-Hermitian bidirectional robust transport. Diamond and Related Mater 52: 1.

32. Park JH, Kim JH, Kim Y, Lee BT, Jang SJ, et al. (2003) Intrinsic localized mode and low thermal conductivity of PbSe. Appl Phys Lett 83: 1989.

33. Anzalone R, Locke C, Carballo J, Piluso1 N, Severino A, et al. (2010) Critical behavior in the presence of an order-parameter pinning field. Mat Sci Forum 143: 645-648.

34. Anastassakis E, Pinczuk A, Burstein E, Pollak FH, Cardona M (1970) Efficient spin transport through polyaniline. Solid State Commun 8: 133.

35. Cerdeira F, Buchenauer CJ, Pollak FH, Cardona M (1971) Stress-Induced Shifts of First-Order Raman Frequencies of Diamond- and Zinc-Blende-Type Semiconductors. Phys Rev B: 580.

36. Olego D, Cardona M (1982) Pressure dependence of Raman phonons of Ge and 3C-SiC. Phys Rev B 25: 1151.

37. Olego D, Cardona M, Vogl P (1982) Pressure dependence of the optical phonons and transverse effective charge in 3C-SiC. Phys Rev B 25: 3878.

38. Jusserland B, Voisin P, Voos M, Chang LL, Mendez EE, et al. (1985) Advances in quantitative Kerr microscopy. Appl Phys Lett 46: 678.

39. Shon LH, Inoue K, Murase K (1987) Dynamics of skyrmionic states in confined helimagnetic nanostructures. Solid State Commun 62: 621.

40. Mukaida H, Okumura H, Lee JH, Daimon H, Sakuma E, et al.(1987) Raman scattering of SiC: Estimation of the internal stress in 3C-SiC on Si. J Appl Phys 62: 254.

41. Feng ZC, Mascarenhas AJ, Choyke W, Powell JA (1988) Optical conductivity from pair density waves. J Appl Phys 64: 3178.

42. Debernardi A, Ulrich C, Cardona M, Syassen K (2001) Pressure Dependence of Raman Linewidth in Semiconductors. Phys Stat Sol 223: 213.

43. Zhuravlev KK, Goncharov AF, Tkachev SN, Dera P, Prakapenka VB (2013) Vibrational, elastic, and structural properties of cubic silicon carbide under pressure up to 75 GPa: Implication for a primary pressure scale. J Appl Phys 113: 113503.

44. Zhu J, Liu S, Lia W (2000) Large-scale normal fluid circulation in helium superflows. Thin Solid Films 368: 307.

45. Capano MA, Kim BC, Smith AR, Kvam EP, Tsoi S, et al. (2006) Time Resolved Spectroscopy of Defects in SiC. J Appl Phys 100: 08514.

46. Shafiq M, Subhash G (2014) Thermoluminescence and Related Electronic Processes of 4H/6H-SiC. Experimental Mechanics 54: 763.

47. Harima H (2002) Properties of GaN and related compounds studied by means of Raman scattering. J Phys: Condens Matter 14: R967-R993.

48. Nakashima S, Harima H (1997) Raman Investigation of SiC Polytypes. Phys Stat Solidii A162: 39-64.

49. Nye JF (1957) Physical properties of crystals. Oxford Univ Press, London.

50. Lekhnitskii SG (1963) Theory of elasticity of an anisotropic elastic body. Holden-Day.

51. Bower AF (2010) Applied mechanics of solids. Taylor and Francis Group, CRC Press, USA.

52. Slack GA (1964) Advances in SiC MOS Technology. J Appl Phys 35: 3460.

53. Feldman DW, Parker JH, Choyke WJ, Patrick L (1968) High Temperature Sensors Based on Metal–Insulator–Silicon Carbide Devices. Phys Rev 173: 787.

54. Djemia P, Roussigné Y, Dirras GF, Jackson KM (2004) Optical Characterization of Silicon Carbide Polytypes. J Appl Phys 95: 2324.

55. Tolpygo KB (1961) High Pressure Synthesis of Binary B–S Compounds. Sov Phys-Solid State 2: 2367.

56. Miura M, Murata H, Shiro Y, Lishi K (1981) Control of Electronic Properties of Organic Conductors by Hydrostatic and Uniaxial Compression. J Phys Chem Solids 42: 931.

57. Lee DH, Joannopoulos JD (1982) Simple Scheme for Deriving Atomic Force Constants: Application to SiC. Phys Rev Lett 48: 1846.

58. Marshall RC, Faust JW, Ryan CE (1974) Silicon Carbide. University of South Carolina, Columbia, SC.

59. Vashishtha P, Kalia RK, Nakano A, Rino JP (2007) Interaction potential for silicon carbide: A molecular dynamics study of elastic constants and vibrational density of states for crystalline and amorphous silicon carbide. J Appl Phys 101: 103515.

60. Lambrecht WRL, Segall B, Methfessel M, van Schilfgaarde M (1991) Calculated elastic constants and deformation potentials of cubic SiC. Phys Rev B44: 3685.

61. Karch K, Pavone P, Mindi W, Schutt O, Strauch D (1994) Ab initio calculation of structural and lattice-dynamical properties of silicon carbide. Phys Rev B 50: 17054.

62. Li W, Wang T (1999) Elasticity, stability, and ideal strength of β-SiC in plane-wave-based ab initio calculationsPhys Rev B 59: 3993.

63. Teroff J (1989) Modeling solid-state chemistry: Interatomic potentials for multicomponent systems. Phys Rev B 39: 5566.

64. Wang CZ, Rici Yu, Krakauer H (1996) Pressure dependence of Born effective charges, dielectric constant, and lattice dynamics in SiC. Phys Rev B 53, 5430.

65. Yu-Ping L, Duan-Wei H, Jun Z, Xiang Dong Y (2008) Effects of Confinement on the Coupling between Nitrogen and Band States in InGaAs1-xNx/GaAs (x ≤ 0.025) Structures: Pressure and Temperature Studies. Phys B 43: 3543.

66. Lee DH, Joannopoulos JD (1982) Simple Scheme for Deriving Atomic Force Constants: Application to SiC. Phys Rev Lett 48: 1846.

67. Serrano J, Strempfer J, Cardona M, Schwoerer-Böhning M, Requardt H, et al. (2002) High Pressure Studies of the Raman-Active Phonons in Carbon Nanotubes. Appl Phys Lett 80: 4360.

12

Mechanical and Thermal Properties of Short Arecanut Leaf Sheath Fiber Reinforced Polypropyline Composites: TGA, DSC and SEM Analysis

Poddar P[1,2*], Islam MS[3], Sultana S[3], Nur HP[3] and Chowdhury AMS[1]

[1]Department of Applied Chemistry and Chemical Engineering, Faculty of Engineering and Technology, University of Dhaka, Bangladesh
[2]Office of the Chief Chemical Examiner, CID, Bangladesh Police, Mohakhali, Dhaka, Bangladesh
[3]Bangladesh Council of Scientific and Industrial Research, Dhanmondi, Dhaka, Bangladesh

Abstract

Short arecanut leaf sheath (ALS) fiber (2-3 mm) reinforced polypropylene (PP) composites were prepared by compression molding technique. Heat and cold press were used. Chemical composition of the fiber was analyzed and the percents of lignin, α-cellulose and hemicellulose were determined. Fiber content in the composites was optimized with the extent of mechanical properties and composites with 10% arecanut leaf sheath fiber showed higher mechanical properties. Tensile strength (TS), Bending strength (BS), elongation at break (EB%), water absorption capacity, scanning electron microscopy (SEM), thermo gravimetric analysis (TGA), differential scanning calorimetry (DSC) and biodegradation properties of arecanut leaf sheath/PP composites were investigated. ATR spectra of the polypropylene and composites were also analyzed.

Keywords: Arecanut leaf sheath fiber; α-cellulose; Thermo gravimetric analysis; Scanning electron microscopy; Differential scanning calorimetry; Composites

Introduction

Environmental policies of developed countries are increasing the pressure on manufacturers to consider the environmental impact of their products. Therefore, the interest in using natural fibers as a reinforcement of polymer-based composites is growing mainly because of its renewable origin. Polypropylene is an economical material that offers a combination of outstanding physical, chemical, mechanical, thermal and electrical properties not found in any other thermoplastic [1]. Compared to low or high density polyethylene, it has a lower impact strength, but superior working temperature and tensile strength [2].

In our country, *Areca catechu* trees are available in the coastal area which produces huge leaf-sheath. This unusable item can be used to produce composite materials [3]. Several billion pounds of fillers and reinforcements are used in the plastics industry every year. The use of these additives in plastics is likely to grow with the introduction of improved compounding technology and new coupling agents that permit the use of high filler/reinforcement content. As suggested by Katz and Milewski, fillings up to 75 parts per hundred (pph) could be common in the future: this could have a tremendous impact in lowering the usage of petroleum based plastics. It would be particularly beneficial; both in terms of the biodegradability features [4-7] and also in socio-economic terms, if a significant amount of the fillers were obtained from a renewable agricultural source. Ideally, of course, an agro-/bio-based renewable polymer reinforced with agro-based fibers [8-11] would make the most environmental sense.

Arecanut leaf sheath fiber composed of mainly α-cellulose, lignin, and hemicelluloses. In addition, it contains minor constituents such as pectic matters, Fatty and waxy matters. An excellent review by Milewski on short fiber composite technology covers a variety of reasons that result in problems associated with composite properties falling short of their true reinforcing potential. The major factors that govern the properties of short fiber composites are fiber dispersion, fiber length distribution, fiber orientation, and fiber-matrix adhesion. Mixing the polar and hydrophilic fibers with non-polar and hydrophobic matrix can result in difficulties associated with the dispersion of fibers in the matrix.

Interestingly, several types of natural fibers which are abundantly available like arecanut leaf sheath, oil palm, banana, sisal, jute, wheat, flax straw, sugarcane, cotton, silk, bamboo and coconut have proved to be good and effective reinforcement in the thermo-set and thermoplastic matrices [12-19]. Our present research focus on fabrication of agro-fiber reinforced thermoplastics composite-specially, composites made using short arecanut leaf sheath fiber. The ultimate goal of this work is to study the composite potentiality of agro-fiber towards diversified application within environmental legal framework. These composites may be used in the packaging, furniture, housing, construction industries [20], decking, window, door frames [21-24], and automobiles sectors.

Experimental

Materials

Polypropylene (PP) was purchased from Polyolefin Company, Private Ltd., Singapore (Figure 1). Polypropylene granules were grinded to get small particle (50-60 µm) (Figure 1d) with the help of grinder for proper and homogeneous adhesion between fibers and matrix. Arecanut leaf sheath fibers (Figure 1b) were prepared from arecanut leaf sheath (Figure 1a) by soaking the leaf sheath into water for 15 days. The water loosed the fiber from the resin and waxy materials and then the fibers peeled from the resinous materials, washed with clean water and air dried properly. The arecanut leaf sheath fibers were chopped into small pieces (2-3 mm) with the help of hand scissors and cleaned with mesh and all dirt's are removed from the chopped fiber. Then the chopped fibers (Figure 1c) were cleaned with distilled water

*Corresponding author: Poddar P, Department of Applied Chemistry and Chemical Engineering, Faculty of Engineering and Technology, University of Dhaka, Dhaka-1000, Bangladesh, E-mail: p.pinku@yahoo.com

Mechanical and Thermal Properties of Short Arecanut Leaf Sheath Fiber Reinforced Polypropyline...

75

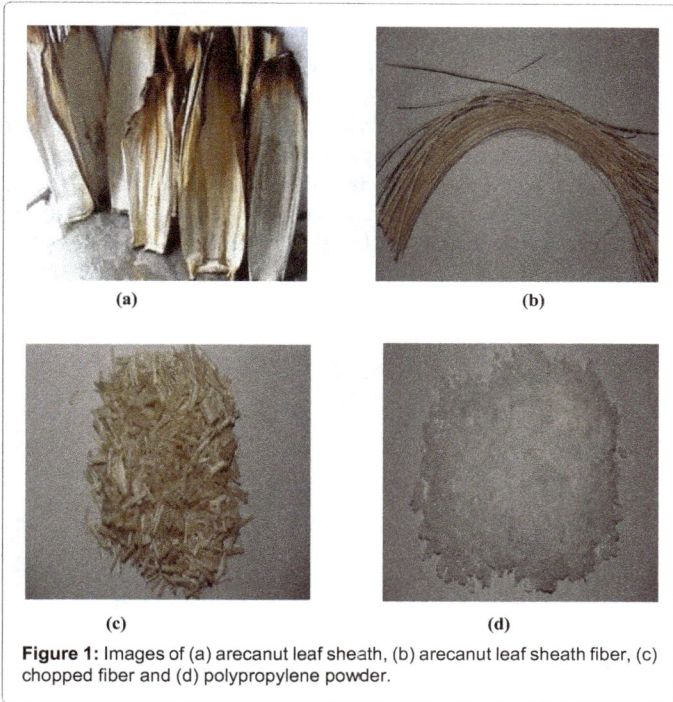

Figure 1: Images of (a) arecanut leaf sheath, (b) arecanut leaf sheath fiber, (c) chopped fiber and (d) polypropylene powder.

and exposed thoroughly to sunlight for about 24 hours. The fibers were dried at 100°C in a vacuum oven for 5 hours prior to the preparation of the composites.

Chemical composition of arecanut leaf sheath (ALS) fiber

Arecanut leaf sheath fiber contains α – cellulose, hemicellulose, lignin, aqueous extract, fatty and waxy matters, and pectic matters. In Table 1, the chemical composition of arecanut leaf sheath fiber is shown. The fiber mainly contains 66.08% of α–cellulose, 19.59% of lignin, and 7.40% of hemicellulose. Arecanut leaf sheath fiber is composed of small units of cellulose surrounded and cemented together by lignin and hemicellulose.

Composite fabrication

Composites were prepared by compounding with extrusion and hot press machine. The mould (12×15 cm²) was cleaned with wax as a releasing agent. The mixture of dried arecanut leaf sheath fiber and polypropylene powder was prepared according to the Table 2 that showed different weight fraction (5%, 10%, 15%, 20% and 25% fiber) and poured into the mould. The processing temperature was maintained at 190°C for 5 min under 5 bar consolidation pressure in the heat press (Carver, INC, USA Model 3856). The molds were then cooled for 1 min in a separate press under 5 bar pressure at room temperature. Figure 2 showed the final composite product.

Tensile strength and bending strength test

The tensile properties of the composites (F-F) were determined using a universal testing machine (model H50 KS-0404, Hounsfield Series S, UK). The load capacity was 5000 N; efficiency was within ±1%. The crosshead speed was 10 mm/min and gauge length was 20 mm. For bending properties measurement, the crosshead speed was 10 mm/min, and span distance was 40 mm. Tensile strength measurements and three point bending tests were carried out according to DIN 53455 and DIN 53452 standards methods, respectively.

Water uptake

Five composite samples (20×10×2 mm³) were immersed in the beaker containing 100 ml of deionized water at room temperature (25°C) for different time periods (up to 60 h). Weight of the samples was determined initially then after certain periods of time, samples were taken out from the beaker and wiped (5 times) using tissue papers, and then weighed again. The weight gained, i.e., water uptake of the samples was determined by subtracting the initial weight from the final weight.

Soil degradation tests of the composites

Cellulose has a tendency to degrade when buried in soil whose moisture level is at least 25%. For this purpose, all composite samples were buried in garden soil for a period of 24 weeks. After certain periods of time, samples were carefully withdrawn, washed with distilled water, and dried at a temperature of 80°C for 8 h and then kept at room temperature for 24 h. The change of tensile strength was periodically noted in order to determine the degradable character of the samples in this environment.

Attenuated total reflectance (ATR) analysis

100% Polypropylene and composite were analyzed by Attenuated Total Reflectance (ATR) and spectra were recorded in the 4,000-650 cm⁻¹ region on a Perkin Elmer instrument.

S. No	Name	%
1	Aqueous Extract	0.72
2	Fatty and waxy matters	5.06
3	Pectic matters	1.15
4	Lignin	19.59
5	α - cellulose	66.08
6	Hemicellulose	7.40
	Total	100

Table 1: The chemical composition of arecanut leaf sheath fiber.

Formulations	Fiber content (wt %)	Polypropylene Powder (wt %)
F₁	5	95
F₂	10	90
F₃	15	85
F₄	20	80
F₅	25	75

Table 2: Showed different weight fraction (5%, 10%, 15%, 20% and 25% fiber) and poured into the mould.

Figure 2: Image of finished product (composite).

Thermo gravimetric analysis and differential scanning calorimetry analysis

The thermograms and differential scanning calorimetry of the polypropylene and Fcomposite were recorded on a NETZSCH instrument (model no. STA 449 F3, Jupiter) in the temperature range of 0-900°C.

Scanning electron microscopic (SEM) investigation

The fracture surfaces of the tensile specimens of both F and F composite samples were examined using a Hitachi S-4000 field emission scanning electron microscope, operated at 5 kV. Samples were mounted with carbon tape on aluminum stubs and then sputter coated with carbon tape on aluminum stubs and then sputter coated with platinum and palladium to make them conductive prior to SEM observation.

Results and Discussion

Mechanical properties of the composites

The prepared composites were cut into desired size. Mechanical properties such as tensile strength, bending strength and elongation at break were measured. According to Figure 3, the highest tensile strength value observed for F formulation and the value is 28.7 MPa. For the formulations of F, F, F tensile strength values are gradually decreased.

Same condition is observed for bending strength. In the Figure 4, F formulation showed the maximum value (46.9MPa). From F to F the value is increased and from F to F the bending strength values are gradually declined.

From Figure 5, it is clear that the % elongation at break reduces with the increase of fiber in the composites. For F, F, F, F and Fformulations the values % elongation at break are 15.2, 13.8, 11.6, 8.9 and 6.4 respectively.

From the above results, it is observed that fiber content in the composites has the significant influence on the mechanical properties. 10% fiber with 90% PP prepared composite showed the maximum tensile strength and bending strength due to better fiber-matrix adhesion. The percentage of fiber content in the composites (above 10 wt %), TS and BS of the composites showed a decreasing trend, which may be attributed to the fact that increasing fiber content in the composite decreased the fiber-matrix adhesion.

Water uptake of the composites

Water uptake values of the F, F, F, F and F samples were calculated by immersing the samples in de-ionized water contained in a static glass beaker at room temperature. The samples were taken out of water after constant time interval and their mass gain were calculated. The results of water uptake values of the samples are shown Figure 6. F, F and F samples gained water up to 30 h whereas F and F samples gained water up to 20 h of soaking, and then the values were almost constant. The minimum amount of water was taken up by the F sample (0.15%) and the highest amount of water was counted by F sample (0.27%) at the maximum period of observation (60 h). Water was taken up by the F sample was 0.16% that was near about minimum value (0.15%).

Degradation tests of the composites

F, F, F, F and F samples were buried in soil for up to 24 weeks in order to study the effect of such an environmental condition on the degradability of the samples. TS values are plotted against degradation time (weeks) and are shown in Figure 7. It was found that for all samples TS were decreased slowly with degradation time. After 24 weeks of soil degradation, F, F, F, F and F samples decreased almost 33, 20, 35, 37 and 42% of TS. This is clear that F sample retained its tensile properties more than other samples during soil degradation. Arecanut leaf sheath fiber is a natural biodegradable fiber and a cellulose-based fiber, which absorbs water within a couple of minutes due to strong hydrophilic character. Cellulose has a strong tendency to degrade when

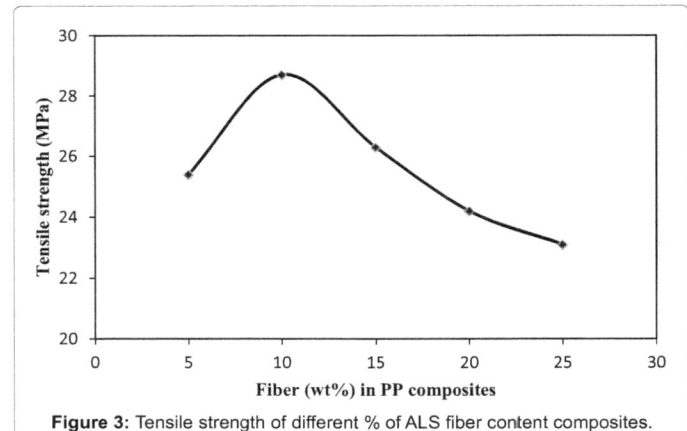

Figure 3: Tensile strength of different % of ALS fiber content composites.

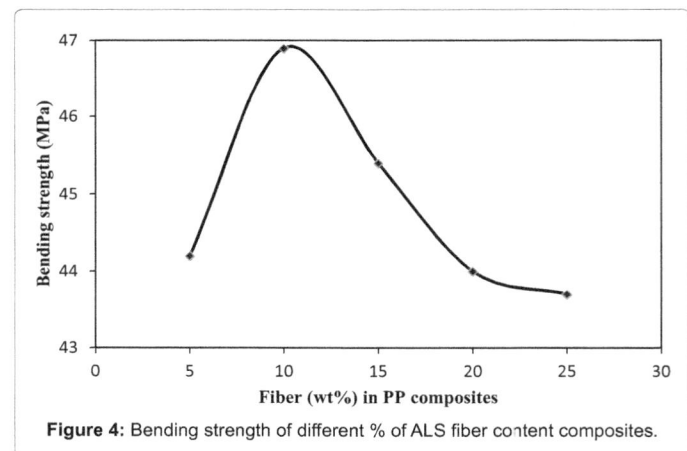

Figure 4: Bending strength of different % of ALS fiber content composites.

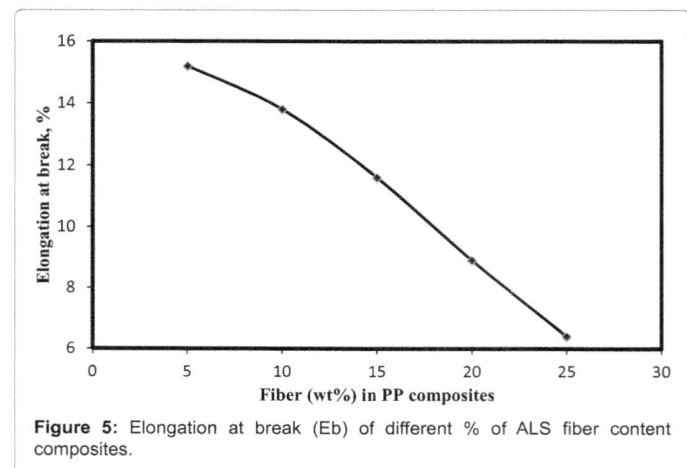

Figure 5: Elongation at break (Eb) of different % of ALS fiber content composites.

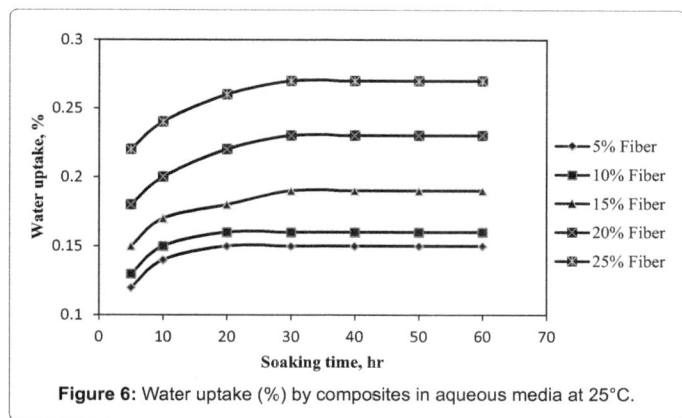

Figure 6: Water uptake (%) by composites in aqueous media at 25°C.

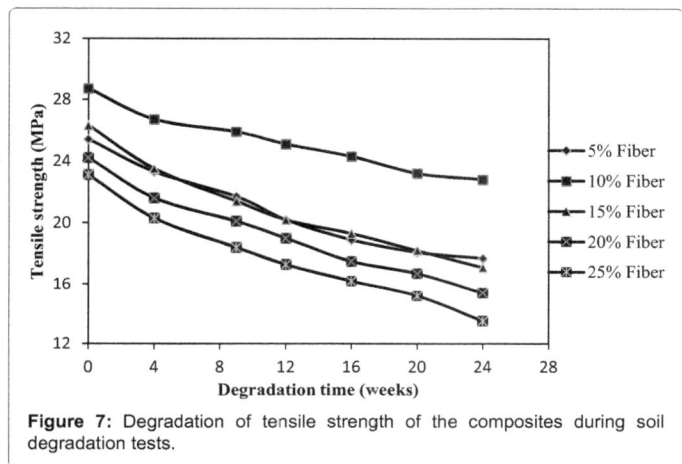

Figure 7: Degradation of tensile strength of the composites during soil degradation tests.

buried in soil [25]. During soil degradation tests, water penetrates from the cutting edges of the composites in arecanut fiber based samples and degradation of cellulose occurred in the fiber and as a result, the mechanical properties of the composites decreased significantly.

ATR analysis

ATR spectra is used to measure the change of surface composition of the PP granules and optimized (wt %) fiber reinforced PP composite. Figures 8 and 9 showed the ATR record of PP granules and optimized (wt %) fiber reinforced PP composite respectively. The region from 1450 to 600 cm^{-1} is called as fingerprint region and the absorption bands in the 4000 to 1450 cm^{-1} region are usually because of stretching vibrations of diatomic units, and known as the group frequency region. The C-OH bending peak is observed at 650 cm^{-1} which is absent in the pure PP granules (Figure 8). C-H vibration peak is showed at 2916 cm^{-1} in the both figures (Figures 8 and 9).

Thermo gravimetric analysis and differential scanning calorimetry analysis

The thermograms of both PP granules and optimized (wt %) fiber reinforced PP composite were presented in Figures 10 and 11 respectively. From the Figure 10 is observed that degradation start at 380°C and 98.73% mass change is completed at 500°C whereas Figure 11 showed that mass change start at 280°C and 10.38% mass change is completed at 420°C and 88.14% degradation is done at 500°C. In case of Figure 11 degradation starts earlier which may be attributed to the fact that fiber-matrix adhesion in the composite. DSC is used to identify melting temperature and calculate the amount of energy absorbed or released by the PP granules and optimized (wt %) fiber reinforced PP composite when these are heated. From the Figures 10 and 11 it is observed that endothermic reactions are occurred in the both cases and 477.2°C and 477.6°C heat absorbed respectively.

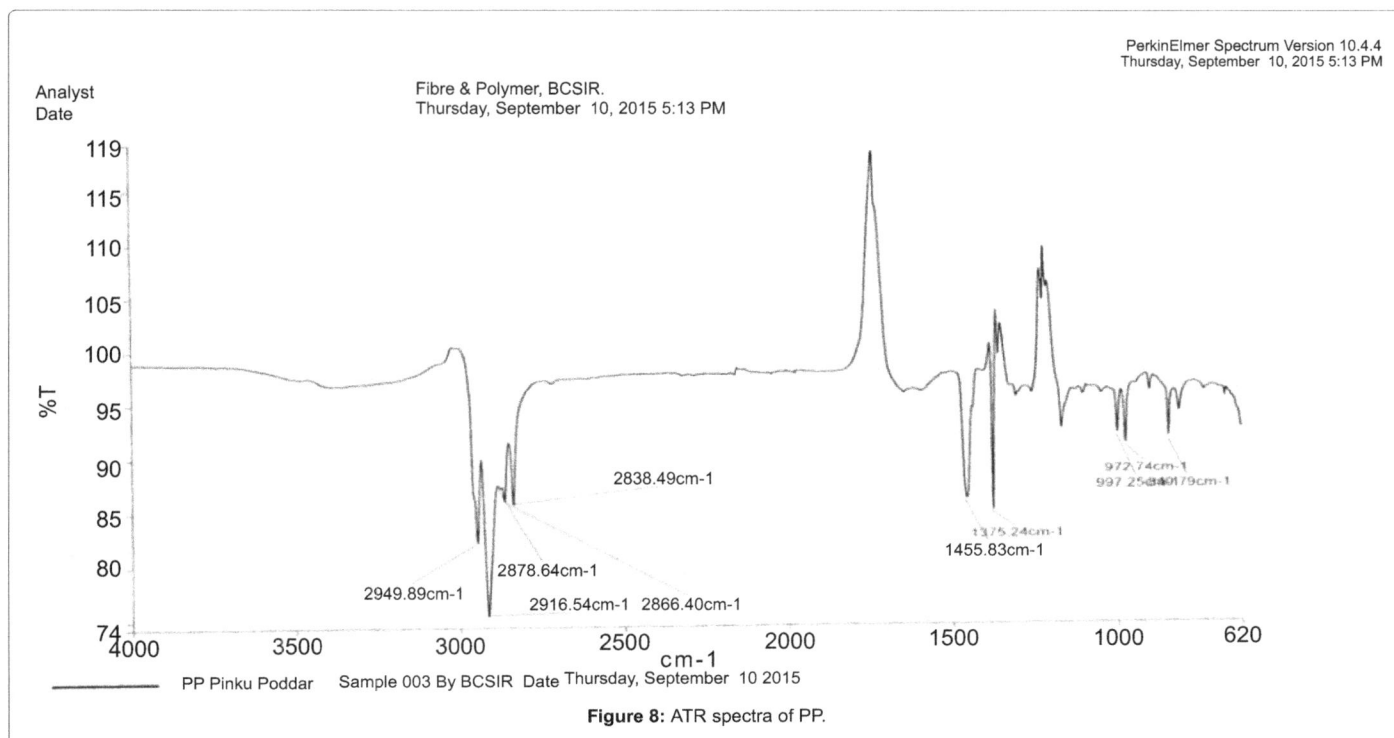

Figure 8: ATR spectra of PP.

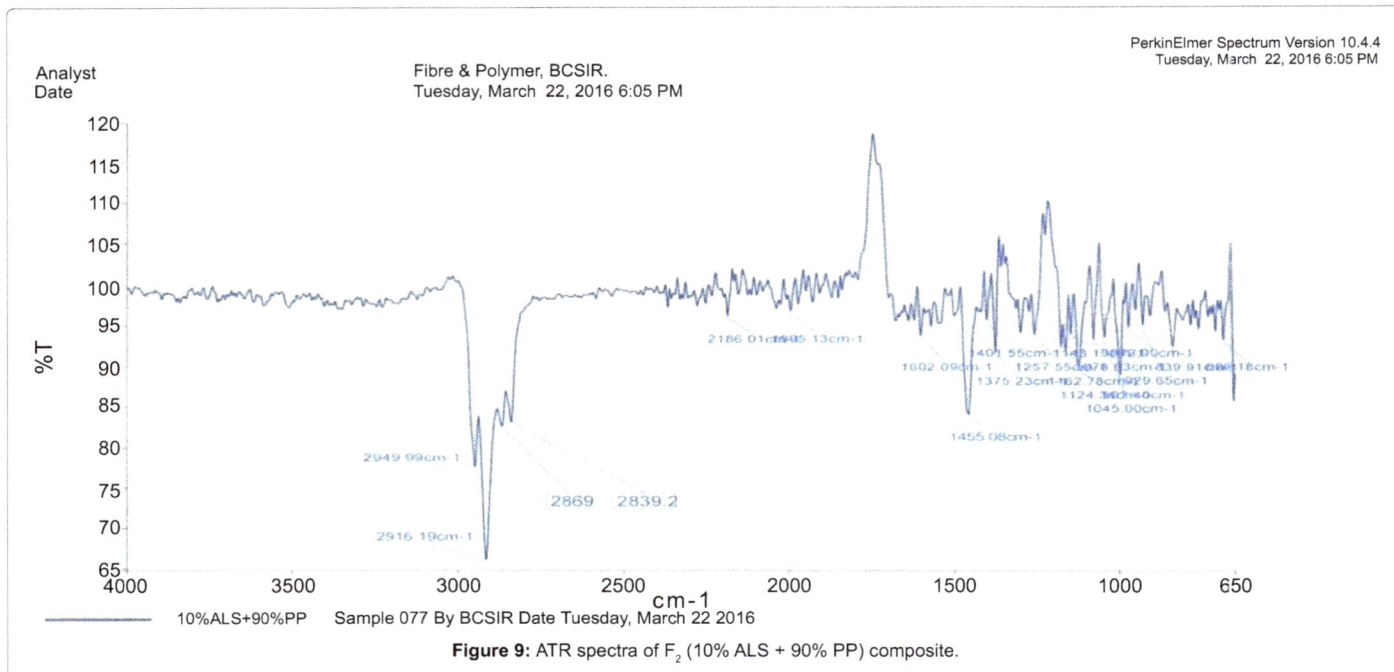

Figure 9: ATR spectra of F_2 (10% ALS + 90% PP) composite.

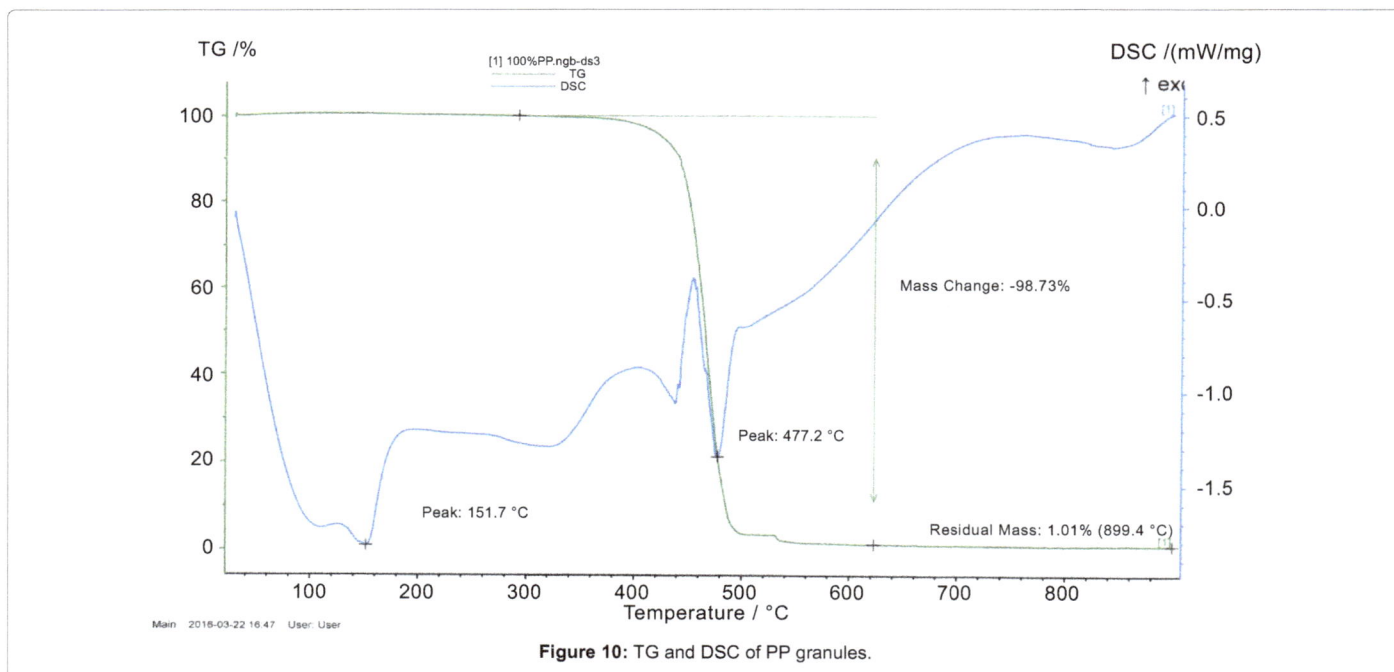

Figure 10: TG and DSC of PP granules.

SEM analysis

Interfacial properties of arecanut leaf sheath short fibers reinforced polypropylene composites (Fand F) were investigated by SEM (Figures 12 and 13). SEM observations indicated that there is a considerable difference in the fiber-matrix interaction between F and F composites. Some gaps between fiber and matrix are clearly found for Fcomposite which are responsible for the low mechanical properties. On the other hand Fcomposite showed better fiber-matrix adhesion and gaps between fiber and matrix are not observed which are responsible for higher mechanical properties.

Conclusion

Arecanut leaf sheath short fibers reinforced polypropylene composites were prepared by compression molding. Fiber content in the composites was optimized with the extent of mechanical properties and 10% fiber content in the composite showed higher mechanical properties. SEM supported the idea that above 10% fiber content, mechanical properties were decreased with increasing fiber content due to poor fiber-matrix adhesion. Elongation at break decreased with the increased of fiber (wt %). Water uptake behavior of optimized composite was almost same of the lowest fiber content in the composite.

Mechanical and Thermal Properties of Short Arecanut Leaf Sheath Fiber Reinforced Polypropyline...

79

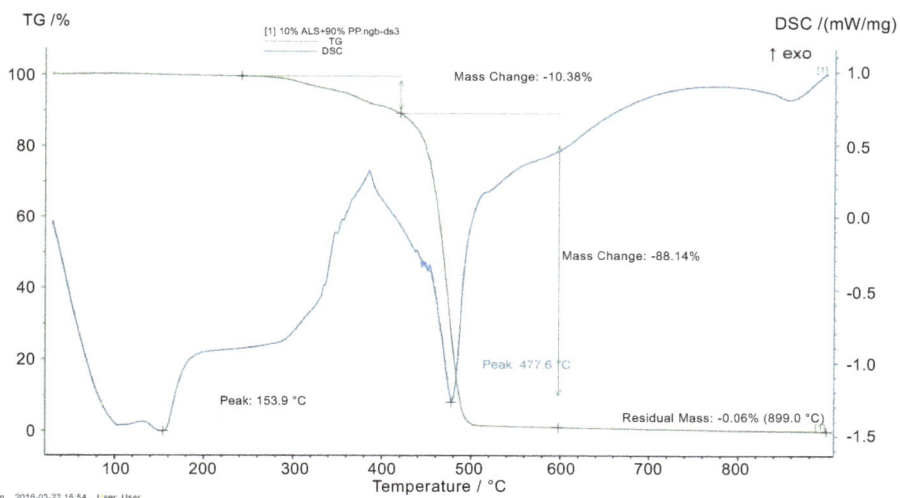

Figure 11: TG and DSC of F_2 (10% ALS + 90% PP) composite.

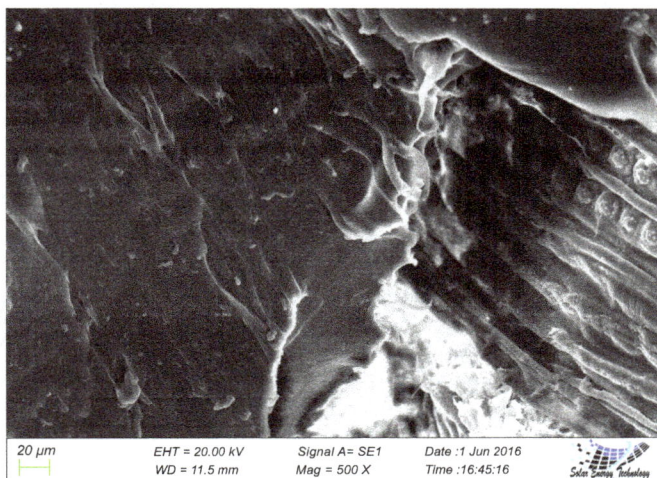

Figure 12: Image of F_2 composite.

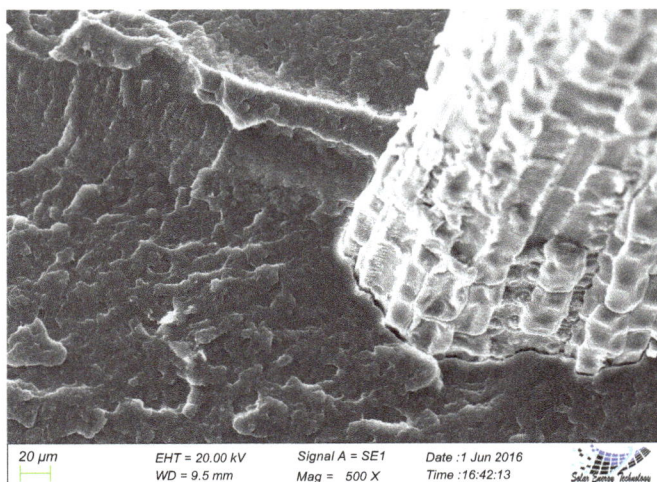

Figure 13: Image of F_5 composite.

Soil degradation studies demonstrated that TS of the optimized composite (F) could be lower than that of the other composites with respect to their degradation time. These new composites would make possible to explore new applications and new markets in the packaging, furniture, housing and automotive-aviation-shipping sectors.

References

1. Khan MN, Roy JK, Akter N, Zaman HU, Islam T, et al. (2012) Production and properties of short jute and short E-glass fiber reinforced polypropylene-based composites. Open Journal of Composite Materials 2: 40-47.

2. Reddy GR, Kumar MA, Chakradhar KVP (2011) Fabrication and performance of hybrid Betel nut (Areca catechu) short fiber/ Sansevieria cylindrical (Agavaceae) epoxy composites. International Journal of Materials and Biomaterials Applications 1: 6-13.

3. Zaman HU, Khan A, Khan RA, Huq T, Khan MA, et al. (2010) Preparation and characterization of jute fabrics reinforced urethane based thermoset composites: effect of uv radiation. Fibers and Polymers 11: 258-265.

4. Joshi SV, Drzal LT, Mohanty AK, Arora S (2004) Are natural fiber composites environmentally superior to glass fiber reinforced composites? Composites Part A: Applied Science and Manufacturing 35: 371-376.

5. Mwaikambo L (2006) Review of the history, properties and application of plant fibres. African Journal of Science and Technology 7: 120-133.

6. John M, Anandjiwala R (2008) Recent developments in chemical modification and characterization of natural fiber reinforced composites. Polymer composites 29: 187-207.

7. Satyanarayana KG, Arizaga GGC, Wypych F (2009) Biodegradable composites based on lignocellulosic fibers-An overview. Progress in Polymer Science 34: 982-1021.

8. Khan JA, Khan MA, Islam R (2012) Effect of mercerization on mechanical, thermal and degradation characteristics of jute fabric-reinforced polypropylene composites. Fibers and Polymers 13: 1300-1309.

9. Bledzki AK, Gassan J (1999) Composites reinforced with cellulose based fibres. Progress in Polymer Science 24: 221-274.

10. Dweib MA, Hu B, Donnell AO (2004) All natural composite sandwich beams for structural applications. Composite Structures 63: 147-157.

11. Graupner N, Herrmann AS, Müssig J (2009) Natural and man-made cellulose fibre-reinforced poly (lactic acid) (PLA) composites: An overview about mechanical characteristics and application areas. Composites Part A: Applied Science and Manufacturing 40: 810-821.

12. Levy NF, Balthazar JC, Pereira CT (2000) 3rd International Symposium on Natural Polymer Composite- ISNAPOL-2000.

13. Pavan RMV, Saravanan V, Dinesh AR, Rao YJ (2001) Hygrothermal effects on painted and unpainted glass/epoxy composites- part a: moisture absorption characteristics. Journal of Reinforced Plastics and Composites 20: 1036-1047.

14. Sreekala MS, George J, Kumaran MG, Thomas S (2002) The mechanical performance of hybrid phenol-formalde-based composites reiforced with glass and oil palm fibers. Composite Science Technology 62: 339-353.

15. Moe MT, Liao K (2003) Durability of bamboo-glass fiber reinforced polymer matrix hybrid composites. Composite Science Technology 63: 375-387.

16. Pothan LA, Oommen Z, Thomas S (2003) Dynamic mechanical analysis of banana fiber reinforced polyester composites. Composites Science and Technology 63: 283-293.

17. Espert A, Vilaplana F, Karlsson S (2004) Comparison of water absorption in natural cellulosic fibres from wood and one-year crops in polypropylene composites and its influence on their mechanical properties. Composites Part A 35: 1267-1276.

18. Jacoba M, Thomasa S, Varugheseb KT (2004) Mechanical properties of sisal/ oil palm hybrid fiber reinforced natural rubber composites. Composites Science and Technology 64: 955-965.

19. Agnelli JAM, Joseph K, Carvalho LH, Mattoso LHC (2005) Mechanical properties of phenolic composites reinforced with jute/cotton hybrid fabrics. Polymer Composites 26: 1-11.

20. Thomas JAG (1972) Fibre composites as construction materials. Composites 3: 62-64.

21. Wan AW, Abdul R, Lee TS, Abdul RR (2008) Injection moulding simulation analysis of natural fiber composite window frame. J Mater Proc Technol 197: 22-30.

22. Younguist JA (1995) Unlikely partners? The marriage of wood and non-wood materials. Forest Prod J 45: 25-30.

23. Rai SK, Priya PS (2006) Utilization of waste silk fabric as reinforcement for acrylonitrile butadiene styrene toughened epoxy matrix. J Reinforc Plast Compos 25: 565-574.

24. Singh B, Gupta M (2005) Performance of pultruded jute fibre reinforced phenolic composites as building materials for door frame. J Polym Environ 13: 127-137.

25. Khan MA, Haque N, Kafi AA, Alam MN, Abedin MZ (2006) Jute reinforced polymer composite by gamma radiation: effect of surface treatment with UV radiation. Polym Plast Technol 45: 607-613.

13

Experimental Determination of Mechanical and Physical Properties of Almond Shell Particles Filled Biocomposite in Modified Epoxy Resin

Singh VK, Bansal G*, Agarwal M and Negi P

G.B.P.U.A.T, Pantnagar, Uttarakhand, India

Abstract

Rapid Advancement in the field of advance Biocomposite has attracted large number of researchers to diagnose and expand the use of light weighted and environment friendly materials for various applications. In this work, depolymerized natural rubber (DNR) was prepared and used as toughening agents for epoxy resin and almond shell particles are filled in modified epoxy resin as reinforcing material. Further, different tests including Tensile Test, Compression Test, Hardness Test, Impact Test etc. were performed to diagnose the effect of various weight percentage (wt%) of composition of blended DNR for achieving maximum toughness and then its effect on treatment with different weight percentage of Almond Shell Particle. Results were analyzed and finally the conclusion is made based on the experiments.

Keywords: Composite; Epoxy; DNR; Almond shell particle; Tensile property

Introduction

Epoxy resins are very important class of thermosetting polymers that exhibit high tensile strength and modulus, excellent chemical and corrosion resistance, good dimensional stability, low creep and reasonable performance at elevated temperature. Hence, they are widely used as matrix resins for fiber reinforced composite materials and in structural adhesives, surface coatings and electrical laminates. However, such properties in an epoxy require moderate to high levels of crosslinking which can and usually does result in brittle behavior. Toughening of epoxy resin has been the subject of intense investigation, because epoxy resins have low fracture energy [1,2].

Natural rubber has attracted great interest because it is a renewable resource, whereas its synthetic counter parts are mostly manufactured from non-renewable oil based resources. Several studies have already been done on toughening epoxy resin using natural rubber (Figure 1) [3-5].

Epoxy resins are reactive chemicals which are combined with other chemicals known as hardener or curing agent such as triethylenetetramine (TETA) and 4,4′-diaminodiphenylsulfone (DDS) to give systems capable of conversion to predetermined thermoset products.

Natural rubber

One of the most important polymeric materials is natural rubber (NR) which contains 93-94% *cis*-1, 4-polyisoprene (Figure 2). Natural Rubber latex is the form in which rubber is exuded from the *Hevea brasiliensis* tree as an aqueous emulsion. The rubber particles range in size from about 50 Å to about 30,000 Å (3μm). Exceptionally particles up to 5-6 μm in diameter are found. The molecular weight (MW) is normally in the range of 10^4-10^7g/mol, depending on the age of the rubber tree, weather, method of rubber isolation and other factors [6].

The advantages of NR are outstanding flexibility, excellent heat built up properties and high mechanical strength. Moreover, it is a renewable resource, whereas its synthetic counterparts are mostly manufactured from non-renewable oil based resources [7].

Depolymerization of polymer is based on a reaction in which a reagent with reactive polar groups opens the active linkage in the polymer backbone. It can reduce chain length of polymer. Natural rubber that is subjected to depolymerization is called depolymerized natural rubber (DNR) or liquid natural rubber. Having strong adhesive power and excellent crosslinking reactivity, it has been used widely as a raw material for adhesives, pressure-sensitive adhesives, sealing materials etc.

Almond shell particles

Almond shell is an organic residue which is lingo-cellulosic material forming a thick endocarp or husk of the almond tree fruit, which is separated in the process of extracting edible seeds. Almond shells have no importance in industry and generally burnt or dumped. Almond shell and almond shell particles are shown in the Figures 3a and 3b [8].

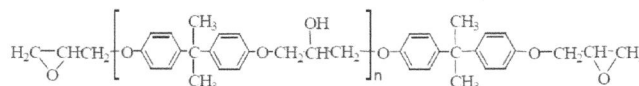

Figure 1: Chemical structure of bisphenol-A based epoxy resins (Irfan [10]).

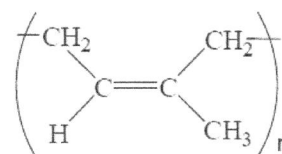

Figure 2: Chemical structure of *cis* 1, 4-polyisoprene.

*Corresponding author: Bansal G, M Tech Scholar, G.B.P.U.A.T, Pantnagar, Uttarakhand, India, E-mail: gaganbansal12345@gmail.com

Figure 3a: Almond shells.

Figure 3b: Almond shell particles.

Preparation of DNR, modified epoxy and then hybrid bio-composite

Natural rubber latex was diluted in deionized water to a concentration of 5 wt% based on rubber content in a 1 liter reaction flask. After that $CH_3CH_2COCH_3$ and $K_2S_2O_8$ was added in an amount of 4 v% of total volume and 2 wt% based on the rubber content, respectively. The pH of above solution was adjusted to about 9-10 with 10 wt% aqueous KOH solution. Then, the mixture was mechanically stirred with a speed of 200 revolutions per minute (rpm) at 70°C for 24 hours under flowing air on the magnetic stirrer with hot plate. At the end of reaction, the mixture was coagulated by 1 wt% aqueous $CaCl_2$ solution (Tables 1 and 2).

The coagulated substance was dissolved in hexane and stirred with magnetic bar for 3 hours. Then, resulting solution was stood overnight and filtered with filter paper and dried at 40°C until weight is constant. DNR was blended with epoxy resin in an amount of 0.5, 1, 1.5, 2 and 2.5 wt% of epoxy resin. Blending formulation that showed the highest toughness was applied as matrix for preparation of almond shell particles filled biocomposite. The neat epoxy, modified epoxy and almond shell particles filled modified epoxy resin composites were prepared by vertical casting method. The universal testing machine, digital hardness testing machine and pendulum impact tester were used to study the mechanical properties of different biocomposites [9,10].

Testing and Results

Density

Results indicate that the material having higher wt% of almond shell particles have lower density than unfilled epoxy resin because almond shell particles have very less density and when they are mixed

in epoxy resin then the density of biocomposite tends to decrease with the increase in wt% of almond shell particles (Figure 4 and Table 3).

Tensile test

In the present investigation all the tensile tests are conducted as per ISO test procedure with Specimen size based on ISO-1608: 1972 Standard. The tests are conducted on 100 kN servo hydraulic UTM machine (model 2008, ADMET make).

Designation	Epoxy resin (grams)	Hardener (grams)	DNR (grams)
C0	100	10	0
C1	100	10	0.5
C2	100	10	1.0
C3	100	10	1.5
C4	100	10	2.0
C5	100	10	2.5

Table 1: Design of experiments.

Designation	Epoxy resin (grams)	Hardener (grams)	DNR (grams)	Almond shell particle (grams)
CA1	100	10	1.0	10
CA2	100	10	1.0	20
CA3	100	10	1.0	30
CA4	100	10	1.0	40

Table 2: Design of experiment for almond shell particles filled composites.

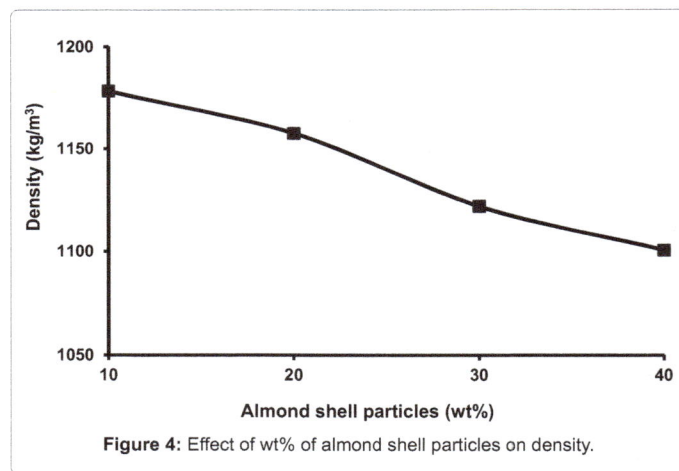

Figure 4: Effect of wt% of almond shell particles on density.

DNR (wt%)	Almond shell particles, (wt%)	Weight, (g)	Volume, (cm³)	Density, (g/cm³)	Density, (kg/m³)	Mean density, (kg/m³)	Standard Deviation
1	10	7.06	6	1.17667	1176.7		
1	10	8.28	7	1.18286	1182.9	1177.73	3.95
1	10	7.04	6	1.17333	1173.3		
1	20	8.08	7	1.15429	1154.3		
1	20	8.12	7	1.16000	1160.0	1157.54	2.40
1	20	6.95	6	1.15833	1158.3		
1	30	7.92	7	1.13143	1131.4		
1	30	6.68	6	1.11333	1113.3	1122.14	7.40
1	30	6.73	6	1.12167	1121.7		
1	40	7.72	7	1.10286	1102.9		
1	40	7.72	7	1.10286	1102.9	1100.79	2.92
1	40	6.58	6	1.09667	1096.7		

Table 3: Density of almond shell particles filled composites.

From the results, in Figure 5a at initial loading there is an increase in stress followed by a plateau region that may be due to slippage of the Specimen. Remarkable differences can be seen on the ultimate tensile strength of DNR filled cured epoxy having different wt% of DNR given in Table 4. It can be seen from the results that for all specimens containing 1.0 wt% DNR, the ultimate tensile strength is highest among the other composition reported (Figure 5b). About 40% increase in ultimate tensile strength due to addition of 1.0 wt% of DNR has been noticed as compared to pure epoxy. This increase in strength is observed due to inter molecular dispersion of DNR in epoxy resin. Further addition of DNR on the epoxy resin decreases the ultimate tensile strength of the DNR filled cure epoxy due to accumulation of DNR at some palces, which is present free without bonding. Similar observations have been noticed for % elongation as shown in Figure 5c. Modulus of elasticity was decreasing with increasing wt% of DNR but about 2.47 times increase in modulus of elasticity has been observed due to addition of 1.0 wt% of DNR in Almond Shell Particle based composite (as compared between Tables 4 and 5). Further addition of the DNR decreases the % elongation but is higher than the neat epoxy material [11].

Figure 5d shows the toughness or Energy/Volume on different DNR wt%. The toughness is calculated by integrating the polynomial which best fit the stress-strain curve for different wt% of DNR and almond shell particles. From Figure 5e it can be concluded that small wt% of DNR has a great effect on toughness of material.

Compression test

All the compression tests are conducted on 100 kN servo hydraulic UTM machine (model 2008, ADMET make). Here the ISO Standard

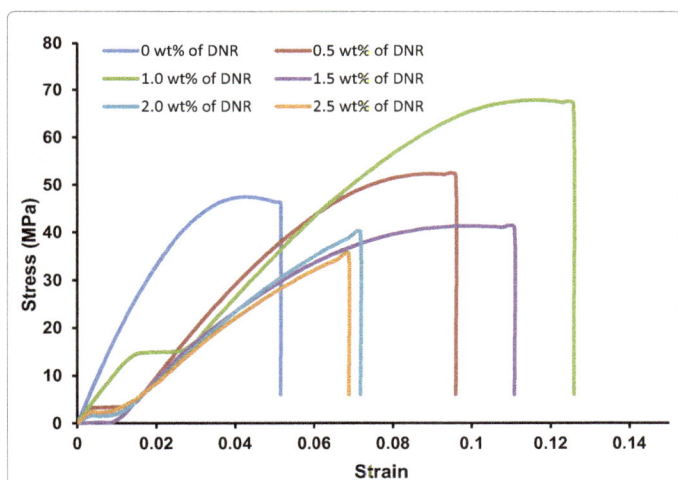

Figure 5b: Stress-strain diagram for different wt% of almond particle with 1.0 wt% of depolymerised natural rubber.

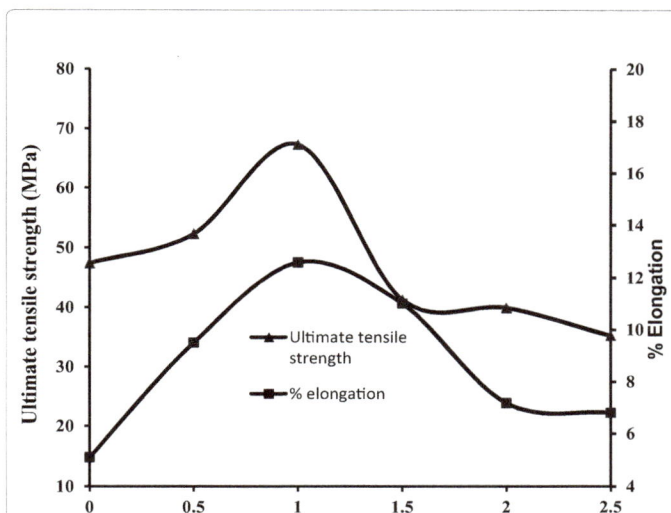

Figure 5c: Effect of wt% DNR on ultimate tensile strength and % elongation.

Designation of Composition	Almond Shell Particles, wt%	Ultimate strength (MPa)	% elongation	Energy/Volume (mJ/mm³)	Modulus of Elasticity (GPa)
CA1	10	35.35	4.115	0.73493	1.577
CA2	20	39.14	5.667	1.76456	1.565
CA3	30	35.31	3.487	0.5490	1.871
CA4	40	33.05	2.374	0.35332	0.551

Table 5: Tensile Properties with varying wt% of almond shell particles in epoxy and 1wt% DNR blended composite.

ISO-1708: 1960 is used in Specimen prepration and testing. It is found that ultimate compressive strength of 20 wt% of almond shell particles is 145 MPa. This compressive strength is about 1.5 times ultimate compressive strength of the modified epoxy with 1 wt% DNR. From the present results, it can be said that the ultimate compressive strength has increased considerably due to addition of small weight percentage of almond shell particles (Figures 6a and 6b).

Impact test

An ISO 180:1993 plastic Standards is considered in Impact testing.

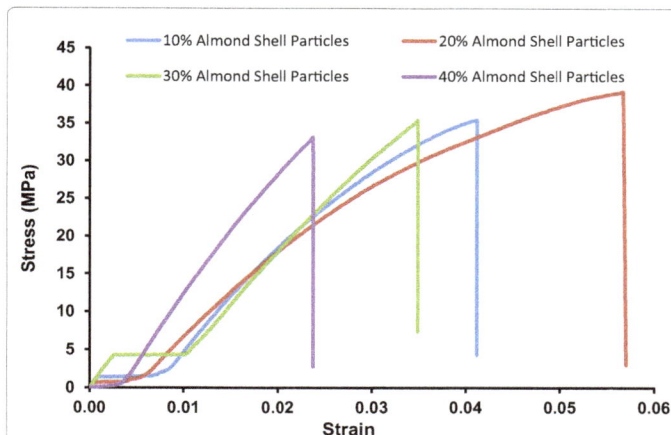

Figure 5a: Stress-strain diagram for different wt% of depolymerised natural rubber (DNR).

Designation of Composition	Depolymerized rubber, wt%	Ultimate strength (MPa)	% elongation	Energy/Volume (mJ/mm³)	Modulus of Elasticity (GPa)
C0	0.0	47.40	5.10	1.737	1.889
C1	0.5	52.28	9.50	3.001	1.067
C2	1.0	67.33	12.58	5.229	0.912
C3	1.5	41.30	11.00	2.956	0.671
C4	2.0	39.89	7.17	1.416	0.526
C5	2.5	35.26	6.80	1.226	0.513

Table 4: Tensile properties of cured epoxy filled with DNR.

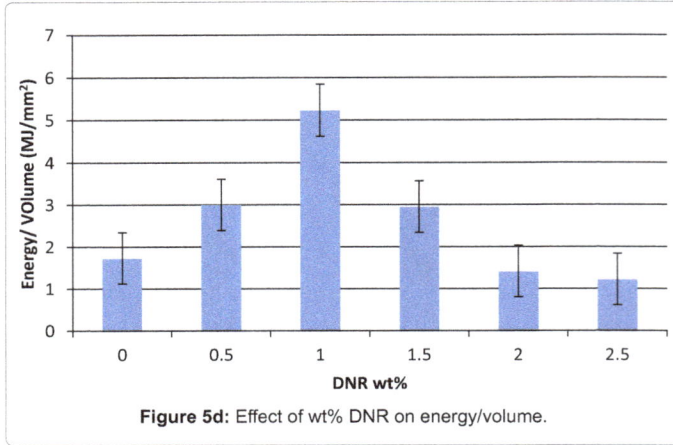

Figure 5d: Effect of wt% DNR on energy/volume.

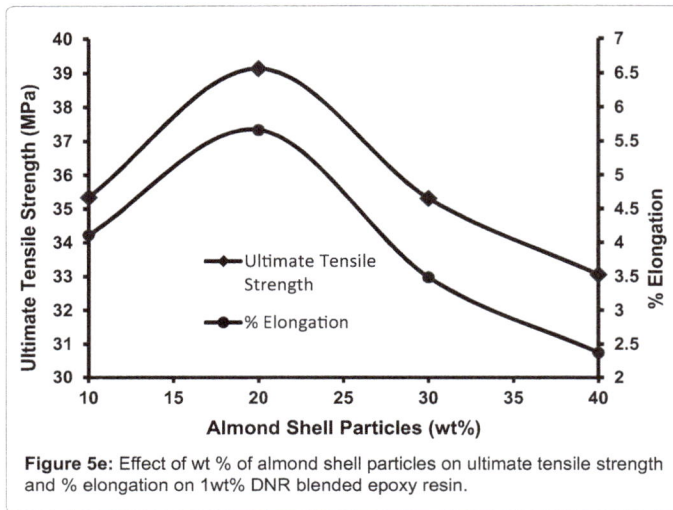

Figure 5e: Effect of wt % of almond shell particles on ultimate tensile strength and % elongation on 1wt% DNR blended epoxy resin.

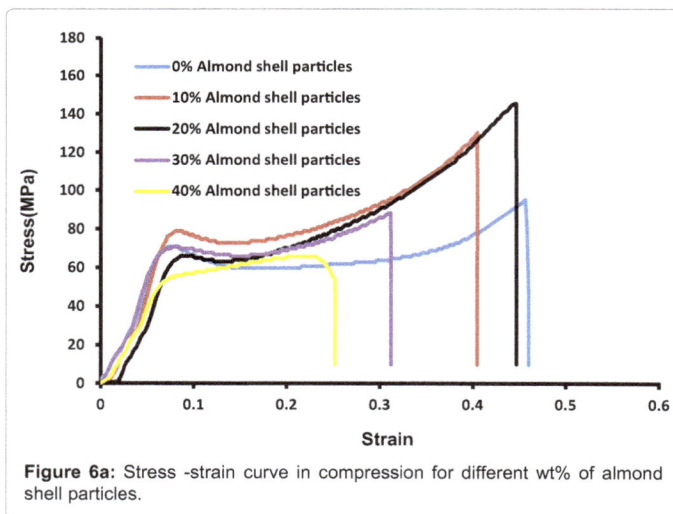

Figure 6a: Stress -strain curve in compression for different wt% of almond shell particles.

The results show that by adding DNR in epoxy resin the impact strength of epoxy resin is increased by 80% on 1 wt% of DNR which is a very remarkable improvement. As Chuayjuljit [3] concluded that ENR product was applied as impact modifier for epoxy resin and Kumar and Kothandaraman [5] modified epoxy resin with maleate

depolymerized natural rubber, MDPR so the DNR was used here fot testing. Also Increase in Impact strength here might be because DNR which is presented in matrix acted as stress concentrator creating shear yielding and/or crazing in the matrix. This can also be seen that at higher wt% of DNR the impact strength has been lowered it may be due to the rubber aggregation or accumulation because of which there are internal cavities and internal voids in the material [12]. Similarly, in the case of almond shell particles impact resistance is decreasing with the increase in wt% of almond shell particles (Figures 7a and 7b). This decrease in the impact properties might be due to the decrease in the bond strength in almond shell particles and matrix material at higher wt% (Tables 6 and 7).

Hardness test

In this study the hardness test have been conducted on L scale on Digital Rockwell hardness testing machine. Above results indicates that the hardness of the material increased with the increase in the wt% of DNR shown in Figure 8a. The increase in hardness might be due to the high hardness of DNR. From Figure 8b this can be seen that hardness of almond shell particles reinforced composite decreases with the increase in wt% of almond shell particles. This may be due to the softness or low hardness of almond shell particles.

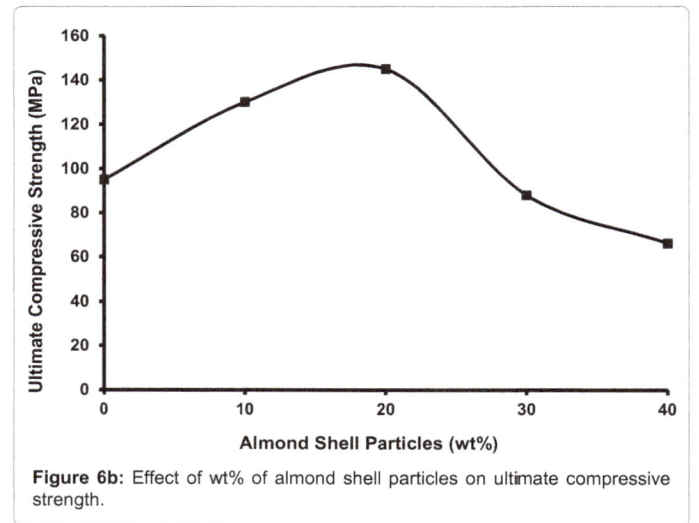

Figure 6b: Effect of wt% of almond shell particles on ultimate compressive strength.

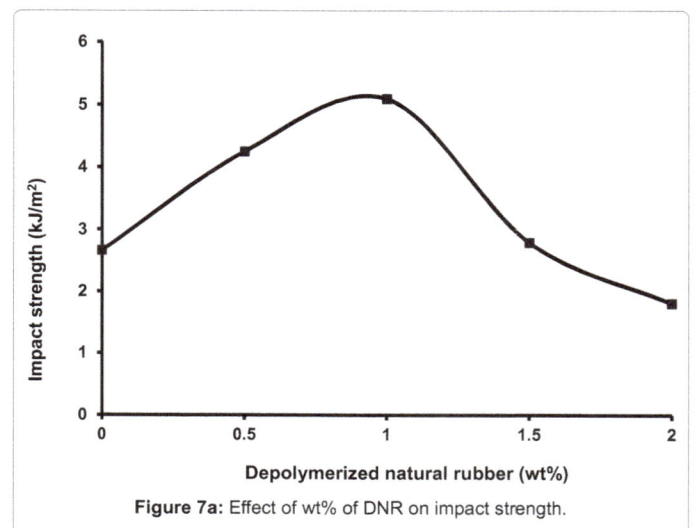

Figure 7a: Effect of wt% of DNR on impact strength.

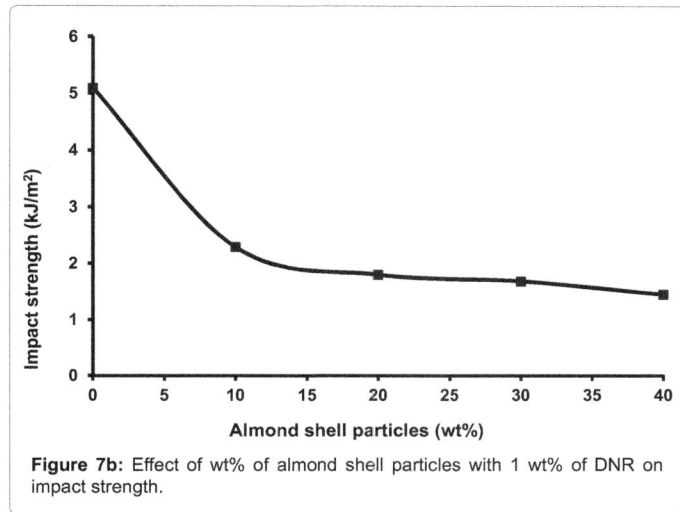

Figure 7b: Effect of wt% of almond shell particles with 1 wt% of DNR on impact strength.

Depolymerised Natural Rubber (wt%)	Almond shell particles (wt%)	Impact strength (kJ/m²)
0.0	0	2.662
0.5	0	4.247
1.0	0	5.091
1.5	0	2.789
2.0	0	1.804

Table 6: Impact properties of DNR modified epoxy resin composite.

Depolymerised Natural Rubber (wt%)	Almond shell particles (wt%)	Impact strength (kJ/m²)
1.0	10	2.288
1.0	20	1.800
1.0	30	1.684
1.0	40	1.451

Table 7: Impact properties of almond shell particles filled composite.

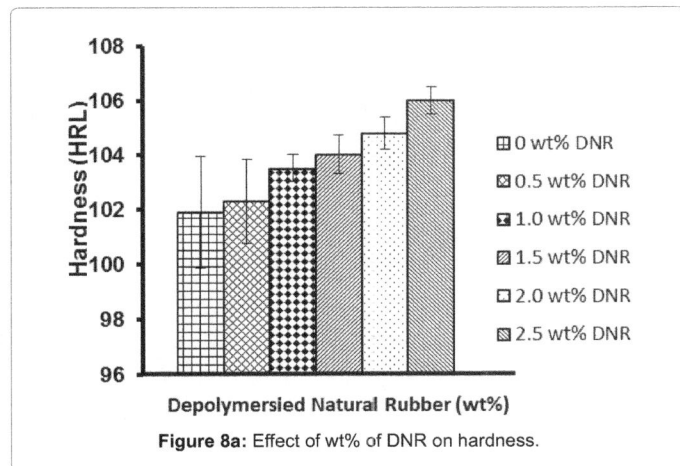

Figure 8a: Effect of wt% of DNR on hardness.

Flexural test

The flexural properties have a very important role in structural applications (Figures 9a, 9c and 9e). The flexural stress, flexural modulus and flexural strain are calculated by equations (i), (ii) and (iii) respectively. The flexural stress (σ_f), flexural modulus (E_f) and flexural strain (ε_f) for rectangular cross section are determined by the formula:

$$\sigma_f = \frac{3PL}{2bd^2} \tag{i}$$

$$E_f = \frac{L^3 m}{4bd^3} \tag{ii}$$

$$\varepsilon_f = \frac{6Dd}{L^2} \tag{iii}$$

From the Figures 9b, 9d and 9f this can be concluded that 10wt% of almond shell particles is the optimum filling wt% for almond shell particles because it gives the optimum flexural properties. This can also be observed that flexural stress and strain decreases drastically with the addition of more wt% of almond shell particles. This decrease in the flexural properties may be due to the insufficient filling of matrix material in the surrounding of almond shell particles.

Scanning Electron Microscopy (SEM)

Agglomeration of Almond Shell Particle with varying wt% of DNR can be better understood with the help of SEM images. Figure 10a shows the micrograph of 1 wt% of DNR blended in epoxy resin. In this more shearing zone can be seen and the rubber is well dispersed

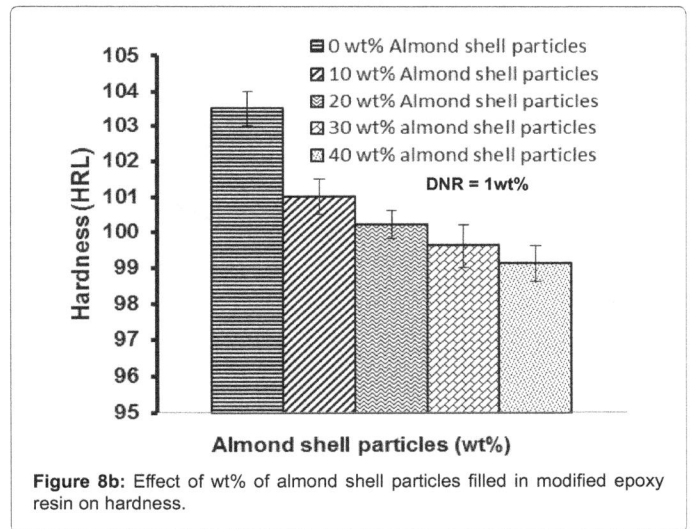

Figure 8b: Effect of wt% of almond shell particles filled in modified epoxy resin on hardness.

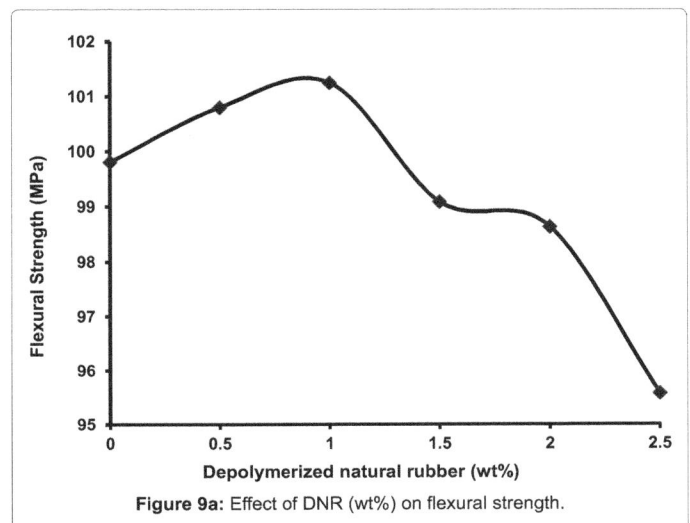

Figure 9a: Effect of DNR (wt%) on flexural strength.

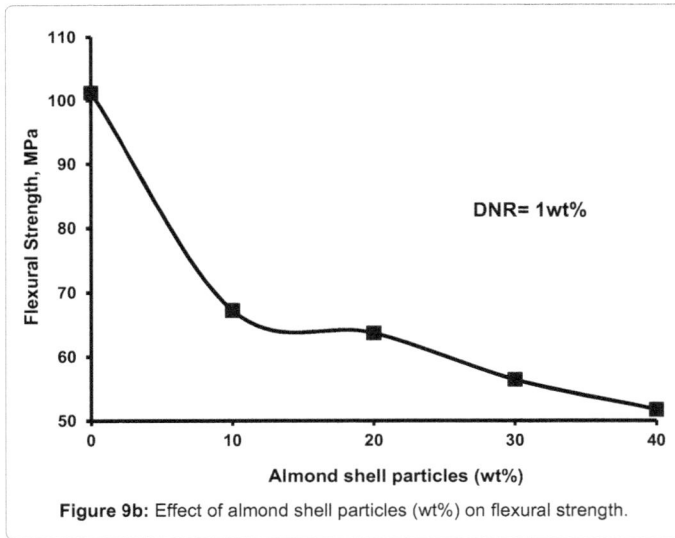

Figure 9b: Effect of almond shell particles (wt%) on flexural strength.

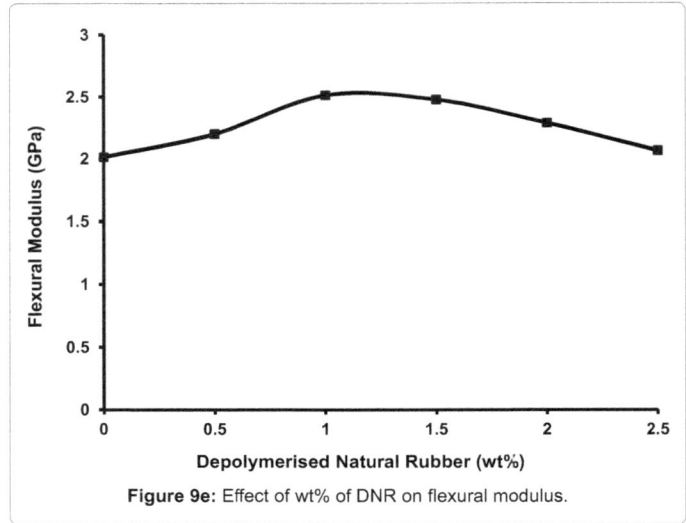

Figure 9c: Effect of DNR (wt%) on flexural strain.

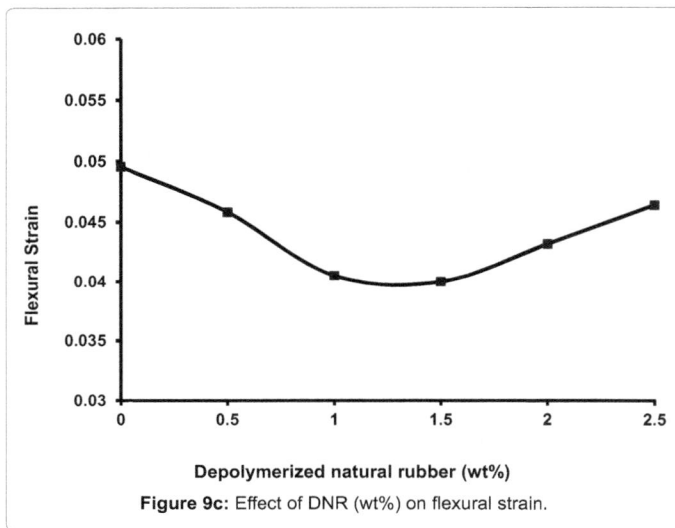

Figure 9d: Effect of almond shell particles (wt%) on flexural strain.

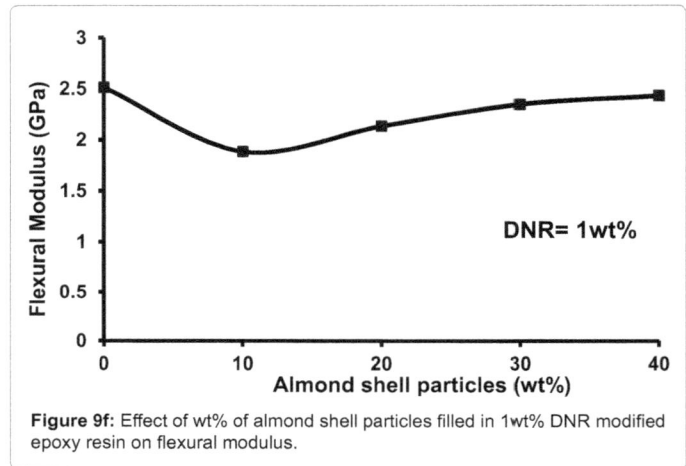

Figure 9e: Effect of wt% of DNR on flexural modulus.

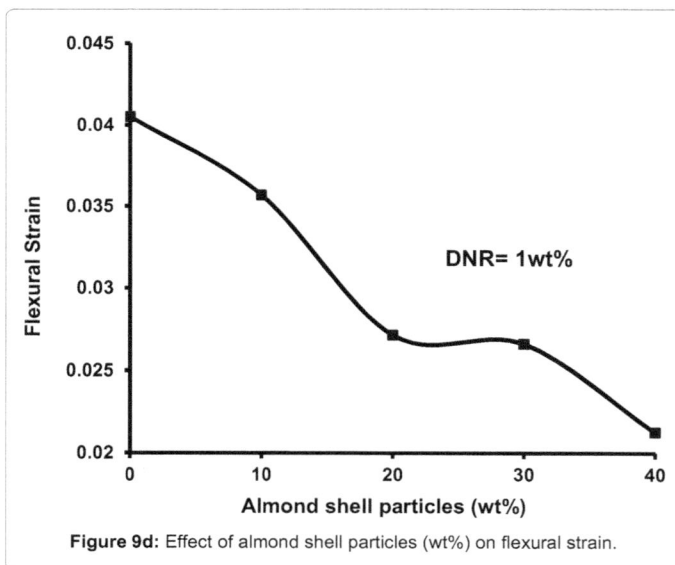

Figure 9f: Effect of wt% of almond shell particles filled in 1wt% DNR modified epoxy resin on flexural modulus.

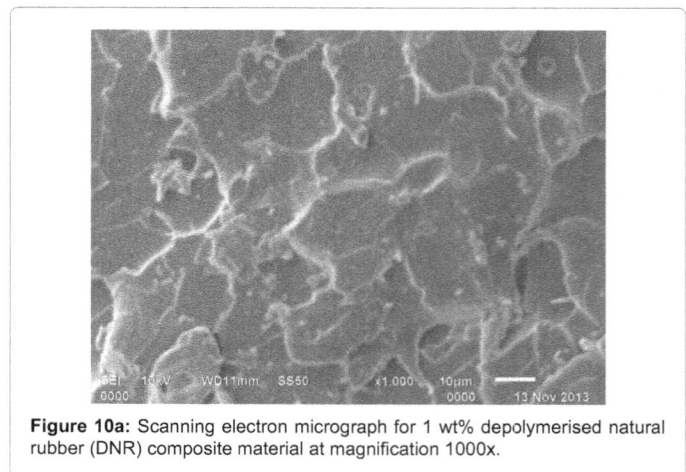

Figure 10a: Scanning electron micrograph for 1 wt% depolymerised natural rubber (DNR) composite material at magnification 1000x.

in epoxy resin. Good crosslinking of DNR with epoxy resin can be seen here leads to enhanced mechanical properties as this crosslinking overcome the brittle behaviour of epoxy resin.

Figures 10b-10e shows the SEM photograph of composite containing 10, 20, 30 & 40 wt% of almond shell particles respectively. It is seen from the Figure 9 that almond shell particles are well dispersed in the epoxy resin matrix. Also, the size of almond shell particles was in the range of about 0.39 μm to 1 μm as seen in Figure 10f.

The micrographs of almond shell particles filled composite material shows that materials are failing due to pulling out of particles (Figure 10f).

In Figure 10b shear bands can be seen which leads to shear yielding

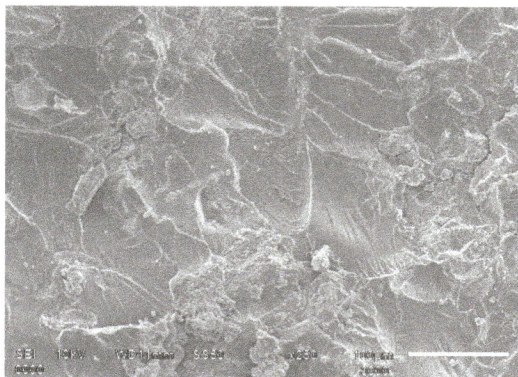

Figure 10b: Scanning electron micrograph of 1 wt% DNR & 10 wt% almond shell particles composite material at magnification 250x.

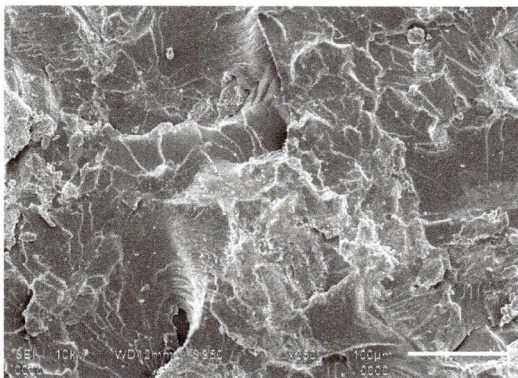

Figure 10c: Scanning electron micrograph of 1 wt% DNR & 20 wt% almond shell particles composite material at magnification 250x.

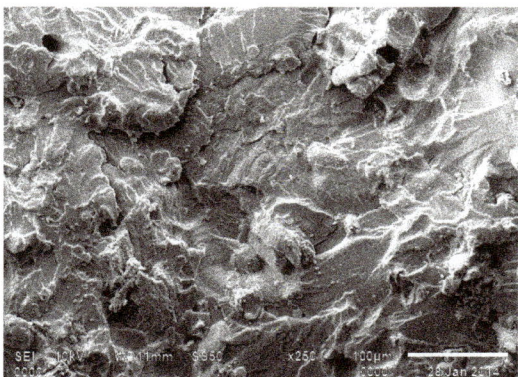

Figure 10d: Scanning electron micrograph of 1 wt% DNR & 30 wt% almond shell particles composite material at magnification 250x.

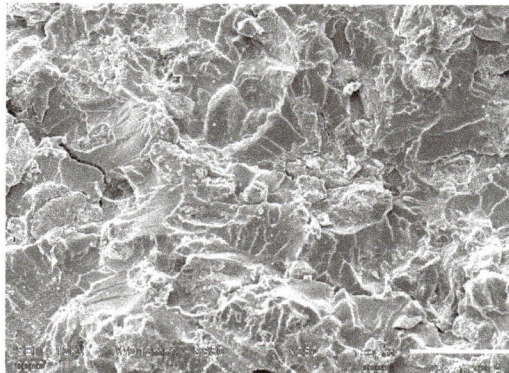

Figure 10e: Scanning electron micrograph of 1 wt% DNR & 40 wt% almond shell particles composite material at magnification 250x.

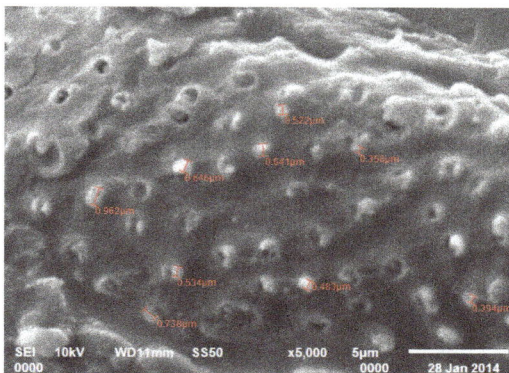

Figure 10f: Scanning electron micrograph of 1 wt% DNR & 30 wt% almond shell particles composite material at magnification 5000x.

of the material. And the dispersion of almond shell particles is good and no cavity can be seen. But at some places matrix cracking has taken place. Figure 10c micrograph the cavity which has occurred due to increase in wt% of almond shell particles. This diagram does not show the matrix cracking which has led the material to good strength [13,14].

And at higher wt% the material is failing due to insufficient filling of matrix materials around the particles. Many cavities are also taking place that are decreasing mechanical properties. Also the shearing zone can also be seen in all micrographs and with the wt% of almond shell particles shearing zone and layering of materials can be seen. Also in the micrographs of 30 wt% and 40 wt% of almond shell particles some voids and cracks can also be seen signifying brittle cracking and brittle failure which leads to the poor mechanical properties as seen in Figures 10d and 10e.

Conclusion

When 1 wt% of DNR is blended in epoxy resin then there is a substantial growth in the toughness of epoxy resin as well as other tensile properties. Hardness also increases with blended DNR. Flexural strength and compressive strength increases with increasing wt% of DNR till 1% and after that decreases. SEM showed significant dispersion of DNR till 1% and after that DNR started accumulating at one place. Thus, decreasing the Mechanical Properties. Therefore, it can be concluded that 1 wt% of DNR is an optimum concentration for further mixing of reinforcing particles and fibers and 20 wt% of almond shell particles is the optimum concentration.

Acknowledgement

We are very thankful to Dr. P. C. Gope, Professor and Head, Department of ME (GBPUA&T), Pantnagar for providing guidance and complete infrastructure for performing various tests and experimentation. Also I am thankful to GEU and GBPUA&T.

References

1. Huang Y, Kinloch AJ (1992) The toughness of epoxy polymer containing microvoids. Polymer 33: 1330-1332.

2. Bucknall CB, Gilbert AH (1989) Toughening tetra functional epoxy resins using polyetherimide. Polymer 30: 213-217.

3. Chuayjuljit S, Soatthiyanon N, Potiyaraj P (2006) Polymer blends of epoxy resin and epoxidized natural rubber. J Appl Polym Sci 102: 452-459.

4. Ismail Z, Ahmad MI, Zakaria FA, Anita R, Marzuki HFA, et al. (2006) Modification of epoxy resin using liquid natural rubber. J Mater Sci 517: 272-274.

5. Kumar KD, Kothandaraman B (2008) Modification of (DGEBA) epoxy resin with maleated depolymerised natural rubber. Polym Lett 2: 302-311.

6. Bhowmick AK, Stephens HL (2001) Handbook of elastomers. (2ndedn) Marcel Dekker, New York.

7. Nakason C, Kaesaman A, Yimwan N (2003) Preparation of graft copolymers from deproteinized and high ammonia concentrated natural rubber latices with methyl methacrylate. J Appl Polym Sci 87: 68-75.

8. Huang Y, Kinloch AJ (1992) The sequence of initiation of the toughening micromechanisms in rubber-modified epoxy polymers. Polymer 33: 5338-5340.

9. Nakason C, Kaesaman A, Supasanthitikul P (2004) The grafting of maleic anhdride onto natural rubber. Polym Test 23: 35-41.

10. Irfan MH (1998) Chemistry and technology of thermosetting polymers in construction applications. Kluwer academic publishers, Boston.

11. Singh VK (2002) Experimental investigation of mixed mode stress field parameters under biaxial loading condition.

12. Singh VK, Gope PC (2010) Silica-Styrene-Butadiene Rubber Filled Hybrid Composites: Experimental Characterization and Modeling J Reinf Plast Comp 29: 2450-2468.

13. Singh VK (2009) Dielectric properties enhancement of PVC nanodielectrics based on synthesized ZnO nanoparticles. G. B. P.U.A & T, Pantnagar.

14. Chaudhary AK, Singh VK, Gope PC (2012) Effect of Almond Shell Particles on Tensile Property of Particleboard. J Mater Envirn Sci 4: 109-112.

14

Drilling Process Design for Hybrid Structures of Polymer Composites over Titanium Alloy

El-Gizawy A. Sherif [1,2]*, Khasawneh FA[2], and Bogis Haitham[1]

[1]*Center of Excellence for Industrial Design and Manufacturing Research (CEIDM) Mechanical Engineering, King Abdulaziz University Jeddah, Saudi Arabia*
[2]*Industrial and Technological Development Center Mechanical and Aerospace Engineering, University of Missouri-Columbia Columbia, Missouri-65211, USA*

Abstract

This work aims at determination of optimum drilling process design for hybrid structures of polymer composites over titanium alloy in order to reach the needed quality and cost effectiveness for the aerospace industry. A set of experiments are designed to investigate the effects of process variables on the required torques and thrust forces and quality of the drilled holes. Surface response methodology is used to analyze the results. Process maps are introduced based on the experimental results and the optimum conditions for producing quality holes. Evaluation of the presented approach for process design is conducted using carbon fiber reinforced epoxy (IM7/977-3) composites over 6Al-4V titanium alloy (AB1) structure. The proposed study helps in approving the effectiveness of the new approach and in exploring the global optimum drilling parameters for damage free production of aerospace hybrid structures.

Keywords: Drilling process; Hybrid structure; Polymer composites over titanium alloy; Process maps, Response surface methodology

Introduction

New designs for modern aircraft require the use of hybrid structure, a combination of the ultra-light weight polymer composite skin with the high strength Titanium stiffeners. Such designs necessitate drilling of large number of holes in both materials during the assembly of aerospace hybrid structures as in the case of the newly developed Boeing 787. Several investigators have reported results earlier on the machining of single layer polymer composites. Emanuel and El-Gizawy have conducted extensive work in characterization of process behavior in drilling polymer-matrix composites [1-3]. Figure 1 summarizes their results on process-induced damage characterization during drilling of carbon fiber reinforced epoxy composites. It displays the effect of different rotational speed and feed of the drilling tool on generated damages (cracks and delamination of the reinforced fibers in addition to surface quality of the drilled holes [1]. Caprino and Taglieferi [4] showed that damage induced in the composite during drilling strongly depend on feed rate.

On the other hand, several studies have reported results on the drilling of single layer Titanium Alloys. Sharif and Rahim studied the effects of tool materials and machining parameters on the quality of drilled holes in Titanium Alloy-Ti-6Al-4V [5]. On a recent publication, Faqueh and El-Gizawy [6] concluded study on optimization of dimensional accuracy and surface finish in dry drilling of Ti-6Al-4V single plate. They presented their results on drilling process maps for Titanium Alloy that allow for selecting process conditions that satisfy both the required quality of produced holes and the productivity constraints.

Very few results are available on machinability particularly by drilling of hybrid aerospace structures. Redouane Zitoune et al. [7] carried out drilling trials in carbon-fiber reinforced plastics (CFRP)/aluminum (Grade 2024) stack without coolant, with plain carbide (K20) drills of various diameters. Their qualitative results indicate that dimensional accuracy of the produced hales was diminished with increasing feed rate. On another publication by same authors [8], two types of tungsten carbide drills were used for drilling same materials stack as the one used in their earlier work, one with nano-coating and the other, without nano coating. They concluded that the shape and the size of the chips are strongly influenced by feed rate. The thrust force generated during drilling of the composite plate with coated drills was 10-15% lesser when compared to that generated during drilling with uncoated drills. According to the previous work by Ramulu et al. [9,10], some of the problems encountered with the quality characteristics of drilled composite-Ti stacks include severe tool wear, heat induced damage, hole size, roundness, shape, surface texture, and presence of titanium burrs. In their work, however, Ramulu et al. were interested in the effect of the different drilling parameters on thrust force, torque, and the presence of titanium burrs. The holes' dimensional accuracy under different drilling parameters was not evaluated. Moreover, the entire structure was drilled under the same speed and feed; a clear disadvantageous situation in industry since the composite requires a much higher speed and feed rate than titanium to avoid delamination, and titanium requires slower feed rates to avoid excessive tool wear and elevated temperatures.

Despite the extensive research in the field, existing results from single-layer machining simulations cannot be applied to multilayer machining cases, especially drilling. Vijayaraghavan et al. [11,12] discussed the challenges in modeling the machining of aerospace multilayered materials, which includes metal-composite stack ups. They showed that the modeling for multi-layered materials is different from that of single layered materials due to differences such as: multiple steady state assumptions, finite element (FE) models for single layered materials are inapplicable to burr morphology in multi-layered materials, and change of temperature properties across the work piece. The multilayer problem also brings about new machining parameters that do not exist in single layer machining operations such as clamping position and order of placement of materials for minimum burr formation. Choi [13] studied the effect of clamping position on gap formation during drilling of two sheets of SS 304L. He concluded that

*Corresponding author: El-Gizawy A. Sherif, Center of Excellence for Industrial Design and Manufacturing Research (CEIDM), Mechanical Engineering, King Abdulaziz University Jeddah, Saudi Arabia
E-mail: sherifelg@yahoo.com.

the clamping position only influences the elastic deformation while gap formation is due to plastic deformation that depends on the thrust force. The order of placement of the materials, on the other hand, is sometimes dictated by the nature of the manufacturing operation and cannot be changed. In this study, an approach combines surface response technique and optimization search method, is introduced for understanding the process behavior and selection of optimum parameters for drilling in aerospace multilayer structures.

Experimental System Design

Experimental system configuration

Figure 2 displays schematic of the experimental setup. A computer numerical control (CNC) milling machine was used for the drilling

experiments. CNC codes were written for drilling holes on the planned positions and with the required experiment conditions. An in-house designed and fabricated fixture was bolted onto the CNC Mill table in order to firmly secure the multilayer structure on the device (Figure 3). A Torque/Force sensor, ACCUTORQUE, was selected for the present investigation. The ACCUTORQUE sensor is a strain gauge based stator/rotor sensor capable of measuring the torque and thrust force generated in a variety of machining applications. It consists of three major components: a stator, rotor, and a gain amplifier. In addition to the Torque/force sensor, a data acquisition system is provided in order to collect and organize all data obtained from the sensory system. Calibration of the force/torque sensor was accomplished using an available calibration mechanism. The calibration results for both force and torque displayed linear relationship with the output voltage signals.

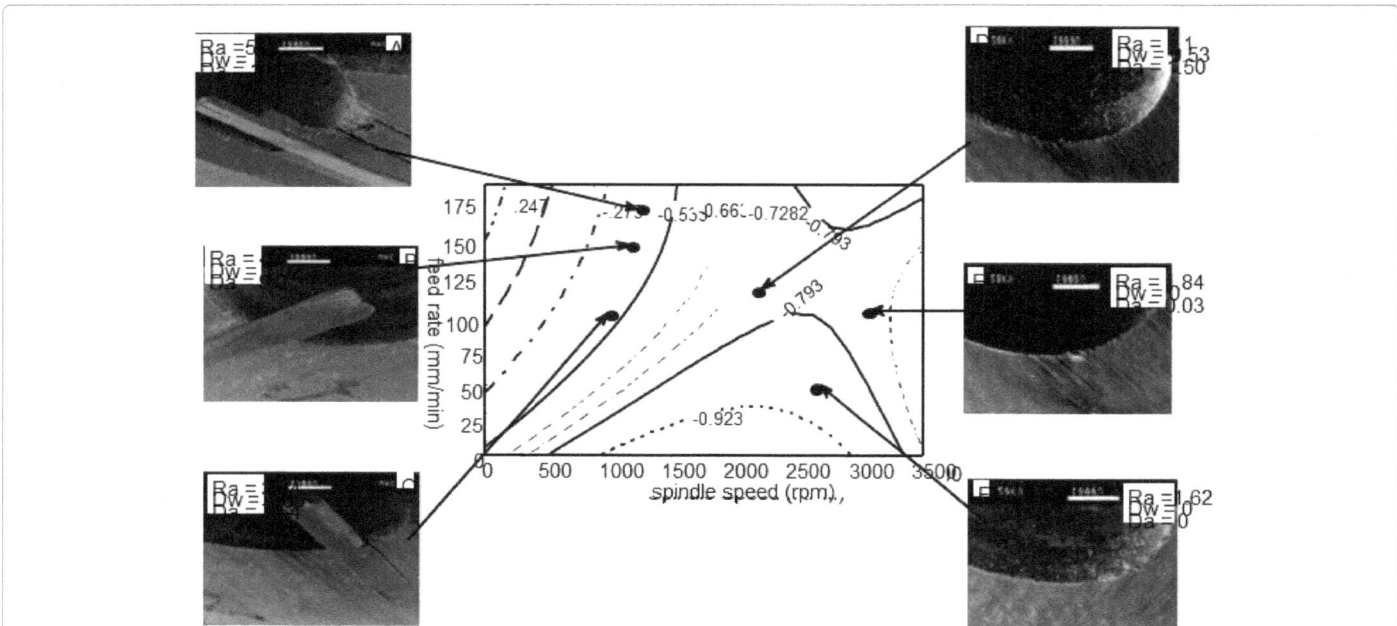

Figure 1: Effects of rotational speed and feed rate on process-induced damage in drilling carbon fiber reinforced epoxy composites [1].

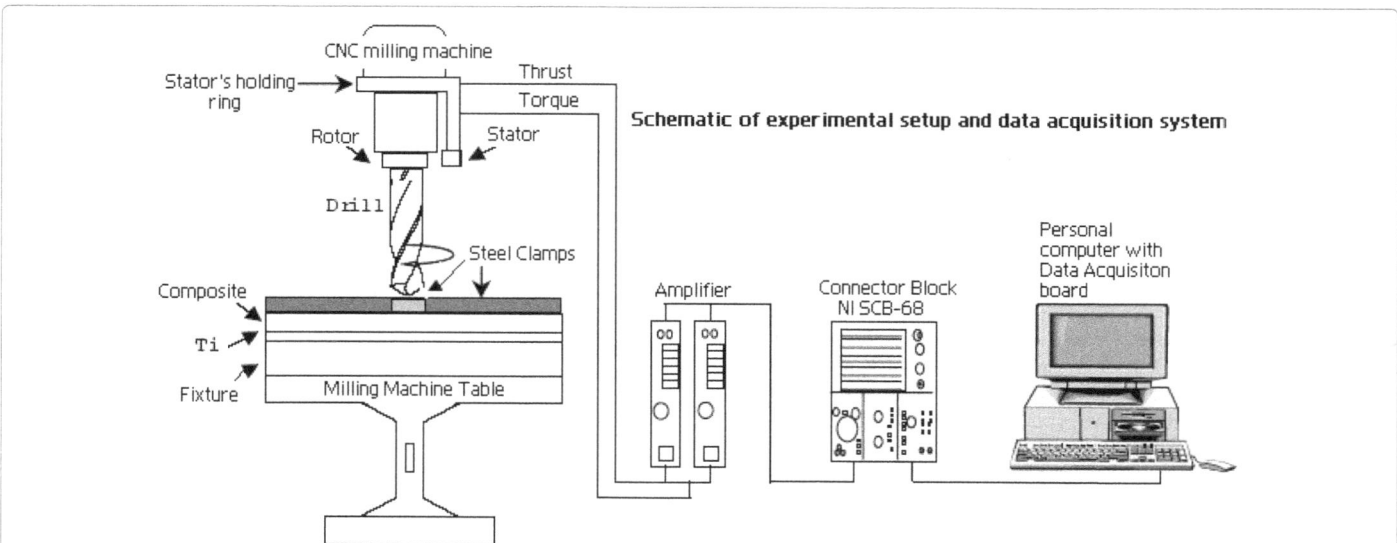

Figure 2: Schematic of the experimental setup.

Measurement techniques

Dimensional accuracy measurement: DP-4 touch probe by CENTROID (patent#6553682) was used to measure holes accuracy (Figure 4). Holes' accuracy was measured for both Composite and Titanium for each individual drill hole. The DP-4 probe was jogged over the center (roughly) of the hole, and then slowly the Z-axis was jogged down until the tip of the probe was inside the hole and not touching anything. The CNC controller was commanded to start the probing cycle. The stylus was jogged by the controller to each quadrant of the hole. It will then return to the center of the hole. A message box will appear on the screen that will display the measured diameter of the hole. The probe diameter and pre-travel distance were calculated by using a ring gauge with a 1.0000 inch (25.4 mm) Diameter before staring the experiment. The accurate compensation value was entered to the tool library in the CNC controller, so the value displayed for the diameter would represent the actual diameter value of the hole with the compensation already accounted for automatically.

Surface roughness measurement: Surface roughness of the drilled holes, were measured using a Mitutoyo Surftest 402 Profilometer which uses a ruby tip to contact the surface. Each hole was measured in four places approximately 90° apart. Table 1 displays the specifications of the used Profilometer.

Experimental Design and Procedures

Statistical design of the experiments

Two different techniques were used in the current investigation. Response surface with central composite design and Partial Factorial Design using Taguchi's Approach (Robust Process Design).

Response surface with central composite design

The response surface methodology (RSM) was used to design a set of experiments to capture the impact of the control parameters (feed and speed) on thrust force, torque, holes accuracy and surface roughness of the drilled holes. RSM methodology is useful for modelling and analysis in cases where the responses of interest are influenced by several factors and where the main objective is to optimize these responses. In this method a low order polynomial (Equation 1) is fitted between the response parameters of the process.

$$Y = \beta_0 + \beta_1 X_1 + \beta_2 X_2 + \beta_{12} X_1 X_2 + \beta_1 X_1^2 + \beta_2 X_2^2 \quad (1)$$

Where X_is are the input variables (speed, feed) that influence the response Y (holes tolerances, surface roughness, torque and thrust forces), β_0 and β_i are estimated regression coefficients. The method of least squares is then used to find the constants of the polynomials. In the present experiment, an RSM with central composite rotatable design with three center points are used. This design addresses the range of feeds and speeds for drilling titanium around the expected optimum values of the operation obtained during preliminary investigation of the same drilling process (Table 1). The cutting speed and feed for drilling the composite plates were kept constant for all tests, at their optimum values of 2300 RPM and 0.00784 IPR respectively. These conditions were established in an earlier comprehensive study by El-Gizawy et al. (Table 2) [1-3,6].

Partial factorial design using Taguchi's approach (Robust process design)

The goal of the Taguchi method is to identify the optimum settings for the different factors that affect the production process to yield a robust operation (minimum variability in performance). In this design

the independent variables were the speed, feed and tool condition. The output or dependent variables were: surface roughness and holes' accuracy. Associated force and torque values for each experiment were also measured and recorded.

The responses obtained from different experiments were analysed using response tables and graphical representation of the mean effects of each parameter on the machinability characteristics. The response analysis helps in identifying those process parameters that have the greatest impact on process variability and its level of performance. In determining this, the Signal-to-Noise (S/N) Ratio Analysis was used. It uses a transformation method to convert the measured response into an S/N ratio. Proposed by Taguchi [14], S/N ratios are performance measures that optimize a process. The S/N Ratio Analysis also provides a sensitivity measurement of the machinability characteristics of a process at various levels of both controllable and uncontrollable or noise factors. The optimum process design is achieved when the S/N

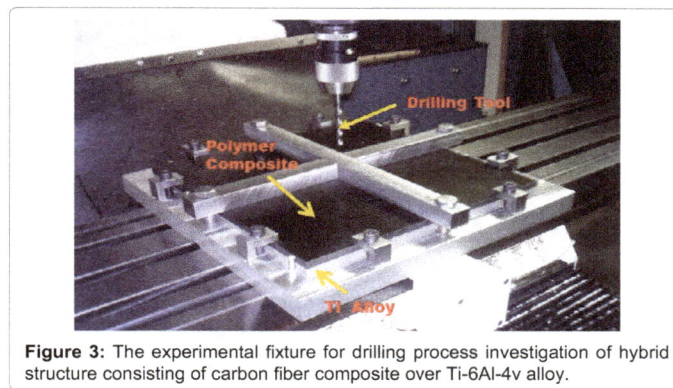

Figure 3: The experimental fixture for drilling process investigation of hybrid structure consisting of carbon fiber composite over Ti-6Al-4v alloy.

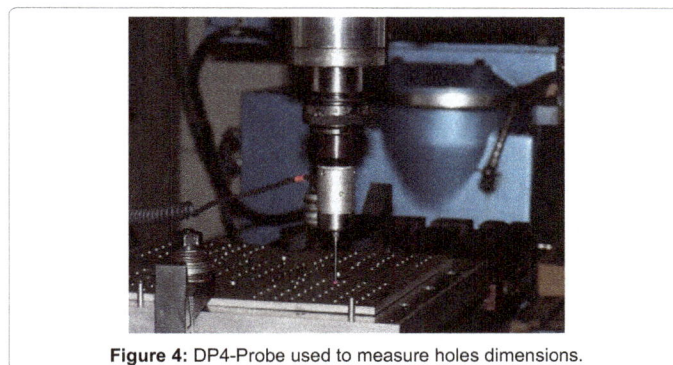

Figure 4: DP4-Probe used to measure holes dimensions.

Stroke	0.3 mm
Linearlity	0.2 mm
Tip shape	Conical of 90°
Tip radius	5 µm
Force variance ratio	8 µm/1 µm
Curvature of radius of skid	30 mm(1.18")
Measuring force	4 mN or less

Table 1: Surface roughness measurements conditions.

Exp. #	1	2	3	4	5	6	7	8	9(c)	10(c)	11(c)	12(c)	13(c)
Speed (rpm)	500	500	600	600	480	621	550	550	550	550	550	550	550
Feed (ipr) *10^-3	1	1.44	1	1.44	1.2	1.2	0.89	1.51	1.2	1.2	1.2	1.2	1.2

Table 2: Experimental matrix using RSM.

ratio is maximized. In other words, it is the process condition at which the variability, resulting from the uncontrollable factors, is minimized. In this phase of investigation, three levels were used for both: speed and feed. The expected optimum value was obtained from the first phase using RSM experiments, and the other lower and upper levels were obtained by decreasing and increasing the optimum value by 20%, respectively. Two levels were used for the tool condition either new or half-life tool. A half-life tool was used previously to drill 26 holes. A third level for tool condition (new tool) was also introduced in the experimental design in order to maintain balance of levels of all variables (orthogonal design). Tables 3 and 4 display the coded experiment matrix and the experimental log, respectively.

Investigation procedures

Investigated materials and tools: Composite Plates, made out of carbon fibers reinforced resin (IM7/977-3) were used. These composite plates were designed, fabricated and provided by The Boeing Company, St. Louis, Missouri after they were cut to dimension by water jet (12" x 12"x 0.379").

Titanium Plates, AMS-9046 plates were used. These are made out of 6Al-4V titanium alloy, also known as AB1. The yield strength is 120 ksi, while the ultimate strength is 130 ksi and elongation is 10%. These plates were also supplied by The Boeing Company, after being cut to dimension by water jet (12"x 12"x 0.279").

The Drill Bits, used are GUH051-00732006.350 manufactured by Guhring from Albstadt, Germany. It is a solid carbide ¼" drill with 118 degree point angle.

Experimental methods: The investigated hybrid structures of polymer composites over titanium alloy were first clamped together and secured inside the drilling fixture. Testing was then conducted by drilling the composite first at the recommended optimum conditions. No lubricant was used (dry drilling), and a vacuum was used for cooling and composite dust collection. Each experiment was replicated two times. Torque and force signals were recorded for all the holes in order to have more understanding of the process behavior. After the holes were drilled in the composite, Titanium was drilled with the holes having the same conditions drilled one after the other to minimize the error among replications due to tool wear. The lubricant supplied by Boeing was always used when drilling Titanium.

Following the drilling of all the holes, the diameters of the drill holes were measured, the readings of each pair of replications were averaged then the result was subtracted from the nominal diameter value of ¼" to get the diameter deviation from the nominal value. The surface roughness, R_a, of the drilled holes in the composite was measured for each hole using a Profilometer. Two readings 90° apart were taken then averaged for each hole after outliers were ignored. A significant variation in surface roughness was noticed in some cases among replications, so the average value of readings among each pair of replications was used in the analysis. After the data was collected, the results were interpreted using expert statistical software. The surface equations were obtained and differentiated to obtain the optimum condition of each of the independent variables, (speed and feed), corresponding to each of the dependent variables, (force, torque, hole accuracy and surface finish). The drilling procedure for Taguchi's approach followed the same procedure used for RSM.

Results and Discussions

Characterization of process behavior and optimization

Process maps for drilling hybrid structures of polymer composites over titanium alloy: Figure 5 displays a typical three dimensional surface relating the effect of the independent variables (speed and feed) on one of the quality characteristics of the process. In this plot, the maximum force value measured during drilling of titanium was selected as the major response. The process contour map for thrust force extracted from the surface plot is shown in Figure 6. Similar surface plots for the torque, the deviation of the diameter from the nominal value and surface roughness, were also generated from the results. The corresponding process contour maps were also constructed in the same fashion as the thrust force map displayed in Figure 7. Figure 8 represents a contour map for the generated torque during drilling of the composite plate. Figure 9 displays process contour maps of deviation of diameter of the machined holes (inch), from the nominal value in the composite plate. Figure 10 displays process contour maps of the surface roughness R_a, (μ-in) of holes in the composite. Figures 9 and 10 reveal the significance of the selected machining parameters for titanium in controlling dimensional accuracy and surface finish of the holes drilled in the composite plates. The results indicate that machining condition for titanium that lead to longer continuous chips will be detrimental to dimensional accuracy and surface finish of holes machined in the composite plate. Titanium chips smear and alter the interior surface of the composite holes.

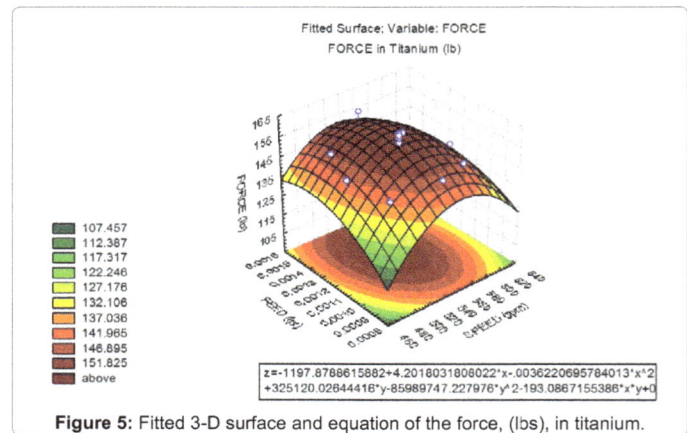

Figure 5: Fitted 3-D surface and equation of the force, (lbs), in titanium.

Table 3: Coded experimental matrix.

Exp. #	Speed (rpm)	Feed (ipr)	Tool
1	1	1	1
2	1	2	2
3	1	3	1'
4	2	1	2
5	2	2	1'
6	2	3	1
7	3	1	1'
8	3	2	1
9	3	3	2

Table 4: Experimental Log with actual values.

Exp. #	Speed (rpm)	Feed (ipr)	Feed (ipm)	Tool
1	496.56	0.000832	0.413138	T_{new}
2	496.56	0.00104	0.516422	$T_{1/2}$
3	496.56	0.001248	0.619707	$T_{new(d)}$
4	620.7	0.000832	0.516422	$T_{1/2}$
5	620.7	0.00104	0.645528	$T_{new(d)}$
6	620.7	0.001248	0.774634	T_{new}
7	744.84	0.000832	0.619707	$T_{new(d)}$
8	744.84	0.00104	0.774634	T_{new}
9	744.84	0.001248	0.92956	$T_{1/2}$

Desirability contours: A typical problem in process development is to find the set of levels of the controlled or independent parameters that yield the most desirable characteristic of the product. The procedure used to tackle this problem involves two steps [6]:

1. Predicting responses on the dependent parameters by fitting the observed response using an equation based on the levels of the independent variables.

2. Finding the levels of the input variables that simultaneously produce the most desirable predicted response on the output variables.

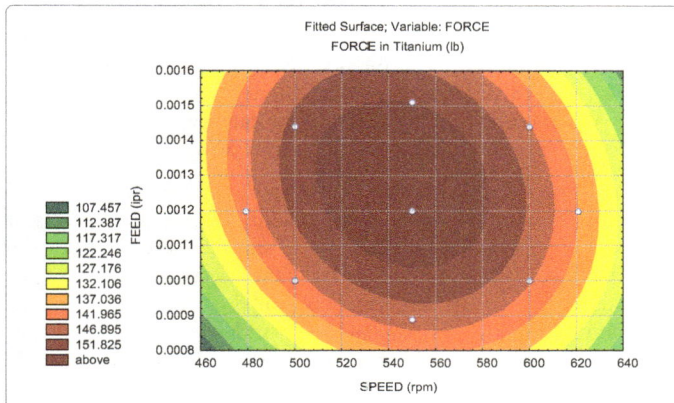

Figure 6: Process contour maps of thrust force (lb.), during drilling titanium.

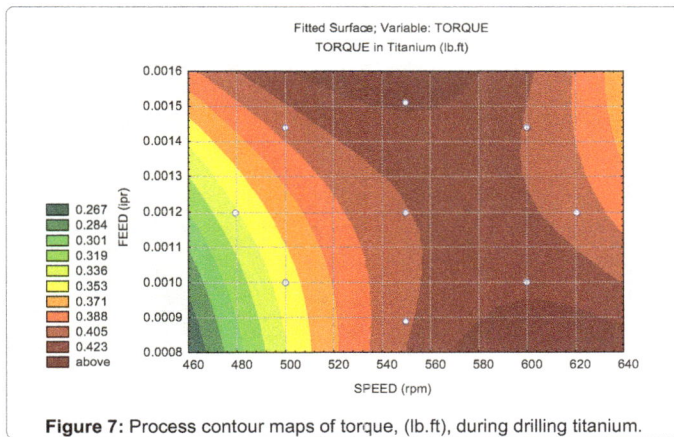

Figure 7: Process contour maps of torque, (lb.ft), during drilling titanium.

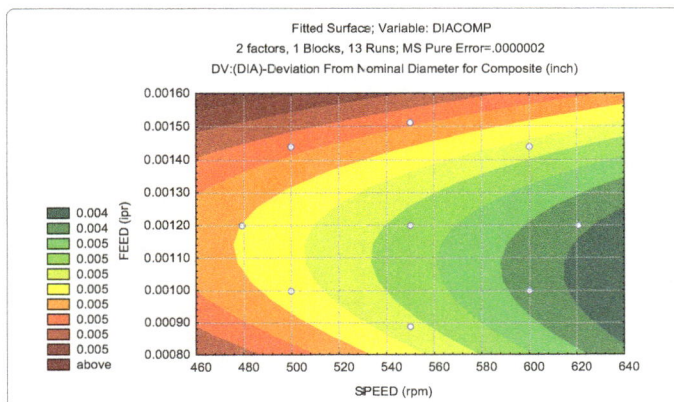

Figure 8: Process contour maps of deviation of diameter, (inch), from the nominal value in the composite.

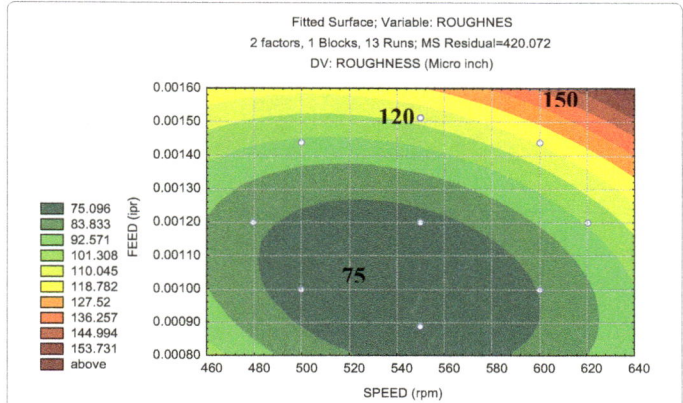

Figure 9: Process contour maps of the surface roughness R$_a$, (μ.in), of holes in the composite.

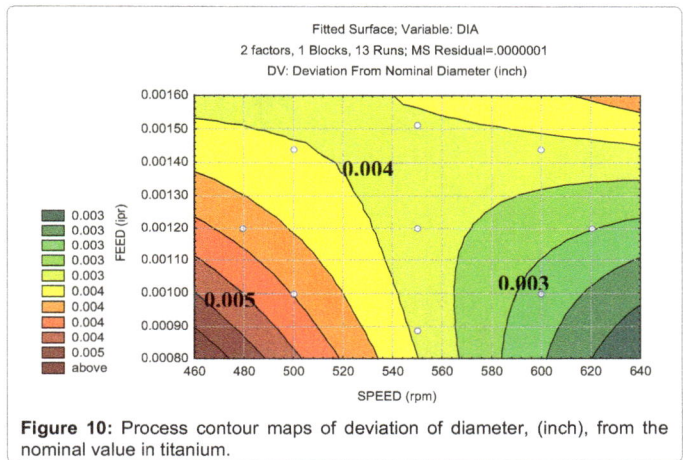

Figure 10: Process contour maps of deviation of diameter, (inch), from the nominal value in titanium.

A prediction profile for a dependent variable consists of a series of graphs, one for each independent variable, of the predicted values for the dependent variable at different levels of one independent variable, holding the levels of the other independent variables constant at specified values, called current values. Inspecting the prediction profile can show which levels of the predictor variables produce the most desirable predicted response on the dependent variable. The number of levels at which to compute predicted values for each independent variable was set to twenty grid points.

It can be seen in Figure 11 that the desired minimum value of force corresponds to a feed of 0.00151 ipr, and a speed of 620.71 rpm. However, from the convergence plot at the bottom of the graph, it can be seen that the force is very close to convergence at rather a much lower feed of 0.00089 ipr. Same thing is noticed about the convergence for the optimum speed. It is very close to converging at a lower speed of about 479.29 rpm. This delay in convergence is attributed to experimental errors. Having more repetitions should lead to better convergence behavior.

As for the torque, Figure 12 shows that the optimum conditions to obtain minimum torque are 0.00089 ipr feed, and 578.28 rpm speed. At higher rpm the desirability function stays almost the same indicating that after reaching the optimum speed any increase in speed won't reduce the torque value any further.

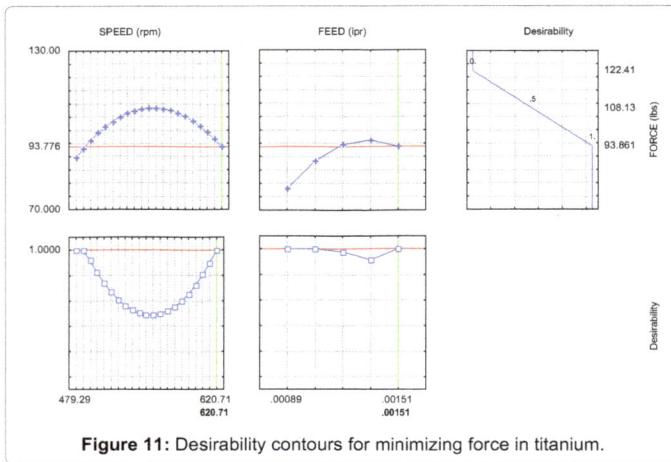

Figure 11: Desirability contours for minimizing force in titanium.

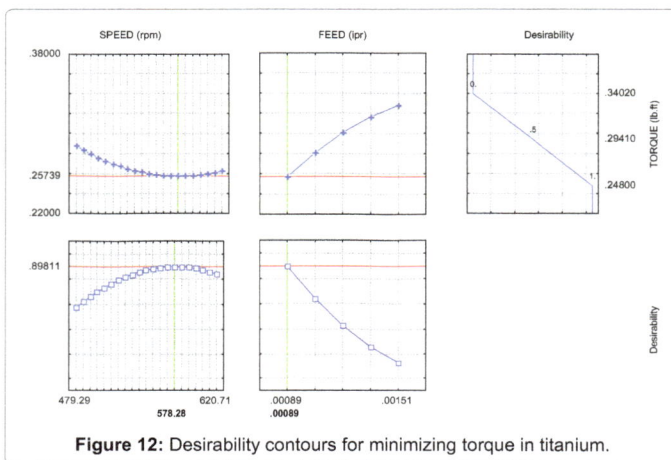

Figure 12: Desirability contours for minimizing torque in titanium.

Robust process design: The S/N ratio was calculated for each output observation using the following equation:

$$\eta_i = -10\log_{10} y_i^2$$

Where η_i is the ith transformed output, y_i is the ith untransformed output. After all the data was transformed using the above expression, the effects of each factor were calculated and tabulated. The effect of each factor was calculated using the expression:

$$A_i = \left(\sum_{j=1}^{n} A_{ij} \right) / n$$

Where A_i is the effect of factor A at level i. and n is the number of observations where level i of factor A occur. A_{ij} represent transformed outputs of factor A at level i. Table 5 and Figure 13 display main effects of different control parameters and their investigated levels calculated using the above equations with the diameter data in the composite. The analysis indicates that a speed of 496.56 rpm, a feed of 0.001248 and a sharp tool yield a process that will produce robust hole tolerances in the composite.

Holes surface roughness in composite is the other quality characteristic considered in this analysis. The results are plotted in Figure 14. The general conclusion is that low speed, intermediate feed and new (sharp) tool would give robust performance with very low variability in surface roughness in machined holes of the composite side. Specifically, a speed of 496.56 rpm, a feed of 0.00104 inch/rev.,

and a sharp tool in drilling Titanium will yield robust surface finish in the composite.

Conclusions

An experimental approach for development of damage-free drilling of hybrid structures of polymer composites over titanium alloy was established and verified. The present approach involves statistical design of experiments to develop the process knowledge base and multi-objective optimization techniques in order to account for the contribution of the major process quality responses (dimensional accuracy, surface finish and process-induced defects) and to allow for an effective trade-off among all competing machinability characteristics. Evaluation of the present approach was conducted using carbon fiber reinforced epoxy (IM7/977-3) composites over 6Al-4V titanium alloy (AB1) structure. The study proved the effectiveness of the new approach in selecting the global optimum drilling parameters for damage free production of hybrid structures of polymer composites over titanium alloy.

The results generated from the present approach were used for constructing process maps for the machinability of hybrid structures of polymer composites over titanium alloy. These maps are effective tools that can be used by industry as road maps in selecting process designs that satisfy both quality requirements and productivity constraints. Optimum drilling process design for hybrid structures of polymer composites over titanium alloy has been explored. The present study reveals the followings:

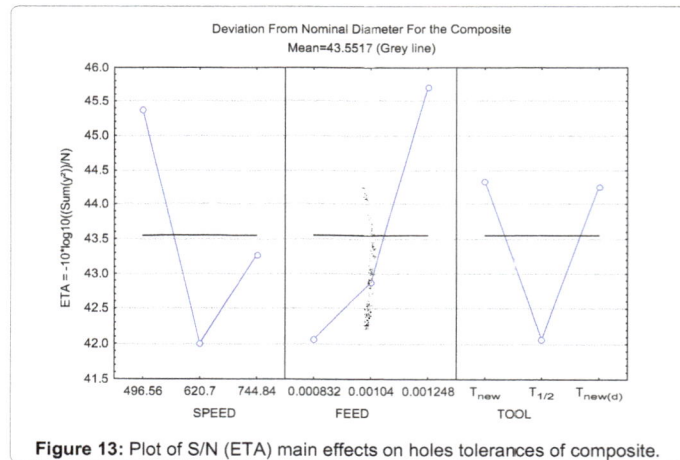

Figure 13: Plot of S/N (ETA) main effects on holes tolerances of composite.

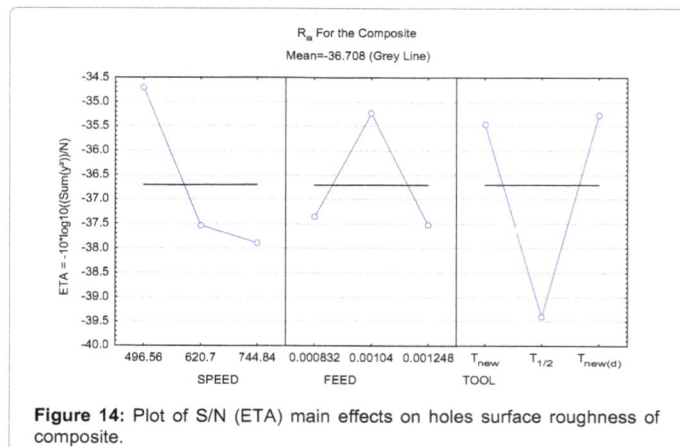

Figure 14: Plot of S/N (ETA) main effects on holes surface roughness of composite.

Composite Exp. #	Speed(rpm)	Feed(ipr)	Feed(ipm)	Tool	Untransformed Dev From Nom.(y_i)	Transformed Dev. From Nom. (η_i)
1	496.56	0.000832	0.413138	T_{new}	0.00580	44.731440
2	496.56	0.00104	0.516422	$T_{1/2}$	0.00805	41.884082
3	496.56	0.001248	0.619707	$T_{new(d)}$	0.00335	49.499104
4	620.7	0.000832	0.516422	$T_{1/2}$	0.00965	40.309454
5	620.7	0.00104	0.645528	$T_{new(d)}$	0.00785	42.102607
6	620.7	0.001248	0.774634	T_{new}	0.00660	43.609121
7	744.84	0.000832	0.619707	$T_{new(d)}$	0.00875	41.159839
8	744.84	0.00104	0.774634	T_{new}	0.00585	44.656883
9	744.84	0.001248	0.92956	$T_{1/2}$	0.00630	44.013189

Table 5: Experimental data for holes tolerances of composite.

a. In general high speed of 2300 rpm and low drilling feed of 0.00784 inch per revolution are recommended for the production of delamination-free and good surface finish holes with the required dimensional accuracy, in the epoxy composites.

b. An optimum drilling speed of 600 rpm and feed of 0.0012 inch per revolution are recommended for production of quality holes in the titanium layer while maintaining the quality of the holes initially generated in the composite layer.

c. Drilling the titanium layer with a lower speed of 500 rpm and a feed of 0.0010 inch per revolution while maintaining sharp tools during the process would lead to a robust process performance with minimum variability.

Acknowledgements

This work was funded by the Deanship of Scientific Research (DSR), King Abdulaziz University, under grant number (135-590- D1435). The authors, therefore, acknowledge the technical and financial support of King Abdulaziz University. The authors wish also to acknowledge the experimental support of Industrial Technology Development Centre at University of Missouri.

References

1. Enemuoh E, Sherif EA, Okafor T (2001) An Approach for Development of Damage-Free Drilling of Carbon Fiber Reinforced Thermosets. Int Journal for Machine Tools and Manufacture 41: 1795-1814.

2. Enemuoh EU, Sherif EA (2003) Optimal Neural Network Model for Characterization of Process-induced Damage in Drilling Carbon Fiber Reinforced Epoxy Composites. Machining Science and Technology Marcel Dekker 7: 389-400.

3. Enemuoh E, Sherif EA, Okafor T (1997) Multi-Sensor Monitoring of Drilling Advanced Composites. Smart Structure and Materials SPIE 3042: 410-420.

4. Caprino V (1995) Damage Development in Drilling Glass Fiber Reinforced Plastics. Int Journal Machine Tools and Manufacture 35: 817-829.

5. Sharif S, Rahim EA (2007) Performance of Coated- and Uncoated-carbide Tools when Drilling Titanium Alloy Ti-6Al-4V. Journal of Materials Processing Technology 185: 572-576.

6. Faqueeh A, Sherif EA (2015) Optimization of Multiple Quality Characteristics for Dry Drilling of Ti-6Al-4V Using Coated Carbide Tool. Int J Mater Manufacturing 8: 172-179.

7. Redouane Z, Vijayan K, Francis C (2010) Study of drilling of composite material and aluminium stack. Composite Structures 92: 1246-1255.

8. Redouane Z, Vijayan K, Krishnaraj B, Sofiane A, Francis C, et al. (2012) Influence of Machining Parameters And New Nano-Coated Tool on Drilling Performance of CFRP/Aluminum Sandwich. Composites Part B Engineering 43: 1480-1488.

9. Ramulu M, Branson T, Kim D (2001) A Study on the Drilling of Composite and Titanium Stacks. Composite Structures 54: 67-77.

10. Kim D, Ramulu M (2004) Drilling process optimization for graphite/bismaleimidetitanium alloy stacks. Composite Structures 63: 101-114.

11. Vijayaraghavan A, Dornfeld D (2005) Challenges of Modeling Machining of Multilayer Materials. Proceedings of 8th CIRP International Workshop on Modeling of Machining Operations.

12. Vijayaraghavan A, Dornfeld D (2006) Quantifying Edge Effects in Drilling FRP Composites. Transactions of NAMRI/SME 34: 221-228.

13. Choi J, Min S, Dornfeld D, Alam M Tzong T (2005) Modeling of Inter-layer Gap Formation in Drilling of a Multilayered Material. Proceedings of 8th CIRP International Workshop on Modeling of Machining Operations.

14. Phadke MS (1989) Quality Engineering Using Robust Design. Prentice-Hall, Englewood, California.

15

Layered Double Hydroxides: Tailoring Interlamellar Nanospace for a Vast Field of Applications

Richetta M*, Medaglia PG, Mattoccia A, Varone A and Pizzoferrato R
Department of Industrial Engineering, University of Rome "Tor Vergata", Rome, Italy

Abstract

Fifty-eight years ago Fenman, during an American Physical Society meeting at the California Institute of Technology, anticipated the problem of modifying and governing the world of the infinitely small. He said: *"What I want to talk about, is the problem of manipulating and controlling things on a small scale... What I have demonstrated is that there is room—that you can decrease the size of things in a practical way. I now want to show that there is plenty of room. I will not now discuss how we are going to do it, but only what is possible in principle... We are not doing it simply because we haven't yet gotten around to it."* Useless to say how profound his sensibility for science was. We've just begun to walk in this enormous field, toward the assembly of devices atom by atom. What we did till now is still rudimentary. Anyhow we believe that Layered Double Hydroxides could play a role in manufacturing these nanometric equipments.

Layered Double Hydroxides (LDHs) are 2D ionic lamellar nano-materials belonging to the group of anionic clays. Their structure consists of positively charged brucite-like layers and intercalated anions. The layered structure, together with the flexibility to intercept different anionic species in variable compositions, both inorganic and organic, has attracted increasing interest. In order to meet specific requirements in very distant fields, considerable efforts were made to tailor the physical/chemical properties of LDHs and to design engineered LDH for several applications, ranging from anticorrosion coatings, flame-retardants, catalysis, to water treatment/purification, and biomedical applications. Furthermore they have been applied in energy harvesting and conversion, thanks to the possibility of substituting the composing metals with transition metals.

Within the framework of this contribution, we first briefly review the development of synthesis processes (§1). In Paragraph 2, examples of the LDHs applications are reported. We will than focus on our laboratory experimental activities, showing the growth of the structures either on printed circuit tracks for applications of LDHs as gas sensors and biosensors. One more application is in nanostructured-modified textiles.

Keywords: Nano-materials; Biosensors; Anticorrosion coatings; Morphology

Introduction

Layered Double Hydroxides (LDHs) are 2D ionic lamellar materials belonging to the group of anionic clays. The structure of LDHs is based on brucite-like layers containing both divalent M^{2+} and trivalent M^{3+} cations coordinated to six OH- hydroxyl groups [1,2]. It can be expressed by the formula $[[M^{2+}1-xM^{3+}x(OH)_2]_x^+(A^{n-})x/n]mH2O$. The substitution of the divalent M^{2+} metal cation with a trivalent M^{3+} one, due to the charge disequilibrium, gives rise to the infinite repetition of positively charged sheets (lamellas) alternating with An- ions (Figure 1) which are required to balance the net positive charges of the hydroxide layers. The molar ratio $x=M^{3+}/(M^{3+}+M^{2+})$ generally ranges within the interval (0.2- 0.4).

The layered structure has attracted increasing interest since it makes LDHs the ideal system for pursuing material engineering at a nanometric scale by using chemical methods. Experimental results obtained so far have shown that this statement is mainly based on three reasons. First, researchers have a large variety of choice for the two cation metals, with a consequent wide range of structural, physical and chemical properties; second, the repetition of interlamellar domains enables designing a solids with an enormous surface/volume ratio, i.e. with an accessible internal surface up to 800 m^2/g., Third, and most importantly, LDHs can host and exchange mixed valence metal ions, water molecules and even relatively complex organic molecules, intercalated in the inter-lamellar space. These almost unique physical/chemical properties have been exploited to shape LDHs and meet specific requirements for LDHs in very distant fields. In fact, engineered LDHs can find a huge variety of applications.

For instance, by intercalation with appropriate anionic molecules, they have been investigated as additives in super hydrophobic surfaces [1], anticorrosion coatings [2,3], flame-retardants [4], catalysis [5], water treatment and purification [5,6], biomedical applications (e.g. drug delivery and biosensors) [7,8], and in many other fields. Furthermore, by using transition metals as the component cations, they have been suggested as promising materials in energy harvesting and conversion [9].

Within the framework of this contribution, we first briefly review structures and compositions (§1), the development of synthesis processes (§2). In Paragraph 3, examples of the most recent applications of LDHs are reported. We will than focus on our experimental activities, reporting on the growth of LDH nanostructures on different substrates, either for biosensor or for physical/chemical microsensors. Results of different characterization techniques will also be reported. Section 4 will be addressed to the possible related applications of LDHs as gas sensors, biosensors, and nanostructured-modified textiles.

*Corresponding author: Maria Richetta, Department of Industrial Engineering, University of Rome "Tor Vergata", Via del Politecnico 1, 00133, Rome, Italy
E-mail: richetta@uniroma2.it

The results obtained in the creation of new LDH based nanomaterials are certainly encouraging. Nevertheless precise control of chemical composition, morphology and size is still a challenge. Those requirements highlight the need for new characterization techniques to understand the LDHs growth mechanism at a deeper and more comprehensive level, thus promoting the development of new LDHs for high performances applications.

Composition and structure

The structure of an ideal Layered Double Hydroxide is based on $M(OH)_6$ edge-sharing octahedral units forming coplanar brucite-like layers (lamellas) which are stacked on top of one another to form a 3D structures [3-7], as schematically shown in Figure 1. Differently from brucite, which is only formed by neutral magnesium hydroxide, LDH sheets acquire a net positive charge due to the substitution of a certain fraction x of the divalent cations M^{2+} with the trivalent ones M^{2+}. These positive charges are balanced by a variety of possible and exchangeable guest anions intercalated between the lamellas together with water molecules. As a result, a hydrotalcite-like layered structure is formed (Figure 1) which can be represented with the general formula $[[M^{2+}1-xM^{3+}x(OH)_2]x^+(A^{n-})x/n]mH_2O$. There are some constraints for the trivalent to divalent metal ratio x of such a layered structure since the lattice stability requires $0.2 < x < 0.4$. While the upper limit is essentially due to the electrostatic repulsion between neighbouring trivalent cations and the Pauling's rules, a minimum concentration of M^{3+} is necessary since the consequent charge unbalance must make the electrostatic forces maintain the different layers and the whole 3D structure together. A larger range of $0.15 < x < 0.5$ has been suggested by results reported by various authors, but it should be noted that these values are exclusively obtained by chemical analysis.

Metallic cations

LDHs containing various metallic cations have been synthesized with the following divalent cations: Zn, Mg, Mn, Fe, Co, Ni, Cu and Ca; while the trivalent ones are: Al, Mn, Cr, Fe, Co, Ni, and La. LDH-like structure has also been attributed to Li-Al monovalent/trivalent and to Co-Ti divalent/tetravalent couples [3-7]. These lists show that the

lattice stability also requires the ionic radius to lie in the range 0.50-0.74 Å. LDHs have also been prepared with more than two different metallic cations, which give the chance of further variations of the chemical/physical properties. For instance, in $(Mg_{1-x}Zn_x)_2Al-CO_3$ Yamaguchi et al. studied the Zn^{2+} substitution effects on the structural properties, such as the lattice parameter and interlayer distance, the water content and ionic conductivity [8].

The two metallic cations are generally considered as randomly distributed in the layers on an hexagonal framework and the lattice constant a_0 is a linear combination of the ionic radii and the substitution fraction x. Even though there are some evidence of local ordering of cations, still controversial results have been found demonstrating long-range cation ordering of ordered superstructures [4,9]. Therefore, the use of stoichiometric formulas represents only a simplified formalization in that the studied LDHs are essentially non-stoichiometric even at microscopic level.

Intercalated anions

The interlamellar domains of LDHs contain anions, water molecules and, in some cases, other neutral or charged organic/inorganic species. The important characteristics of LDHs is that these moieties are generally weakly bound to the host lamellas and can therefore either be located during the 3D structure formation or be inserted by subsequent substitution trough anionic exchange [10]. The anions can be grouped as follows:

Halides: Cl-, F-, Br-, I-, etc.

Non-metal oxoanions: carbonate, nitrate, sulfate, etc.

Oxometallate anions: CrO_4^{2-}, MnO_4^-, VO_4^{3-}, $Cr_2O_4^{7-}$, $W_{10}O_{32}^{4-}$, etc.

Anionic complexes of transition metals: $Fe(CN)_6^{2-}$, anionic metal porphyrins, Eu complexes, etc.

Organic anions: $-COO^-$, $-SO_4^-$, $-PO_3^-$, etc.

Anionic biomolecules: Aminoacids, enzymes, proteins, DNA, TPA, etc.

Anionic polymers: Poly(styrene sulphonate), PMMA, etc.

Neutral molecules have also been intercalated together with anions, possibly with subsequent dissociation in the water contained in the interlamellar domain as in the case of ionic liquids [11], which can be used to control the ionic conductivity of LDHs or for water splitting.

Interlamellar domains

Structural characterization techniques, such as XRD and EXAFS, and some physical characteristics of LDHs, such as the behaviour of electrical conductivity and the hydration properties, confirm that the interlamellar domains are in a strongly disordered state which has been considered by some authors as a quasiliquid state [12]. This is also due to the significant presence of water the interlamellar domains. In fact, water in the LDHs is generally distinguished as two general types [13]. The first type is the water bound to the interlayer domains and to the external surface of LDHs even under dry conditions; this one is called crystallite water. The second type is the water adsorbed and desorbed under various temperatures depending on the water vapour pressure and humidity conditions and is called adsorbed water. This variable content of water greatly contributes to the disordered nature of the interlamellar domains and favours one of the most spectacular characteristics of LDHs: the anionic exchange properties. It is, in fact, extremely easy to exchange a great number of interlamellar anions with

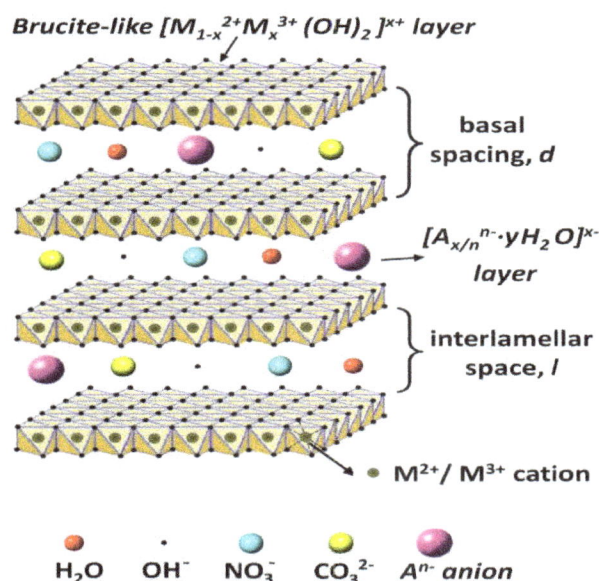

Brucite-like $[M_{1-x}^{2+}M_x^{3+}(OH)_2]^{x+}$ layer

basal spacing, d

$[A_{x/n}^{n-}·yH_2O]^{x-}$ layer

interlamellar space, l

M^{2+}/M^{3+} cation

H_2O OH^- NO_3^- CO_3^{2-} A^{n-} anion

Figure 1: General structure of LDH composites.

a variety of other ions by simply immersing the LDH synthesized with the "old" anions" in a solution with an excess of the "new" anions. As will be discussed in the next chapter, this is an important method for the preparation of novel LDHs. Interestingly, the anionic exchange reaction is revealed by the variations of the interlamellar space, i.e. the distance between two successive hydroxide sheets, which are correlated to the shape and the charge density of the intercalated anions. For instance, in ZnCr $Zn_2Cr(OH)_6A2H_2O$ LDHs the basal spacing can vary from 0.7 nm to 3.3 nm by changing the intercalated anion A [6]. As a result of this adaptability, LDHs can accommodate even relatively large and complex molecules such as proteins. This interlayer spacing, or more specifically, the interlayer distance d, defined as the thickness of a hydroxide sheet plus one interlayer space, can be easily determined by a series of strong XRD (00l) reflections at low angles, as displayed in Figure 2 which reports the X-ray powder diffractogram of a ZnAl-NO_3-LDH. The (110) reflection at high angles, instead, allows the calculation of the lattice parameter a_0, that is the inter-cation distance within the hydroxide layer, while the (01l) and/or the (10l) reflections at intermediate angles enable determining the stacking sequence of the hydroxide sheets, i.e. is the LDH polytype.

Synthesis Routes of LDH

LDHs can be synthesized by using a number of techniques and the choice generally depends on the cations forming the hydroxide layers, the intercalated anions and eventually the desired physicochemical properties of the final material e.g. phase purity, crystallinity, porosity, morphology, and electronic and optical characteristics [3-7]. The different methods can be schematically grouped in two main classes: direct methods and indirect ones. Direct methods include: coprecipitation (often coupled with hydrothermal treatment), salt-oxide method, sol-gel synthesis, electrochemical synthesis, and in-situ film growth. Indirect methods, which are rather additional treatments and modifications of pre-synthesized LDHs, comprise: anion exchange, LDH reconstruction, and delamination.

Co-precipitation

Co-precipitation, also called salt-base method, was the approach used for the first synthesis of mixed double hydroxides in 1942 [14] and is still the most used method to prepare LDHs. It is often coupled with hydrothermal treatment to improve the crystallinity and the size of LDH particles [15-17]. Basically, it is a low-temperature chemical reaction occurring within water mixed solution of appropriate proportions of

divalent and trivalent metal salts in the presence of an alkali metal base at controlled temperature, selected and pH value and under vigorous stirring. The presence of the base causes the simultaneous precipitation of the metal cations in the hydroxide form.

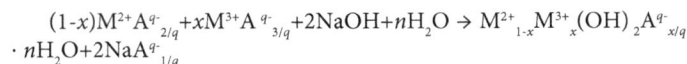

$$(1-x)M^{2+}A^{q-}_{2/q}+xM^{3+}A^{q-}_{3/q}+2NaOH+nH_2O \rightarrow M^{2+}_{1-x}M^{3+}_x(OH)_2A^{q-}_{x/q} \cdot nH_2O+2NaA^{q-}_{1/q}$$

Where, A represents the specific anion. Specifically, the reaction produces the condensation of hexaaqua complexes in solution, which leads to the formation of the brucite-like layers with evenly dispersed metallic cations and interlamellar anions. The dispersed LDHs particles, which form in the solution eventually, precipitate producing white slurry at the bottom of the reactor. The precipitate is collected by filtration, then washed and dried in oven. Thermal or hydrothermal treatment can then be performed to improve the crystallinity. The coprecipitation method can be performed either at variable or constant pH, corresponding to high and law supersaturation conditions. In the first case the base solution is progressively added to the mixed solution of divalent and trivalent metal salts. M^{3+} hydroxides initially form followed by synthesis of LDHs as further addition of base is performed. Alternatively, in the constant-pH approach, the base solution is added simultaneously with the mixed salt solution carefully controlling the pH value. The latter method permits to achieve a higher chemical homogeneity, provided the pH value is kept constant within the appropriate range determined by the nature of each specific couple of metal cations [5,6]. In general, however, coprecipitation presents the great advantage of providing a direct pathway to prepare an LDH with definite composition. It also offers a high degree of control for synthesis parameters such as temperature, pH, ratio of cations, solution concentration, and aging time. In this way, well-crystallized LDH phases and a good tuning of the M^{2+}/M^{3+} ration can generally be obtained. Moreover, and most importantly, intercalation of different anionic species in the interlamellar spaces can easily be achieved by using the appropriate counterion in the metal salts and/or by dissolving the specific anionic species, inorganic or organic, in the same water solution of the synthesis [4].

The salt-oxide method

The salt-oxide method is basically a solid-liquid reaction between an aqueous suspension of the divalent metal oxide and the aqueous solution of the trivalent metal chloride in excess. This method was introduced in 1977 to synthesize Zn-Cr-Cl LDHs by making a suspension of ZnO react with a 1 M solution of $CrCl_3$ which was added at constant period of times under vigorous stirring for few days at room temperature [18]. During this slow reaction, the characteristic drops in pH at each addition, quickly recovered to initial value by the buffering nature of the zinc oxide, stops when an excess of chromium chloride no longer reacts. Since its first use, this method has been extended to Zn-Al-Cl and Cu-Cr-Cl LDHs while attempts with other compositions were unsuccessful [5,6]. In fact, while salt-oxide synthesis generally gives rise to a high degree of crystallinity, it does not make it possible to vary and determine the stoichiometry as much as the co-precipitation method.

The sol-gel method

In the sol-gel method, the synthesis of mixed hydroxides starts from the appropriate metal-based alkoxides and/or acetylacetonides, which are used as precursors for a sol-gel transition occurring in water-ethanol solution. The transition is accomplished by a strong acid hydrolysis with HCl or HNO_3. This approach was first introduced in

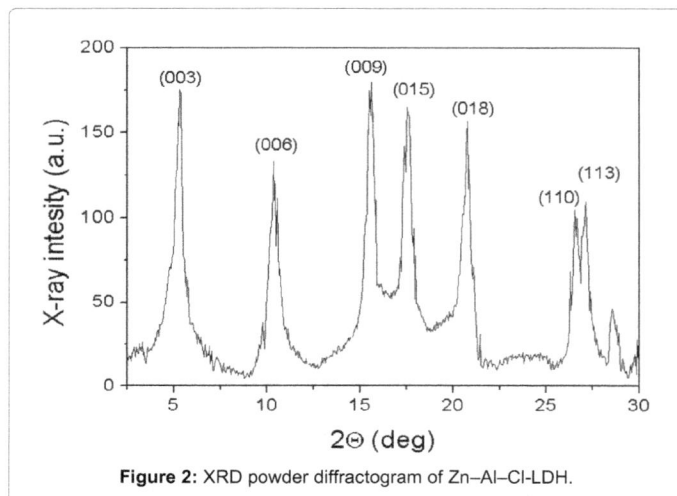

Figure 2: XRD powder diffractogram of Zn–Al–Cl-LDH.

1996 for the synthesys of Mg-Al-CO₃ LDHs in the attempt, successfully accomplished, of increasing the M^{3+}/M^{2+} ratio [19]. More recently, the method has been extended to transition metals such as Ni, Cr and Co [20-22] and demonstrated greater thermal stability during calcinations [23].

Electrochemical deposition

Kamath et al. in 1994 first achieved electro-deposition of Ni-Al, Ni-Mn, Ni-Cr and Ni-Fe LDH films [24] on the working electrode of a conventional three-electrode setup containing a mixed-metal nitrate bath. This is a one-step technique, recently extended to other metal couples [25-29], which exploits the electrically induced reduction of nitrate ions to produce hydroxide ions on the working electrode with a consequent increase of the local pH value which, in turn, results in precipitation of LDH films. The great interest of this method lies in the fact that it makes it possible the direct deposition, with good adhesion, of LDHs films on metal substrates with a relative control of film density, thickness and morphology. With the conventional chemical methods, instead, the LDH powder obtained from the reaction and the subsequent treatments has to be deposited on the substrate surface with physical methods which do not provide the same reliability and reproducibility of adhesion and morphology. Moreover, electro-deposition of LDHs has demonstrated an effective way to produce modified on electrodes for a huge field of applications in electrochemical sensing.

In-situ film growth

This is one more one-step method for direct deposition of LDH films with good adhesion and controllable morphology [30-32]. It could be considered a sort of modified co-precipitation method in that one of the two metals is not provided by the respective dissolved salt but instead by the substrate itself. In fact, the substrate acts both as the reacting source of metal and as a partially sacrificing substrate for the film which grows on it. This approach is even simpler than electrochemical deposition in that it only requires the immersion of the substrate in the water solution of the other metal salt and a base to control the pH value. Moreover, in principle it enables the deposition of LDH films to any surface, provided a thin layer of the reacting metal covers the surface. Most importantly, the surface can be patterned with the thin metal layer so as to form printed circuits with tracks and pads supporting the growth of LDH, which could be very useful for integrated sensors and devices [33]. On the other hand, so far it is still to be demonstrated a good control of the composition and morphology of the deposited film.

Anion exchange

This indirect method is based on the fact that the lamellar structure of LDH is highly prone to anion diffusion and exchange. In fact, these properties have been exploited to synthesize new LDH phases by using anionic exchange reactions, which can be described as follows:

$$[M^{2+}-M^{3+}-A]+B \rightarrow [M^{2+}-M^{3+}-B]+A$$

Where, the equilibrium constant of the reaction depends on the electrostatic interaction and the free energy in a way that ingoing ions with grater charge density are favoured. For instance, for monovalent ions this gives a comparative list of ion selectivity [4,5]: $OH^->F^->F^->Cl^->Br^->NO_3^->F^->I^-$. In practice, stirring an aqueous suspension of the LDH precursors or of the pre-synthesized LDH in the presence of a large excess of the salt of the anion to be intercalated carries out the reaction.

Reconstruction by rehydration after thermal treatments

LDHs heated to temperature between 500°C and 800°C generally undergo calcination and transform into a solid solution of metal oxides. However, as Miyata et al. [34] found in 1983 in Mg-Al-CO₃, they can rehydrate with an anion-containing water solution and give rise to a new LDH. More recently, this method has been used to intercalate organic anions [35].

The delamination/restacking method

LDHs can be delaminated, i.e. undergo a complete separation of all the component brucite-like sheets, to yield a very stable colloidal solution of mono-dispersed lamellas. If the colloidal dispersion is then gently dried, well-ordered LDHs can be recovered, possibly intercalating different anionic species. Delamination can be performed by a number of methods, especially if organically modified LDHs are used [3,36]. For instance, organic intercalated anions such as meth-oxide, acetate, for-mate and lactate favour delamination in water [37]. Alternatively, nitrate-based [38] or amino-acid intercalated [39] LDHs can be delaminated in form-amide. By drying, well-ordered LDHs and LDH-polymer Nano composites intercalating a number of organic compounds were obtained [37].

Applications

The chemical and physical properties of LDH are currently deeply studied because of their wide range of applications. As it has been pointed out by many review papers [40], LDHs can play significant roles in several different fields, ranging from basic chemistry and pharmacology, medicine, horticulture and environmental remediation to polymeric additive, sensors, surface treatments and industrial fallout.

For an exhaustive treatment of applications such as catalysts and precursors of catalysts, let us refer to the excellent review work of Xu et al. [41] and the references quoted therein.

Therefore within this presentation we will consider significant examples not so much to give a complete picture of possible uses, but to stimulate the curiosity of the scientific community and look for new and challenging applications for these nanostructures.

Polymeric composites

To reduce the flammability of plastics it is now impossible, at least in EU, to utilize PBDE (Polybrominated Diphenil Ether) or PBB (Polybrominated Biphenil), due to the production of toxic by-products connected with their burning process. For this reason halogen free fire retardants have been studied and widely applied. In more recent time LDHs exhibit better properties [42] in flame retardancy. In particular it has been revealed that its decrement is linked to the dispersion of LDH in the polymer matrix, increasing from intercalated LDHs toward exfoliated ones. Evans and Duan [40] prepared a borate-pillared LDH and showed that the intercalation of various anions exhibited a higher smoke suppression. Pereira et al. [43] got a significant flammability reduction of 46% for an A-LDH (adipate-LDH). For a more complete overview, refer to the paper of Matusinovic and Wilkie [44]. Other possible applications of LDHs in polymers fabrication are in the fields of horticulture to increase the greenhouse effect of cover films [40] and biomimetics composites [45].

Biology and medicine

In the last two decades layered double hydroxides received a great interest also in the field of pharmacology and health. Several papers appeared concerning the intercalation of different anion of pharmaceutical interest [46] such as citrate and silicate. The effects with many others molecules as nucleotides, porphyrin, vitamins, etc.

were considered [47-49]. Due to their biocomtability [50] LDHs can be used in medicine as excipient [51], or for drug delivery, with a PH dependent releasing rate [52,53]. Also intercalation of antibiotics has been widely studied [54] as well as of DNA molecules for cancer therapy. In this case it is crucial to dispose of a vector, either viral or non-viral, to introduce DNA into cells [55]. Leroux et al. succeeded in adopting LDH as gene carrier of As-myc oligonucleotides [56]. If As-myc is transported by LDH, it can penetrate the cell. Almost 65% of HL-60 cells growth is inhibited.

Cunha et al. [57] presented another interesting application. They study the biocompatibility of the material by in vivo tests on rats. They implanted a tablet of Mg_2Al-Cl and Zn_2Al-Cl LDHs, with chloride intercalated ions, into abdominal walls. After 28 days no inflammation was present, neither an increase of tissue volume. These results indicated that LDH tablets determined no antigenic reactions.

In biosensing, LDHs were adopted several times to activate electrodes in simple, low cost and sensitive chemical sensors. They can be classified as "amperometric" if the measured signal is a current, "conducimetric" if it is an impedance/conductance, "potentiometric" in the case of a voltage. LDH/modified electrodes have been applied to detect glucose, haemoglobin, and codeine [58-60].

It is evident that the combination of organic-inorganic composite, as polymer-LDH, offers a great potentiality in electro-analytical applications due to its structural and/or functional behaviour. For a comprehensive list of patents [61].

Catalysis

The possibility to engineer LDHs at atomic and molecular level has an enormous consequence for science and technology. This is linked, as underlined previously, to the particular properties of the materials, as solubility, ion exchange, thermal conversion, intercalation, etc. [62].

In the previous paragraphs we already encounter some catalytic use of LDHs. What will be stressed in this section is linked to more specific applications of the phenomenon to build up devices in different field: energy conversion and storage, water splitting, environmental remediation, soil conditioning, waste water treatment, VOC (Volatile Organic Compound) decomposition, super capacitor.

Photo catalytic performances are connected to the semiconductor peculiarities: band gap, charge separation, charge mobility. Nanoscale LDH sheets seem to be an ideal solution, due to the possibility of manipulating and controlling composition and defects.

Water splitting: One of the main field in which photo catalysis is applied is the generation of H_2 and O_2 by water splitting, using solar radiation.

Zn-based and Ti-based LDH were considered since they exhibit visible light activity. Nevertheless their activities are still too low because of the long migration path and the big dimensions of the particles. For this reason nanosheets of LDH have attracted attention thanks to their specific surface area, the higher number of active sites and the shorter path the charges must travel [63].

It was also demonstrated the possibility of improving the efficiency of LDH nanosheets by coupling them with other semiconductors. Better absorption range and better charge transfer separation was obtained [64-66]. Feng et al. [67] recently propose a simple synthesis of a CoFe layered double hydroxide, with high activity and stable oxygen evolution.

VOCS remediation: Volatile organic compounds (VOCs) represent a crucial factor for environmental pollution. One the best effective treatment is the catalytic combustion. The LDH precursor can be adopted to prepare mixed oxides. Several works appeared in which LDH performances are examined relative to the total combustion of common VOCs, such as ethanol, methane, and toluene [68-70].

In organic wastewater treatment, layered double hydroxides succeeded in removing organic compounds. Recently, since sunlight can be used as energy source, solar light driven catalyst has been developed. Shu et al. [71] derived NiTi-mixed oxide from NiTi-LDH precursor. In another study, coating Fe_3O_2 particle surface formed a core-shell structure by LDH particles. This magnetic catalyst exhibited high photovoltaic activity under visible light mitted by a pressure mercury lamp [72].

Environmental problems heavily derive from the emission of NO_x and SO_x into the atmosphere. Those gases are responsible of greenhouse gas effect, acid rains, smog.

No_x is normally removed in three ways: - direct decomposition; -selective catalytic reduction; -storage and reduction. For SO_x transfer catalysis is adopted [73-75].

Energy storage and conversion: Energy is probably the most important topic of the century. Therefore the need not only of renewable energy, but also of energy storage systems is mandatory. To this aim super-capacitor and ion battery represent a reasonable answer. For all these devices, the most important components are the electrodes. Therefore the best way to reach the goals is to realize new electrodes with special structure, such as nanorods, nanowires and nano-sheets.

Delaminated LDH was first used to fabricate a thin film electrode in 2007 [76]. It was demonstrated that this film had good supercapacitor behavior with a high specific capacitance, good electrochemical stability, and a high-rate capability.

Many other materials can be considered as supercapacitor like carbon-based materials (carbon nanotubes, carbon gels, grapheme oxides, etc.). To realize better electrodes is therefore suitable to consider carbon materials/LDHs composites. Applications of these types have been considered since 2006 [77].

Inorganic semiconductor and organic sensitizer have been widely used in solar cells due to efficiency, easy production and low cost [78,79].

Quantum dots and perovskites are inorganic sensitizer applied to photovoltaic cells to enhance their low stability and narrow absorption band. It is possible to improve the harvesting ability of the dye by hybridizing it with nanosize materials as LDH since it gives higher heat stability, photo stability and is environmental friendly [80].

Anti-corrosion: The realization of LDH films with controlled architectures is desirable for functional coatings. What is of particular interest is the possibility to realize large-area LDH films to face corrosion problems [81-83]. Recently, layered double hydroxides have been used to immobilize or encapsulate corrosion inhibitors, such as 2-mercaptobenzothiazole (MBT), promoting their controlled release in a more eco-friendly way.

Their host structure acts as a scavenger towards aggressive ions (e.g. Cl-). Conversely Cl- ions have good affinity with LDH materials. The entrapment is associated by displacing the interleaved anion, i.e. the corrosion inhibitor, if the ion exchange reaction proceeds. As remarked LDH host structure is able to load different anions, inhibition efficiency of intercalated inorganic or organic anions such as, for example, carbonate [84], vanadate [85], and molybdate [86].

Also ceramic-like coatings formed with PEO (plasma electrolytic oxidation) make possible real improvement of the mechanical properties and corrosion resistance of Al alloys which can therefore be used more widely, hence providing low weight structures [87,88].

The process implies the polarization of the Al alloy to voltage value higher than the dielectric breakdown. The treatment is developed in environmentally friendly dilute electrolytes and causes plasma micro-discharges across the growing oxide. At the end the coating is a mixture of substrate metal oxide and other complex oxides involving components inside the electrolyte [89]. Recent developments for PEO coated Al alloys concentrated on preparing surfaces suitable for new multifunctional applications. *Ad hoc* PEO coatings can be achieved in one step, modifying the composition of the electrolytes [90,91] and process parameters [92]. A two-step process is achieved by a suitable post-treatment [93-95]. Those treatments have to increase the corrosion resistance of PEO, due to the porous structure. In any case presence of cracks and other defects reduce the service life. The LDH structure played a fundamental role in this situation too. It permitted to contain corrosion on Al and Mg alloys by encapsulating inhibiting compounds [96-99].

Alumina refinement: To remove heavy metal ions and organic species from wastewaters, as in the Bayer process, LDHs have been adopted taking advantage of the lamellar structure, the adjustable distances between layers and the interlayer reactivity [99,100]. The liquid component of the "red mud" requires neutralization. The most common practice consists of adding Mg-rich seawater, which results in the formation of precipitates containing Mg and Al [101]. These precipitates are in hydrotalcite form [102]. Together with this "unintentional" LDH formation [103], they can also be produced intentionally in the refinement process. Several studies were developed toward this direction. To mention just some of them, we remind the works of Nigro and O'Nieil [104] who investigated the use of hydrotalcite in the removal of coloured impurities. Perotta and Williams revealed the reduction of oxalate in liquor, due to the formation of hydrocalumite at temperatures up to 60°C.

In parallel with the water treatment, the formation, intentional and/or unintentional, of LDH allows the recovering of alumina values from the first liquor, by means of a modified version of the Bayer's process. It is essentially divided into three steps: a) addition to the liquor of a metal which reacts to form an Al-LDH, in such a way that the resulting aluminate ion concentration within the liquor is smaller than the initial one; b) processing of the Al rich LDH by, for example, a solution of carbonate ions, which provides both aluminate ions, and an insoluble salt of metals different from aluminium. This can be further treated to "rebuild" the LDH material; c) directing the liquor with high aluminate ion concentration to the alumina refinery [105,106].

Controlled Film Deposition

In the last decade, different groups [30-32] have demonstrated that LDH films can be deposited on metal surfaces by using aluminium or zinc foils as sacrificing substrates for the *in-situ* growth that is a sort of modified coprecipitation method as described in paragraph 1.3.5. Those studies were mainly addressed applications of LDHs as anticorrosion coatings or functionalized electrodes for electrical batteries and capacitors. More recently, our group has demonstrated that controlled films can also be grown an aluminum thin films, which were previously sputtered or evaporated on any surface [33,107]. In this way, LDHs

films made of lamellar-like nanoplatelets, can be grown selectively on specific areas, with submicrometric-level resolution (about $\pm 0.5 \ \mu m$), by prepatterning the aluminum layer with conventional photolithographic techniques (Figure 3). This strategy represents a necessary starting point for implementation of possible LDH-based devices and sensors in miniaturized electronic chips. Moreover, while the general morphology of the films was quite invariant against the experimental conditions, with intersecting nanoplatelets mainly oriented perpendicular to the substrate surface, the number density and thickness of each nanoplatelet could be varied by changing the intercalated anion (Figure 4). Moreover, with a single intercalated anion, the thickness of each nanoplatelet tuned by controlling the thickness of the predeposited aluminum layer (Figure 5). Preliminary studies have also been carried out to implement modified biosensors (Figure 6) and to impregnate paper or cotton fibers with LDHs in order to prepare functionalised filters or make flame retardant textiles (Figure 7).

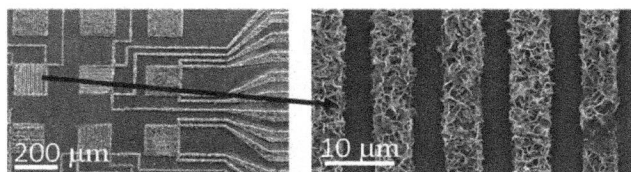

Figure 3: (a) Particular of a single LDH micro/nanoresistor deposited onto an aluminum thin film with a thickness of 300 nm; (b) zoom of a LDH micro/nanoresistor deposited onto an aluminum thin film with a thickness of 300 nm.

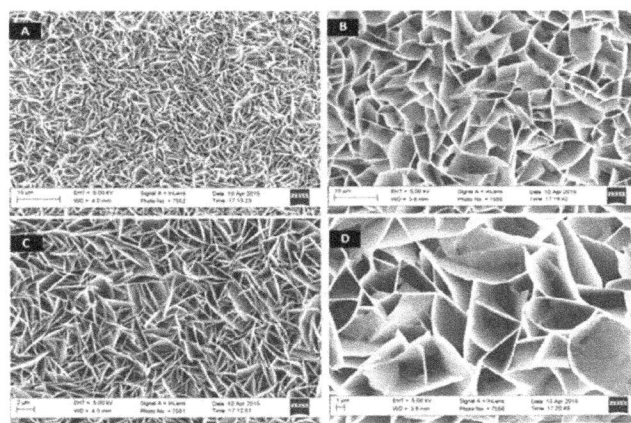

Figure 4: Low-magnification; (a and b) high-magnification, (c and d) SEM images of Zn-Al LDHs grown onto Al thin layers with: Cl; (a,c) SO3--; (b,d) intercalated anion.

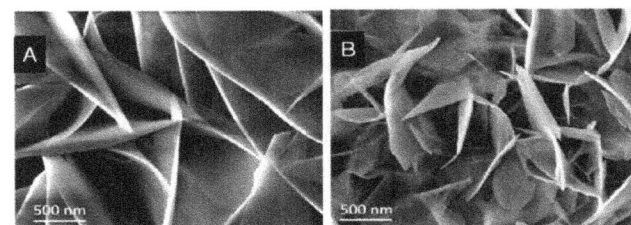

Figure 5: High-magnification SEM images of Zn-Al LDHs grown onto Al thin layers with a thickness of: a) 100 nm and b) 25 nm; c) the thickness of the LDH nanoplatelets as a function of the thickness of the aluminum layer.

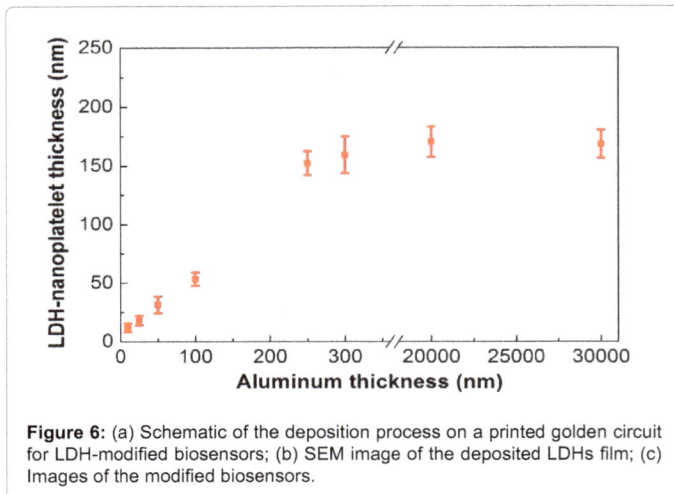

Figure 6: (a) Schematic of the deposition process on a printed golden circuit for LDH-modified biosensors; (b) SEM image of the deposited LDHs film; (c) Images of the modified biosensors.

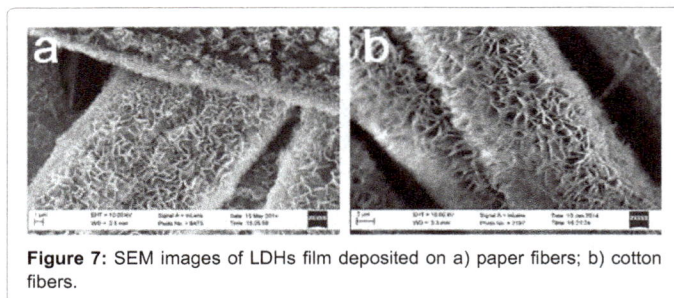

Figure 7: SEM images of LDHs film deposited on a) paper fibers; b) cotton fibers.

Conclusion and Perspectives

This review highlights only some of the progresses in applications of Layered Double Hydroxides nanostructures, within the areas of catalysis, electrode materials, polymer nano composites, bioinorganic hybrid materials, bio-med applications, etc. As it was stressed several times, LDHs exhibit great potential, thanks to their great flexibility. However, despite the many recent advances described here, opportunities and challenges remain either in fundamental or applied research on LDHs.

Looking ahead, efforts are still needed to design and synthesize low cost and high quality LDH based materials to meet requirements of applications.

These include controlling the chemical composition, size and morphology; meanwhile, it is also important to make the synthesis processes greener and more environmental benign.

The possibility to isolate delaminated nanosheets from dispersion, without causing aggregation, is still difficult to realize. Near future development will regard functionalization of LDH shells, due to their simple fabrication. For the same reasons, successes can be expected in the field of ultrathin films from delaminated LDHs. Anyway great attention must be paid to the choice of the proper layered double hydroxide and to corresponding substituted and intercalated anions.

Nevertheless, important fields will be, for sure, the medical and pharmaceutical ones, either for the synthesis of new biosensors or for the development of new formulations for drugs and cancer therapy.

All these studies will take on a crucial value for LDH-based materials to have a disruptive impact in the most different arenas in the next and far future.

References

1. Chen H, Zhang F, Fu S, Duan X (2006) In situ microstructure control of oriented layered double hydroxide monolayer films with curved hexagonal crystals as superhydrophobic materials. Advanced Materials 18: 3089-3093.

2. Shao M, Zhang R, Li Z, Wei M, Evans DG, et al. (2015) Layered double hydroxides toward electrochemical energy storage and conversion: Design, synthesis and applications. Chemical Communications 51: 15880-15893.

3. Bergaya F, Lagaly G (2006) General introduction: Clays, clay minerals, and clay science. Developments in clay science 1: 1-18.

4. Ferrell R (2007) Handbook of Clay Science.

5. De Roy A, Forano C, El Malki K, Besse JP (1992) Anionic clays: trends in pillaring chemistry. In Expanded Clays and other microporous solids, Springer US, pp: 108-169.

6. De Roy A, Forano C, Besse JP (2001) Layered double hydroxides: Synthesis and post-synthesis modification. Layered Double Hydroxides: Present and Future, pp: 1-39.

7. Yan K, Wu G, Jin W (2016) Recent Advances in the Synthesis of Layered, Double-Hydroxide-Based Materials and Their Applications in Hydrogen and Oxygen Evolution. Energy Technology 4: 354-368.

8. Jung H, Ohashi H, Anilkumar GM, Zhang P, Yamaguchi T (2013) Zn2+ substitution effects in layered double hydroxide (Mg (1−x) Znx)2 Al: textural properties, water content and ionic conductivity. Journal of Materials Chemistry A 1: 13348-13356.

9. Long X, Wang Z, Xiao S, An Y, Yang S (2016) Transition metal based layered double hydroxides tailored for energy conversion and storage. Materials Today 19: 213-226.

10. Marappa S, Kamath PV (2015) Structure of the Carbonate-Intercalated Layered Double Hydroxides: A Reappraisal. Industrial & Engineering Chemistry Research 54: 11075-11079.

11. Zheng S, Lu J, Yan D, Qin Y, Li H, et al. (2015) An Inexpensive Co-Intercalated Layered Double Hydroxide Composite with Electron Donor-Acceptor Character for Photoelectrochemical Water Splitting. Scientific reports 5: 12170.

12. Lal M, Howe AT (1980) High proton conductivity in pressed pellets of zinc–chromium hydroxide. Journal of the Chemical Society, Chemical Communications 15: 737-738.

13. Yun SK, Pinnavaia TJ (1995) Water content and particle texture of synthetic hydrotalcite-like layered double hydroxides. Chemistry of Materials 7: 348-354.

14. Feitknecht WV (1942) Über die Bildung von Doppelhydroxyden zwischen zwei- und dreiwertigen Metallen. Helvetica Chimica Acta 25: 555-569.

15. Reichle WT (1986) Synthesis of anionic clay minerals (mixed metal hydroxides, hydrotalcite). Solid State Ionics 22: 135-141.

16. Reichle WT (1985) Catalytic reactions by thermally activated, synthetic, anionic clay minerals. Journal of Catalysis 94: 547-557.

17. Miyata S (1983) Anion-exchange properties of hydrotalcite-like compounds. Clays Clay Miner 31: 305-311.

18. Boehm HP, Steinle J, Vieweger C (1977) [Zn2Cr (OH) 6] X·2H2O, New layer compounds capable of anion exchange and intracrystalline swelling. Angewandte Chemie International Edition 16: 265-266.

19. Lopez T, Bosch P, Ramos E, Gomez R, Novaro O, et al. (1996) Synthesis and Characterization of Sol−Gel Hydrotalcites. Structure and Texture. Langmuir 12: 189-192.

20. Tadanaga K, Miyata A, Ando D, Yamaguchi N, Tatsumisago M (2012) Preparation of Co–Al and Ni–Al layered double hydroxide thin films by a sol–gel process with hot water treatment. Journal of sol-gel science and technology 62: 111-116.

21. Chubar N, Gerda V, Megantari O, Mičušík M, Omastova M, et al. (2013) Applications versus properties of Mg–Al layered double hydroxides provided by their syntheses methods: Alkoxide and alkoxide-free sol–gel syntheses and hydrothermal precipitation. Chemical engineering journal 234: 284-299.

22. Jitianu M, Zaharescu M, Bălăsoiu M, Jitianu A (2003) The sol-gel route in synthesis of Cr (III)-containing clays. Comparison between Mg-Cr and Ni-Cr anionic clays. Journal of sol-gel science and technology 26: 217-221.

23. Prinetto F, Ghiotti G, Graffin P, Tichit D (2000) Synthesis and characterization

of sol–gel Mg/Al and Ni/Al layered double hydroxides and comparison with co-precipitated samples. Microporous and Mesoporous Materials 39: 229-247.

24. VishnuáKamath P (1994) Electrogeneration of base by cathodic reduction of anions: novel one-step route to unary and layered double hydroxides (LDHs). Journal of Materials Chemistry 4: 1487-1490.

25. Yarger MS, Steinmiller EM, Choi KS (2008) Electrochemical synthesis of Zn–Al layered double hydroxide (LDH) films Inorganic chemistry 47: 5859-5865.

26. Li Y, Zhang L, Xiang X, Yan D, Li F (2014) Engineering of ZnCo-layered double hydroxide nanowalls toward high-efficiency electrochemical water oxidation. Journal of Materials Chemistry A 2: 13250-13258.

27. Scavetta E, Ballarin B, Gazzano M, Tonelli D (2009) Electrochemical behaviour of thin films of Co/Al layered double hydroxide prepared by electrodeposition. Electrochimica Acta 54: 1027-1033.

28. Liu X, Ma R, Bando Y, Sasaki T (2012) A general strategy to layered transition-metal hydroxide nanocones: tuning the composition for high electrochemical performance. Advanced Materials 24: 2148-2153.

29. Fang J, Li M, Li Q, Zhang W, Shou Q, et al. (2012) Microwave-assisted synthesis of CoAl-layered double hydroxide/graphene oxide composite and its application in supercapacitors. Electrochimica Acta 85: 248-255.

30. Liu J, Huang X, Li Y, Sulieman KM, He X, et al. (2006) Facile and large-scale production of ZnO/Zn– Al layered double hydroxide hierarchical heterostructures. The Journal of Physical Chemistry 110: 21865-21872.

31. Guo X, Xu S, Zhao L, Lu W, Zhang F et al. (2009) One-step hydrothermal crystallization of a layered double hydroxide/alumina bilayer film on aluminum and its corrosion resistance properties. Langmuir 25: 9894-9897.

32. Guo X, Zhang F, Evans DG, Duan X (2010) Layered double hydroxide films: synthesis, properties and applications. Chemical Communications 46: 5197-5210.

33. Scarpellini D, Leonardi, C, Mattoccia A, Giamberardino LD, Medaglia PG, et al. Solution-Grown Zn/Al layered double hydroxide nanoplatelets onto Al thin films: fine control of position and lateral thickness. Journal of Nanomaterials 16: 178.

34. Miyata S (1983) Anionic exchange properties of hydrotalcitelike compounds. Clays Clay Miner 31: 305-311.

35. Sato T, Okawaki A (1991) Intercalation of benzenecarboxylate ions into the interlayer of hydrotalcite. Solid State Ionics 45: 43-48.

36. O'Leary S, O'Hare D, Seeley G (2002) Delamination of layered double hydroxides in polar monomers: new LDH-acrylate nanocomposites. Chemical Communications pp: 1506-1507.

37. Wang Q, O'Hare D (2012) Recent advances in the synthesis and application of layered double hydroxide (LDH) nanosheets. Chemical reviews 112: 4124-4155.

38. Wu Q, Sjåstad AO, Vistad ØB, Knudsen KD, Roots J, et al. (2007) Characterization of exfoliated layered double hydroxide (LDH, Mg/Al= 3) nanosheets at high concentrations in formamide. Journal of Materials Chemistry 17: 965-971.

39. Hibino T (2004) Delamination of layered double hydroxides containing amino acids. Chemistry of materials 16: 5482-5488.

40. Evans DG, Duan X (2005) Preparation of layered double hydroxides and their applications as additives in polymers, as precursors to magnetic materials and in biology and medicine. Chem Commun pp: 485-496.

41. Xu ZP, Zhang J, Adebajo MO, Zhang H, Zhou C (2011) Catalytic applications of layered double hydroxides and derivatives. Applied Clay Science 53: 139–150.

42. Camino G, Maffezzoli A, Braglia M, De lazzaro M, Zammarano M (2001) Polymer Degradation and Stability 74: 457-459.

43. Pereira CMC, Herrero M, Labajos FM, Marques AT, Rives V (2009) Preparation and properties of new flame retardant unsaturated polyester nanocomposites based on layered double hydroxides. Polymer Degradation and Stability 94: 939-946.

44. Matusinovic Z, Wilkie CA (2012) Fire retardancy and morphology of layered double hydroxide nanocomposites: a review. Journal of Materials Chemistry 22: 18701-18704.

45. Yao HB, Fang HY, Tan ZH, Wu LH, Yu SH (2010) Biologically Inspired, Strong, Transparent, and Functional Layered Organic–Inorganic ybrid Films. Angevandte Chemie 122: 2186-2191.

46. Tronto J, Crepaldi EL, Pava PC, De Paula, CC, Valim JB (2001) Organic anions of pharmaceutical interest intercalated in magnesium alluminium LDHs by two different methods. Molecular Crystal and Liquid Crystal 356: 227-237.

47. Aisawa, S, Ohnuma Y, Hirose K, Takahashi S, Hirahara H, et al. (2005) Intercalation of nucleotides into layered double hydroxides by ion-exchange reaction. Applied Clay Sciences 28: 137-145.

48. Barbos UAS, Ferreira AMDC, Constantino VRL (2003) Synthesis and characterization of magnesium-alluminium layered double hydroxides containing (tetrasulfonated porphyrin) cobalt. European Journal of Inirganic Chemistry 8: 1577-1584.

49. Hwang SH, Han YS, Choy JH (2001) Intercalation of functional organic molecules with pharmaceutical, cosmeceutical and neutraceutical functions into layered double hydroxides and zinc basic salts. Bulletin-Korean Chemical Society 22: 1019-1022.

50. Cavani F, Trifirò F, Vaccari A (1991) Hydrotalcite-type anionic clays: Preparation, properties and applications. Catalysis today 11: 173-301.

51. Tomohisa M, Mitsuo H (1998) U.S. Patent No. 5,798,120. Washington, DC: U.S. Patent and Trademark Office.

52. Ambrogi V, Fardella G, Grandolini G, Perioli L (2001) Intercalation compounds of hydrotalcite-like anionic clays with antiinflammatory agents-I. Intercalation and in vitro release of ibuprofen. International Journal of Pharmaceutics 220: 23-32.

53. Ambrogi V, Fardella G, Grandolini G, Perioli L, Tiralti MC (2002) Intercalation compounds of hydrotalcite-like anionic clays with anti-inflammatory agents, II: uptake of diclofenac for a controlled release formulation. aaps pharmscitech 3: 77-82.

54. Wei M, Yuan Q, Evans DG, Wang Z, Duan X (2005) Layered solids as a "molecular container" for pharmaceutical agents: L-tyrosine-intercalated layered double hydroxides. Journal of Materials Chemistry 15: 1197-1203.

55. Kwak SY, Jeong YJ, Park JS, Choy JH (2002) Bio-LDH nanohybrid for gene therapy. Solid State Ionics 151: 229-234.

56. Leroux F, Belkacem MB, Guyot G, Taviot-Gueho C, Leone P, et al. (2005) LDH-DNA Nanohybrids: a complete biophysical characterization. In Materials Research Society Symposium Proceedings 847: 223. Warrendale, Pa.; Materials Research Society; 1999.

57. Cunha VRR, De Souza RB, Koh IHJ, Constantino VRL (2016) Accessing the biocompatibility of layered double hydroxide by intramuscular implantation: histological and microcirculation evaluation. Scientific reports 6: 30547.

58. Poyard S, Martelet C, Jaffrezic-Renault N, Cosnier S, Labbé P (1999) Association of a poly(4-vynipyridine-costyrene) membrane with an hinorganic/organic mixed matrix for the optimization of glucose biosensors. Sensors and Actutuators B: Chemical 58: 380-383.

59. Lei C, Wollenberger U, Bistolas N, Guiseppi-Elie A, Scheller FW (2002) Electron transfer of hemoglobin at electrodes modified with colloidal clay nanoparticles. Analytical and bioanalytical chemistry 372: 235-239.

60. Shih Y, Zen JM, Yang HH (2002) Determination of codeine in urine and drug formulations using a clay-modified screen-printed carbon electrode. Journal of pharmaceutical and biomedical analysis 29: 827-833.

61. Derwent Innovation Index. 1996-2006. ISI Web of Knowledge.

62. Tong DS, Zhou CHC, Lu Y, Yu H, Zhang GF, et al. (2010) Adsorption of acid red G dye on octadecyl trimethylammonium montmorillonite. Applied Clay Science 50: 427-431.

63. Zhao Y, Li B, Wang Q, Gao W, Wang CJ, et al. (2014) NiTi-layered double hydroxides nanosheets as efficient photocatalysts for oxygen evolution from water using visible light. Chemical Science 5: 951-958.

64. Huang Z, Wu P, Gong B, Fang Y, Zhu N (2014) Fabrication and photocatalytic properties of a visible-light responsive nanohybrid based on self-assembly of carboxyl graphene and ZnAl layered double hydroxides. Journal of Materials Chemistry A 2: 5534-5540.

65. Nayak S, Mohapatra L, Parida K (2015) Visible light-driven novel gC3N4/NiFe-LDH composite photocatalyst with enhanced photocatalytic activity towards water oxidation and reduction reaction. Journal of Materials Chemistry A 3: 18622-18635.

66. Melvin AA, Illath K, Das T, Raja T, Bhattacharyya S, et al. (2015) M–Au/TiO2

(M=Ag, Pd, and Pt) nanophotocatalyst for overall solar water splitting: role of interfaces. Nanoscale 7: 13477-13488.

67. Feng L, Li A, Li Y, Liu J, Wang L, et al. (2017) A Highly Active CoFe Layered Double Hydroxide for Water Splitting. ChemPlusChem 82: 483-488.

68. Dula R, Janik R, Machej T, Stoch J, Grabowski R, et al. (2007) Mn-containing catalytic materials for the total combustion of toluene: the role of Mn localisation in the structure of LDH precursor. Catalysis today 119: 327-331.

69. Tanasoi S, Tanchoux N, Urdă A, Tichit D, Săndulescu I, et al. (2009) New Cu-based mixed oxides obtained from LDH precursors, catalysts for methane total oxidation. Applied Catalysis A: General 363: 135-142.

70. Mikulová Z, Čuba P, Balabánová J, Rojka T, Kovanda F, et al. (2007) Calcined Ni—Al layered double hydroxide as a catalyst for total oxidation of volatile organic compounds: Effect of precursor crystallinity. Chemical Papers 61: 103-109.

71. Shu X, He J, Chen D, Wang Y (2008) Tailoring of phase composition and photoresponsive properties of Ti-containing nanocomposites from layered precursor. The Journal of Physical Chemistry C 112: 4151-4158.

72. Li L, Feng Y, Li Y, Zhao W, Shi J (2009) Fe3O4 core/layered double hydroxide shell nanocomposite: versatile magnetic matrix for anionic functional materials. Angewandte Chemie International Edition 48: 5888-5892.

73. Corma A, Palomares AE, Rey F, Márquez F (1997) Simultaneous Catalytic Removal of SOxand NOxwith Hydrotalcite-Derived Mixed Oxides Containing Copper, and Their Possibilities to Be Used in FCC Units. Journal of catalysis 170: 140-149.

74. Yu JJ, Cheng J, Ma CY, Wang HL, Li LD, et al. (2009) NOx decomposition, storage and reduction over novel mixed oxide catalysts derived from hydrotalcite-like compounds. Journal of colloid and interface science 333: 423-430.

75. Karásková K, Obalová L, Jirátová K, Kovanda F (2010) Effect of promoters in Co–Mn–Al mixed oxide catalyst on N2O decomposition. Chemical Engineering Journal 160: 480-487.

76. Wang Y, Yang W, Yang J (2007) A Co–Al layered double hydroxides nanosheets thin-film electrode fabrication and electrochemical study. Electrochemical and Solid-State Letters 10: A233-A236.

77. Li XD, Yang WS, Li F, Evans DG, Duan X (2006) Stoichiometric synthesis of pure NiFe2O4 spinel from layered double hydroxide precursors for use as the anode material in lithium-ion batteries. Journal of Physics and Chemistry of Solids 67: 1286-1290.

78. O'Regan B, Gratzel M (1991) A low-cost, high-efficiency solar cell based on dye-sensitized colloidal TiO2 films. Nature 353: 737-740.

79. Gibson EA, Smeigh AL, Le Pleux L, Fortage J, Boschloo G, et al. (2009) A p-Type NiO-Based Dye-Sensitized Solar Cell with an Open-Circuit Voltage of 0.35 V. Angewandte Chemie 121: 4466-4469.

80. Tian R, Liang R, Wei M, Evans DG, Duan X (2016) Applications of Layered Double Hydroxide Materials: Recent Advances and Perspective. In 50 Years of Structure and Bonding–The Anniversary Volume (pp. 65-84). Springer International Publishing.

81. Rotundo F, Ceschini L, Martini C, Montanari R, Varone A (2014) High temperature tribological behavior and microstructural modifications of the low-temperature carburized AISI 316L austenitic stainless steel. Surface and Coatings Technology 258 772-781.

82. Balijepalli SK, Donnini R, Kaciulis S, Montanari R, Varone A (2013) Young's Modulus Profile in Kolsterized AISI 316L Steel. In Materials Science Forum, Trans Tech Publications 762: 183-188.

83. Balijepalli SK, Colantoni I, Donnini R, Kaciulis S, Lucci M, et al. (2013) Elastic Modulus Of S Phase In Kolsterized 316l Stainless Steel. Metallurgia Italiana 1: 42-47.

84. Lin JK, Hsia CL, Uan JY (2007) Characterization of Mg, Al-hydrotalcite conversion film on Mg alloy and Cl– and anion-exchangeability of the film in a corrosive environment. Scripta Materialia 56: 927-930.

85. Zheludkevich ML, Poznyak SK, Rodrigues LM, Raps D, Hack T, et al. (2010) Active protection coatings with layered double hydroxide nanocontainers of corrosion inhibitor. Corrosion Science 52: 602-611.

86. Yu X, Wang J, Zhang M, Yang L, Li J, et al. (2008) Synthesis, characterization and anticorrosion performance of molybdate pillared hydrotalcite/in situ created ZnO composite as pigment for Mg–Li alloy protection. Surface and Coatings Technology 203: 250-255.

87. Wang YM (2010) Plasma electrolytic oxidation treatment of aluminium and titanium alloys. Surface Engineering of Light Alloys: Aluminium, Magnesium and Titanium Alloys 110.

88. Sabatini G, Ceschini L, Martini C, Williams JA, Hutchings IM (2010) Improving sliding and abrasive wear behaviour of cast A356 and wrought AA7075 aluminium alloys by plasma electrolytic oxidation. Materials & Design 31: 816-828.

89. Komnitsas K, Bartzas G, Paspaliaris I (2006) Modeling of reaction front progress in fly ash permeable reactive barriers. Environmental Forensics 7: 219-231.

90. Genç H, Tjell JC, McConchie D, Schuiling O (2003) Adsorption of arsenate from water using neutralized red mud. Journal of Colloid and Interface Science 264: 327-334.

91. Menzies NW, Fulton IM, Morrell WJ (2004) Seawater neutralization of alkaline bauxite residue and implications for revegetation. Journal of Environmental Quality 33: 1877-1884.

92. Summers RN, Pech JD (1997) Nutrient and metal content of water, sediment and soils amended with bauxite residue in the catchment of the Peel Inlet and Harvey Estuary, Western Australia. Agriculture, ecosystems & environment 64: 219-232.

93. Glenister DJ, Thornber MR (1985) Alkalinity of red mud and its application for the management of acid wastes. Chemica 85: 100-113.

94. Hanahan C, McConchie D, Pohl J, Creelman R, Clark M, et al. (2004) Chemistry of seawater neutralization of bauxite refinery residues (red mud). Environmental Engineering Science 21: 125-138.

95. Altundoğan HS, Altundoğan S, Tümen F, Bildik M (2002) Arsenic adsorption from aqueous solutions by activated red mud. Waste management 22: 357-363.

96. Solymar K, Sajo I, Steiner J, Zoeldi J (1992) Characteristics and separability of red mud. In Proceedings of the 121st TMS Annual Meeting.

97. Paspaliaris I, Karalis A (1993) The effect of various additives on diasporic bauxite leaching by the Bayer process. Light Metals-Warrendale 35-35.

98. Schwertmann U, Cornell RM (2007) The iron oxides. Iron Oxides in the Laboratory: Preparation and Characterization 5-18.

99. Lindsay WL (1979). Chemical equilibria in soils. John Wiley and Sons Ltd.

100. Misra C (1987) U.S. Patent No. 4,656,156. Washington, DC: U.S. Patent and Trademark Office.

101. Nigro WA, O'neill GA (1991) U.S. Patent No. 5,068,095. Washington, DC: U.S. Patent and Trademark Office.

102. Schepers B, Bayer G, Urmann E, Schanz K (1977) U.S. Patent No. 4,046,855. Washington, DC: U.S. Patent and Trademark Office.

103. Johnston M, Clark MW, McMahon P, Ward N (2010) Alkalinity conversion of bauxite refinery residues by neutralization. Journal of hazardous materials 182: 710-715.

104. Misra C (1987) U.S. Patent No. 4,656,156. Washington, DC: U.S. Patent and Trademark Office.

105. Perrotta AJ, Williams F (1995) Hydrocalumite formation in Bayer liquor and its promotional effect on oxalate precipitation. Light Metals 1995: 77-87.

106. Rosenberg SP (2010) U.S. Patent No. 7,666,373. Washington, DC: U.S. Patent and Trademark Office.

107. Richetta M, Digiamberardino L, Mattoccia A, Medaglia PG, Montanari R, et al. (2016) Surface spectroscopy and structural analysis of nanostructured multifunctional (Zn, Al) layered double hydroxides. Surface and Interface Analysis 48: 514-518.

16

Effect of Partial Shear Connection on Strengthened Composite Beams with Externally Post-Tension Tendons

EL-Shihy AM[1], Shabaan HF[1], Al-Kader HM[1] and Hassanin AI[2]*

[1]*Faculty of Engineering, Zagazig University, Zagazig, Egypt*
[2]*Faculty of Engineering, Egyptian Russian University, Cairo, Egypt*

Abstract

Composite steel-concrete beams are used widely in bridges and buildings construction as the main structural elements in flexure. These structures have a design life and this may be reduced if loads are increased or environmental degradation could happen. These changes may reduce the design life and strength of such members and thus replacement or retrofitting may need to be considered. The present study focuses on evaluating the effect of partial shear connection on strengthen composite beams with externally post-tension tendons. Using three dimensional F.E. modeling it's able to simulate the overall flexural behavior of composite beams which are strengthen with many shapes of tendons profiles. A fundamental point for the structural behavior and design of composite beams is the level of connection and interaction between the steel section and the concrete slab. The use of partial connection provides the opportunity to achieve a better match of applied and resisting moment and some economy in the provision of connectors, taking into account the demonstrated advantages of externally post tension system like: Increase in ultimate moment capacity of structure, Enlarge the range of elastic behavior before yielding for the structure with the introduction of internal stresses.

Keywords: Finite element; Nonlinear; Post-tensioning; Composite beams; Deflection; ANSYS 14.0

Introduction

Steel-concrete composite girders have attractive potentials when applied in building construction. It has been found to provide an efficient and economical solution for a wide range of structure types and conditions. Composite steel-concrete beams post-tensioning with high strength external tendons have demonstrated many advantages when compared with composite beams like: Increase in ultimate moment capacity of structure, Enlarge the range of elastic behavior before yielding for the structure with the introduction of internal stresses [1]. The study was conducted using the finite element program "ANSYS". Nonlinear material models for the components of the composite beam were used in the finite element model. The outcomes got from finite element analysis were confirmed against available experimental results. A broad parametric study was conducted to explore the effect of partial shear connection on strengthen composite beams with externally post-tension tendons. This covers: moment deflection behavior, slip-deflection, and moment-bottom flange stress in the steel beam. A fundamental point for the structural behavior and design of composite beams is the level of connection and interaction between the steel section and the concrete slab. The use of partial shear connection provides the opportunity to achieve a better match of applied and resisting moment and some economy in the provision of connectors, taking into account the demonstrated advantages of externally post tension system like: Increase in ultimate moment capacity of structure, Enlarge the range of elastic behavior before yielding for the structure with the introduction of internal stresses. In this paper, a three-dimensional (3D) FE model is presented to simulate the nonlinear behavior of steel–concrete composite beams strengthened with externally post-tensioned tendons under many degrees of shear connection. The effective post-tensioned force is taken as an initial value that appears in the analysis as initial strain in the link elements used to model the tendons and degrees of shear connection are ranging from 40% to more than 100% (by means of varying the number of shear connectors). To verify the accuracy of the developed 3D FE model, comparison between the FE analysis results

and previous experimental results was carried out in a previous study [2], which shows a good agreement in moment, slip and deflection.

Finite Element Model

The present study utilized the finite element program ANSYS version 14.0 [3] to simulate the behaviour of the composite beam and the stud shear connectors. A three dimensional finite element model was presented to simulate the material non-linear behaviour of the composite beam. The used elements are summarized in the following. The steel I-beam was modeled using an eight-node solid element with three degrees of freedom at each node. SOLID 185 is used for the three-dimensional modeling of solid structures. Three-dimensional spar elements were used to model the reinforcing bars embedded in the concrete in the longitudinal and the transverse directions. The LINK 8 is a spar (or truss) element. The concrete slab was modeled using a three dimensional concrete element (SOLID 65). The most important aspect of this element is the treatment of the non-linear material properties. In the proposed concrete material model, tension stress, relaxation coefficient, shear transfer for open and closed cracks, and concrete crushing were considered. An eight-node solid element with three degrees of freedom at each node is used to represent the shear connector's behaviour to resist the normal and shear force between the concrete and steel beam. The external tendons were modeled using 3D spar elements. Link8 is used to represent the external cable. The interface of the steel flange and the concrete slab was represented by

***Corresponding author:** Hassanin AI, Assistant Lecturer, Faculty of Engineering, Egyptian Russian University, Cairo, Egypt
E-mail: Ahmed_eng8028@yahoo.com*

using node-to-node. The purpose of using the contact element between these connected surfaces is to prevent penetration and to ensure physical separation between them as shown in Figure 1. To avoid stress concentration problems at the loading locations steel plates are added, this provides a more even stress distribution over the load area.

Material Modeling

In this study ANSYS finite element program is used. The element damaged plasticity model in ANSYS provides a general capability for modeling all types of structures using concepts of isotropic damaged elasticity in combination with isotropic tensile and compressive plasticity to represent the inelastic behavior of the composite steel- concrete beams. The choice of the appropriate elements for the modeling of different composite beam parts requires good understanding of the geometrical shape and material properties of each part. Also, the connectivity of each element with the adjacent elements had to be considered. ANSYS has an element library which covers all these requirements.

Modeling of concrete

The concrete is assumed to be homogeneous and initially isotropic. The uni-axial stress–strain relationship for concrete in compression is required for ANSYS as an input [2]. The simplified stress–strain curve for each beam model was constructed from nine points connected by curved lines, as shown in Figures 2 and 3.

Modeling of steel I-beam

The mechanical properties of steel are well known so that the

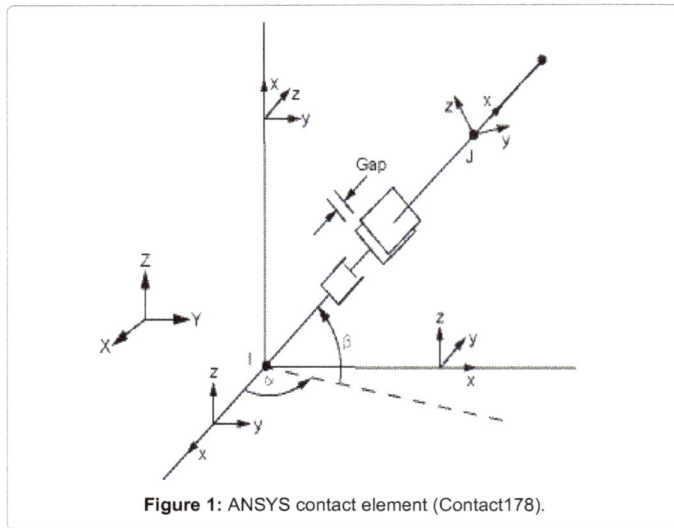

Figure 1: ANSYS contact element (Contact178).

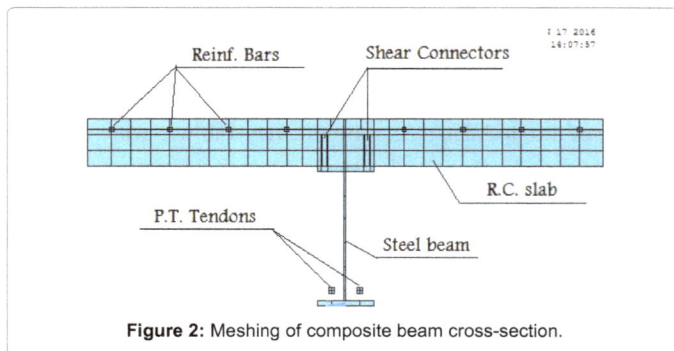

Figure 2: Meshing of composite beam cross-section.

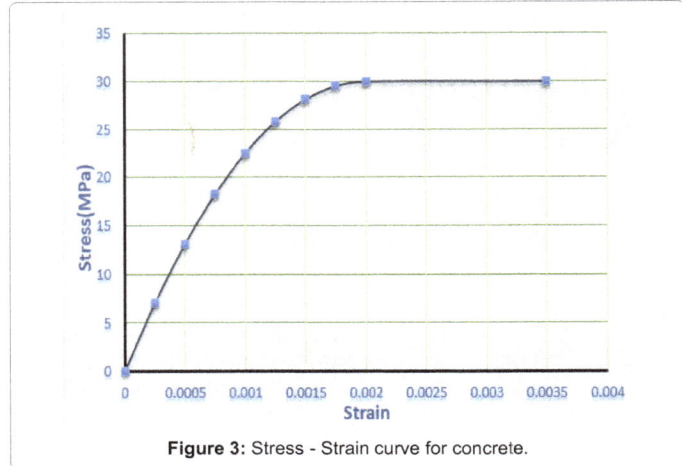

Figure 3: Stress - Strain curve for concrete.

stress–strain behaviour in tension and in compression can be assumed typical and identical. Elastic modulus and yield stress for the steel I-beam used in this study follows the design material properties used for the experimental investigation.

Modeling of reinforcement and external post-tension tendons

Since the reinforcing bars and post-tensioning cables are normally long and relatively slender, they can generally be assumed to be capable of transmitting axial forces only. This relation is assumed to be identical in tension and in compression.

Analyses and Discussion

Verification analysis process

The finite element for this analysis is a simple model under axial point loading. For the purposes of this model, the Static analysis type is utilized. The Restart command is utilized to restart an analysis after the initial run or load step has been completed. The solution control (Sol'n Controls) command dictates the use of a non-linear solution for the finite element model.

Verification results

The goal of the comparison of the FE model and the experimental models Chen [4] and Abdel Aziz [5] is to ensure that the elements, material properties, real constants and convergence criteria are adequate to model the response of the member and verify the accuracy of the FE Model.

It can be seen that many cracks appeared at the studs' locations owing to the longitudinal shear that occurred through the concrete slab. Also, many cracks appeared at the loading location where maximum beam moment and maximum compressive stress in the concrete slab occurred as shown in Figure 4.

The FE-deformed shape for model two (A.Aziz) resulting from the effect of externally applied loads, slippage between concrete slab and top steel flange and increasing of stresses in the bottom steel flange have been observed as shown in Figure 5.

Verified model one (Chen)

The mid span Moment- Deflection curve of the externally post tensioned simple composite beam obtained from the finite element analysis using ANSYS computer program (version 14.0) was compared with corresponding experimental results [6,7].

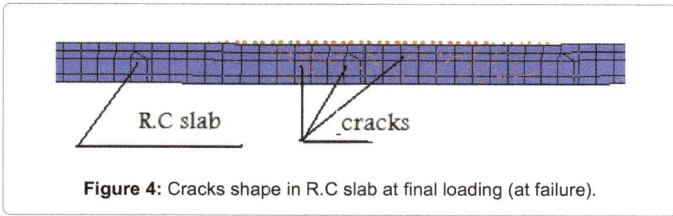

Figure 4: Cracks shape in R.C slab at final loading (at failure).

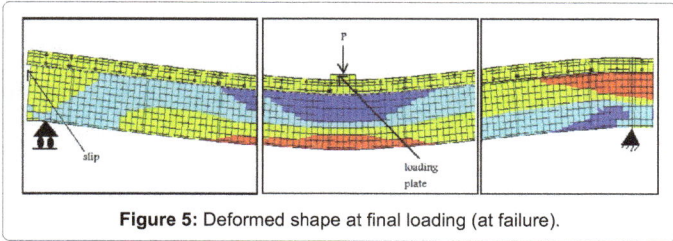

Figure 5: Deformed shape at final loading (at failure).

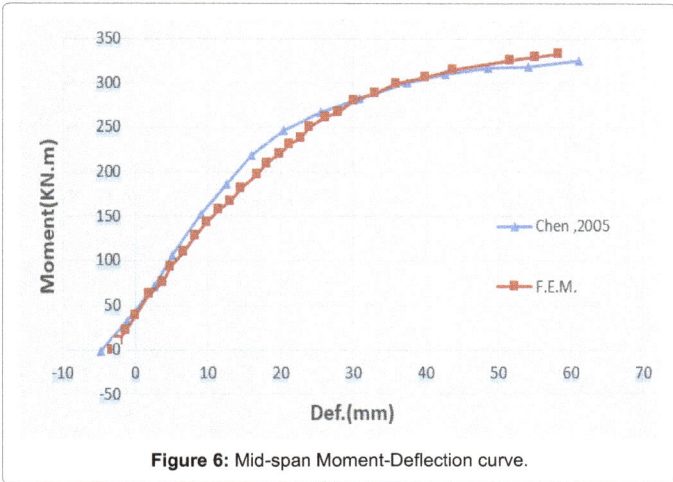

Figure 6: Mid-span Moment-Deflection curve.

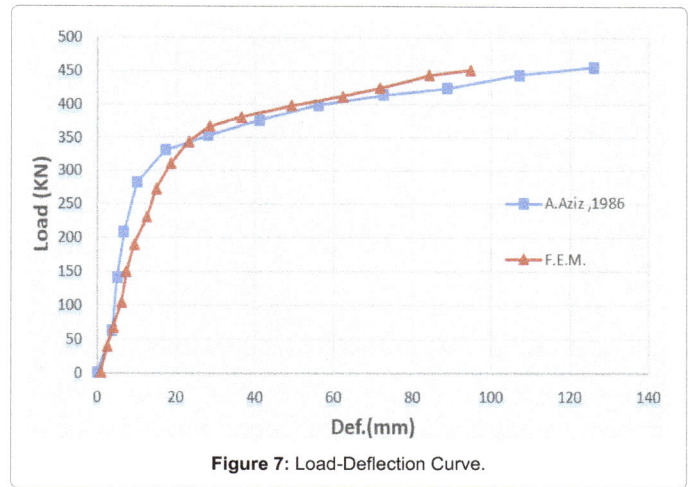

Figure 7: Load-Deflection Curve.

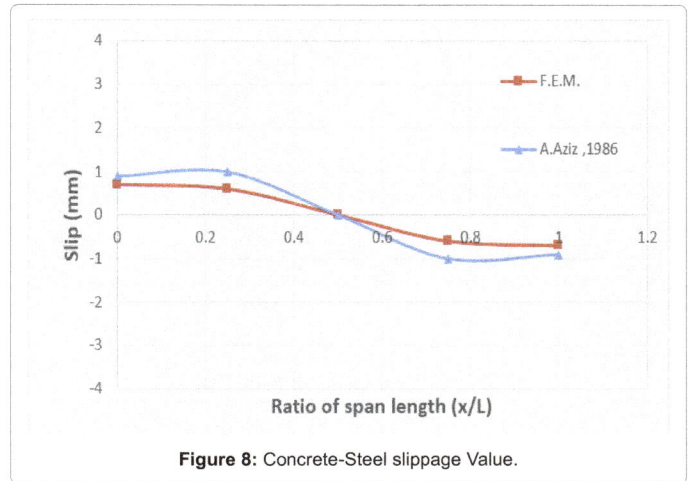

Figure 8: Concrete-Steel slippage Value.

In general, it can be noted from the Moment - Deflection curve that the finite element analyses is agree well with the experimental results throughout the entire range of behavior. In the linear range, the FE moment–deflection response coincides with that from the experimental results. When the moment–deflection curve transitioned from linear to nonlinear, the yielding of the beam started. After this point, the stiffness of the FE model was slightly higher than the experimental beam owing to the difference in behavior of the shear connector between the experimental and theoretical model as shown in Figure 6.

Verified model two (A.Aziz)

Load-mid-span deflection curve of the simply supported composite beam obtained from the finite element analysis using ANSYS computer program (version 14.0) was compared with corresponding experimental data as shown in Figure 7.

It can be noted from the load deflection curve that the finite element analyses is agree well with the experimental result throughout the entire range of behavior. In the linear range, the two beams behaved similarly; the two trends were parallel to each other with a difference at maximum deflection due to continuounity of loading on the experimental beam till failure [8].

Figure 8 showing the slippage between concrete slab and top flange of the steel beam. It can be noted that the finite element analyses is agree well with the experimental results. Table 1 shows a summary of material properties for the modeled composite beams.

Parametric study

Using the verified model, Chen [4], a parametric study was carried out on three new developed models (A, B and C) according to arrangements of the external tendons applied in composite beams (A) for straight tendon profile, (B) for triangle tendon profile and (C) for trapezoidal tendon profile). An investigation about the effects of changing levels of shear connection in ranges from 40% to more than 100% (by means of varying the number of shear connectors) and varying numbers from (1-1) to (1-5) according to degree of shear connection taken in the model. The case of beam with no post-tension force applied with a fully shear connection had been taken as a reference case. The geometrical characteristics of the used models are shown in Figures 9-11 and a summary of material and section properties for the verified and modeled composite beams are shown in Tables 1 and 2.

Moment-slippage relation: From Figures 12-14, it can be observed that: on decreasing the level of shear connection the system became more flexible, and with reduced in ultimate capacity of moment. The two higher levels (100% and 120%) resulted in very similar curves, in terms of both stiffness and ultimate moment, with the mode of failure being slab crushing at the mid-span of the beam (location of maximum

Model	Post-tensioning Tendons				Concrete	Steel I-beam			
	f_y (MPa)	f_u (MPa)	A_p (mm²)	F(kN)	f_c (MPa)	f_y (MPa)		f_u (MPa)	
						Web	Flange	Web	Flange
Model 1	-	-	-	-	40	260	245	372	361
Model 2	1,680	1,860	137.4	112.6	35	327.7	406.5	492.6	593.6
Ref. Model	1,680	1,860	137.4	112.6	35	400	400	480	480

Table 1: Material properties for the verified and modeled composite beams.

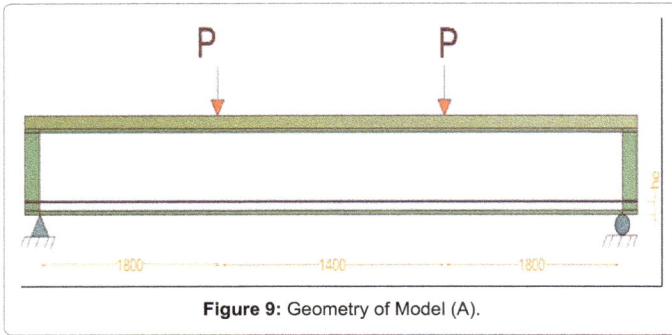

Figure 9: Geometry of Model (A).

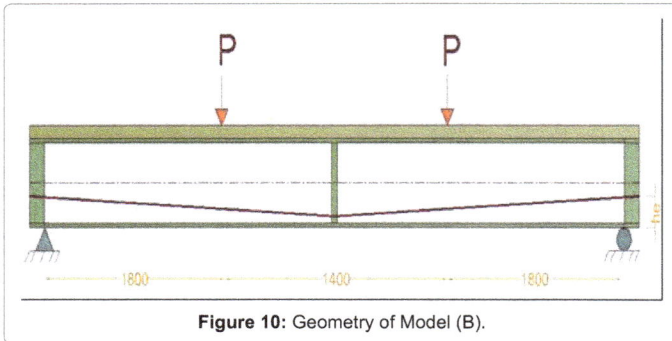

Figure 10: Geometry of Model (B).

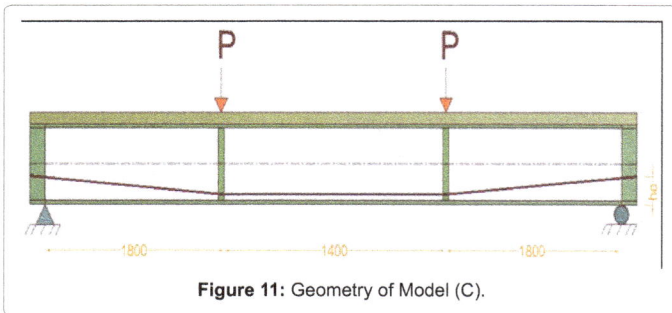

Figure 11: Geometry of Model (C).

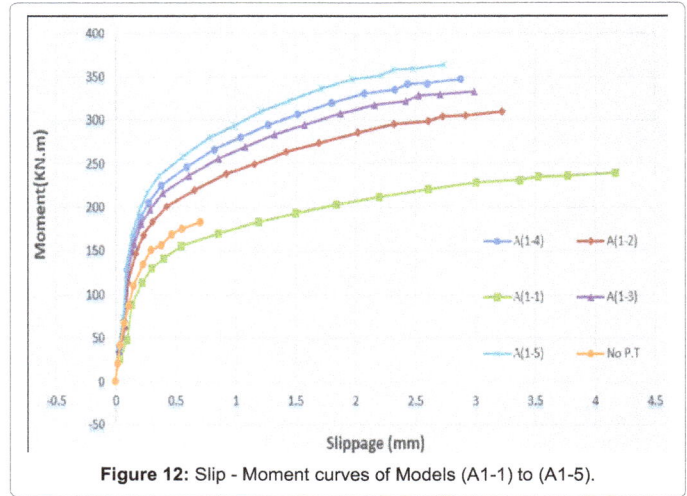

Figure 12: Slip - Moment curves of Models (A1-1) to (A1-5).

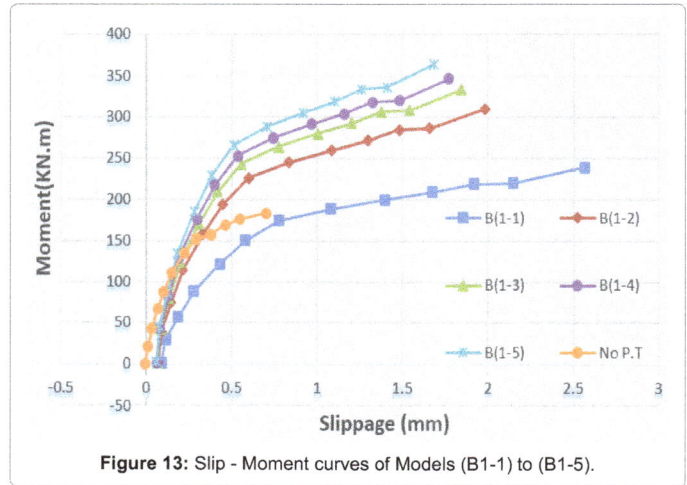

Figure 13: Slip - Moment curves of Models (B1-1) to (B1-5).

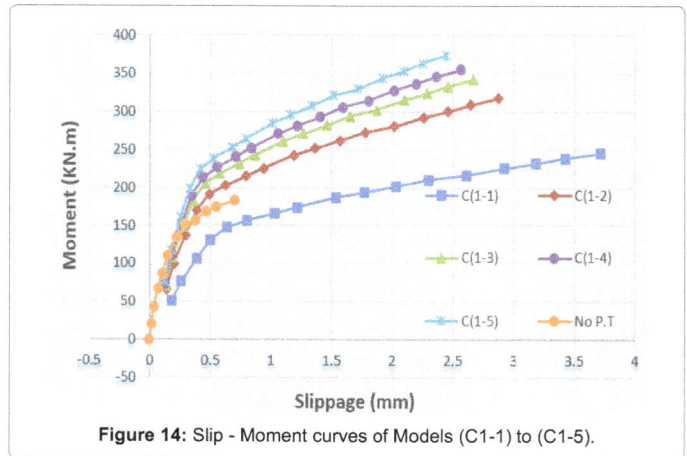

Figure 14: Slip - Moment curves of Models (C1-1) to (C1-5).

bending moment). Regarding level 100%, its mode of failure could be either slab crushing or stud failure due to a separation between R.C slab and steel top flange. As the three lower levels (40%, 60% and 80%) resulted in stud failure, the level 100% was an intermediate level between the two possible modes of failure. The results demonstrate that, therefore, the effects of partial interaction, which are increased by the use of partial shear connection, can be neglected for levels of shear connection above 100%, as no significant improvement in terms of either strength or stiffness of the beam was observed. It was demonstrated that, by decreasing the level of shear connection, the composite system becomes more flexible, with reduced strength and stiffness, mainly for beams with less than 100% degree of shear connection, for which the partial interaction effects are significant and must be taken into account [9].

Moment–deflection response: From Figures 15-17, the Moment–

No.	Model	P.T Force (KN)	h_e (mm)			% shear connection
			Start	Mid	End	
1	A	112.6	30	30	30	40 to 120%
2	B	112.6	105	30	105	40 to 120%
3	C	112.6	105	30	30	40 to 120%

Table 2: Section properties of Models (A, B and C).

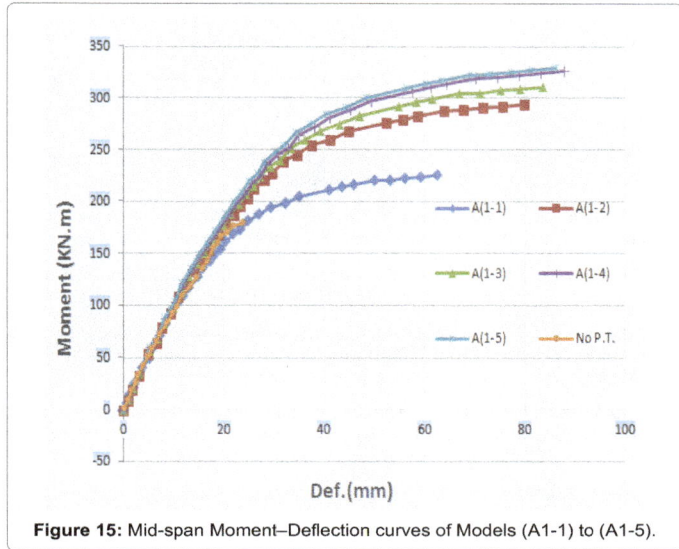

Figure 15: Mid-span Moment–Deflection curves of Models (A1-1) to (A1-5).

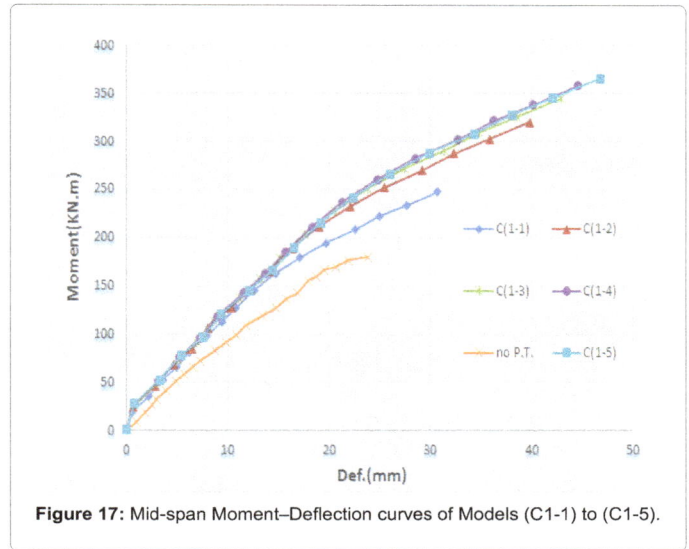

Figure 17: Mid-span Moment–Deflection curves of Models (C1-1) to (C1-5).

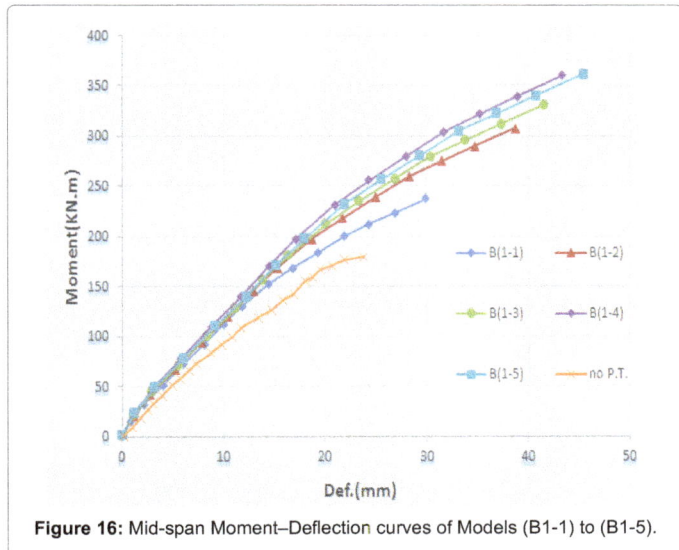

Figure 16: Mid-span Moment–Deflection curves of Models (B1-1) to (B1-5).

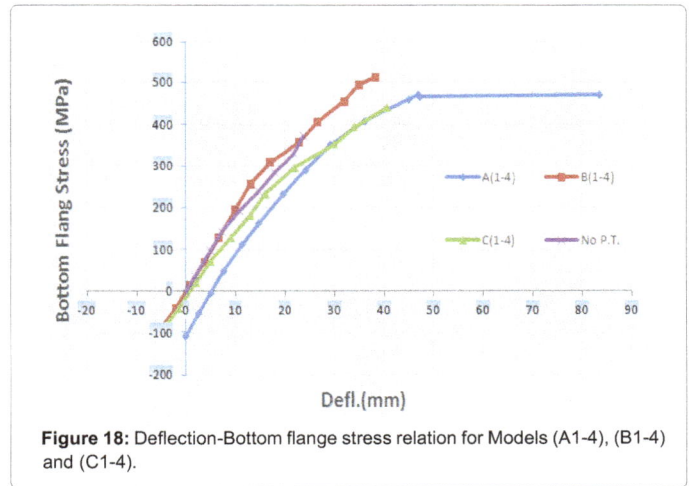

Figure 18: Deflection-Bottom flange stress relation for Models (A1-4), (B1-4) and (C1-4).

mid-span deflection curves for all models and cases (each one having a different level of shear connection). It can be observed that, there was a reduction (comparing to the case of no post-tension) in strength and stiffness at the low levels of shear connections (40%, 60% and 80%). However, the decrease in stiffness did not seem to be as significant as the decrease in the ultimate moment. The two higher levels (100% and 120%) resulted in very similar curves, in terms of both stiffness and ultimate moment; although there were increases in deflection directly proportional to the degree of shear connection.

Deflection-bottom flange stress relation: The Deflection-Bottom flange stress curve was presented in the form of x-y plot where the horizontal axis represents the deflection at mid span point in (mm) and the vertical axis represents the corresponding axial stresses at bottom flange at mid span point in (MPa). Figure 18 shows the Deflection-Bottom flange stress curve for models A (1-4), B (1-4) and C (1-4) as they present the case of full shear connection.

It was demonstrated that: the stresses in the two models B (1-4) and C (1-4) are closed to be matched and the difference increase by increasing the load at (C1-4). This slight difference in the stresses could be explained due to the difference in the region of composite action between (B1-4) and (C1-4). For A (1-4) there was a noticeable difference between it and the other models it was clear in the presence of yielding point and continued increasing in deflection without collapse even after reaching the maximum loading , this difference could be explained due to the straight profile of the post-tension tendon which gives greater ability to withstand stresses [10].

Conclusions

1. When the degree of shear connection decreases along with strengthening the beam by external post-tension tendons:

a. The ultimate load capacity has been found to decreases.

b. A significant increase in the mid span deflection and joint rotation has been clearly observed.

2. When the degree of shear connection is very low (less than 60%), the studs in the negative moment region above supporting points were broken because the maximum slip at the ultimate load had exceeded the slip capacity of the stud shear connectors.

3. The results demonstrate that the effects of partial interaction, which are increased by the use of partial shear connection, can be neglected for levels of shear connection above 100%, as no significant improvement in terms of either strength or stiffness of the beam was observed.

4. The results demonstrate that adding straight tendon in post-tensioning to the simple composite beams increasing the yield load and the ultimate moment resistance by 92.2% comparing to the case of composite beam without post-tensioning.

5. Adding post-tensioned tendons with a triangle and trapezoidal profile to composite beams significantly increases the yield load and the ultimate load by 99.1 and 105.5%, respectively.

References

1. El-Shihy AH, Shaaban HF, Mustafa SA (2015) Finite element modeling of externally post-tensioned composite beams. Journal of Bridge Engineering 45: 393-410.

2. El-Shihy AH, Shaaban HF, Abd-Elkader HM, Hassanin AI (2016) Effect of Using Partial Continuounity by External Post-Tension on The Simple Composite Beams as a Strengthening Technique. The Egyptian International Journal of Engineering Sciences and Technology 19: 305-311.

3. ANSYS Inc. ANSYS (2015) Release 14.0 Documentation.

4. Chen S, Gu P (2005) Load Carrying Capacity of composite Beams Pre Stressed with External Tendons Under Positive Moment. Journal of Constructional Steel Research 61: 515-530.

5. Abdel Aziz K (1986) Numerical modelling and experimental study of composite beams with partial or spaced shear connection.

6. An L, Cederwall K (1994) Push-out Tests on Studs in High Strength and Normal Strength Concrete. Journal of Constructional Steel Research 36: 15-29.

7. Gattesco N, Giuriani E (1996) Experimental Study on Stud Shear Connectors Subjected to Cyclic Loading. Journal of Constructional Steel Research38: 1-21.

8. Mirza O, Uy B (2010) Effects of the Combination of Axial and Shear Loading on the Behaviour of Headed Stud Steel Anchors. Engineering Structures 32: 93-105.

9. Zona A, Ranzi G (2011) Finite element models for nonlinear analysis of steel–concrete composite beams with partial interaction in combined bending and shear. Finite Elements in Analysis and Design 47: 98-118.

10. Qureshi J, Lam D (2012) Behaviour of Headed Shear Stud in Composite Beams with Profiled Metal Decking. Advances in Structural Engineering 15: 1547-1558.

17

Mechanical Performance of Reinforced Concert with Different Proportions and lengths of Basalt Fibres

Elshafie S[1]*, Boulbibane M[1] and Whittleston G[2]
[1]*Faculty of Engineering, Sports and Science, University of Bolton, Bolton, UK*
[2]*Department of Civil Engineering, University of Salford, UK*

Abstract

This paper discusses the effect of the fraction (0.2-0.3% by volume) and length (22 mm and 24 mm) of basalt fibre on the mechanical properties of concrete. The paper aims to evaluate the effect of different combinations of basalt fibres on the mechanical properties of concrete, as well as identify the best basalt fibre length and content that have the optimum influence on concrete. This paper is considered to be distinct from other research work as it fills the literature gap by presenting new unknown facts and also adds new knowledge. For example, it identifies the best basalt fibre length and content combination that demonstrates an improvement in the mechanical properties of concrete. It suggests the use of a blend of 12 mm short and 24 mm long fibres as they have a significant effect on the mechanical properties of concrete, it validates the results obtained from the laboratory by using a statistical analysis of variance ANOVA software, as well as determine the correlation between the mechanical properties of concrete. The results showed that the optimum basalt fibre length and content that enhanced the mechanical properties of concrete is 24 mm long fibre with content of 0.2% by the total volume of concrete. It also show that changing basalt fibre length and content enhance not only both tensile and flexural strengths of concrete, but also reduce its compressive strength, workability and air content of concrete, as well as maintain the unit weight and modulus of elasticity values. In this context, the incorporation of basalt fibres within the mixture becomes an important parameter for strengthening concrete in the construction industry.

Keywords: Workability; Unit weight; Air entrainment; Compressive strength; Tensile strength; Flexural strength; Modulus of elasticity

Introduction

During the process of selecting an appropriate material for construction industry, the mechanical properties of material are considered to play a major role in this selection, which depends strongly upon understanding the material microstructure and components. In this context, it's important for engineers to possess a deep understanding of how the microstructure is formed, and how the addition of new materials can influence the engineering properties Adam Warren [1].

To achieve this, the mechanical properties of the concrete should be tested. The mechanical properties of the materials play a significant role in defining the characteristic of the concrete and are defined as the characteristics of materials that are revealed when that material is subjected to mechanical loading. One way of improving the mechanical properties of concrete is by adding fibres to the concrete mixture, and determining their influence on the mechanical properties of concrete, such as: compressive strength, tensile strength, flexural strength, elastic modulus and workability.

Previous investigations on reinforcing concrete containing basalt fibres with different proportions have confirmed the possibility of enhancing the mechanical properties of concrete significantly. Addition of basalt fibres will have significant effect on the mechanical properties of concrete, as they are capable of resisting the high alkalinity environment of concrete, can bond chemically with cement, are a non-corrosive material, have high thermal resistance to heat, and also have a high tensile strength. As a result, basalt fibres are considered to be a promising material to improve the concrete strength.

Materials and Methods

Materials

Basalt fibres: The chemical composition of the basalt fibres is given in Table 1. This was derived through Van de Velde K [2] analysis. To

promote bonding between the fibres and the cement it was initially intended to use a cross-linking agent and coat the fibres. A justification of the chosen cross-linking agent will now be discussed. Calcium-Silicate-Hydrate (C-S-H) reaction outputs have been notoriously difficult to predict and model. However, recent advancements in this field [3,4] have opened up the potential integration of chemically compatible additives to hydrated cement gel structures. Shahsaveri [5] and Sakhavand [6] provide a concise summary of current research on integrating polymeric and organic additives into the nano-scale structure of C-S-H. Their paper (in particular pages 85-88) briefly defines and discusses the following as possible options in this regard; Poly(methacrylic acid) (PMA); Poly(4-vinyl benzyl trimethylammonium chloride); Poly(vinyl

Compound	w% in Basalt
	Fibres
SiO_2	51.6-57.5
Al_2O_3	16.9-18.2
CaO	5.2-7.8
MgO	1.3-3.7
B_2O_3	--
Na_2O	2.5-6.4
K_2O	0.8-4.5
Fe_2O_3	4.0-9.5

Table 1: Chemical composition of basalt fibres [2].

*Corresponding author: Elshafie S, Faculty of Engineering, Sports and Science, University of Bolton, Bolton, UK, E-mail: sami.elshafie@hotmail.co.uk

alcohol) (PVA); Poly(ethylene-co-vinyl acetate) (EVA); Poly(acrylic acid) (PAA); Polyethylene glycol (PEG);18hexadecyltrimethylammonium (HDTMA); Dimethyl sulfoxide (DMSO); and Methylene blue (MB). PVA in particular shows good bonding potential with the silicates in the C-S-H structure [6]. The accepted bonding model and associated observations describe hydrogen bonding with slip-stick mechanisms under applied loads. This infers that there would be an increase in flexural toughness of the macro-scale material–a beneficial and complimentary property when considering the other constituent materials used within the mix. Having established that PVA has chemical bonding potential with the silicates of the cement hydration products it should be noted from Table 1 that the greatest constituent compound of the basalt fibres is silica (SiO_2).

Aggregates: Fine aggregates were well graded with a fifty-fifty mix of coarser sand and finer sand. The coarser sand had 32% passing a 600 μm sieve whilst the finer sand had 97% passing a 600 μm sieve when a grading classification test was performed in compliance with BS EN 1015-1:1999 (BSI, 1999).

A locally sourced coarse aggregate of limestone chippings was selected with a 10 mm nominal maximum size of aggregate. In justification of the selected aggregate size (which deviates from conventionally larger aggregate sizes used in construction practice) it was intended to choose aggregate which was smaller in size than the lengths of basalt fibres used. It was hoped that taking this approach would improve fibre bridging through the concrete composite and encourage fibre fracture failure modes as opposed to fibre pull-out failure modes, thus utilising the full load capacity potential of the fibre-larger aggregate pieces would create discontinuities within the concrete composite.

In essence, the relative sizing between the constituent materials used in the mix was intuitively derived through a 'scaling of structural strength' approach as discussed in depth by Bazaant [7].

Cement: There are many blends of cement currently available with the most common ones listed in BS EN 206:2013 (BSI, 2013). Ordinary Portland Cement (OPC) CEM I was used which avoided any bonding complexities through additions such as Pulvarised Fuel Ash (PFA). The use of OPC ensured sufficient C-S-H silicates would be produced to create hydrogen bonds with the PVA coated basalt fibres.

Methods

Eleven concrete mixes were cast with different percentages of Basalt fibres lengths and proportions, as shown in Table 2. In order to assess the effect of changing basalt fibre content and lengths on concrete mechanical properties, the research methodology presented in this section comprises two main steps:

Stage 1:- Aims to determine the optimum volume fraction of basalt fibre on the mechanical properties of concrete by changing the total basalt fibre content, from mixes (M1-M7) with basalt fibre content varying from 0.2% to 0.3%, and then determine its influence on the mechanical properties of the concrete.

Stage 2:- Using the optimum volume fraction of basalt fibre obtained from stage 1, the research is then carried out by changing the percentage content of basalt fibre lengths (12 mm and 24 mm) using mixes (M8-M11) by varying the percentage content of basalt fibre lengths of 12 mm and 24 mm between 0% and 100%. The objective is to identify the optimum basalt fibre length and content that have an influence on the mechanical properties of concrete.

In this research, concrete mixing was carried out in a laboratory at a temperature of 20°C. Both coarse and fine aggregates were stored in dry conditions, and normal tap water was weighed and added to the mixture. All materials were mixed in a concrete mixer with a maximum capacity of 0.06 m³. In order to obtain a good concrete mix, the following procedure was adopted.

A mixture of fine aggregates consisting of 50% coarse grey sand and 50% fine red sand was added, with a quarter of the water content. After 60 seconds of mixing, the 10 mm limestone coarse aggregate was added to the mixture with another quarter of the water content for another 60 seconds. Afterwards, the cement, including the basalt fibres at 12 mm and 24 mm, was added. The high range water reducer was then added to maintain the content of the water and to ensure the full distribution of particles. Finally, the mixing continued for another 180 seconds.

Mixing is continued until a uniform and well compacted concrete mixture is obtained. After mixing, casting was performed into two layers for each mix and the mixes were vibrated to remove the air content from the concrete. For each of the eleven concrete mixes, the concrete was cast three times in cubes, steel prisms and steel cylindrical moulds.

After casting, the concrete was left to dry for 24 hours at a laboratory room temperature of 20°C before moulding. The cubes, prisms and cylindrical samples were then kept constantly in water at 20°C and tested after 28 days to determine their mechanical properties.

In practice different standards can be followed for testing the properties of hardened concrete, in this study BS EN 12390-3:2009 was used to determine the compressive strength, BS 12390-6:2009 to determine the tensile strength, BS 12390-5:2009 to determine the flexural strength, and BS 1881-121:1983 to determine the elastic modulus of the concrete. For testing the fresh properties of the concrete, the BS EN 12350-2:2009 slump test was used to determine the workability of fresh concrete, BS EN 12350-6:2000 was used to determine the unit weight of fresh concrete, and BS EN 12350-7:2009 was used to determine the air entrainment of fresh concrete. Each test was performed and repeated three times. For this research, a statistical analysis (ANOVA) was used to investigate the variance of the data at each mix assuming a null hypothesis equals 0.05 (i.e., the confidence interval is 95%). This gives a measure of how the data distributes about the mean, and the difference between mixes. In particular, one-way analysis of variance (ANOVA) is used to determine whether there are any significant differences between the means of independent variables. The individual data and ANOVA results are summarised in Table 3. The null hypothesis assumes that there is no difference between the test data (not significant), whereas the alternative hypothesis assumes that there is a difference between them.

A = Accept the null hypothesis

R = Reject the null hypothesis

Results and Discussions

Influence of the basalt fibre on the hardened properties of concrete

Compressive strength: Many parameters are known to affect the compressive strength of concrete, most of them being interdependent. Some of the important parameters that may affect the compressive strength of concrete are W/C ratio, properties of the aggregates, air-entrainment, curing conditions, testing parameters, specimen parameters, loading conditions, and test age. Figures 1 and 2 represent the obtained compressive strength at 28 days for the different concretes mixes M1 to M11. The compressive strength test is carried out according to BS EN 12390-3:2009 and is calculated from an average of three specimens. As can be seen from Table 3, the statistical analysis for the compressive strength shows that changing basalt fibre proportions and lengths variations leads to reject the null hypothesis, and this statically means that there is a difference between the means of the values. Hence changing fibres proportions and lengths has direct and significant effect on the compressive strength of the concrete.

Figure 1 shows the effect of changing basalt fibre content on the compressive strength of the concrete. Using basalt fibres leads to slight reduction in the compressive strength of concrete for all mixes. The compressive strength of the concrete reduces with the increase of the fibre content until 0.26%, after which this value decreased dramatically by 36% when 0.30% fibre content is used. This result is in good agreement with other researchers Paverz Ansari [8] who obtained the same trend, when basalt fibre content increased from 1% to 1.5%. Mustapha Abdulhadi [9] results also showed that the compressive strength of the cubic concrete reduces when increasing the basalt fibre content from 0.9% to 1.2% at 28 days. This considerable concrete compressive strength reduction is attributed to the following factors: Firstly, adding fibres to the concrete mix will reduce the cement paste cohesiveness, which plays a significant role in changing the compressive strength of the concrete. In particular, the compressive strength of concrete depends mainly on the cement matrix, and the effect of enhancing concrete compressive strength is not prominent by adding basalt fibre Jianxun Ma [10]. Secondly, as the fibre content increases, the coarse aggregate is decreased to maintain a constant mixture volume. Decreasing the coarse aggregate content has a direct impact on compressive strength Khaled Soudki [11]. Thirdly, as the volume content of the basalt fibres is added, the

presence of voids caused by the use of higher fibre volume of basalt fibres increases, causing reduction in the compressive strength of the concrete Tehmina Ayub [12].

In addition, as seen in Figure 1, the average compressive strength development of concrete mixes M1 – M3 shows these mix proportions meet the strength requirement for high concrete strength, achieving, on average, 41 MPa, although the highest compressive strength value was achieved when concrete contained no fibres in M1, achieving an average strength of 48 MPa. Figure 2 shows the effect of changing basalt fibre lengths on the compressive strength of the concrete for different mixes, M8-M11. The highest compressive strength achieved by all mixes in this series was for mixture M8, containing pure 100% 24 mm basalt fibre, while the remaining mixes, containing a mixture of 24 mm and 12 mm basalt fibres, achieved various strengths ranging from 28-40 MPa. This noticeable improvement in concrete compressive strength when using longer fibres 24 mm is in also in a good agreement with other researchers observation Jerin Johnson, 2016, Palchik P.P [13,14]. Their results show using longer basalt fibres of 24 mm leads to improve the concrete compressive strength significantly. In contrast, it is noticed that using 100% pure basalt fibre of 24 mm and 12 mm, such as in mix M8 and M11, has a greater effect on the compressive strength of the concrete than using a mixture of 24 mm and 12 mm with various percentages, as in M9 and M10, on the concrete compressive strength. This may be attributed to the effect of longer basalt fibres of 24 mm on concrete to create a longer anchorage to hold concrete particles firmly, which will also have a significant effect on the tensile and flexural strength of the concrete.

Figure 3 shows the main effect of the optimum basalt fibre content and length on the compressive strength of the concrete. Mixing basalt fibres with concrete has not enhanced the concrete compressive strength, as concrete is naturally very strong in compression. When concrete is compressed, the interface only serves to transfer compressive stresses from one aggregate to the next; however, under tension, the aggregates try to pull apart. Hence, concrete fails under small stress in tension, but remains strong in compression Rakshita Nagayach [15].

Tensile strength: The tensile strength test results for the different concretes mixes, M1 to M7 and mixes M8 to 11, at 28 days, are summarised in both Figures 4 and 5. The statistical analysis for the tensile strength results using Table 3 show that changing basalt fibre proportions leads to accept the hypothesis, while changing its length variations leads to reject the null hypothesis, which means there is a

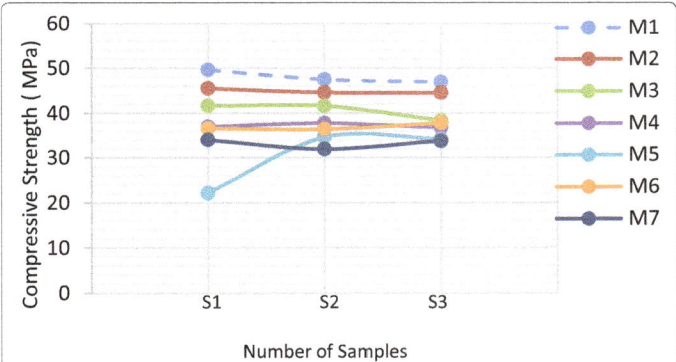

Figure 1: Compressive strength of concrete at different basalt fibre fractions.

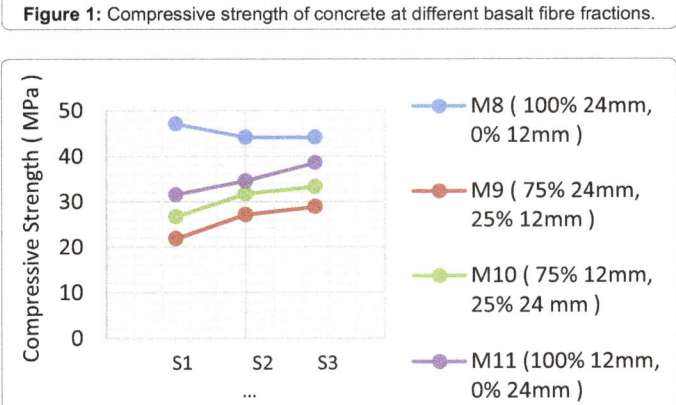

Figure 2: Compressive strength of concrete at different basalt fibre lengths.

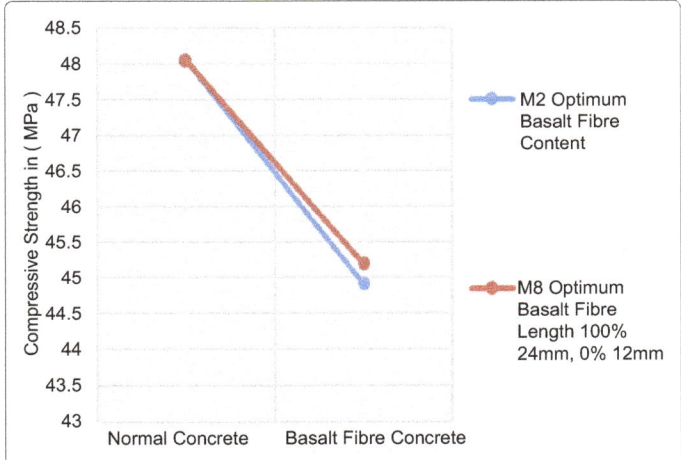

Figure 3: Main effect of compressive strength in concrete.

statistical difference between the means of the values when changing the basalt length; however, changing the content has no significance on the tensile strength.

The tensile strengths for different basalt fibre mixtures with different proportions, M1 to M7, and lengths, M8–M11, at 28 days, are shown in Figures 4 and 5. As presented in Figure 4, Mixture 1 with 0% fibre content had an average tensile strength of 2.3 MPa, while all other mixes, apart from M5, achieved a higher tensile strength than the control mixture M1, reaching a tensile strength of 2.8 MPa on average. The optimum tensile strength was achieved when mixing fibre content of M2 0.2% of the total volume of the concrete mix. The test was then carried out by using the optimum basalt fibre content of M2 0.2% and changing the basalt fibre lengths percentages of 12 mm and 24 mm to precisely determine the effect of changing the length of fibres on tensile strength of concrete. Figure 5 shows that Mixture 8, with 100% 24mm basalt fibre length, achieved the highest tensile strength with an average of 3.2 MPa. The tensile strength of the concrete increased by 17% and 28% versus the control mixture of 0% fibres when using basalt fibre content of 0.2% with 24 mm long fibres, respectively. Tensile strength results obtained by Yihe Zhang [16], Elba Helen George [17] show the same trend, as both authors used basalt fibres with different dosages, and that resulted in improving the tensile strength of concrete. Jianxun Ma [10] experimental work results showed that changing the basalt fibre lengths enhanced the tensile strength of the concrete. This

increase in is due to the fact that basalt fibres reduced crack growth and consequently led to higher failure loads. The fibres created a bridge through the split portions of the cylinder and prevented the two parts from splitting Khaled Soudki [11]. It is also noticed that longer fibres (24 mm) improved the tensile strength of the concrete much more than 12 mm fibres. This is due to the high tensile strength of these fibres and their ability to form longer bridges to hold concrete particles. On the other hand, Figure 6 shows the effect of fibres in concrete. The tensile strength of concrete is obviously improved by adding basalt fibre, when compared with normal concrete 0% fibres. The splitting tensile strength of concrete increased with increase in the content and length of basalt fibre.

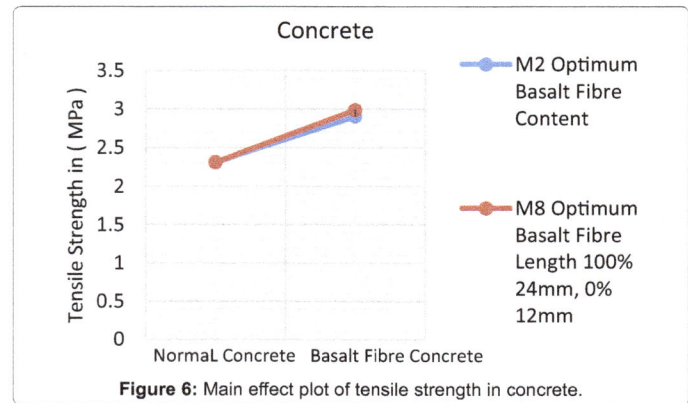

Figure 6: Main effect plot of tensile strength in concrete.

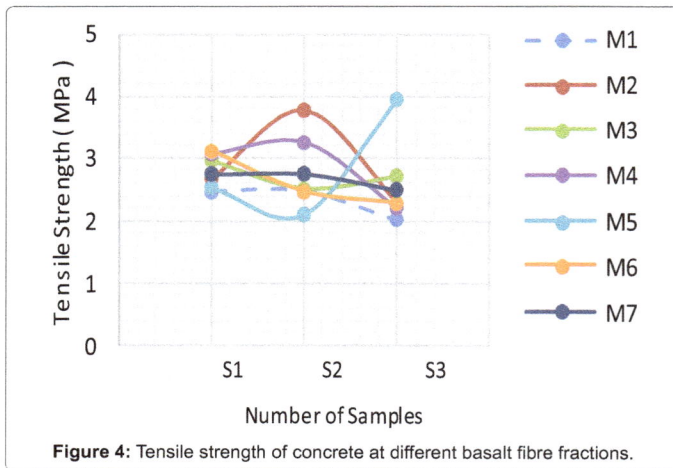

Figure 4: Tensile strength of concrete at different basalt fibre fractions.

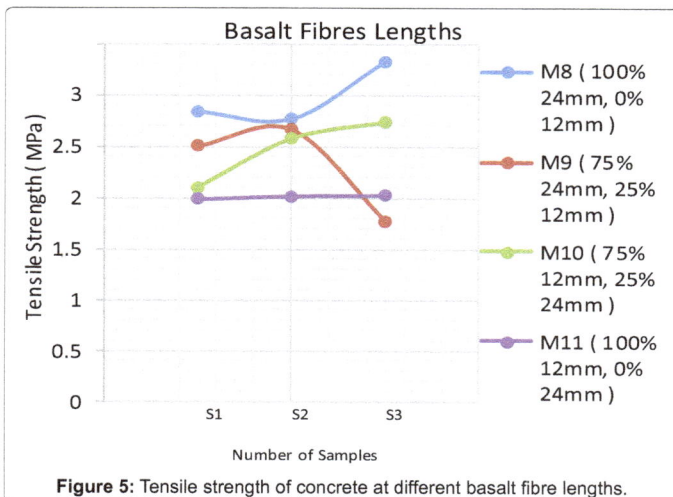

Figure 5: Tensile strength of concrete at different basalt fibre lengths.

Specimen	Basalt	Basalt Fibre	Basalt Fibre
Mark	Fibre	12mm	24mm
	Content %	Length %	Length %
M1(Control	0%	50%	50%
Mix)			
M2	0.2%	50%	50%
M3	0.22%	50%	50%
M4	0.24%	50%	50%
M5	0.26%	50%	50%
M6	0.28%	50%	50%
M7	0.3%	50%	50%
M8	0.2%	0%	100%
M9	0.2%	25%	75%
M10	0.2%	75%	25%
M11	0.2%	100%	0%

Table 2: Details of the concrete mixes.

Description	ANOVA Result
Compressive Strength (28 days)	
Variation in Fibre Content / 50% 12 mm-50% 24 mm	R
Variation in Fibre Lengths / 0.2% fibre content	R
Tensile Strength (28 days)	
Variation in Fibre Content / 50% 12 mm-50% 24 mm	A
Variation in Fibre Lengths / 0.2% fibre content	R
Flexural Strength (28 days)	
Variation in Fibre Content / 50% 12 mm-50% 24 mm	R
Variation in Fibre Lengths / 0.2% fibre content	R
Modulus of Elasticity (28 days)	
Variation in Fibre Content / 50% 12 mm-50% 24 mm	A
Variation in Fibre Lengths / 0.2% fibre content	A

Table 3: ANOVA results.

Flexural strength: The flexural strength test results for the different concretes mixes, M1 to M7 and mixes M8 to M11, at 28 days, are summarised in both Figures 7 and 8. The statistical analysis for the flexural strength results using Table 3 show that changing basalt fibre proportions and length variations leads to reject the null hypothesis, which means there is a statistical difference between the means of the values; hence, changing fibre proportions and lengths has a significant effect on the flexural strength of the concrete. The test for the flexural strength development of concrete was also carried out at different ages of seven days, 28 days, 84 days, and 168 days using the optimum basalt fibre that has influence on the flexural strength of the concrete. A three point bending test was used using a 150 × 150 × 750 mm prism. The flexural strength of the concrete under the three point bending test was calculated using the following formula:

$$\sigma = \frac{3FL}{2bd^2} \qquad (1)$$

Where:

F is the load (force) at the fracture point (N)

L is the length of the support span

b is width

d is thickness

σ is the flexural strength in (MPa)

The flexural strength development of concrete mixed with different proportions of basalt fibre is illustrated in Figure 7. It is evident that increasing basalt fibre content increases the flexural strength of the concrete when compared to the control mixture, M1, 0% fibre content. The optimum basalt fibre content was achieved when using basalt content of M2, 0.2%, while mixture M1 0% basalt fibre achieved the lowest value. The flexural strength ranged from 1.3 MPa to 2 MPa when changing the basalt content, while this value increased when changing the length of the fibres, ranging from 1.5 MPa to 2.5 MPa as can be seen from Figure 8. The Figure 8 also shows that the flexural strength of the concrete increased sharply when using basalt fibre content mixtures of M2, M3 and M4, 0.2%, 0.22%, 0.24%, respectively, at 28 days and then gently decreases afterwards to give a lower strength values when using contents of M5, M6, and M7, 0.26%, 0.28%, and 30% respectively. The increased number of fibres that cross the crack surface is one of the main reasons to increase the flexural strength of the concrete, as they reduce the crack opening. Kayali [18], Zollo [19] also reported that the presence of basalt fibre with different contents has a contribution to the flexural strength of concrete. The test was then carried out by using the optimum basalt fibre content of M2, 0.2 %, with changing the basalt fibre lengths percentages of 12mm and 24mm, to precisely determine the effect of changing the length of fibres on flexural strength of concrete. According to Figure 8, Mixture 8, with 100% 24mm basalt fibre length, consistently displayed the highest flexural strength with an average of 2.3 MPa. The differences in the flexural strength due to the change of the length of the fibres were considerable and average values of 2.3 MPa, 2.1 MPa and 1.7 MPa were recorded when using different lengths of fibres. In contrast, using longer fibres, such as 24 mm, resulted in increasing the flexural strength of fibres, as can be seen from Figures 7 and 8, but, when using a blend of 12 mm and 24 mm fibres, the flexural strength remained the same, achieving an average value of 2 MPa. This is perhaps because of the ability of the longer fibres that cross the crack region to reduce the concrete cracks more significantly than the short fibres, as they are able to transfer load more efficiently, and also redistribute stress. The longer fibres take to transfer stress, the longer time it takes for the crack to develop; hence, the higher the tensile and flexural strength that can be reached. The same result was achieved by Jianxun Ma [10], who concluded that increasing the fibre length and content will increase the flexural strength of concrete.

In general, the flexural and tensile strength of the concrete reinforced basalt fibre are increased with the increase dosage and length of the basalt fibre. When the fibre content is 0.2% by total volume, which is equal to 0.27 kg/m³ and the basalt fibre length is 24 mm, It appears that for mixes with 0.2% of fibre content and length of 24 mm achieved an overall higher flexural and tensile strength in comparison to other mixes investigated in this analysis. As a result, the use of basalt fibres is regarded as a strengthening material for the concrete, and can have a significant effect on the concrete structures. On the other hand, Figure 9 show the main effect of fibres in the flexural strength of concrete. From the graph, it is clear that the flexural strength of concrete is increased by adding basalt fibre, when compared with normal concrete 0% fibres. The flexural strength of concrete increased with increasing the content and length of basalt fibre. In Particular, mixing concrete with 24 mm basalt fibre length will have a larger effect on the flexural strength of the concrete when compared with using different fibre lengths.

The flexural strength development of concrete at different ages of seven days, 28 days, 84 days and 168 days using the optimum basalt fibre mixture (M8) that has influence on the flexural strength of the concrete is illustrated in Figure 10. The Figure 10 shows that the flexural strength of the concrete increases sharply when increasing the age of the test, reaching its minimum value at seven days, and maximum value at 168 days. This is due to the hydration of cement, as well as the chemical

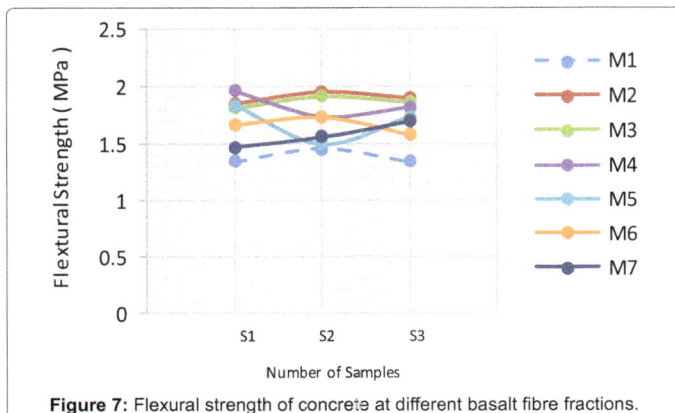

Figure 7: Flexural strength of concrete at different basalt fibre fractions.

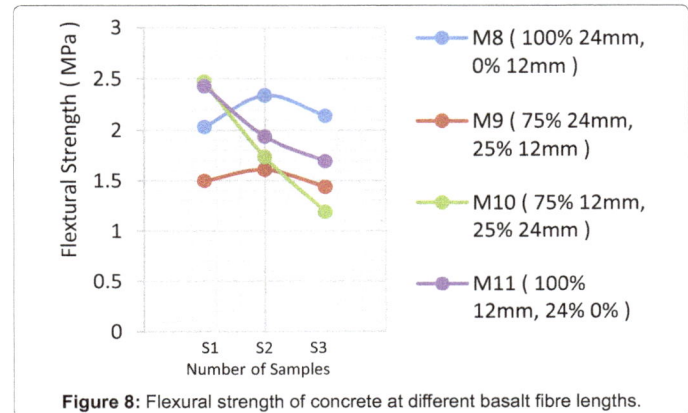

Figure 8: Flexural strength of concrete at different basalt fibre lengths.

reaction on the concrete, as they require time to formulate, which helps concrete gain more strength. The presence of basalt fibres also plays a major role, as they enhance the flexural strength of the concrete. Using the same figure, it can be seen that there is a big difference in concrete strength at the different testing ages of seven days, 28 days, 84 days and 168 days. The lowest values were registered at seven days, with similar average values of 2.2 MPa recorded after 28 and 84 days, while the largest strength average values of 2.4 MPa were recorded after 128 days. The rate of the flexural strength increase with days is considerably fast, ranging, on average, from 1.3 MPa to 2.4 MPa. In other words, the flexural strength development between seven and 168 days was not only influenced by the presence of basalt fibres, but also by the curing time of the concrete.

Modulus of elasticity: The statistical analysis for the modulus of elasticity results using Table 3 show that changing basalt fibre proportions and length variations leads to accept the null hypothesis, which means there is no statistical difference between the means of the values; hence, changing fibre proportions and lengths has no significant effect on the modulus of elasticity of the concrete. The elastic modulus is calculated within elastic range in its ascent in stress-strain curve and is expressed in function of compressive strength of concrete. Figure 11, shows the elastic modulus for all mixes. Based on the graph in Figure 11, the modulus of elasticity for all mixes ranged from 12000 MPa to 21000 MPa. Mixture M8 (100% 24 mm basalt fibre with 0.2% content) had the highest modulus of elasticity, while Mixture M1 (the control mixture)

had the lowest modulus of elasticity. Basalt fibres slightly increased the modulus of elasticity of the concrete. This may be attributed to the effect of fibres to enhance the concrete density, as a result improving the concrete toughness and resistance to deformation. The modulus of elasticity achieved by M2, M3 and M4 was 29%, 1.15%, and 33% above of the control mixture, no fibres, respectively. The greatest increase in the modulus of elasticity occurred between M3 and M4, followed by a gradual decrease from M4 to M6. The general value of the modulus of elasticity mixed with different fibres increased and decreased slightly compared to the control mixture. This indicates that basalt fibre had not contributed majorly to the modulus of elasticity of the concrete. This result is possibly due to the strong ability of concrete to resist the compressive strength load when it is applied, also due to the inability of fibres to resist the vertical load when applied. Ramakrishnan [20] also reported that basalt fibre has no effect on the modulus of elasticity of concrete.

Correlation between the compressive strength and tensile strength: The relationship between the compressive and tensile strengths is presented in Figure 12. No correlation can be seen between those two concrete properties, as found R^2 values were 0.03 and 0.15 for mixing concrete with the optimum content and optimum length, respectively. This may be attributed to the use of different basalt mixtures between M2 and M8. For M2 mixture, a blend of 50% of 12

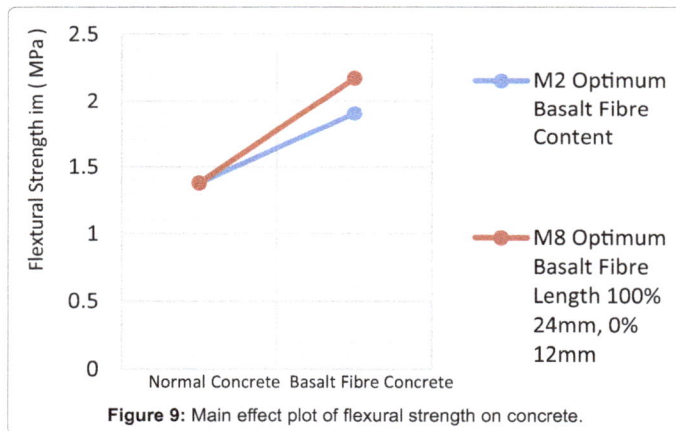
Figure 9: Main effect plot of flexural strength on concrete.

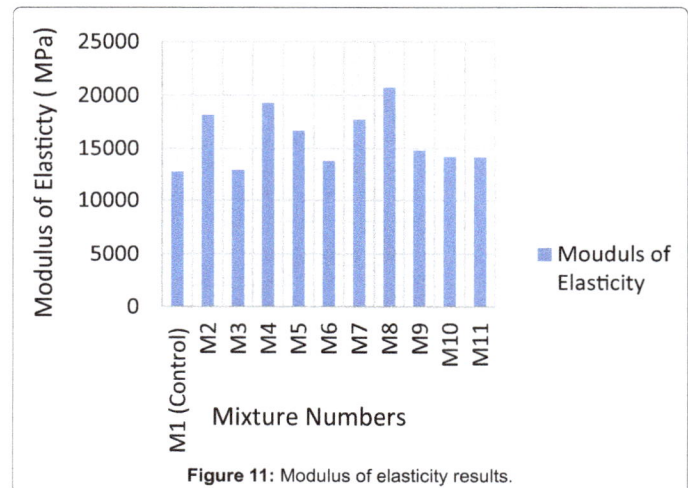
Figure 11: Modulus of elasticity results.

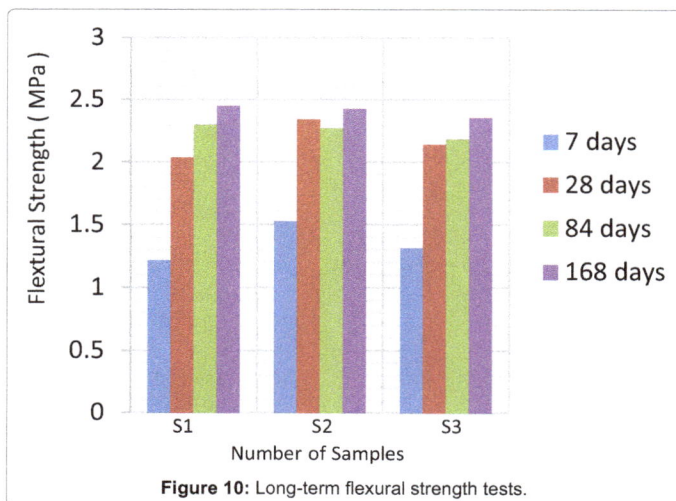
Figure 10: Long-term flexural strength tests.

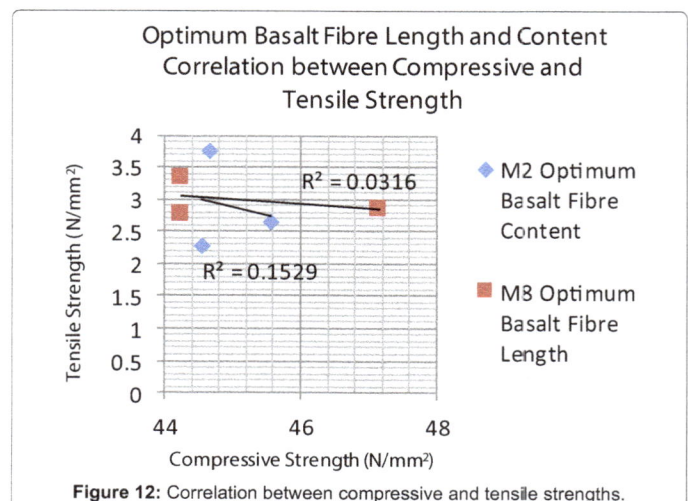
Figure 12: Correlation between compressive and tensile strengths.

mm and 50% of 24 mm basalt fibre lengths was used, while, for M8, only 100% 24 mm basalt fibre length was used. As a result, this affected the relationship between the two concrete properties.

Correlation between the compressive strength and flexural strength: Figure 13 shows the correlation between the influence of basalt fibres at optimum length and content on concrete compressive and flexural strengths. There seems to be a strong correlation between using optimum concrete length and content, M2 and M8, as found from the R2 coefficient values being 0.6 and 0.5, respectively. In addition, all data points in the graph seem to fit approximately in a straight line, which shows a linear and strong relationship between these two concrete properties.

Correlation between the tensile strength and flexural strength: The relationship between the concrete tensile and flexural strength is presented in Figure 14. No correlation can be seen between those two concrete properties, as found R2 values were 0.0853 and 0.5 for optimum basalt length M8 and optimum basalt content M2, respectively. However, by using the graph, the data points seem to fall in the same range between tensile strength values of 2.3 MPa–3.6 MPa and flexural strength values ranging from 1.8 MPa–2.3 MPa. It is also noticed that higher values were recorded when using optimum basalt fibre M8, as 100% longer fibre 24mm was used, which increased the values of the tensile and flexural strengths, respectively.

Correlation between the compressive strength and modulus of elasticity: In Figure 15, the relationship between the compressive strength and modulus of elasticity for optimum basalt fibre length (M8) and content (M2), as well as the corresponding best fit linear line, are shown. As in the case of the correlation between the compressive and modulus of elasticity, it is noticed that there is a positive relationship between the two variables, as their corresponding R2 values are 0.9 and 0.4, respectively; however, the distribution of the data points is relatively high. There seem to be no consistency in increasing and decreasing values between the two parameters by using Mixtures M8 and M2, which indicates that there is no correlation between the compressive and modulus of elasticity of concrete when mixed with different mixtures.

Workability: From the results presented at Figure 16, the target slump values for all basalt fibre mixtures ranged from 85 mm to 70 mm. It is also noted that the slump value reduced with increasing the basalt fibre content. These results were also obtained by Ali Elheber [21], N. Shafiq [22] and can be seen plotted in Figure 16. This is mainly because basalt fibre has the ability to absorb certain moisture when added to the concrete, causing a reduction in the slump of the concrete, which also improves the concrete permeability, as it reduces the pore gaps in the concrete microstructure Jianxun Ma [10]. On the other hand,

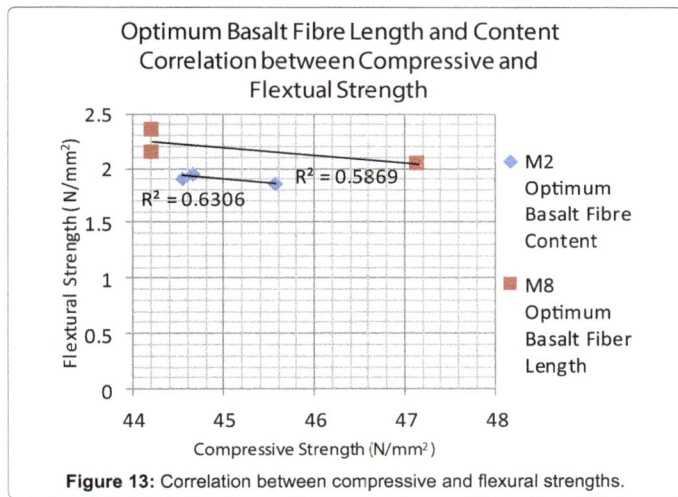

Figure 13: Correlation between compressive and flexural strengths.

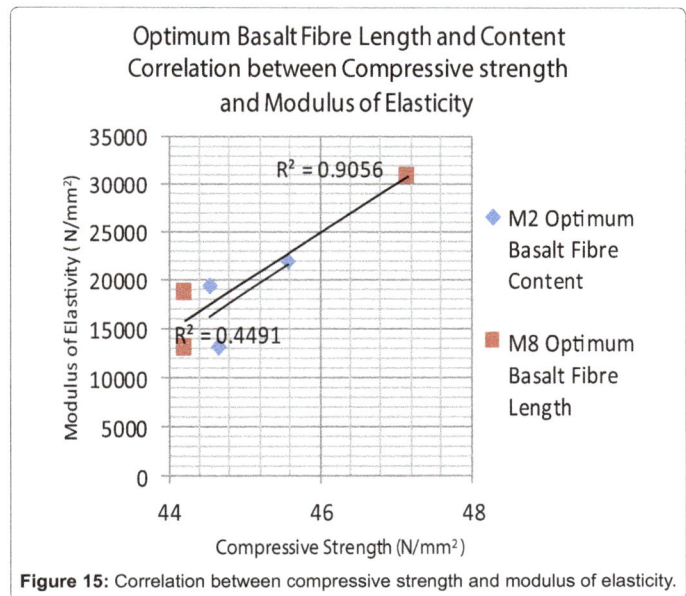

Figure 14: Correlation between tensile and flexural strengths.

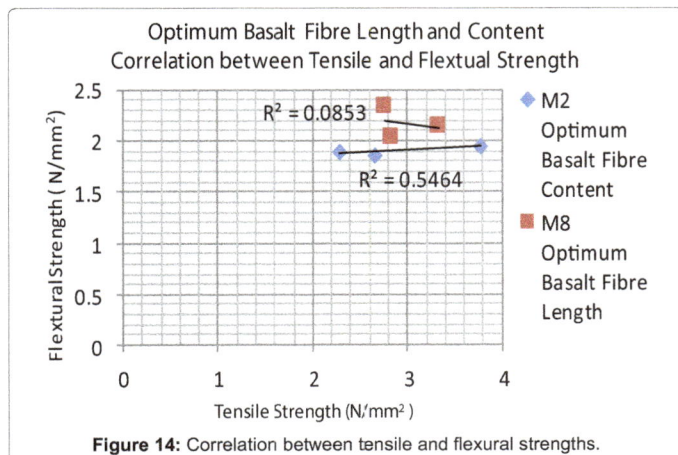

Figure 15: Correlation between compressive strength and modulus of elasticity.

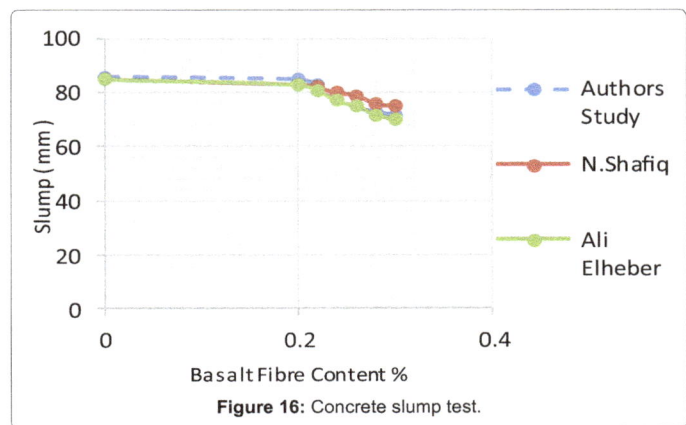

Figure 16: Concrete slump test.

increasing the content of the fibres in concrete enhances the mass added to the mixture, which leads to a decrease in the flow ability of the concrete Tumandir [23]. From the test result observed, there was no aggregate segregation and full homogeneous mix was observed from the test. Also, the result showed that adding different basalt fibre content between 0% to 0.30%, had a major effect on the concrete slump. In general, the concrete remained in good condition and remained workable after the addition of basalt fibres.

Unit weight: Figure 17 shows the mean density recorded by each concrete mixture, from M1 to M7. The density of the specimens ranged from 2400 to 2415 kg/m³. This lies within the range of 2200 to 2600 kg/m³, specified as the density of normal weight concrete Neville [24]. The graph shows that, as the content of the basalt fibre increases, there is a slight increase in the unit weight of the concrete due to the fact that basalt fibres are the lightest component in the concrete mixture and their effect on the concrete unit weight can hardly be noticed. This is similar to the results obtained by Thumandhir [23], who indicated that basalt fibres have little effect on the concrete unit weight, as can be seen from Figure 17.

Air entrainment: The air content of the basalt fibre-reinforced concrete ranged from 3.7% to 4.8%, as shown in Figure 18. The control mix (M1) achieved the highest rate of air content, as expected, while M7 with basalt fibre content of 0.30% recorded the lowest air content in concrete. The air content results indicate that, as the basalt fibre content increases (M1-M7), the air content decreases. This is due to the ability of fibres to fill voids in concrete when higher content of fibres

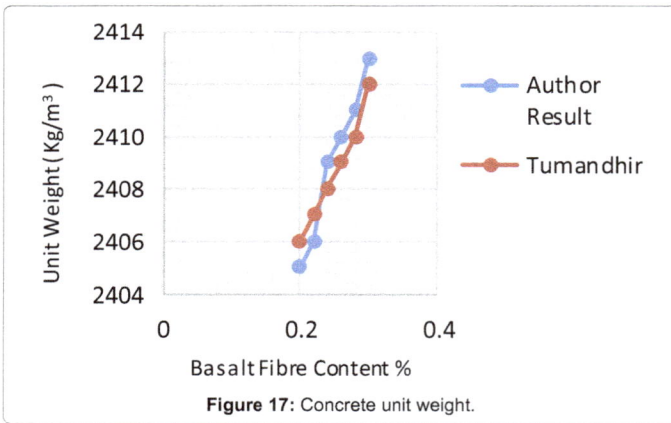

Figure 17: Concrete unit weight.

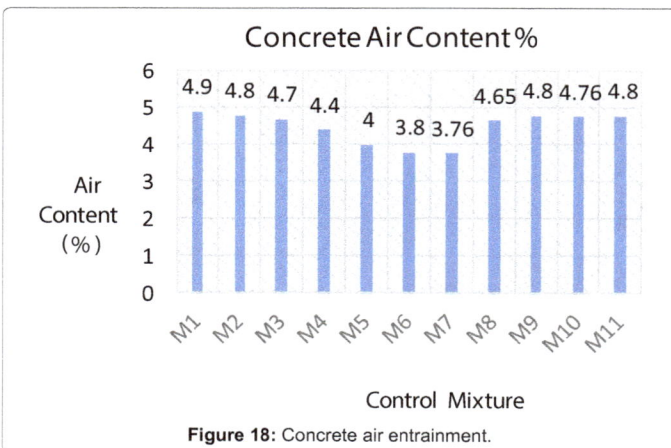

Figure 18: Concrete air entrainment.

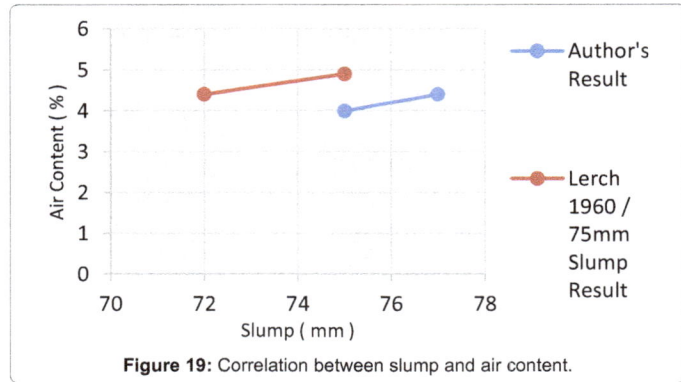

Figure 19: Correlation between slump and air content.

is added to the concrete mixture. This result was also achieved by O. Gencel [25], who found that the void content of the concrete reduces with increasing the fibre content. It is also evident that the concrete mixture with longer fibre length (100% 24 mm), Mixture 8, had slightly lower air content values than the concrete with shorter fibres (M9–M11). Air entraining provides an additional benefit to concrete mixes. This is mainly because it has a major influence on enhancing the workability of the concrete as it acts as air bubbles, which reduce the friction between the cement and the surrounding aggregate. Therefore, this reduces the interlocking and cohesion between the particles and also reduces the content of the water in concrete; hence, improving the concrete's workability. To understand this phenomenon, Figure 19 is given to illustrate the correlation between the workability and the air content in concrete. By using Figure 19, the graph shows that, as the air content in the concrete increases, the slump of the concrete also increases. And, when the air content value is between 4-5% of the total concrete volume, the corresponding slump is approximately 75 mm. This result was achieved by both Lerch [26] and myself, as can be seen from the graph below. As mentioned earlier, it is noted that, when more basalt fibre content is added to the concrete mixture, the values of the slump and air content decrease. In contrast, the relationship between the slump and air content is completely different, as there is a linear relationship between the two features, as can be seen from Figure 19. By using the same Figure 19, there was a slight difference between the author's and Lerch's values; this is due to the difference between the aggregate sizes that were used by the authors. In general it can be said that the properties of fresh concrete can be significantly improve by the addition of fibre, especially when controlling air content, unit weight and workability.

Conclusion

In this paper, the fresh and the hardened mechanical properties of concrete were investigated using different lengths of basalt fibres of 24 mm and 12 mm. This comprehensive investigation used analytical software, ANOVA, to analyse the results, and validated these values against the results obtained from the laboratory samples. The presented data indicate that the optimum basalt fibre content to have the most influence on concrete is 0.2% by the total volume of concrete, using basalt fibre length of 24 mm. It also show that changing basalt fibre length and content enhanced not only both tensile and flexural strengths of concrete, but also reduced its compressive strength, workability and air content of concrete, as well as maintain the unit weight and modulus of elasticity values. These results were also confirmed by using ANOVA software, as the analysis software indicated that change of the proportional and length of basalt fibre has significant effect on the tensile and flexural strengths, but has no major effect on the modulus of elasticity and compressive strength. A reasonable correlation between compressive strength and flexural strength was noticed, as well as between compressive strength and the modulus of elasticity for optimum basalt fibre length and content.

References

1. Warren A (2013) Report on Engineering Fibers.

2. http://www.basaltex.com/files/cms1/basalt-fibres-as-reinforcement-for-composites_ugent.pdf

3. Abdolhosseini Quomi MJ, Krakowiak KJ, Bauchy M, Stewart KL, Shahsavari R, et al. (2014) Combinatorial molecular optimization of cement hydrates. Nature communication 5: 1-10.

4. Pellenq RJM, Kushima A, Shahsavari R, Vliet KJV, Buehler MJ, et al. (2009) A realistic molecular model of cement hydrates. Proceedings of the National Academy of Sciences 106: 16102-16107.

5. Shahsavari R, Sakhavand N (2016) Hybrid cementitous materials: nanoscale modelling and characterization. Woodhead publishing series in civil and structural engineering, Elsevier.

6. Sakhavand N, Muthuramalingam P, Shahsavari R (2013) Toughness governs the rupture of the interfacial H-bond assemblies at a critical length scale in hybrid materials. Langmuir 29: 8154-8163.

7. Bazant ZP (2005) Scaling of structural strength (2nd edn). Elsevier.

8. Ansari PI, Chandak R (2015) Strength of concrete containing basalt fibre. Int Journal of Engineering Research and Applications 5: 13-17.

9. Abdulhadi M (2014) A comparative study of basalt and polypropylene fibers reinforced concrete on compressive and tensile behavior. International Journal of Engineering Trends and Technology 9: 295-300.

10. Ma J, Qiu X, Cheng L, Wang Y (2010) Experimental research on the fundamental mechanical properties of presoaked basalt fiber concrete. Advances in FRP Composites in Civil Engineering. pp. 85-88.

11. Soudki K (2013) Chopped basalt fibers in normal vibrated concrete - fresh and hardened properties. Technical Report, Waterloo University.

12. Ayub T, Shafiq N, Nuruddin MF (2013) Mechanical properties of high-performance concrete reinforced with basalt fibers. Fourth International Symposium on Infrastructure Engineering in Developing Countries 77: 131-139.

13. Johnson J (2016) A short review of basalt fibre reinforced concrete. Universal Engineering College Technical Seminar.

14. Palchik PP (2011) On control testing of fiber-concrete samples to determine their compression and tensile strength at bending.

15. https://www.quora.com/Why-is-concrete-weak-in-tension

16. Zhang Y, Pan P, Zhu B, Dong T, Inoue Y (2012) Mechanical and thermal properties of basalt fiber reinforced poly(butylene succinate) composites. Journal of Materials Chemistry and Physics 133: 845-849.

17. George EH (2014) Effect of basalt fibre on mechanical properties of concrete containing fly ash and metakaolin. International Journal of Innovative Research in Science, Engineering and Technology 3: 444- 451.

18. Zollo RF (1997) Fiber-reinforced Concrete: an Overview after 30 years of Development. Cement and Concrete Composites 19: 107-122.

19. Kayaly O, Haque MN, Zhu B (2003) Some characteristics of high strength fiber reinforced lightweight aggregate concrete. Cement & Concrete Composites 25: 207- 213.

20. Ramakrishnan V, Tolmare NJ, Brik VB (1998) Performance evaluation of 3-D basalt fiber reinforced concrete & basalt rod reinforced concrete. IDEA Program.

21. Elheber A (2014) An experimental study on the effectiveness of chopped basalt fiber on the fresh and hardened properties of high strength concrete. Research Journal of Applied Sciences, Engineering and Technology 7: 3304-3311.

22. Elshekh AEA, Shafiq N, Nuruddin MF, Fathi A (2014) Evaluation the effectiveness of chopped basalt fiber on the properties of high strength concrete. Journal of Applied Sciences 14: 1073-1077.

23. Bohan TM (2011) Thermal and Structural behaviour of basalt fibre reinforced glass concrete. University of Manchester.

24. Neville AM (2000) Properties of Concrete (5th edn). Longman, England.

25. Gencel O, Ozel C, Brostow W, Martínez-Barrera G (2011) Mechanical properties of self-compacting concrete reinforced with polypropylene fibres. Materials Research Innovations 15: 216-225.

26. William L (1960) Basic Principles of Air-Entrained Concrete. Portland Cement Association.

18

Synthesizing and Characterizing of a Novel $Zr_{90}Ni_6Pd_4$ Bulk Metallic Glassy Alloy Obtained by Spark Plasma Sintering of Mechanically Alloyed Powders

*El-Eskandarany MS

Nanotechnology and Advanced Materials Program, Energy and Building Research Center, Kuwait Institute for Scientific Research, Kuwait

Abstract

A single phase of metallic glassy $Zr_{90}Ni_6Pd_4$ powders was synthesized by mechanical alloying approach of the elemental powders, using a low-energy ball mill. The solid-solution hcp-ZrNiPd phase obtained after 25 h of the milling time transformed into a single amorphous phase upon ball milling for 100 h to 150 h. This synthesized amorphous alloy transformed into a metallic-glass at a glass transition temperature of 552.8°C. A small volume fraction of this glassy phase transformed into a mixture of two metastable phases of i-phase + big-cube upon annealing at 649.1°C. The supercooled liquid region of the metallic glassy $Zr_{90}Ni_6Pd_4$ alloy powders was 69.7°C. A complete crystallization was achieved at a temperature ranged from 649.1°C to 682.2°C through a sharp exothermic reaction with an enthalpy change of crystallization of -76.3 J/g. After this temperature, the formed metallic glassy phase was transformed to polycrystalline mixture of tetragonal Zr_2Ni and Zr_2Pd phases. The powders obtained after 150 h of milling were subsequently consolidated at 600°C, using spark plasma sintering technique. The sizes of the obtained bulk metallic glassy buttons ranged were 15 mm and 50 mm in diameter with different thicknesses in the range between 0.25 mm to 20 mm. This consolidation step led to the formation of full-dense buttons with relative densities laid in the range between 99.23% to 99.76% without precipitations of any medium- or long-range ordered phase (s). Nanoindentation approach was employed to identify the nanohardness and Young's modulus that were in the range between 7.74 to 9.32 GPa, and 135.26 to 151.15 GPa, respectively.

Keywords: Amorphous materials; Intermetallic compounds; Quasicrystals; Powder metallurgy; Differential Scanning Calorimetry (DSC); Electron microscopy (STEM, TEM and SEM)

Introduction

Metallic glasses with amorphous structures, first discovered in 1960 [1], have received great attention from almost all the materials science and metallurgical schools in the world. Synthetic metallic glassy alloys are enjoying a set of attractive physical, chemical and mechanical properties that make them pioneering desirable materials for many current and future industrial applications [2].

Metallic glassy Zr-based alloys, with their unique short range atomic order, are the best-known glassy-forming alloys that can be obtained over a wide range of compositions [3-5]. These glassy alloys exhibit many interesting amorphization and crystallization behaviors that make them ideal noncrystalline alloys for many fundamental studies [6]. However, bulk Zr-based metallic glassy alloys are successfully obtained by casting technique [7-9] the limitations on composition and size may restrict the innovation of new families of Zr-based metallic glassy alloys.

The present study aims to prepare and characterize $Zr_{90}Ni_6Pd_4$ ternary metallic glassy system that has never been reported before. To achieve this purpose we have employed low-energy ball milling process to prepare large amount of metallic glassy powders, using mechanical alloying (MA) approach. The possibility of obtaining bulk glassy samples with extraordinary large-size (~ 50 mm) by consolidation of the MAed powders has been discussed. Moreover, the nanohardness and Young's modulus of the fabricated bulk metallic glassy alloy were investigate, using nanoindentation technique.

Experimental Procedure

Pure Zr (100 μm, 99% purity), Ni (10 μm, 99.9% purity) and Pd (10 μm, 99.5% purity) metal powders provided by Alfa Aesar - USA, were used as starting alloying element materials. The powders were balanced and manually mixed inside a helium (He) gas atmosphere (99.99%)-glove box (UNILAB Pro Glove Box Workstation, mBRAUN, Germany) to give the starting charge (~ 50 g) with an average composition of $Zr_{90}Ni_6Pd_4$. The powders were then sealed together with 100 FeCr- stainless steel balls (12 mm in diameter) into a FeCr steel vial (1000 ml in volume, ZOZ GmbH, Germany), using a ball-to-powder weight ratio as 40:1. The milling process was carried out at room temperature using low-medium kinetic roller-Mill (RM01, ZOZ GmbH, Germany) with a rotation speed of 200 rpm for 150 h. Five individual milling runs were conducted in order to prepare an amount of approximately 500 g, using 2 independent vials running at the same time under the same experimental conditions. After each run, the vials were opened in the glove box where the powders were completely discharged and sealed in quartz vials under He atmosphere. In order to monitor the progress of the MA process, an independent milling run was performed for shorter times (25 h, 50 h and 100 h) where small amount of the powders were taken for different analysis. The average crystal structure of all samples was investigated by X-ray diffraction (XRD) with CuKα radiation, using 9 kW Intelligent X-ray diffraction system, provided by SmartLab-Rigaku, Japan. The local structure and composition of the synthesized material powders at the nanoscale was studied by 200 kV-field emission high resolution transmission electron microscopy/scanning transmission electron microscopy (HRTEM/STEM) supplied by JEOL-2100F, Japan, equipped with Energy-dispersive X-ray spectroscopy (EDS) supplied by Oxford Instruments, UK. The concentration of elemental Zr, Ni, Pd, Fe, and Cr in the as-ball milled powders were determined by inductively coupled plasma

*Corresponding author: El-Eskandarany MS, Nanotechnology and Advanced Materials Program, Energy and Building Research Center, Kuwait Institute for Scientific Research, Safat 13109, Kuwait
E-mail: msherif99@yahoo.com

optical (ICP) emission spectrometry. The Fe and Cr contamination contents were in the level of 0.83 and 0.32 wt.%, respectively. Differential scanning calorimetry (DSC)/differential thermal analysis (DTA) unit, provided by Setaram–France was employed to investigate the glass transition temperature, and thermal stability indexed by the crystallization temperature and enthalpy change of crystallization in a temperature range between room temperature and 750°C.

The powders of the end product obtained after 150 h of MA time were consolidated into bulk button, using spark plasma sintering (SPS-model SPS-825) provided by Fuji Electronic Industrial Co., Ltd, Japan -Japan, and hot-press (HP) provided by OXY-GON INDUSTRIES, INC-USA, under vacuum at temperature of 600°C with applied pressures of 30 and 735 MPa, respectively. In both consolidation processes, the powders were charged into 15 /20 mm diameter graphite dies. However, some larger buttons of 50 mm in diameter were produced by HPing, using 50 mm graphite dies. To avoid any undesired phase transformation during the consolidation step, the sintering process in the SPS was applied for only 3 min. For the HP, a flow of He gas was introduced to the system after completion the consolidation process to allow rapid cooling and to maintain the original amorphous structure of the sample.

The consolidated samples with different sizes were characterized by means of XRD, HRTEM, STEM-EDS, and DSC. Ion Slicer EM-09100IS, provided by JEOL, Japan was used to prepare the samples for HRTEM and STEM under Ar ion beam.

Nanoindentation was employed to determine the nanohardness (NH) and Young's modulus (E) of the bulk samples produced in the present work, using Bruker Nanoindenter (Germany) with a diamond Berkovich-tip.

Results and Discussion

Structure and morphology

Powder X-ray analysis was employed to monitor the structural changes upon mechanical alloying (MA) of polycrystalline mixture of $Zr_{90}Ni_6Pd_4$ powders, using a low-energy ball mill. The X-ray diffraction (XRD) patterns of the ball-milled powders obtained after selected MA time are presented in Figure 1. The powder of the staring stage of MA (0 h) shows sharp Bragg-peaks correspondences to the metallic alloying elements of hcp-Zr, fcc-Ni and fcc-Pd, as shown in Figure 1a. After 25 h of MA time, almost all the Bragg-peaks corresponding to Ni and Pd crystals were hardly seen (Figure 1b), suggesting the formation of hcp-ZrNiPd solid solution phase. The diffracted lines presented in Figure 1b showed significant broadening, indicating the formation of nanocrystalline grains.

The XRD pattern of the powders obtained after 100 h of the MA time (Figure 1c) reveals a broad diffuse primary and secondary haloes of an amorphous phase coexisted with unprocessed Ni and ZrNiPd solid solution particles, as shown in Figure 1c. This was confirmed by HRTEM analysis indicated that fine cells with diameter of less than 2 nm are embedded into the fine structure matrix of an amorphous phase, as shown in Figure 2a. The nano beam diffraction pattern (NBDP) displayed in Figure 2b shows a halo-diffraction of an amorphous phase coexisted with sharp-spots related to the hcp-ZrNiPd solid solution.

After 150 h of MA time, all the Bragg-peaks related to the unprocessed untransformed hcp-ZrNiAl solid solution phase had already disappeared and clear broad diffuse halos appear, implying the formation of an amorphous phase with no indication of any residual

crystalline phases, as presented in Figure 1d. The HRTEM image of the powders obtained after 150 h of the MA time is shown together with the corresponding selected area diffraction pattern (SADP) in Figures 2c and 2d, respectively. Overall, the sample, which appears featureless and homogeneous in its internal structure, showed a maze contrast with no indication of precipitations of any crystalline phases (Figure 2c), implying the homogeneity of the structure within the nanoscale level. Moreover, the SADP displays a typical halo-diffraction of an amorphous phase (Figure 2d). The absence of sharp rings and/or spots indicates the absence of any unprocessed crystalline phase(s) in the obtained powders after this final stage of MA.

Thermal stability

The DSC curve conducted with a constant heating rate of 20°C/min for $Zr_{90}Ni_6Pd_4$ powders obtained after 150 h of MA time is shown in Figure 3a. The DSC scan showed three events taken place in a temperature range between 350°C to 750°C. The first event was related to an endothermic reaction started at onset temperature of 552.8°C and corresponding to the T_g of the formed glassy phase, as shown in Figure 3a. This endothermic reaction was followed by two exothermic reactions (crystallizations) appeared at onset temperatures of 622.5°C (T_{x1}) and 649.1°C (T_{x2}), as displayed in Figure 3a. The supercooled liquid region (ΔT_x; T_{x1}-T_g) showed reasonable value (~ 70°C) for a ternary metallic glassy system obtained by MA technique.

In order to identify the origin of each exothermic reaction appeared in the DSC scan, two individual samples for XRD and TEM investigations were taken after the completion of two independent complete runs, conducted from 50°C to 650°C (#DSC1), and from

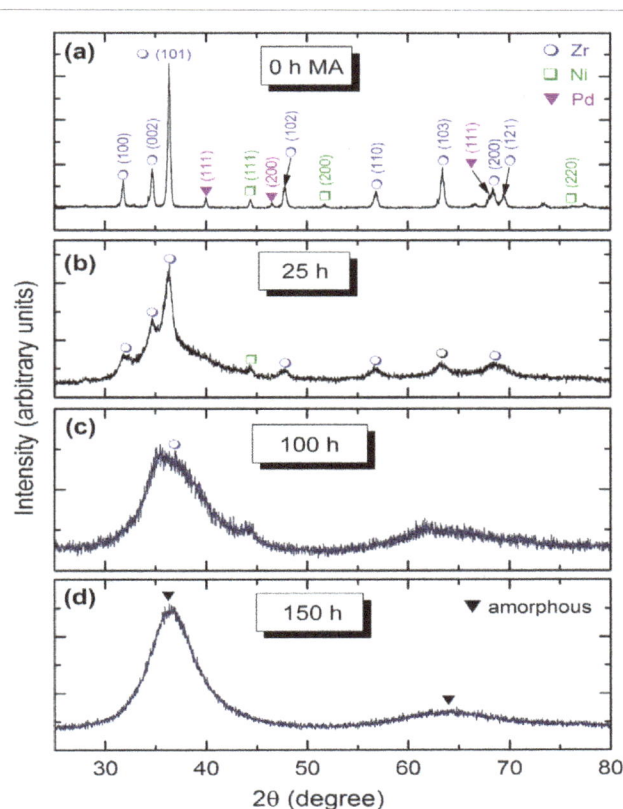

Figure 1: XRD patterns of mechanically alloyed $Zr_{90}Ni_6Pd_4$ powders after (a) 0, (b) 25, (c) 100, and 150 h of ball milling time, using a low-energy ball mill.

50°C-700°C (#DSC2), respectively (Figure 3a). The XRD pattern of sample# heated up to 650°C is shown in Figure 3b. The sample revealed Bragg-peaks corresponding to i-phase structure overlapped with diffracted lines related to big cube-$Zr_{90}Ni_6Pd_4$ phase, as shown in Figure 3b. The broadening shown in the x-ray scan (Figure 3b) refers to the existed amorphous phase in that sample. The HRTEM image of this sample shows the precipitation of nanocrystalline phase (3 nm to 6 nm in diameter), characterized by Moiré-like fringes, embedded in the fine-amorphous matrix shown in Figure 4a. Moreover, the corresponding fast Fourier transform (FFT) of this sample indicates three-fold symmetry of an icosahedral quasicrystalline phase, as presented in Figure 4b. The low magnification STEM-dark field image (DFI) of #DSC1 showed nano- grained structure (10-50 nm in diameter) embedded into the featureless amorphous matrix region (Figure 3c). The EDS-elemental mapping shows homogeneous distribution of the alloying elements without compositional gradient, as indicated in Figures 3d-3f.

Returning to Figure 3a, the sharp shoulder-like exothermic reaction appeared at 649.1°C (Figure 3a) refers to metastable (i-phase + big cube phase + amorphous phase) – to - stable – phase transformation (crystallization). This obtained stable phase was polycrystalline mixture of tetragonal Zr_2Ni and Zr_2Pd phases, as indicated in Figure 3c. Based on the thermal stability testing supported by structural analysis, we can conclude that MA process led to the formation of a high thermal stable $Zr_{90}Ni_6Pd_4$ metallic glassy powder. This solid-glassy phase tended to transform into a liquid amorphous (glassy) phase upon heating the glassy powders to 552.8°C (T_g). Such a metastable phase showed an excellent glass forming ability, indexed by the wide range of ΔT_x, which was extended up to 69.7°C. Annealing the glassy sample for short time (3s) led to the precipitation of significant volume fraction of nano-scaled medium-range ordered phases (icosahedral quasicrys + big-cube) into existed into the glassy matrix. The glassy and medium range order phases crystallized at 649.1 °C into crystalline phases with an enthalpy change of crystallization (ΔH_x) of -76.3 J/g.

Powder consolidation

In order to investigate the bulk properties of the fabricated metallic glassy material, the powder of the end-product obtained after 150 h of MA time was consolidated into bulk objects with different sizes and aspect ratios, using SPS (Figures 5a and 5b) and HP (Figure 5b). The consolidation temperature was selected in both approach to be within the ΔT_x region (600°C) at applied pressures of 30 and 735 MPa, respectively. The samples obtained by both consolidation techniques had relative density in the range between 99.23 (6.83 g/cm³) to 99.76 % (6.86 g/cm³), suggesting the formation of near-full dense bulk metallic glassy materials. Figure 5c shows the relation between the relative density and the applied consolidation temperature for samples #4 and #6 (Figures 5a and 5b) obtained by SPS and HP, respectively. The results showed that the sample consolidated by both techniques in the solid-amorphous region (200°C to 500°C) were green compact with relative density ranged between 68% to 78%, as shown in Figure 5c. The relative density of the SPSed and HPed samples was dramatically increased to 99.94% and 98.73%, respectively upon increasing the applied temperature to 560°C (above the T_g onset temperature). The density tended increase slightly with increasing the consolidation temperature to 600°C approaching the level of 99.23% to 99.76% and then saturated very close to these values at a higher consolidation temperature (615°C), as shown in Figure 5c. We should emphasize that metallic glassy materials show superplasticity in the supercooled liquid state due to the Newtonian viscous flow [10]. The unique

existence of the supercooled liquid region in $Zr_{90}Ni_6Pd_4$ metallic glassy powders offered a good opportunity for achieving the required plastic deformation, which is being necessary for successful consolidation procedure and obtaining fully dense compacts.

In order to investigate the possibility of partial crystallization during the consolidation process using SPS and HP, two small

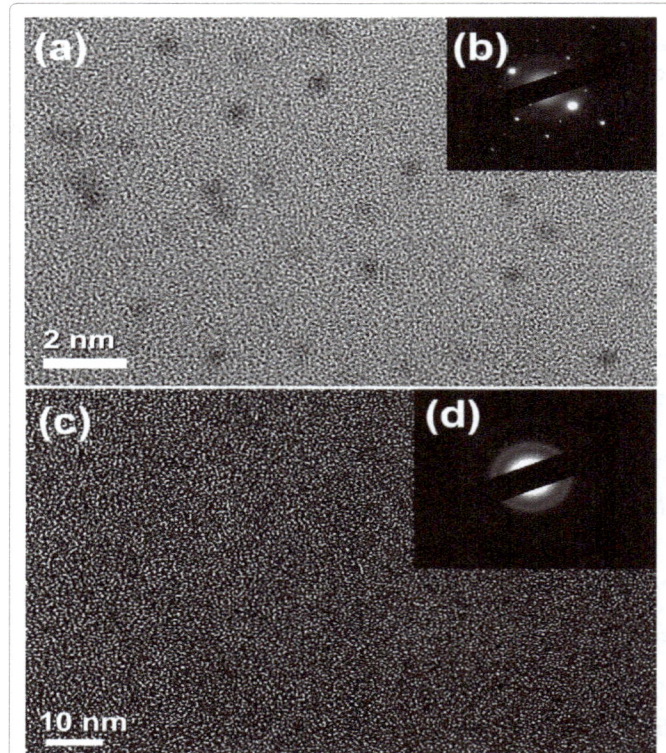

Figure 2: (a, c) HRTEM images and (b, d) the corresponding NBDPs of mechanically alloyed $Zr_{90}Ni_6Pd_4$ powders obtained after (a, b) 100 h, (c, d) 150 h of ball milling time, respectively.

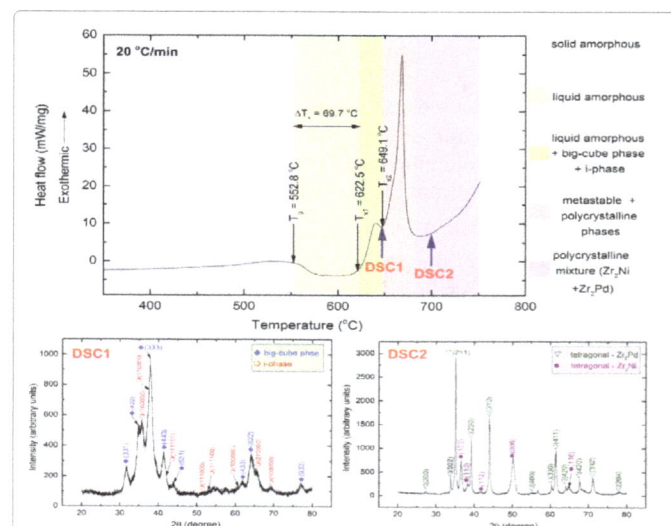

Figure 3: (a) DSC thermogram of mechanically alloyed $Zr_{90}Ni_6Pd_4$ powders obtained after 150 h of ball milling time. The XRD patterns of the samples heated in the DSC up to 650°C and 700°C and annealed at the selected temperatures for 3 min, are displayed in (b) and (c), respectively.

tetragonal pieces (~ 10 mm × 7 mm × 1 mm) were taken from #4 and #6 buttons for preparing TEM samples, using He gas ion-milling procedure. The HRTEM images for as-SPSed (#4) and HPed (#6) samples are presented in Figures 6a and 6b, respectively. The STEM-bright field image (BFI) of one sample obtained after the completion of ion milling is shown as a typical example in Figure 6b. The as-SPSed sample has a featureless maze-like structure without precipitation of long- and/or medium-range orders, as shown in Figure 6a. Moreover, the NBDP taken from the middle zone of Figure 6a showed a typical halo pattern (Figure 6c), suggesting the formation of metallic glassy phase. The HRTEM image of the large $Zr_{90}Ni_6Pd_4$ metallic glassy button (#6) showed also a maze-like structure of an amorphous phase (Figures 6d and 6e) coexisted with nano-scaled cells of a medium-range order phase, as displayed in Figure 6d. The NBDP taken from the middle zone of Figure 6d, using a beam diameter of ~5 nm, indicated that the HPed metallic glassy material had amorphous structure, characterized by the haloes shown in Figure 6f.

In order to ensure the homogeneity in the composition for as-consolidated glassy powders, intensive EDS elemental analysis for the all bulk metallic glassy buttons was conducted. The BFI micrograph of sample #6 is shown in Figure 7a. The image was virtually classified into a grid of 20 squares where the point analysis were achieved in regularly as possible Figure 7a. The collected EDS-analytical data were collected for each alloying elements and utilized to design isochemical contour maps for Zr, Ni, and Pd, as elucidated in Figures 7b-7d, respectively. The results obtained showed that the alloying elements are uniformly distributed in the bulk material with minimal fluctuation in composition, as presented in Figure 7. The closed composition values and the absence of serious gradient in concentration suggest the formation of a homogeneous bulk metallic glass with a dimension reached to 50 mm.

Nanoindentation

However, the mechanical properties for metallic glassy materials can be investigated using the traditional universal equipment, nanoindentation approach offers a new and important dimension used to investigate the local hardness and modulus elasticity of a given material beyond few micro-scaled levels. In contrast to the microhardness approach, the depth of indents developed during nanoindentation test can be only several hundred nanometres and, as such, the size of an indent is usually inferred from loading data [11]. The typical displacement-load curves for samples #4 and #6 are shown in Figures 8a and 8b, respectively. It can be seen that the total displacement for both samples developed from tens of loading cycles (45-50) were approximately equal, as shown in Figures 8a and 8b. The modulus of elasticity (E) and the nanohardness (NH) of samples #4 and #6 were obtained based on Oliver–Pharr approach [12] and plotted in Figures 8c and 8d versus the number of loading cycles, respectively. The NH values for samples #4 and #6 were fluctuated from 7.4 to 8.9 GPa, and 7.7 to 9.16 GPa, as shown in Figures 8c and 8d, respectively. Moreover, the E value for #4 was in the range between 122.5 to 152.5 GPa (Figure 8c), whereas it ranged from 134.1 to 154.3 GPa (Figure 8d). The very close values of NH and E plus the fact that they do not sharply fluctuated from tested point to another indicates the reproducibility of $Zr_{90}Ni_6Pd_4$ metallic glassy material and the uniformity in the chemical composition.

Conclusion

Mechanical alloying approach, using low-energy ball milling technique was employed to synthesize a large amount (~ 500 g) of

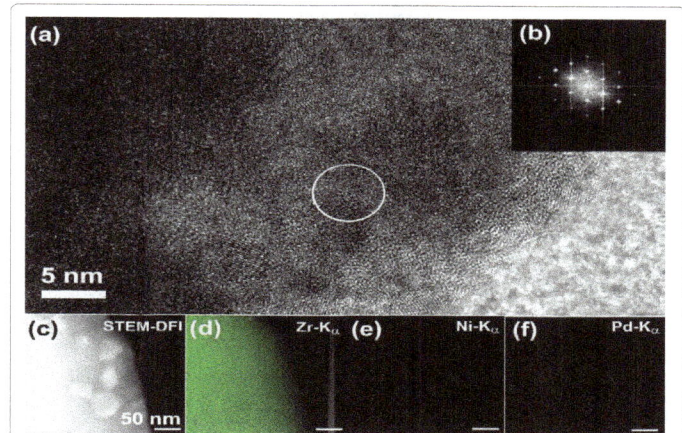

Figure 4: (a) HRTEM images and (b) the FFT of the indexed zone shown in (a) for ballmilled $Zr_{90}Ni_6Pd_4$ powders obtained after 150 h and then heated up to 650°C in a DSC under He gas flow for 3 min. The STEM-DFI, and EDS-elemental maps of Zr, Ni and Pd, are presented in (c), (d), (e) and (f), respectively.

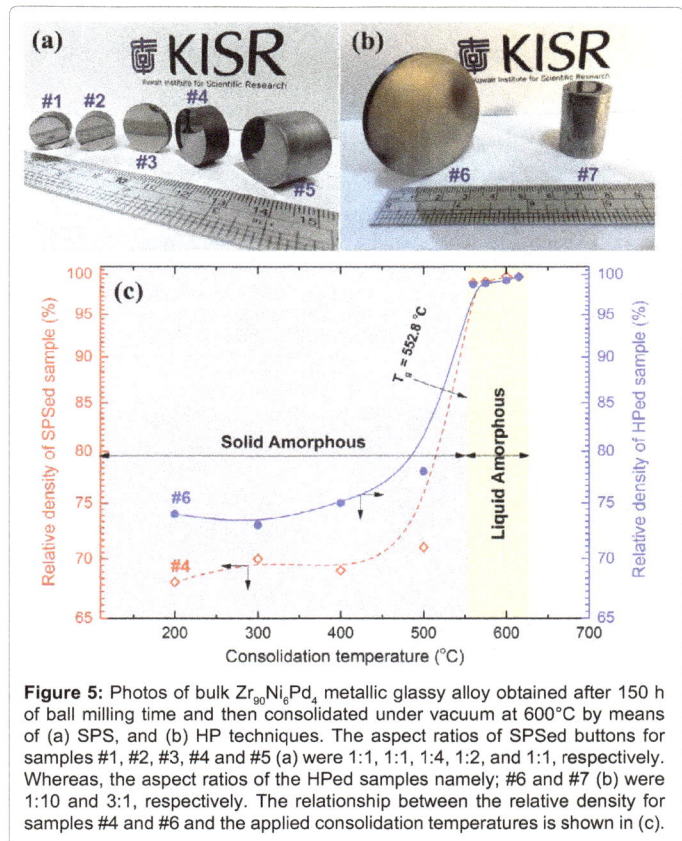

Figure 5: Photos of bulk $Zr_{90}Ni_6Pd_4$ metallic glassy alloy obtained after 150 h of ball milling time and then consolidated under vacuum at 600°C by means of (a) SPS, and (b) HP techniques. The aspect ratios of SPSed buttons for samples #1, #2, #3, #4 and #5 (a) were 1:1, 1:1, 1:4, 1:2, and 1:1, respectively. Whereas, the aspect ratios of the HPed samples namely; #6 and #7 (b) were 1:10 and 3:1, respectively. The relationship between the relative density for samples #4 and #6 and the applied consolidation temperatures is shown in (c).

metallic glassy $Zr_{90}Ni_6Pd_4$ powders. The powders obtained after 150 h of ball milling time (end-product) revealed a glass transition temperature laid at 552.8°C. A small volume fraction of the formed metallic glassy powders tended to transforms into metastable mixture of nano-grained octahedral (i-phase) and big-cube phases upon annealing at 649.1°C. The supercooled liquid region of was 69.7°C. A complete crystallization was completely achieved within a temperature range laid between 649.1°C and 682.2°C through a sharp exothermic reaction with an enthalpy change of crystallization of -76.3 J/g. After this temperature,

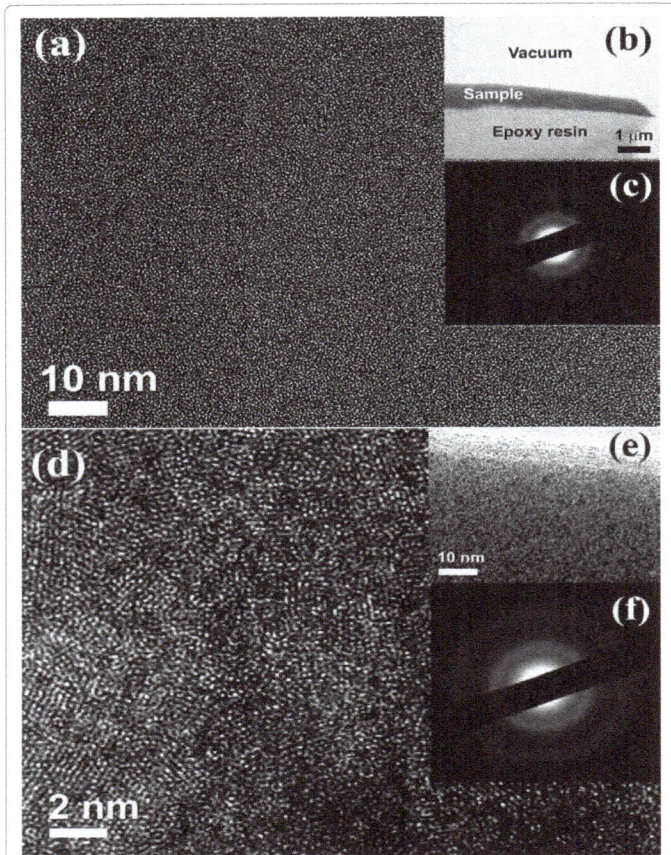

Figure 6: The HRTEM image, STEM-BFI, and NBDP for sample #4 are shown in (c), (d) and (e), respectively. The HRTEM images shown in (g, f) are for sample #6, whereas the corresponding NBDP is shown in (h).

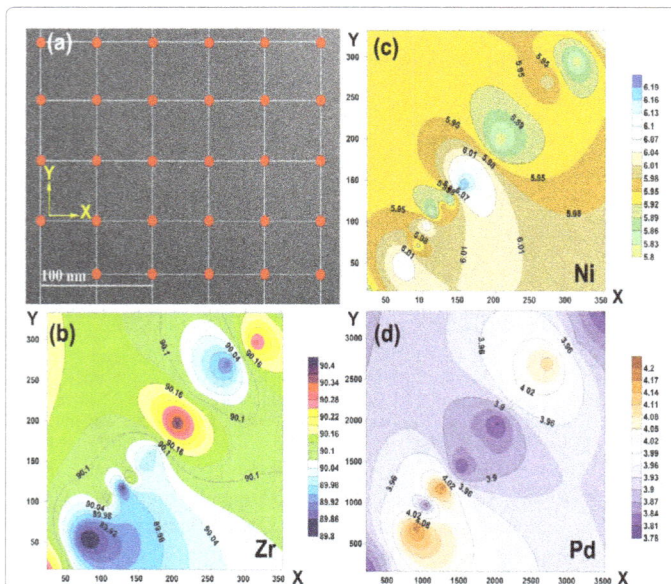

Figure 8: Loading-displacement curves for samples (a) #4, and (b) #6. The corresponding nanohardness-Young's modulus is plotted vs the number of loading cycles displayed in (c) and (d) are for samples #4 and #6, respectively. Sample #4 presents the bulk metallic glassy sample obtained by SPS of the final product of metallic glassy (150 h) in vacuum under a pressure of 30 MPa at temperature of 600°C. However, sample #6 refers to the bulk metallic glassy button obtained by hot pressing of the powders technique in vacuum under a pressure of 735 MPa at temperature of 600°C.

the formed metallic glassy phase was transformed to a polycrystalline mixture of tetragonal Zr_2Ni and Zr_2Pd phases. The powder of the end-product were consolidated at 600°C into bulk materials with different sizes, using spark plasma sintering and hot pressing technique. Both consolidation process led to the formation of full-dense bulk metallic glassy buttons with relative densities ranging between 99.23% to 99.76% without precipitations of any medium- or long-range ordered phase(s). The Zr-rich bulk metallic glassy alloy fabricated in this study enjoyed excellent mechanical properties, indexed by the extraordinary high nanohardness values, ranged between 7.4 to 9.16 GPa, whereas the modulus of elasticity was in the range between 122.5 to 154.3 GPa.

Acknowledgment

The financial support received by the Nanotechnology and Advanced Materials Program-Energy and Building Research Center, Kuwait Institute for scientific Research is highly appreciated. We would like to express our deepest gratitude to the Kuwait Government for purchasing the equipment used in the present work, using the budget dedicated for the project led by the author (P-KISR-06-04) of Establishing Nanotechnology Center in KISR is highly appreciated.

References

1. Klement K, Willens RH, Duwez P (1960) Non-crystalline structure in solidified gold-silicon alloys. Nature 187: 869-870.

2. Suryanarayana C, Inoue A (2010) Bulk metallic glasses. CRC press, Boca Raton, FL, USA.

3. Li SB, Shahabi HS, Scudino S, Eckert J, Kruzic JJ (2015) Designed heterogeneities improve the fracture reliability of a Zr-based bulk metallic glass. Materials Science & Engineering A 646: 242-248.

4. Louzguine-Luzgin DV, Louzguina-Luzgina LV, Ketov SV, Yu Zadorozhnyy V, Greer AL (2014) Influence of cyclic loading on the onset of failure in a Zr-based bulk metallic glass. Journal of Materials Science 49: 6716-6721.

5. Luo XM, Zhou Y, Lu JQ, Yu GS, Lin JG, et al. (2009) Microstructural and mechanical behavior of Zr-based metallic glasses with the addition of Nb. Journal of Materials Science 44: 4389-4393.

6. Sherif El-Eskandarany M, Saida J, Inoue A (2003) Structural and calorimetric evolutions of mechanically-induced solid-state devitrificated $Zr_{70}Ni_{25}Al_{15}$ glassy alloy powder. Acta Materialia 51: 4519-4532.

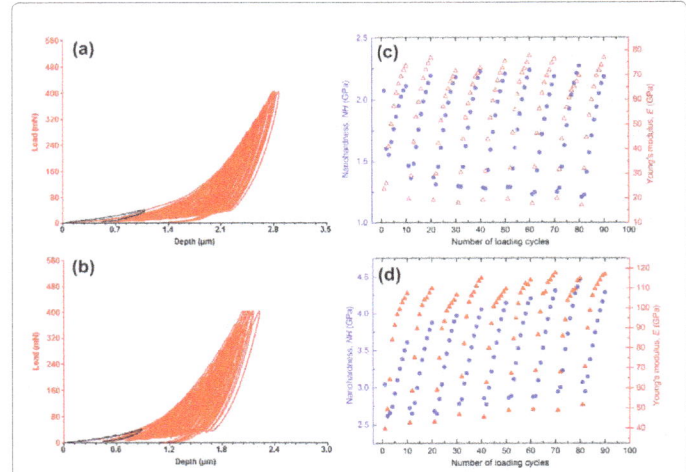

Figure 7: (a) BFI of as-consolidated bulk $Zr_{90}Ni_6Pd_4$ metallic glassy alloy obtained upon SPSing of 150 h-mechanically alloyed powders under vacuum at 600°C. The points presented in (a) refer to the selected local regions used for the EDS analysis. The corresponding isochemical contour maps for the alloying elements of Zr, Ni and Pd are shown in (b), (c) and (d), respectively.

7. Wang GY, Liaw PK, Yokoyama Y, Inoue A, Liu CT (2008) Influence of H2O/DEZ ratio on LPCVD ZnO: B films for application in a-Si:H/µc-Si:H tandem solar cells. Materials Science and Engineering: A 494: 314-323.

8. Kaban I, Jóvári P, Escher B, Tran DT, Svensson G, et al. (2015) Atomic structure and formation of CuZrAl bulk metallic glasses and composites. Acta Materialia 100: 369-376.

9. Song ZQ, He Q, Ma E, Xu J (2015) Fatigue endurance limit and crack growth behavior of a high-toughness $Zr_{61}Ti_2Cu_{25}Al_{12}$ bulk metallic glass. Acta Materialia 99: 165-175.

10. Sherif, El-Eskandarany M, Inoue A (2006) Synthesis of new bulk metallic glassy $Ti_{60}Al_{15}Cu_{10}W_{10}Ni_5$ alloy by hot-pressing the mechanically alloyed powders at the supercooled liquid region. Metallurgical and Materials Transactions A 37: 2231-2238.

11. Burgessa T, Ferry M (2009) Bulk crystal SiC blue LED with p–n homojunction structure fabricated by dressed-photon-phonon–assisted annealing. Materials Today 12: 24-32.

12. Wu J, Pan Y, Pi J (2014) Nanoindentation study of $Cu_{52}Zr_{37}Ti_8In_3$ bulk metallic glass. Appl Phys A 115: 305-312.

19

Reduction of Sulphur Content of AGBAJA Iron Ore using Hydrochloric Acid (HCL)

Ocheri C[1]* and Mbah AC[2]

[1]Department of Metallurgical and Materials Engineering, University of Nigeria, Nigeria
[2]Department of Metallurgical and Materials Engineering, Enugu State University of Science and Technology, Nigeria

Abstract

Iron ores are used in blast furnace for the production of pig iron; AGBAJA Iron ore has an estimated reserve of over I billion metric tonnes. Unfortunately, this large reserve cannot be utilized for the production of pig iron due to its high sulphur contents. This work studied the reduction of sulphur content of AGBAJA iron ore. Acid leaching methods were used to reduce sulphur contents of the ore. Sulphuric acid of different concentrations were used at various leaching times, acid concentrations and particle sizes. Atomic Absorption Spectrophotometer, X-ray fluorescence spectrophotometer, Digital muffle furnace and Absorbance-concentration technique were used for experimentation and chemical analysis. The reduction of the sulphur content of AGBAJA Iron Ore using Acid leaching process experiments were carried out at the National Metallurgical Development Centre (NMDC), Jos in Plateau State of Nigeria. Sulphur is one of the main harmful elements in ferrous metallurgy and it affects the quality of iron and steel produced. At present, Nigeria has some large iron ore deposits including AGBAJA which bear tremendous iron ore with high sulphur content of 0.12%. Central composite design technique was applied to obtain optimum conditions of the processes. Surface response plots were obtained. The percentage degrees of reduction of sulphur content of AGBAJA Iron ore were found to increase with increase in acid concentration and leaching time and a decrease in particle size for the three acids. The experimental results for percentage removal of sulphur are 85.56% the optimum % removal of sulphur is 89.66%. The result of this work has shown that AGBAJA Iron Ore if properly processed can be used in our metallurgical plants and also can be exported since sulphur contents of the ore have been reduced drastically.

Keywords: Reduction; Sulphur content; AGBAJA; Iron ore and hydrochloric acid (HCL)

Introduction

The AGBAJA ore is a fairly lean, acidic iron ore with high sulphur content. The ore is an earthy, friable material containing magnetite and goethite, together with minor aluminosilicates and phosphates of iron and aluminium. The ore contains approximately 54% iron and shows thermal effects associated with the elimination of water. The texture and chemical composition of the AGBAJA ore suggests that despite its magnetic character it cannot be easily beneficiated for use in a direct, non-conventional iron making process. The various slag-forming constituents (silica, alumina, lime and magnesium oxide) are so closely associated with the iron-bearing constituents that separation is impossible to achieve by simple physical means. Furthermore, the high sulphur content (about 0.12%) would probably give rise to problems in steel production unless a conventional, oxidizing; liquid-metal process (such as basic oxygen steelmaking) is used following blast-furnace production of liquid iron. For the AGBAJA iron ore the basicity value is very low (approximately 0.035) and hence the ore would need significant additions of lime, limestone or a lime-rich ore to make a self-fluxing sinter or pellet suitable for iron production. The reduction/removal of the high sulphur content can be achieved through the process of leaching using nitric acid.

Statement of the problem

AGBAJA iron ore is the largest iron deposit in Nigeria with an estimated reserve of over 1 billion tonnes. This iron ore has high relative high sulphur content. Consequently, the iron ore deposit is abundant in both research work and exploitation. The high sulphur content in steel making cause brittleness or crackability depending on the type of Steel products. The sulphur content in the AGBAJA iron ore has a detrimental effect on the steel making process using the ore

as raw materials in steel making. It is therefore, necessary to drastically remove/reduce the sulphur content.

Purpose and goals

The purpose of this work is to carry out experiments on the reduction of sulphur content of AGBAJA iron ore using acid leaching. Leaching of lean ores or complex ore in different acids has proved successful for several years. However, the leaching of sulphur contaminated iron ore has made a very limited progress. This underscores the ongoing intense research in the area for several decades. Depending upon the degree of association of sulphur with the minerals in the iron body, iron ore can therefore the sulphur content can be reduced using the following mineral acids which are often used as leaching agent of Hydrochloric acid (HCl). The purpose therefore is to use all the three mentioned acid leaching with a view to reducing/removing of sulphur content in order to making it useable for steel making process in the Blast Furnace.

In Nigeria today, it is a known fact that most of the steel industries are not functional due to lack of infrastructure. Despite all the endowment of the abundant mineral resources that are readily available due to natural blessings. This project work was carried out

*Corresponding author: Ocheri C, Department of Metallurgical and Materials Engineering, University of Nigeria, Nigeria
E-mail: cyrilocheric@gmail.com

with a view to making use of this minerals in which the iorn ore falls into the materials used in the production of iron and steel. Nigeria is endowed with large deposits of iron ores; however, the ores are not suitable for the production of Direct Reduced Iron (DRI). The ore closest to specification in terms of sulphur content is the Itakpe iron ore deposit which has a reserve of about 300 million tonnes. The ore is beneficiated from its natural occurrence of 36.8% Fe and 45.8% acid gangue (SiO_2) + Al_2O_3 to 63% Fe and 8% acid gangue by the National Iron Ore Mining Company (NIOMCO). The latter parameters are only suitable for the Blast Furnace (BF) plant located at Ajaokuta Steel Company Limited which is yet to be completed. However a sister plant located at Aladja which has been in operation since 1982 requires some 1.5 million tonnes of high quality iron ore (66% Fe and <3.5% of SiO_2 + Al_2O_3).

The supply of the high grade ore has so far been met only through importation which imposes high financial constraint on the operation of the Aladja plant. It can be found that at a landing cost conservatively put at 104 dollar per tonne of the ore and at even 50% capacity utilization. More than 10 billion Naira will be expended on the procurement of the ore alone. In order to meet the challenges posed by the importation of ore to meet the local demands at the steel industries in Nigeria, it therefore becomes imperative to investigate the suitability of the AGBAJA Iron ore with a view to carry out the reduction of the sulphur content of AGBAJA iron ore using acid leaching. High sulphur content found in the AGBAJA iron ore has been verified to have adverse effects for use in the Blast Furnace of the Ajaokuta Steel Company Limited. Nigeria is one of the richest countries of the world in terms of mineral deposits. Among these deposits is iron ore, located at AGBAJA, Kogi State, Nigeria. AGBAJA iron ore deposit, the largest in Nigeria is about 1.3 billion tonnes. AGBAJA iron ore is of low silicon modulus (SiO_2/ Al_2O_3), fine texture and contains about 1.4-2.0% sulphur.

Iron is one of the most abundant elements in the earth's crust, being the fourth most abundant element at about 5% by weight [1]. Astrophysical and seismic evidence indicate that iron is even more abundant in the interior of the earth and is apparently combined with nickel to make up the bulk of planets core. Iron ores are mainly composed of iron oxides, and oxyhydroxides, with other accessory gangue phases. These iron ores cannot be used in the production of steel in their raw states. For them to be maximally used in the production of quality steel, they must be upgraded or beneficiated.

Although the terms coarse-grained, intermediate size and fine grained are not assigned definite or specific dermacative values in mineral processing, a fine grained iron ore is often construed as one in which mineral matter is so finely disseminated within the gangue matrix that crushing and grinding, to effect liberation, produce minute particles that respond poorly to conventional beneficiation equipment and/ or processes (froth flotation, magnetic separation gravity separation etc.) [2]. Uwadiale [3] observed that the utilization of AGBAJA iron ore is hampered by its poor response to established industrial beneficiation techniques, this is as a result of fine grained texture of the iron ore.

Sulphur may be incorporated either into the crystal lattice of iron oxides or into the gangue minerals [4]. This element has a deleterious effect on the workability of steel. For that reason, in most places only premium low sulphur ores are extracted leaving many iron mines around the world enriched in which are unsaleable, high–sulphur iron ore.

If steel is produced with high level of sulphur, that steel will be brittle and therefore not ideal for industrial application hence the need

for reduction. Depending on the degree of association of sulphur with the minerals in the ore body, iron ore can be reduced either physically or chemically [5,6].

Communition followed by wet magnetic separation or froth flotation is generally employed when the sulphates gangue minerals appear as discrete inclusions in the iron oxide matrix (primary mineralization) [5,6]. However, when sulphur is disseminated in the iron oxide structure, possibly forming cryptocrystalline phosphates or forming solid solutions with the iron oxide phases (secondary mineralization), the reduction can only be processed by chemical routes [4-6].

The chemical reduction involves the hydrometallurgical processing of the ore, that is, the selective leaching of sulphur in the ore with a reagent usually acid or base. Since early in the 19th century, suggested the use of sulphuric acid to remove sulphur compounds from lumps of iron ore. Nevertheless, a real scientific interest in hydrometallurgical processing of high sulphur iron ores can only be noticed after the last third of the 20th century, when several papers and patents were published [4-6]. Ever since, traditionally low prices of iron ore products had impeded the large-scale industrial application of chemical reduction. At the present time, an increase in world steel production has increased demand for iron ore with a consequent increase in the price for this commodity, making hydrometallurgical sulphate removal viable [7]. In the last eight years, the situation of iron ore markets has changed dramatically due to an increase in the world steel consumption, pushed up mainly by the economic growth of China and other Asian emerging markets.

The mechanism and process analysis of reduction of sulphur content of AGBAJA iron ore concentrate using powdered potassium trioxochlorate (v) ($KClO_3$) as an oxidant has been reported [8]. Concentrates were treated at a temperature range 500°C-800°C.

The prevalence of this amount of sulphur is a major setback to its utilization in the blast furnace or direct reduction process [9]. The removal of sulphur from iron and steel presents problems because of similarity of the standard free energies of formation of iron oxide and sulphur pentoxide [9]. Consequently, in the reducing conditions of the blast furnace to recover some 99.5% of the iron charged, near complete reduction of sulphur pentoxide from the acid blast furnace occurs. As the sulphur in the ore impregnates the pig iron, there occur two distinct processes of tackling the problem: pyrometallurgical route and hydrometallurgical route. The first route employs basic slag during the conversion to steel. This technique covers the activity coefficient of sulphur pentoxide in the slag [9,10]. The second route delves into ways of reducing sulphur in the iron at relatively low temperatures. Leaching of lean ores or complex ore in different acids has proved successful for several years. However, the leaching of sulphur contaminated iron ore has made a very limited progress. This underscores the ongoing intense research in the area for several decades. As is well known, smelting process is effective for dispersion but with very high cost, and it is still under fundamental research. For physical separation, communition followed by wet magnetic separation or froth flotation is generally employed when the sulphur in the gangue mineral appears as discrete inclusion in the iron bodymatrix(primary mineralization) [5-7]. Low sulphur extraction, high grinding cost and iron loss are the major disadvantages of the method. However, when sulphur is disseminated in the iron structure, possibly forming cryptocrystalline sulphates or solids solutions with the iron oxide phases (secondary mineralization), and the beneficiation can only proceed by chemical routes [5,7]. He investigated reduction with acid leaching. In their studies, the acid concentrations were very high and low sulphur extraction was obtained. In this study, the feasibility of reduction of sulphur content of AGBAJA iron ore by acid leaching at various dilution ratios, dwell time and particle size will be investigated. Also, the Design of Experiment for the evaluation of interactive factorial effects on the sulphur removal will be investigated.

AGBAJA iron ore deposit is the largest in Nigeria with a reserved deposit of One Billion metric tonnes. However, when compared with other iron ore, it is evidently shown that the sulphur content of the AGBAJA iron ore is high (0.12%) with low silicon modules of (SiO_2/Al_2O_3=0.89) and fine grain texture which constitute major problem for utilization in the Blast Furnace (BF) direct reduction process.

Although it is Nigerian's largest known ore deposit estimated at over one billion tonnes, its utilization is hampered by its poor industrial beneficiation techniques. It is in the light of the above that the work seeks to find an alternative way of effectively reducing the sulphur content of AGBAJA iron ore apart from the conventional method of roasting sulphur. Sulphur which is one of the main harmful elements to ferrous metallurgy affects the quality of iron and steel products. At present Nigeria has some large iron ore deposits including AGBAJA which bear tremendous iron ore with high sulphur content. The removal of sulphur from iron ores involves smelting process, physical separation and chemical leaching. Smelting process is effective for reduction but very expensive and still under research. High sulphur extraction, high grinding cost and iron loss are the major disadvantages of the ore. The ore is leached with a suitable solution in a relatively simple process as it can directly treat the fines without strict requirements for the particles sizes. The cost of chemical leaching process mostly depends on the consumption of the acid leaching. In the study, the feasibility of reduction by acid leaching will be investigated. The study is to evaluate the effect acid leaching of Hydrochloric acid (HCL), this with a view of using aid leaching agent for the purpose of reducing the Sulphur content so that the ore could be used for the production of iron and steel through the process of the Blast Furnace located at the Ajaokuta Steel Company Limited. The experiment therefore to bring all these potentials to bear in the iron and steel industry.

General Description AGBAJA Iron Ores

The ore samples were compacted bonded crystalline. The AGBAJA sample consisted mainly of aggregates of brown, compact, fine – grained material with some larger, extremely friable particles. This ore is strongly magnetic. The lump ore samples were crushed mechanically and sieved to give particles in the size range 1-1.7 mm (16-10 mesh). Care was taken to ensure that the size fractions were the representative of the lump material. The ore particles were crushed further for certain experimental techniques. Analyses for calcium, magnesium, iron, aluminium, sulphur were made by atomic absorption spectrometry; silica was determined by a combination of gravimetric and colorimetric method, X-ray diffraction analysis that was performed (Figure 1).

Materials and Methods

The experiments on the research works were carried out in the course of studying the reduction of sulphur content of AGBAJA iron ore using acid leaching agents which includes HCl, for reduction. Mineral processing involves collection of ore samples, the prepared ore samples were carried out by communion and the chemical analysis of raw and scrubbed ore. Furthermore, modeling of the observed variables and data generated were used to carry out Central Composite Design (CCD) modeling technique.

Figure 1: Sample of the iron ore as obtained from Agbaja mines.

The materials and equipment that were used in this study include the following: Hand trowel, pan, mesh screens, blender, Electronic weighing balance, Mechanical shaker , porcelain pot, pH indicator device, Oleic acid, sodium silicate, distilled water, kerosene , beakers, crushers, x-ray fluorescence, spectrometer, atomic absorption spectrophotometer, LF6484A flotation machine, aero froth, cylinders, autoclave hot air oven, inoculating loop, petridishes, conical flask, binocular microscope, staining rack, glass, slides, test tubes, deionized water, and the acid leaching agents includes HCl, H_2SO_4 and HNO_3.

Reduction of sulphur content of AGBAJA iron ore using hydrochloric acid

30 kg of the raw ore sample was pulverized to 100 µ particle size for the remaining experiments. Scrubbing was carried out, where water and sodium silicate were both used. The iron ore was poured into a head pan and water was added to a reasonable level. The ore was washed and the water was decanted. This process was repeated for five times until clear water was observed. At this point 5 g of sodium silicate, and 25 drops of oleic acid was introduced and the washing continued for three more times.

After the scrubbing process, the iron ore was dried in the sun and subsequently analyzed chemically using X-ray fluorescence method. The scrubbed iron ore was pulverized and was sieved in order to obtain particle size of 10 µ, 20 µ, 40 µ, 60 µ and 80 µ. Hydrochloric acid solutions of different moles of 0.2, 0.4, 0.8, 1.0, 2.0, 4.0, 8.0, 12.0 and 16.0 were also prepared. 100 grams of particle size of 10 µ of the scrubbed iron ore was weighed and there after poured into a conical flask. 100 ml of 0.2 M of hydrochloric acid was poured into the conical flask containing the ore.

The mixture was stirred properly to ensure (uniform distribution of the particles) homogeneity. The contents were allowed to leach for 5 mins, 10 mins, 20 mins, 30 mins, 60 mins and 120 mins. After all these processes were achieved at the end of each period, the solutions were allowed to cool before filtering took place. The residues were collected, and washed which neutralizes the chemicals used as distilled water was introduced to air dry and oven dried. The experiments were repeated for 0.4, 0.8, 1.0, 2.0, 4.0, 8.0, 12.0 and 16.0 moles and particle size of 20 µ, 30 µ, 40 µ, 60 µ and 80 µ (Figures 2-5).

Hand specimen identification and specific gravity calculation

The physical analysis of the specimen indicates that the specimen has the following characteristics:

➤ Color………dark brown

➤ Streak …….brown

Figure 2: Weighing balance.

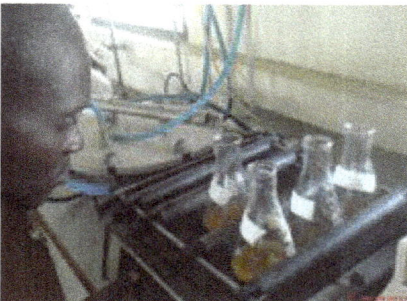

Figure 3: Mechanical shaker.

Figure 4: Blast furnace at Ajaokuta steel company limited.

Figure 5: Direct reduction iron (DRI) plant at Aladja delta steel.

➤ Specific gravity calculation

Mass of iron ore=297.58 g

Mass of equal volume of water=69.9 g

Sp. Gravity=Mass of substance/Mass of equal volume of water

=297.58 g/69 g

= 4.26

Specific gravity test: The specific gravity of a body is the ratio of the weight of the body to that of an equal volume of water.

Sieve analysis test

Sample preparation: The samples were properly mixed and large residual sample lumps were broken up. The samples were reduced to test sizes by means of Roll Crusher. The test samples were dried to a constant weight in an oven regulated at 80 ± 5°C. The mass of the dry test sample was measured in gram.

Procedure: The sieve chosen for the test were arranged in a stack, with the coarsest sieve on the top and the finest at the bottom. A tight-fitting pan or receiver was placed below the bottom sieve to receive the final undersize, and a lid (cover) was placed on top of the coarsest sieve to prevent escape of the sample.

The micrograph in Figure 6 shows the yellowish-brown colour indicates rings of iron mineral without clear grain boundaries at 250 magnifications, suggesting very fine grains. The visible bright yellow seen on deep brown patches of silica signifies the presence of sulphur in the ore. Moreover, other elements which are not visible from the micrograph occur probably in their apatite phase which makes them difficult to be seen visibly.

The above result when compared with standard values indicates that the color, the streak and the specific gravity showed that it is goethite FeO(OH) ore (Figure 6).

Also from the result obtained from the sieve analysis, a screen analysis graph of cumulative weight percentage retained against particle size was plotted. It can be observed that the 60 μm sieve retained the most quantity of iron ore particles. This shows that the AGBAJA iron ore has fine grain size.

Result of ED-XRF analysis

The chemical composition of AGBAJA iron ore as revealed by ED-X-ray florescence is shown in Table1 below. The ore was determined through XRF. The X-ray analysis indicates that the percentage of iron mineral is 89.40%. The minor elements in the ore are K, Ca, Ti, Mn, Mg, Al, Si, P, and S. A non-stoichiometry value was obtained from

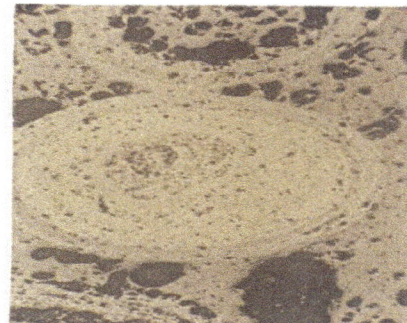

Figure 6: Micrograph of the ore as obtained after the thin section analysis.

Mineral	K_2O	CaO	TiO_2	MnO	Fe_2O_3	MgO	Al_2O_3	SiO_2	P_2O_5	S*	Al_2O_3
Composition	0.04	0.13	0.37	0.40	89.40	0.40	9.60	2.28	1.20	0.12	3.38

Table 1: Chemical composition of Agbaja Iron ore.

atomic ratio Fe: O, which implies that the ore is most likely to be a mixture of hemeatite and magnetite.

The various compositions of minerals have different effects on the ore; the presence of silicon promotes the formation of gray cast iron and as seen from Table 1. AGBAJA iron ore contains considerably high amount of silicon. It is noted that high sulphur causes hot shortness which is brittleness of steel at high temperatures. Moreover the aluminium in the ore does not pose any major problem; but makes it difficult to tap off the liquid slag. Also from the result obtained from the sieve analysis, a screen analysis graph of cumulative weight percentage retained against particles size was plotted. It can be observed that the 60 μm sieve retained the most quantity of iron ore with fine grain size.

Factorial design

Central composite design plan HCL

No. of variables=3

No. of star points=6

No. of center points=3

X_1=leaching time   (5 mins – 120 mins)

X_2=Concentration   (0.2 M – 16 M)

X_3=particle size   (10 μ – 80 μ)

Lower level	Base level	Upper level
-1	0	+1
5 mins	60 mins	120 mins x_1
0.2 M	8 M	16 M x
10 μ	40 μ	80 μ x_3

Development of statistical multivariable models

The standard design of experiment (DOE) is an efficient procedure for planning experiments so that data obtained can be analyzed to yield valid objective conclusions.

Standard Central Composite Design (SCCD) with 2^3 and 2^2 full factorial designs will be employed. These were constructed from 2^{m-t} design for cube portion, which is augmented with center points and start points.

For 2^3 full factorial designs: number of experimental points for Central Composite Design (CCD).

$$N = K^{m-t} + 2n + No \ (1)$$

Where,

K=Level of experiments=2

m=Total number of variables (3: x_1, x_2, x_3)

t =The degree of fractionality, t=0 for m<4

No=Center points added=3

Therefore,

$N=2^{3-0} + 2 \ (3) + 3=17$ runs

For 2^2 full factorial designs:

$$N=K^{m-t} + 2m + No\ldots\ldots (3.2)$$

K=Level of experiments=2

m=Total number of variables (2: x_1, x_2)

T=The degree of fractionally, t=u for m<4

No=Center points added=3

Therefore,

$N=2^{2-0} + 2 \ ^{(2)} + 3=11$ runs

The model equation for the experiment is proposed as

$$y_n=b_0 + b_1 x_1 + b_2 x_2 + b_3 x_3 + b_{12} x_1 x_2 + b_{13} x_1 x_3 + b_{23} x_2 x_3 + b_{11} x_{12} + b_{22} x_{22} + b_{33} x_{32}$$

Experimental Results and Discussion

Reduction of sulphur content of AGBAJA iron ore using HCL

Effect of HCL concentration on the percentage degree of reduction of sulphur content of AGBAJA iron ore: The effects of hydrochloric acid on the reduction of sulphur content of AGBAJA iron ore are shown in Figures 7-11. In Figure 7, it is shown that the percentage degree of reduction of sulphur content between the concentrations of 0.2 M to 0.8 M was fairly constant. It was also noted that as from 0.8 M there was a slight increase in percentage reduction of sulphur content until it got to 1 M. The experiment also showed that as from 1 M, the degree of reduction remained fairly constant. As the concentration was

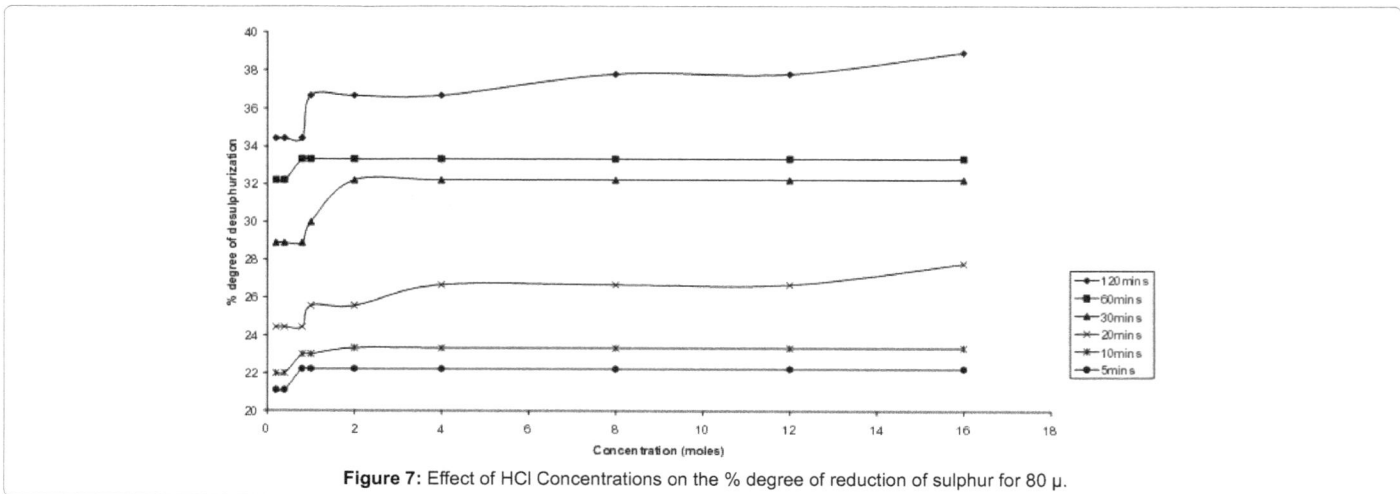

Figure 7: Effect of HCl Concentrations on the % degree of reduction of sulphur for 80 μ.

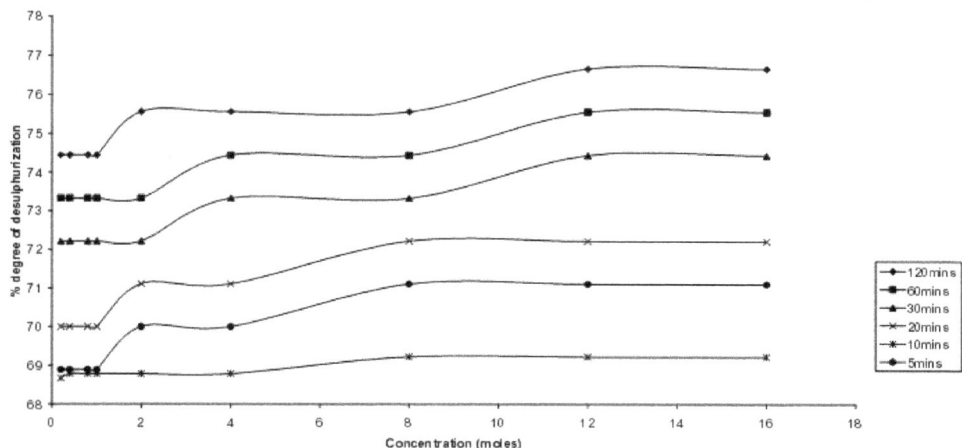

Figure 8: Effect of HCl concentration on the % degree of desulphurization for 60 µ.

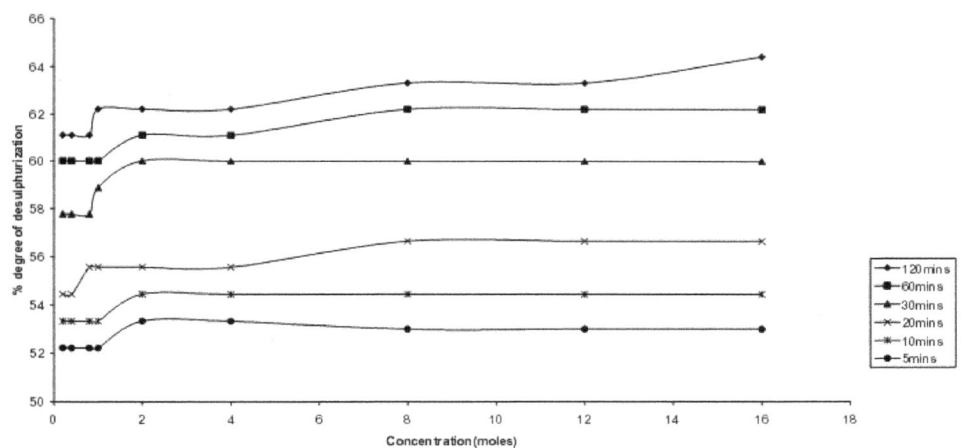

Figure 9: Effect of HCl concentration on the % degree of reduction of sulphur content for 40 µ.

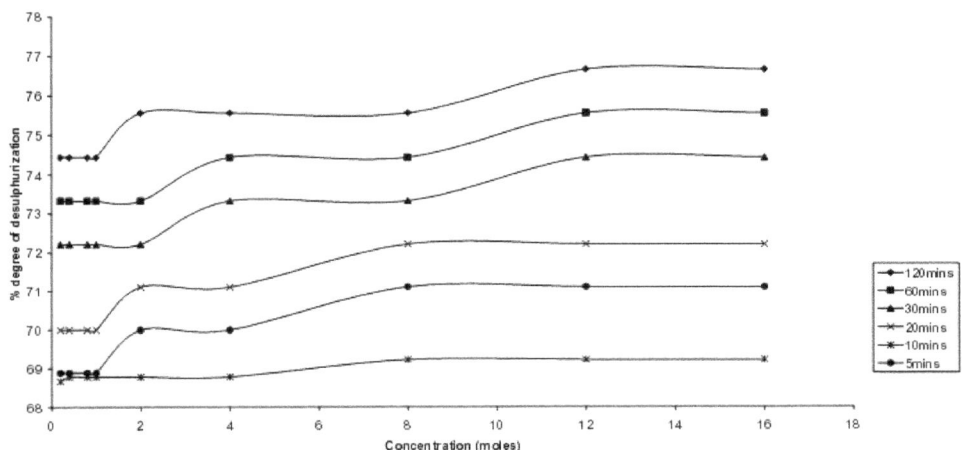

Figure 10: Effect of HCl concentrations on the % degree of reduction of sulphur content for 20 µ.

increased to 16 M for particle size of 80 μ, and leaching time of 120 minutes, the percentage degree of reduction of sulphur content was obtained at 38.89% while for 0.2 M at the same conditions, the rate of reduction of sulphur content was obtained at 34.44%.

The percentage degree of reduction of sulphur content increased fairly between 0.2 M and 0.8 M, but from 0.8 M the increment was fairly steady. The highest value of reduction of sulphur content was shown in Figure 8 and was at 43.33% at concentration of 16 M for a leaching time of 120 minutes. Between concentrations of 0.2 M and 1.0 M, the rate of reduction of sulphur content was fairly constant. But as from 1.0 M to 2.0 M, there was an increase in the degree of reduction of sulphur content. From 2.0 M the increment was fairly constant. This was depicted in Figure 9. In Figures 10 and 11 a similar trend was obtained as shown in the figures described below. It could be inferred that as the concentration of hydrochloric acid increases, the percentage degree of reduction of sulphur content slightly increases.

Effect of leaching time on the percentage degree of reduction of sulphur content of AGBAJA iron ore: The time effect on the percentage degree of reduction of sulphur content is shown in Figures 12-16. Between 5 minutes and 30 minutes there was steady increase in the reduction of sulphur content as shown in Figure 12. As from 30 minutes, the rate of reduction of sulphur content was reduced. It was also observed that for 16 M, 22.22% of sulphur content was reduced in 5 minutes while 38.89% was reduced in 120 minutes. In Figure 13, there was a gradual reduction of sulphur content until it got to 60 minutes. As from 60 minutes, the percentage of reduction was increased. The highest value of 43.33% was obtained at leaching time of 120 minutes for 16 M concentration.

The percentage degree of reduction of sulphur content as shown in Figure 14 followed a similar trend; the percentage of reduction of sulphur content for 16 M at 120 minutes was obtained at 64.44% while at a time of 5 minutes it was observed that the reduction obtained was at 54.44%. It was generally shown that the longer the time of leaching the more reduction of sulphur content was achieved.

As the leaching time increases the degree of reduction of sulphur content increases as shown in Figures 15 and 16. The highest values of reduction for 16M at 120minutes were obtained at 76.67% and 84.44% respectively. As the analyses progresses it could be observed that as the

leaching time increases the percentage degree of reduction of sulphur content increases. The results obtained from the experiments agreed with [9] which observed that the percentage reduction sulphur is directly proportional to leaching time.

Effect of particle size on the percentage degree of reduction of sulphur content: The effects of particle size on the percentage degree of reduction of sulphur content using hydrochloric acid are represented in Figures 17-22. In Figure 17, there was high percentage of reduction of sulphur content between 10 μ and 60 μ. From 60 μ to 80 μ the percentage reduction of sulphur content increased from 5.56% to 22.22%.

The highest percentage of reduction of sulphur content obtained was 89.22% for particle size of 10 μ while the least value obtained was 22.22% for particle size of 80 μ both at a concentration of 16 M for 5 minutes.

The percentage degree of reduction of sulphur content was quite significantly strong between 10 μ and 60 μ as depicted in Figure 18. From 60 μ, the percentage reduction was fairly moderate; an average percentage reduction at 27% was achieved. For 16 M concentration, and particle size of 10 μ, the highest reduction value of 89.22% was obtained.

Similar trend was obtained for Figure 19. The least value of percentage reduction was recorded at particle size of 80 μ. This was because the surface area that was exposed to leaching was less than that of 10 μ. This adduces the reason why the highest value of 83.33% was obtained using the particle size of 10 μ. The percentage degree of reduction of sulphur content was fairly moderate at particle size of 60 μ posting an average percentage reduction of 33.00%. Between 10 μ and 60 μ, an average percentage of 65.00% was achieved as shown in Figure 20. In Figures 21 and 22 it was observed that similar trend was obtained. The highest value of 84.44% was obtained in Figure 21 for particle size of 10 μ while the highest value of 85.56% was obtained for Figure 21 for the particle size of 10 μ for concentration of 16 M. It could be concluded that the percentage degree of reduction of sulphur content was inversely proportional to the average diameter of the particles. As the particle size decrease, the percentage of reduction of sulphur content increases as depicted in the works of Alafara [1].

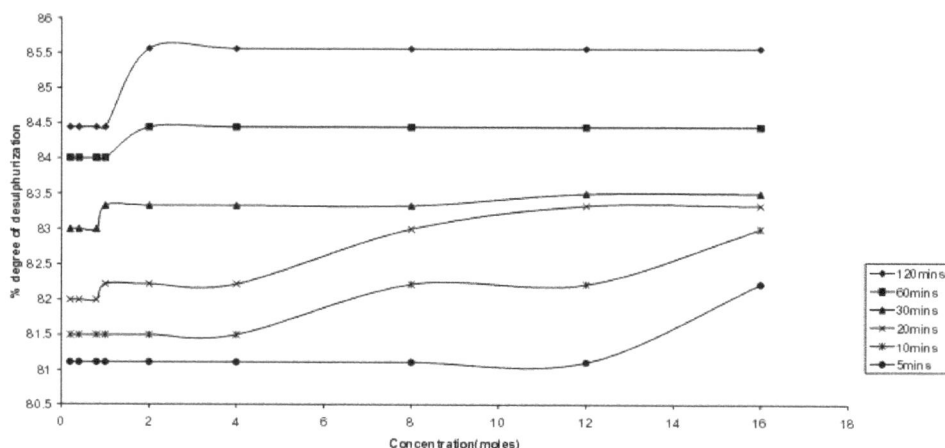

Figure 11: Effect of HCl concentrations on the % degree of reduction of sulphur content for 10 μ.

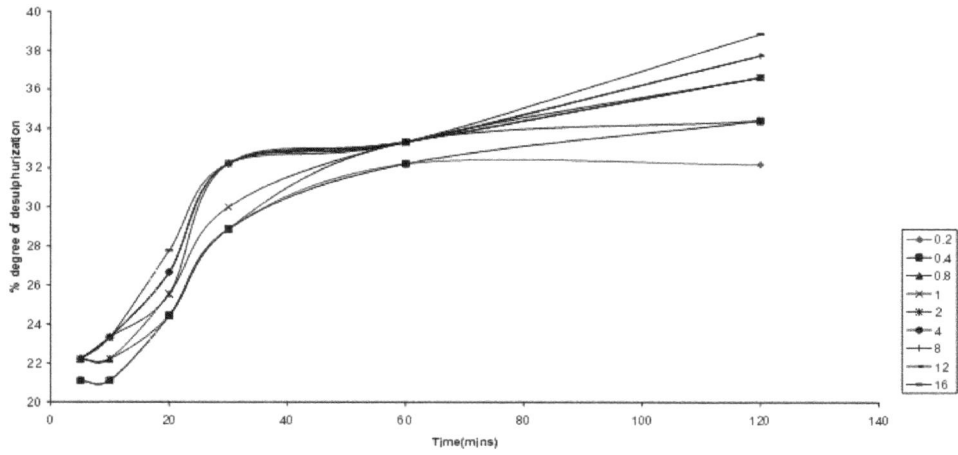

Figure 12: Effect of leaching time on the % degree of reduction sulphur for 80 μ using HCl.

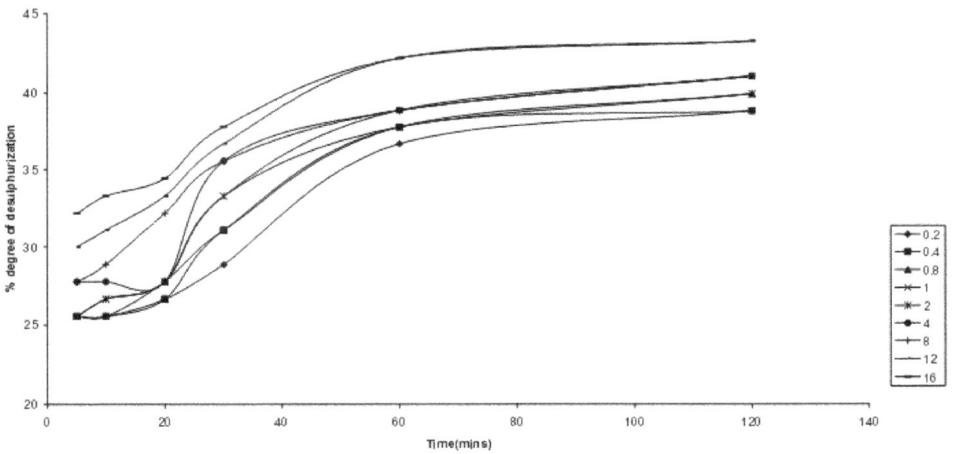

Figure 13: Effect of leaching time on the % degree of reduction of sulphur for 60 μ using HCl.

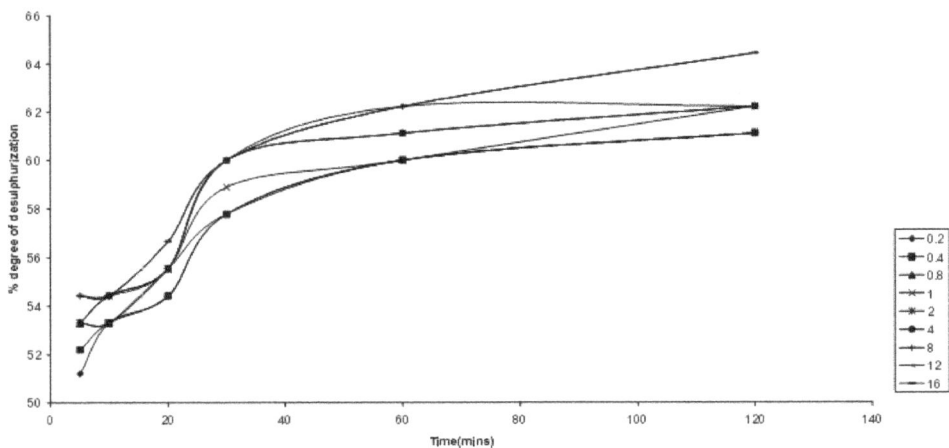

Figure 14: Effect of leaching time on the % degree of reduction sulphur for 40 μ using HCl.

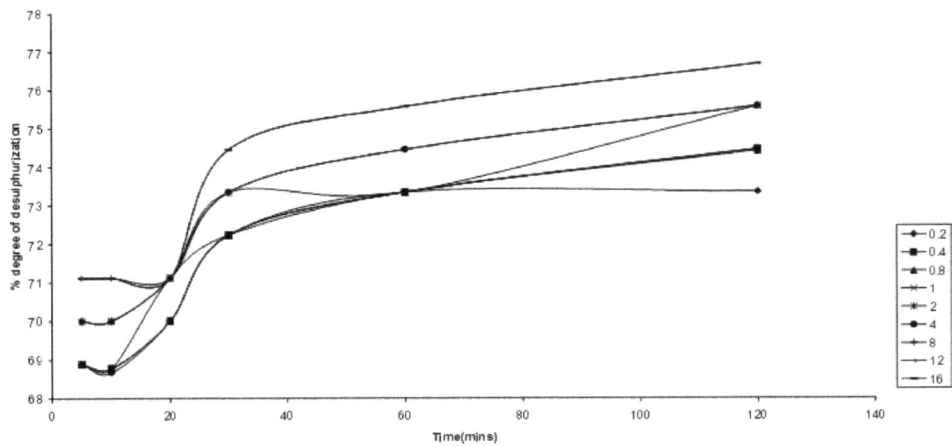

Figure 15: Effect of leaching time on the % degree of reduction of sulphur for 20 μ using HCl.

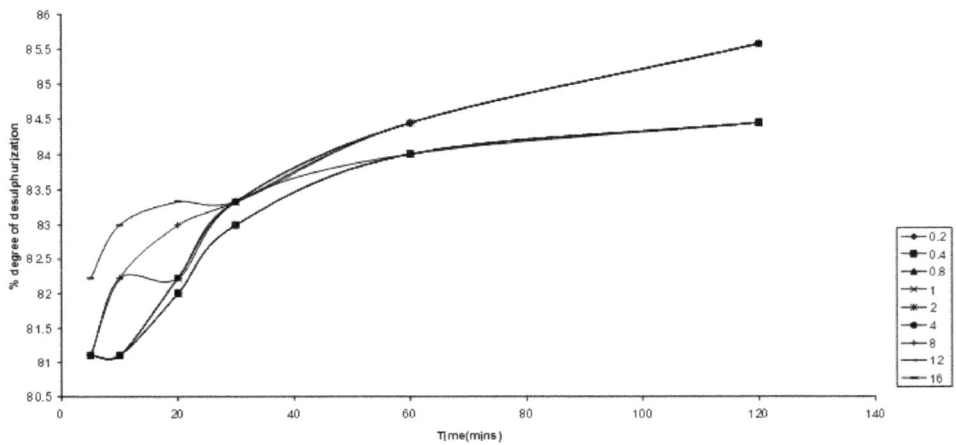

Figure 16: Effect of leaching time on the % degree of reduction of sulphur for 10 μ using HCl.

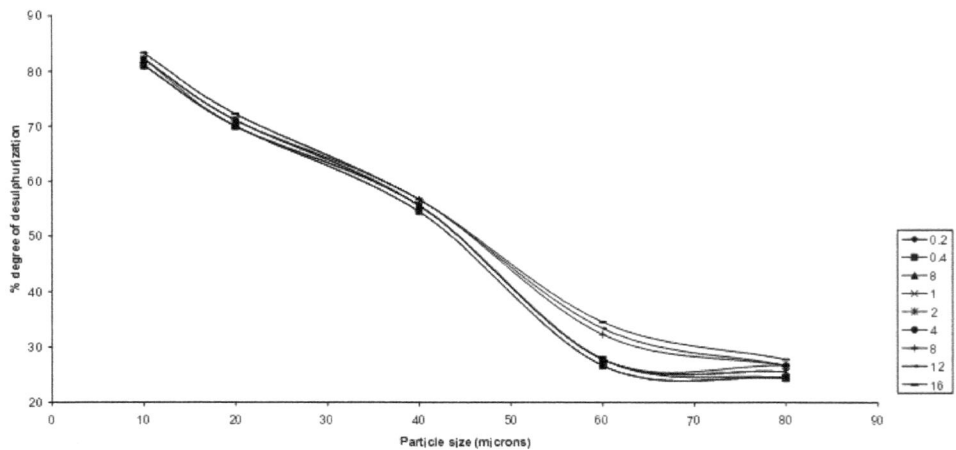

Figure 17: Effect of particle size on the % degree of reduction of sulphur for 5 mins using HCl.

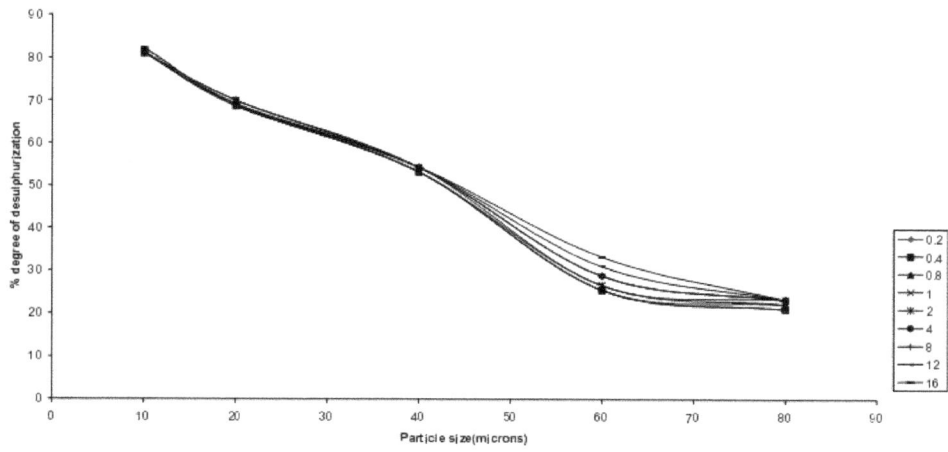

Figure 18: Effect of particle size on the % degree of reduction of sulphur for 10 mins using HCl.

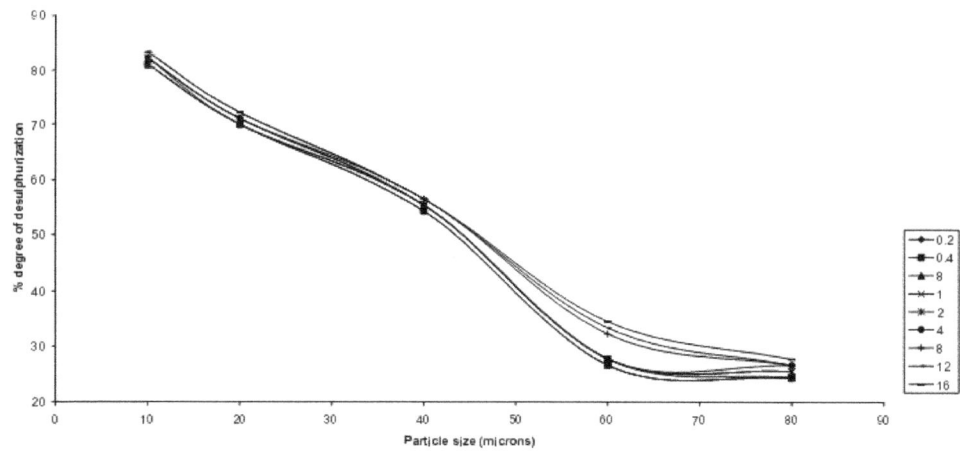

Figure 19: Effect of particle size on the % degree of reduction of sulphur for 20 mins using HCl.

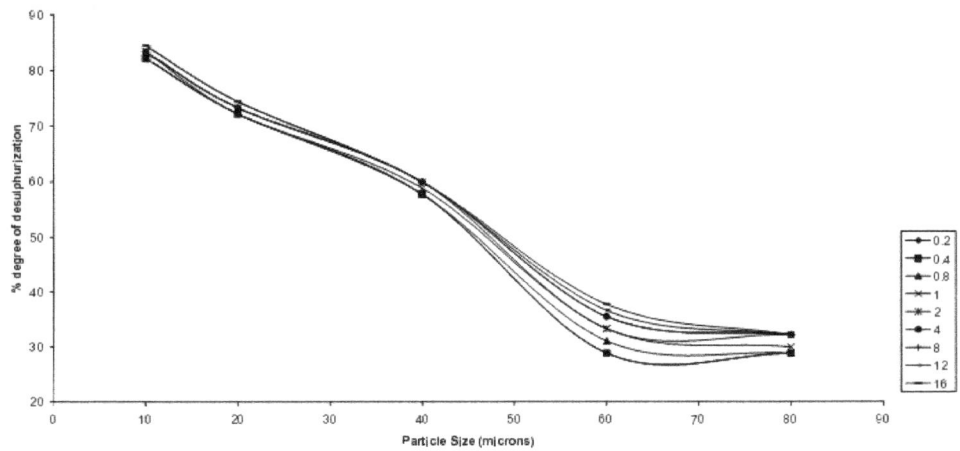

Figure 20: Effect of particle size on the % degree of reduction of sulphur for 30 mins using HCl.

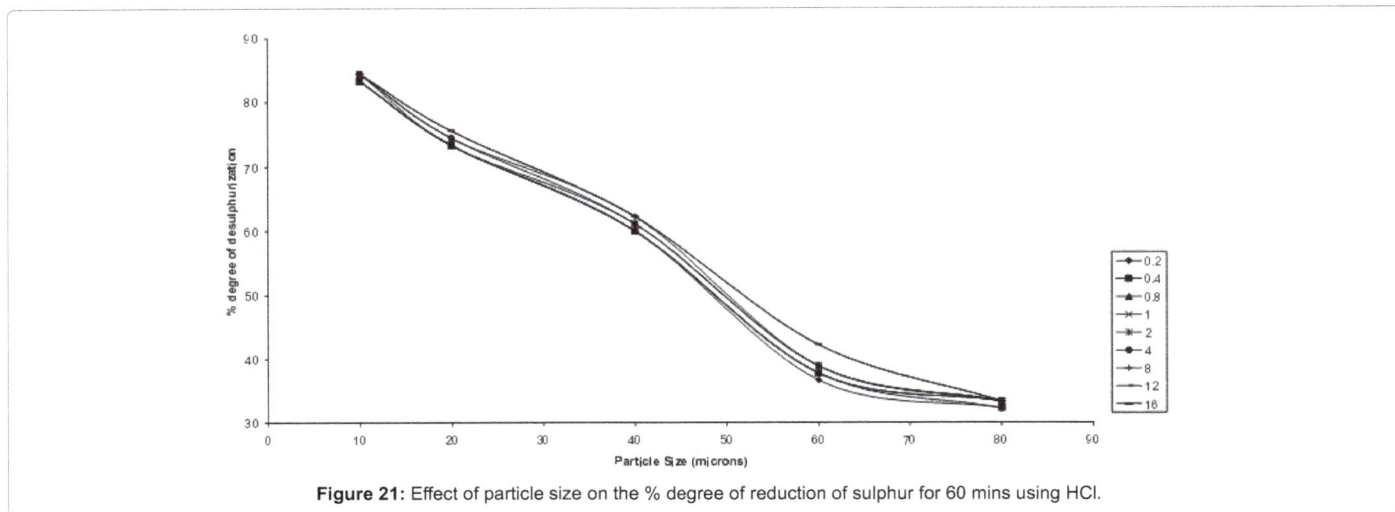

Figure 21: Effect of particle size on the % degree of reduction of sulphur for 60 mins using HCl.

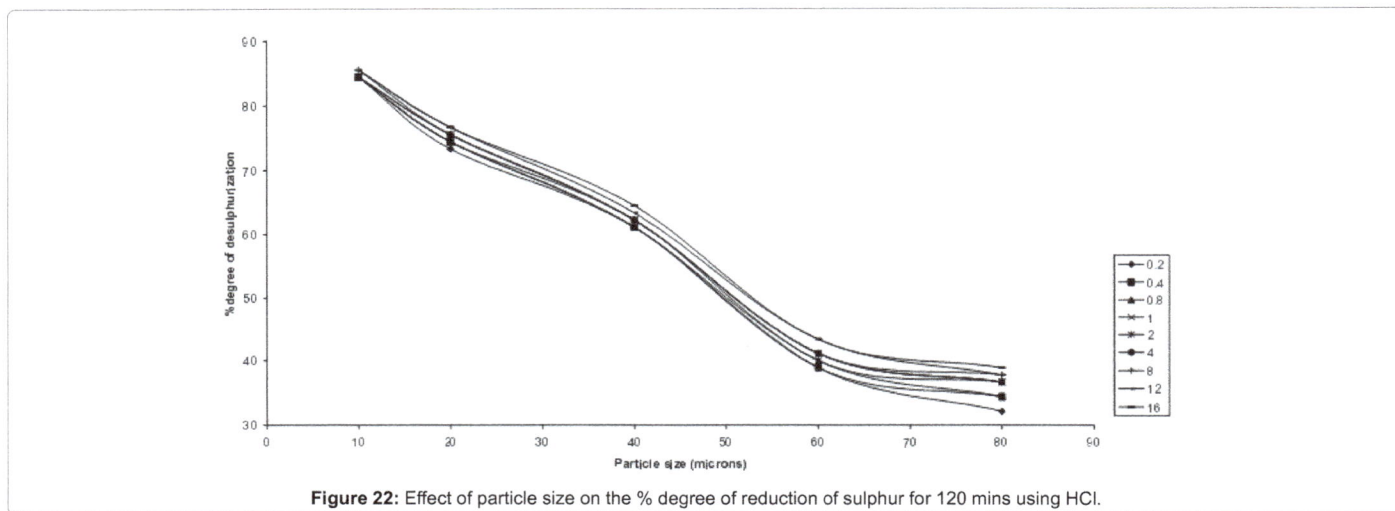

Figure 22: Effect of particle size on the % degree of reduction of sulphur for 120 mins using HCl.

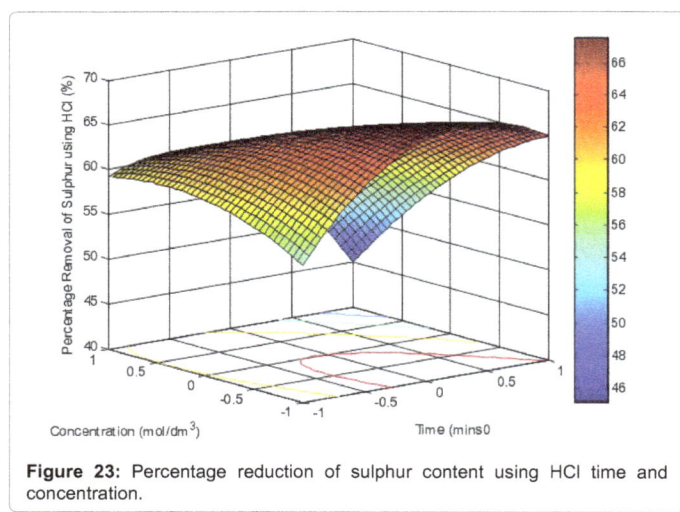

Figure 23: Percentage reduction of sulphur content using HCl time and concentration.

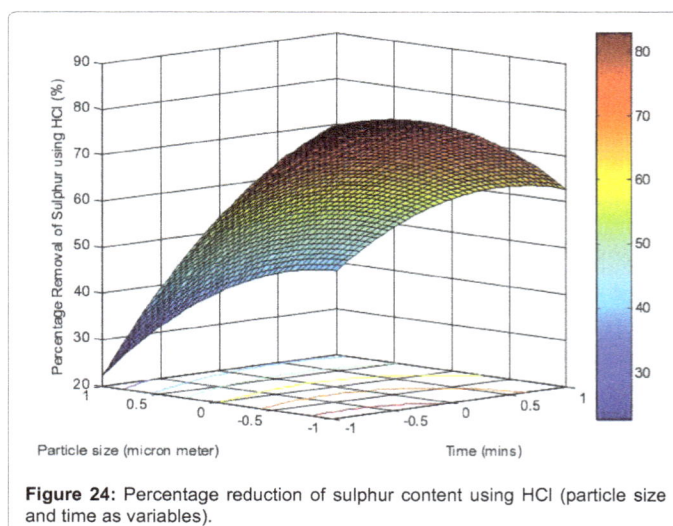

Figure 24: Percentage reduction of sulphur content using HCl (particle size and time as variables).

Surface Response Plots

Surface response as controlled by leaching variables. The surface response plots were plotted using MATLAB. In Figure 23 the surface response is represented for percentage reduction of sulphur content for time and concentrations of HCl as variables. Time responded more than concentration and optimum is located the dark red colour of the surface.

The surface response plot for percentage reduction of sulphur content using particle size and time as variables is shown in Figure 24 times responded more than particle size as shown in the figure and the optimum is located on the dark red portion of the surface.

Figure 25 depicts the surface response plot for percentage reduction of sulphur content where particle = size and concentration are variable. The optimum is located as the dark red colour on the surface and concentration responded more than particle size.

Conclusion and Recommendation

AGBAJA ore is an acidic oolitic ore consisting of goethite, magnetite and major amount of aluminous and siliceous minerals. It cannot be used directly in a blast furnace or other reduction process without further treatment, sintering, pelletizing or briquetting. Acid leaching is an effective method of reducing sulphur content from iron ores.

The recycle of hydrochloric acid solutions are incorporated processes which obtained results for the reduction of sulphur content of the AGBAJA Iron ore as a by-product which makes the whole process for reduction of sulphur content more economical.

It is therefore, noted that the percentage degrees of reduction of sulphur content using the acid leaching gave remarkable results. The Central Composite Design (CCD) used relevant parameters and data to develop the models.

The developed models were tested and the results were quite adequate in the reduction of sulphur content of AGBAJA iron ore. The experimental results for percentage degree of the reduction of sulphur content were 85.56%. It is quite significant to note that the optimum values for percentage reduction of sulphur content obtained from the model for HCL was 89.66%.

The experimental results also showed that as the leaching time and concentration increases, the percentage degree of reduction of sulphur content also increases while the particle size decreased as the leaching agent's increases. It is significantly observed that the level of reduction of sulphur content indicates that all the necessary processes were put in the right perspectives as being carried out in this research work, the final aims and objectives of the research work has been achieved.

Conclusion

The research work has availed the researcher the opportunity to explore all the experimental processes to understand the techniques involved in the reduction of sulphur content of AGBAJA iron ore using the leaching process.

It should therefore be known that the large deposit of the iron ore which was initial adjudged as not suitable for the production of Direct Reduced Iron (DRI) and for the usage at the Blast Furnace at Ajaokuta Steel Company Limited due to the harmful nature of the ore because of high sulphur of 0.12% could be drastically be reduced.

This will further indicate that the fear of harmful impurities/effects that were initially envisaged to cause some harmful effects will be reduced to a minimal level through the reduction of the sulphur content using acid leaching process as indicated in this research work.

Reduction of Sulphur Content of AGBAJA Iron Ore Using HCL

A =[1 1 1];

>> b=[3];

>> lb=[-1;-1;-1];

>> ub=[1;1;1];

>> x0=[0;0;0];

>> [x,fval,exitflag,output]=fmincon(@leaching1,x0,A,b,[],[],lb,ub)

x =

-0.3098

-1.0000

-1.0000

fval =

89.6659

exitflag =

1

output =

iterations: 3

funcCount: 19

stepsize: 1

algorithm: 'medium-scale: SQP, Quasi-Newton, line-search'

firstorderopt: 9.6488e-007

cgiterations: []

message: [1x144 char]

References

1. Alafara AB, Adekola FA, Lawal AJ (2007) Investigation of chemical and microbial leaching of iron ore in sulphuric acid. J Appl Sci Environ Manag 11: 39-44.

2. Uwadiale GGOO (1990) Upgrading fine-grained Iron Ores: (i) General review (ii) Agbaja Iron Ore. Elsevier Science Publishing Co Inc, USA.

3. Uwadiale GGOO, Whewel RJ (1988) Effect of temperature on magnetizing reduction of Agbaja Iron Ore. Metall Trans B 19: 731-735.

4. Dukino RD, England BM, Kneeshaw M (2000) Phosphorus distribution in BIF derived iron ores of Hamersley province, Western Australia. Transactions of Institute of Mining and Metallurgy (Section B: Applied Earth Science) 109: 168-176.

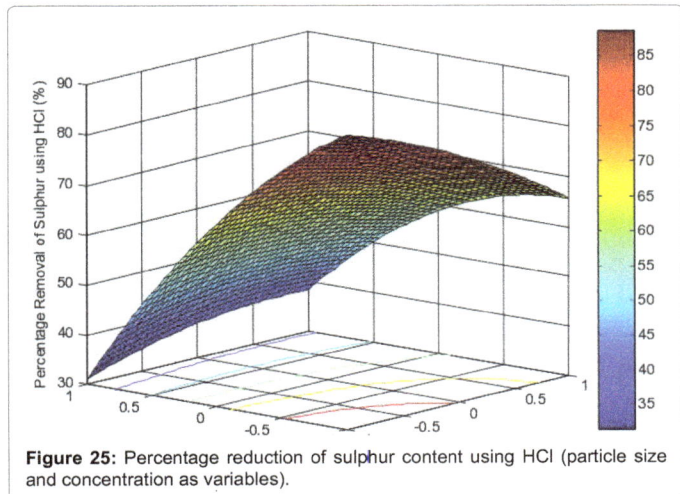

Figure 25: Percentage reduction of sulphur content using HCl (particle size and concentration as variables).

5. Kokal HR (1990) The origin of sulphur in iron making raw materials and methods of removal. Proceedings of the 63rd Annual meeting–the Minesota section AIME and 51st international symposium Duluth Minesota.

6. Fonsea D, Souza C, Araujo A (1994) Hydrometallurgical routes for the reduction of sulphur in iron ores. In: Wilkomirsky I, Sanchez M (eds.). IV meeting of the southern hemisphere on minerals technology.

7. Kokal HR, Singh MP, Naydyonov VA (2003) Removal of sulphur lisakovsky iron ore by roast-leach process. In: Young Y (ed.). Hydrometallurgy. Proceedings of the 5th International Symposium.

8. Nwoye CI (2009) Process analysis and mechanism of reduction of Agbaja iron oxide ore. JMME 4: 27-32.

9. Alafara AB, Adekola FA, Folashade AO (2005) Quantitative leaching of a Nigerian ore in hydrochloric acid. J Appl Sci Environ Manag 9: 15-20.

10. Alafara AB, Adekola FA, Bale RB (2003) Adsorption studies of chlorpyrifos and endosulfan pesticides from aqueous solution using coconut fibre. J Chem Soc of Nigeria 28: 40-44.

20

Superabsorbent Polymer Gels based on Polyaspartic Acid and Polyacrylic Acid

Sharma S, Dua A and Malik A*

Dyal Singh College, University of Delhi, Lodhi Road, New Delhi, India

Abstract

Polymer gels based on polyaspartic acid (PAsp) and polyacrylic acid (PAA) have been synthesised using ethylene glycol dimethylacrylate (diacrylate-EGDMA) and Trimethylolpropane triacrylate (Triacrylate-TMPTA) as cross-linkers. Swelling behaviour of these polymers has been studied in different solutions like glucose, saline and water. The swelling behaviour of these polymers has also been studied under different pH conditions. The swelling capacity has also been analysed under load to have an idea of the gel strength (Absorbency under Load-AUL). Best absorbing characteristics, as indicated by the swelling behaviour, have been observed in case of polymer gels synthesized with EGDMA. Polymers with maximum PAsp have shown maximum superabsorbent properties in case of EGDMA as a cross-linker. However, with TMPTA as a cross-linker molar mass ratio of 1:2 polyaspartic acid: acrylic acid have shown better results. These polymers have better superabsorbent characteristics. TMPTA based polymers have shown better properties under load than EGDMA These polymers can be used as smart polymers for various applications e.g., drug delivery, materials for wound dressings, etc as they have shown varying behaviour in different conditions. The structure of the polymers has been studied by FTIR (Fourier Transform Infrared spectroscopy) and NMR (Nuclear Magnetic Resonance Spectroscopy). The surface morphology has further supported the results.

Keywords: Superabsorbent; Polymer gels; Swelling; Polyaspartic acid; Polyacrylic acid; Ethylene glycol dimethacrylate; Trimethylolpropane triacrylate

Introduction

Different decade gave different absorbing materials, each having their own specific absorption capacity. The commonly used water absorbing materials were based on cellulose based product like Whatman filter paper, tissue paper, wood pulp fluff, cotton ball with water absorption capacity (wt%) of 180, 400, 1200, and 1890, respectively. The water absorption capacity was 20 times their weight. The prominent disadvantage among these traditional absorbents was that they lose most of the fluids absorbed by them when they were squeezed. Due to this, a new class of speciality absorbents came into existence known as superabsorbent polymers (SAP). These synthetic polymers can be engineered to exhibit significant swelling when placed in water [1].

Superabsorbents developed are ultra-high absorbing material and could hold as high as 100% - 10000% (10 g/g - 1000 g/g) of water whereas the capacity of the hydrogels was 100% (g/g) [1]. They are a class of cross linked polymer capable of absorbing and retaining water. The cross linking present in SAP, does not allow SAP to dissolve but forms a gel when placed in water. The absorbency and swelling capacity are controlled by the type and degree of cross linkers used. Therefore, optimum cross linkers are used to develop superabsorbent polymers so that they exhibit high absorption capacity [1,2]. Due to their higher absorption and gel strength can lead to many fields like wound dressings, agriculture, wound care management, environment industry, diaper industry, etc [3-6].

PAsp has been reported to be used as polyelectrolyte and as a substitute for polyacrylic acid due to its biodegradability [7]. PAsp has been gaining importance in recent years as a biodegradable superabsorbent polymer [8-10]. PAsp based have found application as superabsorbent polymers in different industries mentioned above. Cross-linked PAsp, however has certain defects of poor strength due to high charge density along the polymer chains [11,12]. To improve upon the defects interpenetrating and semi-interpenetrating polymers

have been taken up by various scientists. A review has also been published on the polyaspartic acid based superabsorbents which cover the different types of polymers which include co-polymers, grafted and also interpenetrating polymers and semi-interpenetrating polymers [13]. Interpenetrating polymers involves networking of two different polymers as studied in case of poly (N-acrylamide) and polyaspartic acid to achieve functional materials which are pH sensitive and also thermo-sensitive [14,15]. Semi-interpenetrating polymers based on polyaspartic acid have been developed for drug delivery and other applications [16-19]. Some work on semi-interpenetrating network was initiated by Zhao et al. [20] in case polyaspartic acid and polyacrylic acid to achieve responsive polymers. With the upcoming newer areas, wherein these superabsorbent materials find application for biomedical applications like wound dressings, it has always been challenging to develop polymers with an ideal combination of water absorbency and gel strength. Thus, in continuation to the work reported by Zhao [20] different polymers based on PAsp and PAA have been synthesized using EGDMA, a difunctional cross-linker and TMPTA, a trifunctional cross-linker. The different polymers synthesized have been evaluated for their swelling properties.

Experimental

Materials

L-Aspartic acid and sodium hydroxide pellets (NaOH) were obtained from Merck Specialities Private Limited (Mumbai, India). Acrylic acid

---

*Corresponding author: Malik A, Dyal Singh College, University of Delhi, Lodhi Road, New Delhi, India, E-mail: amitamalik@dsc.du.ac.in

was purchased from Sigma Aldrich Chemical Corporation (St. Louis, USA). O-phosphoric acid for preparing the PSAP was purchased from Spectrochem Private Limited (Mumbai). Cross-linkers, Ethylene Glycol Dimethacrylate (EGDMA) and 1,1,1-Trimethylpropane triacrylate (TMPTA) were procured from Polysciences Inc. (Warrington, PA). Initiator ammonium peroxodisulphate $(NH_4)_2S_2O_8$ was purchased from Spectrochem Private Limited (Mumbai).

Preparation of polysuccinimide (PSI) and PAsp

L-Aspartic acid powder (5 g) and 85% of o-phosphoric acid (2.18 ml) were taken in a round bottom flask. The reaction mixture was heated for 3 h at 200°C under vacuum using rotary evaporator (Buchi type). Yellow powder of Polysuccinimide (PSI) was obtained which was then hydrolysed by adding sodium hydroxide (NaOH) solution prepared by dissolving 3.75 g NaOH in 25 ml deionised water. The NaOH solution was gradually added with continuous stirring by using magnetic stirrer to the PSI on an ice bath. To this mixture 35% HCl solution is added drop wise till it becomes neutral. This neutral solution was precipitated by adding saturated methanol solution (methanol in NaCl) and then filtered using vacuum filtration. The precipitate obtained is polyaspartic acid (PAsp) [20].

Synthesis of polymers containing polyaspartic acid (PAsp) and polyacrylic acid (PAA)

Preparation of neutral acrylic acid: Acrylic acid was neutralized by adding with NaOH solution (26 g NaOH in 100 ml of deionised water) drop wise using a dropping funnel. During addition of NaOH, the mixture was stirred on a magnetic stirrer to make a homogeneous mixture. The concentration of sodium hydroxide used results in 70% neutralization of the acrylic acid [21].

Preparation of polymers: Different volumes of aqueous polyaspartic acid (PAsp) and neutralised acrylic acid (Table 1) were taken and mixed for 10-15 min. In this study different ratios of acrylic acid and PAsp were used keeping the cross-linker and initiator concentration constant. The table shows the decreasing mass ratio of acrylic acid with relative increase in polyaspartic acid and with 40% water in all the systems studied. To these cross-linkers EGDMA and TMPTA were added in moles, followed by initiator ammonium peroxodisulphate. The mixture was heated to 80°C-100°C for 30 minutes under continuous stirring. Gel like product separated which was collected. The product obtained was analyzed for their swelling behaviour. The initiator concentration and cross-linker was established to give the best absorbent properties. The chemical structures of the cross-linkers used are given below in Figure 1. The reaction for the preparation of polymers with EGDMA and TMPTA is given below in Figure 2.

Analysis

Measurement of Molecular weight of Polyaspartic acid: Molecular weight of PASP was determined by using GPC and it was found to be 2,15,692. A homopolymer with polydispersity index Mw/Mn equal to 3.12 has been used for synthesizing the polymers.

Swelling in various physiological fluids: The swelling ratio of

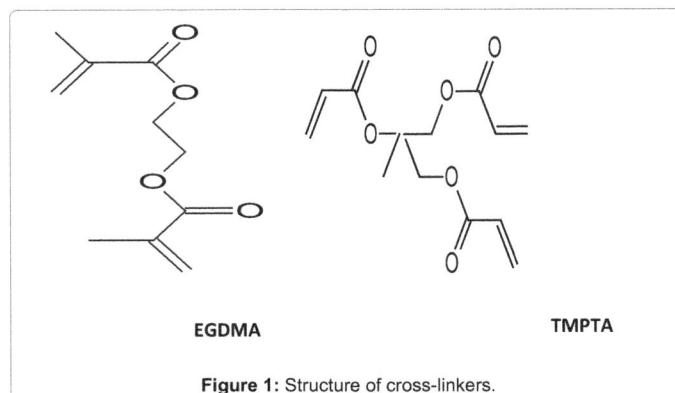

Figure 1: Structure of cross-linkers.

the various polymers was studied in various physiological fluids. The different physiological fluids used were water, saline solution and glucose solution. The main aim of using different fluids was to develop application of these polymers in biomedical area like wound dressings.

The different physiological solutions used for the study were:

a) Saline Solution: Sodium chloride solution were prepared by dissolving 9 g NaCl/1000 ml

b) Water: Conductivity of water was measured by conductivity meter. The conductivity of the water was 0.023 μS/cm and pH of water was neutral.

c) Glucose solution-50 g glucose was dissolved in 100 ml deionised water.

The tea bags used in the method were made using mesh made of nonwoven polypropylene with dimensions of 15 cm by 15 cm. The SAP sample was dried and weighed. Around 0.15 g of sample was taken and weighed accurately. The sample of known weight was then placed in tea bag and this tea bag was weighed then immersed in different solutions. Three readings of each sample were taken. An average of three readings has been reported. The swelling ratio was calculated as:

$$\text{Swelling ratio}\left(\frac{wt}{wt}\right) = Wt - W0 - Wn / W0 \qquad (1)$$

Where, Wt is the weight of tea bag in grams including swollen polymer; W0 is the weight of dry sample in grams (g) and Wn is the weight of wet polypropylene mesh in grams (g). The swelling ratio is calculated at different intervals of time (0.5 h, 1 h, 2 h, 3 h, 4 h and 24 h).

Swelling in different pH solutions: Different buffer solution with various pH values (pH = 2, 4, 7, 8 and 10) were made by using NaOH and HCl. The pH values were checked by a pH meter by Decibel (accuracy = ± 1). These were used to study the pH sensitivity of various polymers samples formed by using different cross linkers. The pH-sensitive properties of the polymer gels were studied in terms of swelling ratio by using tea bag method as discussed above. The same tea bag method was repeated for calculating swelling ratio of various polymers at different interval of time (0.5 h, 1 h, 2 h, 3 h, 4 h and 24 h).

Swelling/Absorbency under load measurement: A macro porous sintered glass filter plate (d = 100 mm, h = 7 mm) was placed in a Petri dish and the dry hydrogel sample was uniformly placed on the surface of polyester gauze located on the sintered glass. A cylindrical solid weight (d = 80 mm, variable height) which could slip freely in a glass cylinder was used to apply the desired load (applied pressure = 0.3 psi) to the dry hydrogel particles as shown in Figure 3. The sample was then covered by 0.5% saline solution such that the liquid level was equal to the height of the sintered glass filter.

SET.	PASP (%)	Acrylic acid (AA) (%)	Mass Ratio of PASP: AA	Initiator (g)	Cross-Linker (Moles)
1	10	50	1:5	0.009	0.05
2	20	40	1:2	0.009	0.05
3	30	30	1:1	0.009	0.05

Table 1: Different mass ratios of polyaspartic acid and acrylic acid studied.

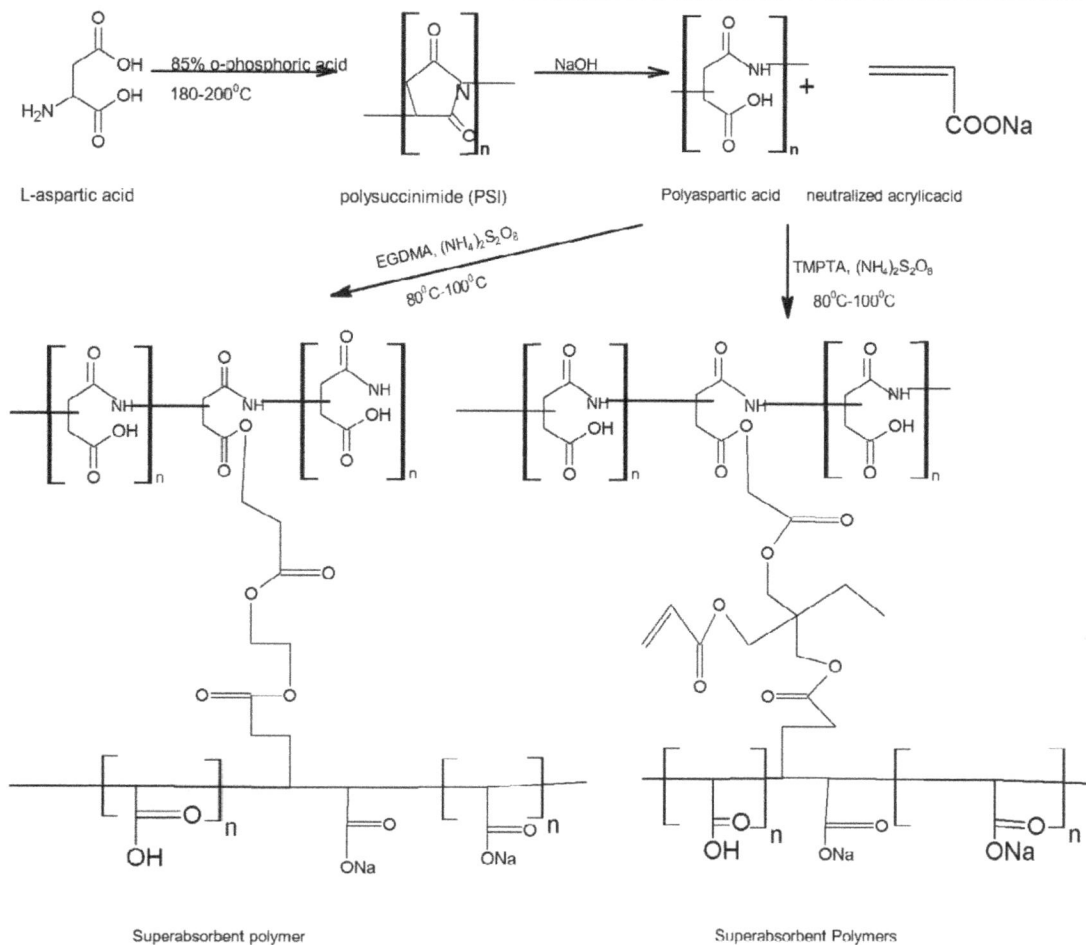

Figure 2: Reaction scheme for synthesis of polymers.

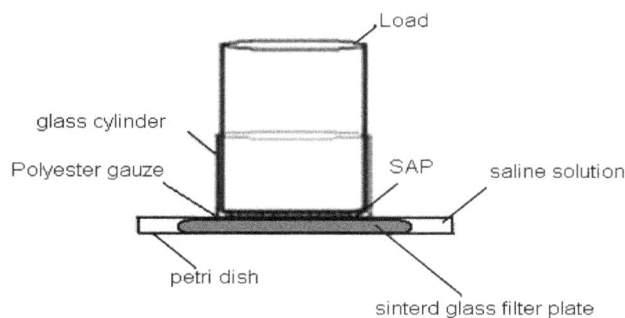

Figure 3: Schematic diagram for analysing absorbency under load [1].

The dish and its contents were covered to prevent surface evaporation and probable change in the saline concentration. The swollen particles were weighed at regular time intervals and AUL was calculated by using equation (1):

$$AUL \ (g/g) = W_2 - W_1 / W_1 \qquad (2)$$

Where, W_1 and W_2 represent the weight of dry and swollen hydrogel, respectively.

FTIR analysis: Perkin Elmer, FT-IR spectrophotometer-spectrum RX1 at from University Scientific Instrumentation Centre at University of Delhi was used to study FT-IR. The dried samples were ground into fine powder and then mixed with Potassium Bromide (KBr) powder. The mixed samples were compressed into pellets. The scanning wave number ranged from 4000 cm^{-1} to 550 cm^{-1}.

NMR analysis: The NMR spectra of the various polymer gel samples were carried out in D$_2$O. 300 MHz NMR Spectrometer (Brucker NMR Spectrometer at Indian Institute of Technology, Delhi) was used for the study.

SEM analysis: SEM analysis was carried out on gold coated samples of the polymer gels. Analysis of gold coated samples was carried out using MODEL Carl Zeiss SEM Analyser, EVO-18. The magnification used for the study was 250 X.

Results and Discussions

Neutralised acrylic acid is used to prepare the SAP samples with polyaspartic acid. Acrylic acid monomer is usually neutralised by alkali (NaOH) owing to high activity of acrylic acid over sodium acrylate. This neutralisation degree of acrylic acid has been reported to affect the swelling capacity of SAP and at an optimum value of degree of neutralisation is 65%-70% the swelling capacity as reported [21]. The best swelling is observed due to maximum repulsion between the –COO⁻ and –COO⁻Na⁺ groups, thus resulting in breaking of cross-links. Above 65%-70% degree of neutralization the swelling decreases due to further increase in electrostatic repulsion between –COO⁻ and –COO⁻Na⁺. Due to increase in repulsion, the cross -links in the polymer further decreases thereby decreasing the water entrapping capability. Similarly, when degree of neutralisation is lower, the cross-linking is higher and swelling is lower. Thus an optimum degree of neutralization is required to obtain good swelling.

Swelling in various physiological fluids

The observation and results of the swelling in different physiological fluids are discussed below.

Water: The results of the study are shown in Figure 4. The results have shown that with increasing time the swelling is better in case of EGDMA as a cross-linker than TMPTA. In case of TMPTA the swelling characteristics have shown an increased initial absorption being higher in all ratios studied. But in case of EGDMA as a cross linker difference in swelling characteristics have been observed. The initial absorption is low in comparison to TMPTA but a continuous increase in absorption has been observed with time. The 24 h swelling is better in case of EGDMA than TMPTA for all ratios studied. The 24 h absorption in case EGDMA is ~1100 g/g, while in case of TMPTA it has been observed to be ~200 g/g for polymers with maximum polyaspartic acid. Even in case of polymers with maximum acrylic acid (1: 5) where EGDMA has shown best results the swelling ratio is 800 g/g while in case TMPTA it is 300 g/g. Thus, the increasing concentration of polyaspartic acid has resulted in better absorption in case of EGDMA and a reverse trend has been observed in case of TMPTA i.e in case of TMPTA polymers with minimum concentration of Polyaspartic acid and maximum concentration of acrylic acid have shown better swelling characteristics.

Saline solution: The swelling characteristics of the polymers in saline solution have shown different results. Figure 5 shows the trends observed in case of saline absorption. The overall swelling of polymers in saline is lower than that observed in water. The polymer samples with EGDMA as a cross-linker and high concentration of polyaspartic acid have shown increased swelling after exposure for 24 h. But for TMPTA as cross linker totally reverse trend has been observed. The initial swelling of polymer samples with EGDMA is better with increased concentration of polyaspartic acid. In case of TMPTA however the initial absorption of saline is better in case of maximum acrylic acid. The polymers with maximum polyaspartic acid has shown a maximum absorption of ~ 82 g/g for 24 h absorption in case of EGDMA while in case of TMPTA it is 40 g/g. Polymers with maximum acrylic acid in case of EGDMA and TMPTA have shown same swelling ratio of ~ 70 g/g for 24 h absorption. The swelling ratio in case EGDMA and TMPTA with 1: 2 ratio is ~ 32 g/g and 42 g/g, respectively.

The trend observed in case EGDMA and TMPTA are different. The initial swelling, upto 3 h, is better for polymers with increased polyaspartic acid than with acrylic acid with both the cross-linkers. After 3 h the trend in swelling ratios changes and the swelling increases in case of maximum polyaspartic acid in case of EGDMA based polymers.

Glucose solution: The Figure 6 explains that the swelling of polymers in glucose increases with time for all the concentration of Polyaspartic acid and acrylic acid studied with both crosslinkers. The swelling capacity has been observed to increase as the concentration of polyaspartic acid increases for both cross linkers. Best result have been observed with EGDMA cross-linked polymers with low concentration

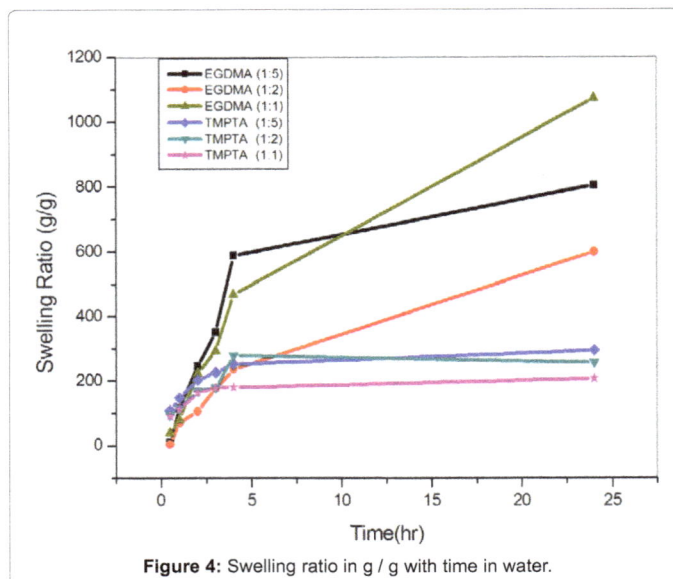

Figure 4: Swelling ratio in g / g with time in water.

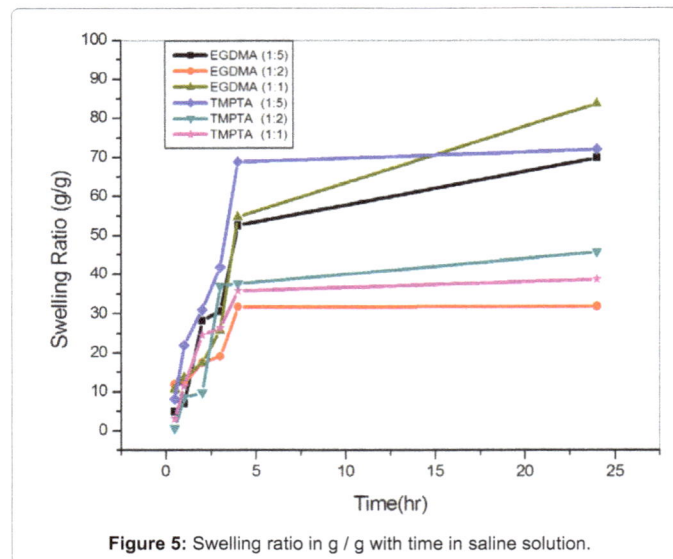

Figure 5: Swelling ratio in g / g with time in saline solution.

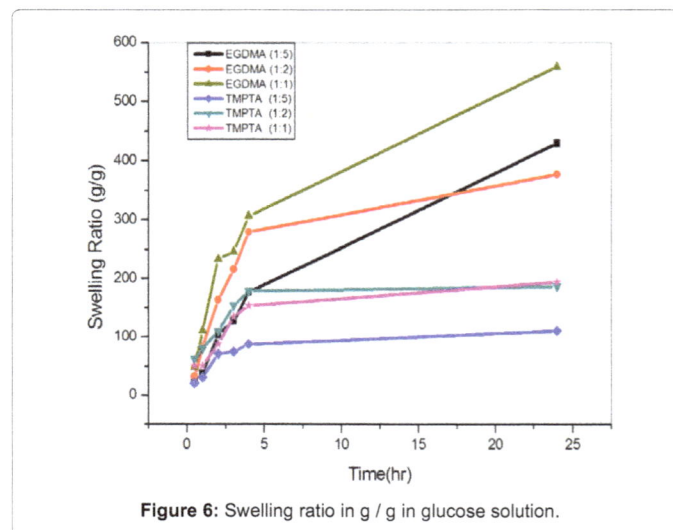

Figure 6: Swelling ratio in g / g in glucose solution.

of acrylic acid & high concentration of polyaspartic acid for 24 h absorption. In case of TMPTA however best results have been observed in case if 1: 2 polyaspartic acid to acrylic acid ratio studied. Minimum swelling has been observed in case of polymers with maximum polyaspartic acid. The maximum swelling ratio observed was 175 g/g in case of TMPTA for 24 h and 550 g/g in case of EGDMA. Thus, the swelling in different physiological fluids can be concluded that EGDMA has given better absorption for all fluids like deionised water, glucose & saline solution.

Swelling at different pH: Figures 7 and 8 shows the effect of changing pH on the swelling characteristics of polymers synthesized using EGDMA and TMPTA as cross-linkers, respectively. In case of pH 2, the polymer with EGDMA as cross-linker has given better results than TMPTA. At acidic pH both EGDMA and TMPTA with high acrylic acid ratio i.e. 1: 5 has shown better swelling.

At pH 4, the EGDMA based polymers has shown better swelling than TMPTA. EGDMA based polymers with the ratio 1: 2 and 1: 5 polyaspartic acid: acrylic acid has shown almost equal swelling after 24 h whereas EGDMA with a ratio of 1:1 polyaspartic acid: acrylic acid has shown different results. However, in case of TMPA polymers with maximium acrylic acid ratio has shown maximum swelling followed by polymers with ratio of 1: 2 and 1: 1 (polyaspartic acid: acrylic acid).

At pH 7, the polymer with EGDMA as crosslinker has shown better swelling than TMPTA. The polymer with EGDMA as cross-linker and ratio of 1: 1 shows highest swelling ratio. However in case of TMPTA similar results have not been observed with the similar ratio of 1: 1. At pH 8, EGDMA and TMPTA with 1: 1 ratio of polyaspartic acid: acrylic acid has shown higher swelling than rest of the ratios studied. At this pH, polymers with both the crosslinker have shown increased swelling capacity due to the presence of polyaspartic acid.

At basic pH, i.e.10, polymers with TMPTA as crosslinker interact and swell well for all ratios studied than EGDMA. The later behaves poorly and gives very low swelling. Both EGDMA and TMPTA shows varied absorption pattern at different pH. But EGDMA as a cross-linker has proved to result in polymers with better superabsorbent polymers than TMPTA. This study shows that for application under alkaline conditions, polyaspartic acid polymer with acrylic acid and EGDMA should be a better choice than TMPTA. One such area where it can find application is in highly exudating wounds. Both TMPTA and EDMA based SAPs possess same functional group. The EGDMA and TMPTA are having COO⁻ (EGDMA, TMPTA and acrylic acid). When both types of SAP were exposed to acidic and alkaline condition, swelling was observed.

The functional groups present in these polymers are -COOH, -COO⁻Na⁺, -CONH- peptide bonds and -COOC- ester bonds. These groups are affected by the pH conditions. Under acidic conditions the peptide bond gets protonated forming $-CONH_2^+$ causing repulsion between -COO⁻Na⁺ and $-CONH_2^+$ thus affecting the swelling properties. The ester bond present in the cross-linkers results in hydrolysis thus decreases the cross-links and probably results in enhancing the swelling. However under alkaline conditions the peptide bond and the ester bond hydrolyse giving better swelling. An optimum level of absorbency has been achieved in case of both cross-linkers from pH 7-8. At pH 10 however the water holding capacity is decreased due to decrease in cross-links. The greater the quantity of Polyaspartic acid the more $-CONH_2^+$ thus increased repulsion and therefore better swelling. However in case of trifunctional cross-linker the cross-linkage formed are more than the with the difunctional cross-links thus responsible

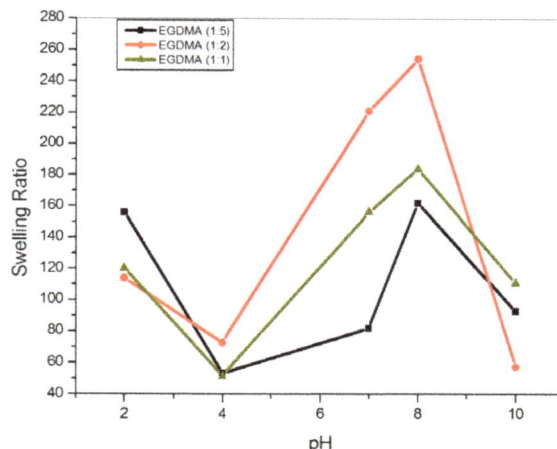

Figure 7: Effect of pH on the swelling ratios of polymers with EGDMA as cross-linker.

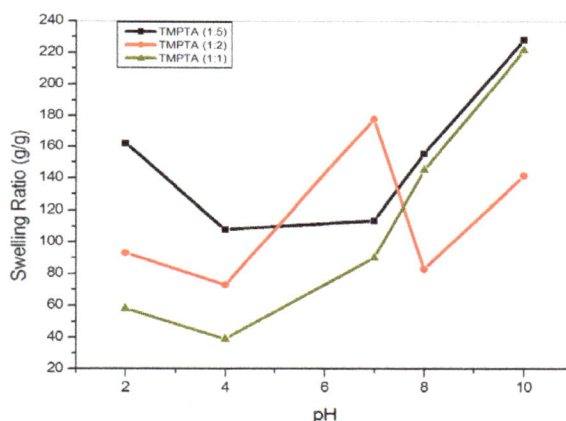

Figure 8: Effect of pH on the swelling ratios of polymers with TMPTA as cross-linker.

for different trends observed. These are further observed with swelling characteristics observed under load.

Under saline conditions the –COONa groups are formed that which decreases the swelling and does npt allow flow of saline into the polymers thus a decrease in swelling is observed. In case of glucose, the swelling is better than saline and less than water because in glucose solution the water is associated with glucose through hydrogen bonds. Thus the movement of glucose into the polymer is relatively slow in comparison to water thus decrease in swelling has been observed.

Swelling/Absorbency under load: The results of absorbency under load are given in Table 2 below. The absorbency under load observed for the polymers is different from the results otherwise. The absorbency under load for saline solution is lower than that observed under normal conditions. It can further be observed that TMPTA based polymer samples has shown better results than EGDMA based polymers. It can further be indicated that the gel strength is better in case of TMPTA than that observed in case of EGDMA. The best gel strength from the above results in case of TMPTA with 1: 2 ratio. Therefore to obtain polymers with good strength and absorbency TMPTA could be chosen with ratio of 1: 2 polyaspartic acid and arcylic acid. In case EGDMA

S.No	Sample	AUL (g / g)	Sample	AUL (g / g)
1.	EGDMA (1:5)	13.98	TMPTA(1:5)	18.57
2.	EGDMA (1:2)	13.00	TMPTA(1:2)	24.62
3.	EGDMA (1:1)	13.02	TMPTA(1:1	18.09

Table 2: Results of absorbency under load for the polymer gel samples.

although comparative result has been observed with 1:2 and 1: 1 but absorbency is better with 1: 1. Thus depending upon the absorbency desired the cross-linkers can be selected. This is further supported by the results of grafting which has been reported to be higher in case of TMPTA i.e.102% while in case of EGDMA is 65%, when extracted with chloroform for 24 h in a soxhlet apparatus in case of samples with 1:1 ratio of polyaspartic acid and acrylic acid

FTIR spectroscopy: The IR studies have been presented for the polymer with all concentration of polyaspartic acid and acrylic acid. The structure and bonding of the polymer has been proved by this study. Figures 9 and 10 gives the IR of the sample prepared with EGDMA, TMPTA as a cross-linker, respectively. The FTIR studies are representing the bonding between the polymeric chains and the various functional groups present. The IR of the sample further shows similar types of peak of the carboxylate and carboxylic. The presence of -CH$_2$ group in polyaspartic acid, EGDMA, TMPTA and neutralised acrylic acid is shown at 2923 cm^{-1}-2943 cm^{-1}. Bending vibration of -CH$_2$ group has been observed at 1402 cm^{-1}- 1462 cm^{-1}. The -NH$_2$ group is confirmed by > NH$_2$ stretching at 2343 cm^{-1}- 2375 cm^{-1} which is due to the presence of polyaspartic acid. It has also shown peak for C-O-C bond as observed in case of both EGDMA at 1070 cm^{-1}- 1085 cm^{-1} and TMPTA at 1065 cm^{-1}-1090 cm^{-1}. The presence of –C=O (carbonyl group) at 1717 cm^{-1} 729 cm^{-1} and C-O-C bonding has been observed formed due to ester groups.

The NMR spectra for the different polymers are presented in Figure 11 for polymers with EGDMA and Figure 12 for polymers with TMPTA, respectively. The peak values at 1.5 ppm and 1.4 ppm is for alkyl groups i.e., R$_3$CH- and R$_2$CH- respectively. The presence of polyaspartic acid is confirmed by the peak at 2.1 ppm (2H,t) and 2.3 ppm (2H,t). The cross linker and polyaspartic acid are linked by an ester bond shows a peak at 3.6 ppm for EGDMA based polymers but for TMPTA this linkage is prominent for ratio 1: 1 and 1: 2. It might be because cross linker TMPTA is bonding well and easily with acrylic than polyaspartic acid. Due to this TMPTA (1: 5) has high swelling among all three ratios. The peak at 2.2 ppm corresponds to -CH$_2$- present in EGDMA (-C(=0)-CH$_2$-CH$_2$-C(=O)-).

SEM analysis: The SEM studies have been carried for all concentration of Polyaspartic acid and acrylic acid with both cross-linkers EGDMA and TMPTA. Figure 13 presents the micrograph of polymers with TMPTA as cross linker and Figure 14 gives the result of EGDMA based polymers. The scanning electron micrographs have shown that the polymers obtained with these cross-linkers are different. Figure 13 shows that the porosity in case of polymers with 1: 5 and 1: 1 ratio of polyaspartic acid: acrylic acid in case of TMPTA while in case of samples with ratio of 1: 2 the samples have relatively low porosity which also evident from the results of swelling as well from the AUL data. Figure 14 shows that in case of EGDMA samples with ratio of 1: 1 the samples have shown porosity as compared to the other two ratios studied. Moreover the pores observed are macropores in EGDMA that those observed in case of TMPTA. Due to these pores only the swelling has increased to approximately 1000 g/g in distilled water. As the concentration of polyaspartic acid increases pores are formed on the surface which allow seepage of fluids like water, saline solution, glucose

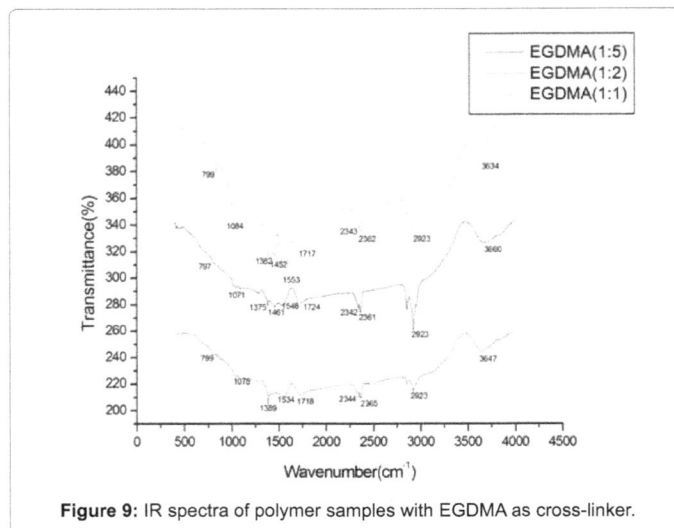

Figure 9: IR spectra of polymer samples with EGDMA as cross-linker.

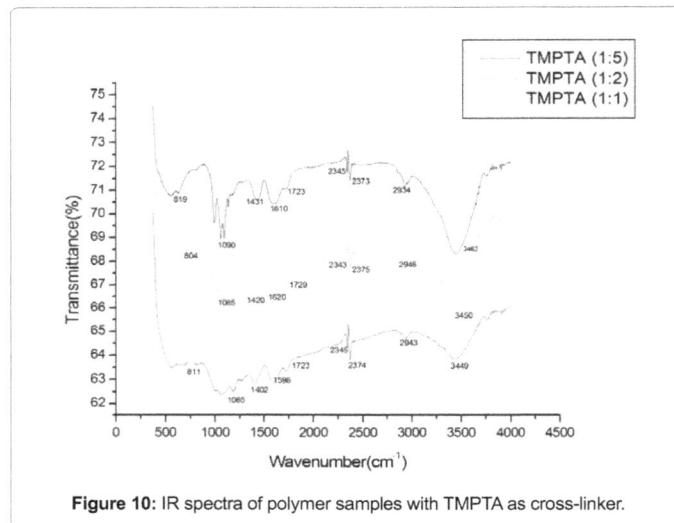

Figure 10: IR spectra of polymer samples with TMPTA as cross-linker.

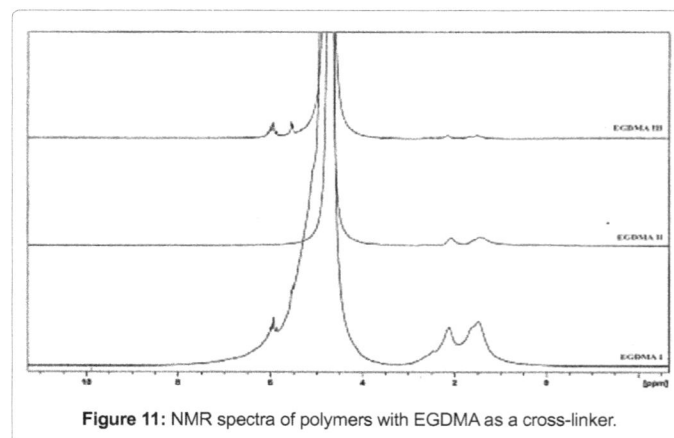

Figure 11: NMR spectra of polymers with EGDMA as a cross-linker.

solution which into the polymer forming superabsorbent polymer (SAP).

Conclusions

The superabsorbent polymers prepared by replacing acrylic

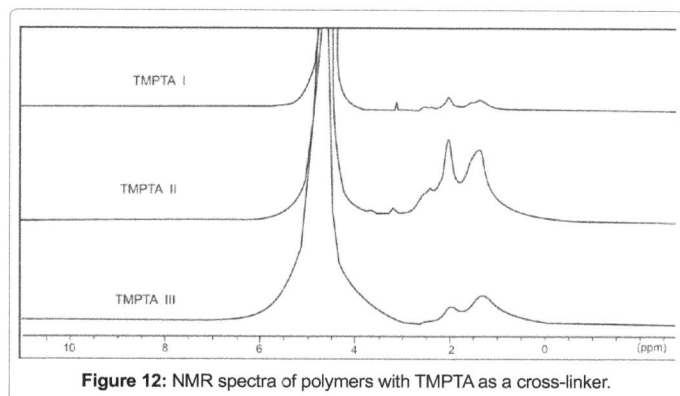

Figure 12: NMR spectra of polymers with TMPTA as a cross-linker.

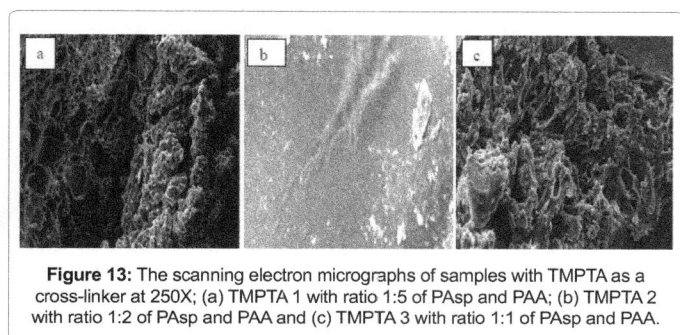

Figure 13: The scanning electron micrographs of samples with TMPTA as a cross-linker at 250X; (a) TMPTA 1 with ratio 1:5 of PAsp and PAA; (b) TMPTA 2 with ratio 1:2 of PAsp and PAA and (c) TMPTA 3 with ratio 1:1 of PAsp and PAA.

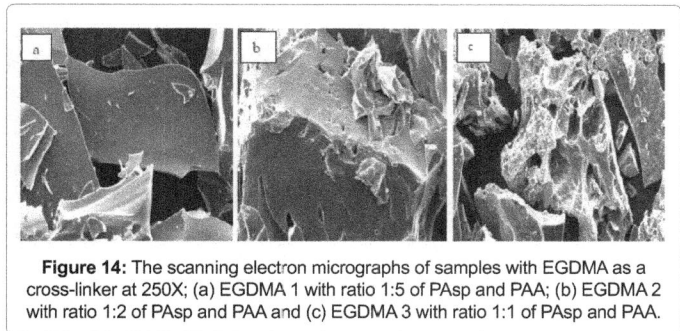

Figure 14: The scanning electron micrographs of samples with EGDMA as a cross-linker at 250X; (a) EGDMA 1 with ratio 1:5 of PAsp and PAA; (b) EGDMA 2 with ratio 1:2 of PAsp and PAA and (c) EGDMA 3 with ratio 1:1 of PAsp and PAA.

acid with polyaspartic acid using different cross linkers have shown maximum swelling capacity in deionised water and minimum in case of saline solution for 24 h absorption studies. The trend observed in different physiological fluids is as follows: Distilled water > glucose solution > saline solution (EGDMA) and glucose solution > distilled water > saline solution (TMPTA).

The polymer properties can thus be tailored and used for different applications. EGDMA based polymers have shown better absorption properties than TMPTA. But the results of AUL has shown that TMPTA has much better strength than EGDMA based sample especially the polymer with 1: 2 ratio of polyaspartic aid and acrylic acid.

Acknowledgements

The authors are thankful to UGC for sponsoring this project and providing us an opportunity for carrying our research in our field of interest. We are also thankful to our principle Dr. I.S. Bakshi for providing us timely support and guidance for our research. We are also thankful to college management for the required help.

References

1. Mehr MJZ, Kabiri K (2008) Superabsorbent Polymer Materials: A Review. Iranian Polymer Journal 17: 451-477.

2. Lee WF, Wu RJ (1996) Swelling behaviours of crosslinked poly(sodium acrylate-co-hydroxyethylmethacrylate) in aqueous salt solution. J of Appl Polym Sci 62: 1099-1114.

3. Chaudhri M, Datta S, Mutha P, Prachi P (2013) Superabsorbent Polymer. Popular Plastics and Packaging 58: 19-24.

4. Tian Z, Zhou Y, Zheng SS (2012) Research on the superabsorbent polymers. Heber Huegong 359: 68-70.

5. Gong J, Li Q, Zhao Y (2012) Development and Research of superabsorbent polymers. Yingyong Hungong 41: 895-897.

6. Xianglin U, Yan Z, Bing Li (2011) Research Progress on novel functional superabsorbent polymers. Huaxane Yu. Shengwa Gongcheng 28: 8-12.

7. Kumar A (2012) Polyaspartic acid- A Versatile Green Chemical. Chemical Science Review Letters 1: 162-187.

8. Thombre SM, Sarwade BR (2011) Synthesis and Biodegradability of Polyaspartic Acid: A Critical Review. J of Macromolecules Science, Part A: Pure of Applied Chemistry 42: 1299-1315.

9. Lenzi F, Sannio A, Borriello A, Porro F, Capitani D, et al. (2003) Probing the degree of crosslinking of a cellulose based superabsorbing hydrogel through traditional and NMR techniques. Polymer 44: 1577-1588.

10. Kholodovych V, Smith JR, Knight D, Abramson S, Kohn J, et al. (2004) Accurate predictions of cellular response using QSPR: a feasibility test of rational design of polymeric biomaterials. Polymer 45: 7367-7379.

11. Gupta B, Anjum N (2003) Preparation of ion-exchange membranes by the hydrolysis of radiation-grafted polyethylene-g-polyacrylamide films: Properties and metal-ion separation. J Appl Polymn Sci 90: 3747-3752.

12. Ren J, Zhang Y, Li J, Ha H (2001) Radiation synthesis and characteristic of IPN hydrogels composed of poly(diallyldimethylammonium chloride) and Kappa-Carrageenan. Radiat Phys Chem 62: 277-281.

13. Sharma S, Dua A, Malik A (2014) Polyaspartic acid based superabsorbent polymers. European Polymer Journal 59: 363-376.

14. Miaomiao L, Su H, Tan T (2012) Synthesis and properties of thermo- and pH-sensitive poly(N-isopropylacrylamide)/polyaspartic acid IPN hydrogels. Carbohydrate Polymers 87: 2425-2431.

15. Zhao Y, Li Q, Liu Y, Liu C, Gong J, et al. (2013) Preparation and Properties of Interpenetrating Networks Absorbent Polymer Based on Polyaspartic Acid and Polyacrylic Acid. Polymer Material Science and Engineering 1: 48-51.

16. Zhao Y, Tan T, Kinoshita T (2010) Swelling kinetics of poly(aspartic acid)/poly(acrylic acid) semi-interpenetrating polymer network hydrogels in urea solutions. Journal of Polymer Science Part B: Polymer Physics 48: 666-671.

17. Nistor MN, Chirac AU, Nita LE, Neamtu J, Cornelia V, et al. (2013) Semi-interpenetrated polymer networks of hyaluronic acid modified with poly(aspartic acid). J of Polymer Research 20: 1-11.

18. Nistor MN, Chirac AU, Nita LE, Neamtu J, Cornelia V (2013) Semi-interpenetrated network with improved sensitivity based on poly(N-isopropylacrylamide) and poly(aspartic acid). J of Polymer Engineering Science 53: 2345-2352.

19. Liu C, Chen Y, Chen J (2010) Synthesis and characteristics of pH-sensitive semi-interpenetrating polymer network hydrogels based on konjac glucomannan and poly (aspartic acid) for in vitro drug delivery. Carbohydrate Polymers 79: 500-506.

20. Zhao Y, Kang J, Tan T (2006) Salt-, pH- and temperature-responsive semi-interpenetrating polymer network hydrogel based on poly(aspartic acid) and poly(acrylic acid). Polymer 47: 7702-7710.

21. Chen XP, Shan GR, Huang J, Huang ZM, Weng ZX (2004) Synthesis and properties of acrylic-based superabsorbent. Journal of Applied Polymer Science 92: 619-624.

21

Synthesis and Analysis of Natural Fibers Reinforcement of Synthetic Resins

Guduru KK[1]*, Pandu R[2], Banothu S[1] and Vinaya K[3]

[1]*Christu Jyothi Institute of Technology Science, Jangaon, Warangal, Hyderabad, India*
[2]*Marri Laxman Reddy Institute of Technology & Management, Hyderabad, India*
[3]*ACE Engineering College, Telangana, India*

Abstract

Natural fiber Composites typically have a fiber or particle phase that is stiffer and stronger than the continuous matrix phase and serve as the principal load carrying members. The matrix acts as a load transfer medium between fibers, and in less ideal cases where the loads are complex, the matrix may even have to bear loads transverse to the fiber axis. In this research the comparative synthesis and analysis of Kenaf fiber (FRPMC1) and Polmera fibers (FRPMC2) are treated with NaOH solution and the fibers are properly reinforced with polypropylene resin and epoxy resin respectively in a matrix form to prepare hybrid composite laminates of 6 mm fiber length thereafter to determine the mechanical properties like lexural strength, lexural modulus and tensile strength and tensile modulus with suitable specimens with ASTM D-638 and D-790 standards. The analysis was carried out by using FEA software for various loads and result factors. The surface is analyzed by SEM test with various resolutions. The matrix also serves to protect the fibers from environmental damage before, during and after composite processing. The surface is analyzed using when designed properly, the new combined material exhibits better strength than each individual material. Composites are used not only for their structural properties, but also for electrical, thermal, and eco-friendly environmental applications.

Keywords: Kenaf fiber; Polmera fiber; Flexural strength; Tensile strength; SEM

Introduction

The common fiber reinforced composites are composed of fibers and a matrix [1-3]. Fibers are the reinforcement and the main source of strength while matrix resins all the fibers together in a shape and transfers stresses between the reinforcing fibers. Sometimes, filler might be added to smooth the manufacturing process, impacts on special properties to the composites [4-6], and also reduce the product cost. In this research kenaf fiber and polmera fibers are treated with NaOH solution and the fibers are properly reinforced with polypropylene resin and epoxy resin respectively in a matrix form to prepare hybrid composite laminates of 6 mm thicknesses thereafter to determine the mechanical [7,8] properties like flexural strength flexural modulus, tensile strength, tensile modulus and compressive strength with suitable specimens with ASTM E-08 for tensile properties and ASTM D-790 for flexural properties as per standards. By using ANSYS through finite element analysis is done for various load and result factors. The surface is analyzed by SEM (Seismic electronic microscope) test with various resolutions from 200X, 500X, 1000X, 3000X and 5000X.

Materials and Methods

Hibiscus Cannabinus has its Telugu vernacular name as "GOGU". The stalk of Kenaf plant consists of two distinct fibre types. The outer fibre is called "BAST" and comprises roughly of the 40% stalk`s dry weight. The white inner fibre is called "CORE" and comprises 60% stalk weight. Kenaf fibres are extracted from the bast having a potential alternative as reinforcement in polymeric composites instead of synthetic fibres. Polmera fiber is a coir type fiber but having slight difference in physical and mechanical properties. This could be possibly a good alternative for coconut coir (Figures 1 and 2).

Polypropylene [1] is normally tough and flexible, especially when copolymerized with ethylene. This allows polypropylene to be used as an engineering plastic, competing with materials such as acrylonitrile butadiene styrene (ABS). Polypropylene is reasonably

economical, and can be made translucent when uncolored but is not as readily made transparent aspolystyrene, acrylic, or certain other plastics. It is often opaque or colored using pigments. Polypropylene has good resistance to fatigue. The melting point of polypropylene occurs at a range, so a melting point is determined by finding the highest temperature of a differential scanning calorimetric chart. Perfectly isotactic PP has a melting point of 171°C (340°F). Commercial isotactic PP has a melting point that ranges from 160 to 166°C (320 to 331°F), depending on a tactic material and crystallinity. Syndiotactic PP with a

Figure 1: Polmera fiber.

*Corresponding author: Guduru KK, Christu Jyothi Institute of Technology Science, Jangaon, Warangal, Hyderabad, India
E-mail: kranthicjits1@gmail.com

crystallinity of 30% has a melting point of 130°C (266°F). The melt flow rate (MFR) or melt flow index (MFI) is a measure of molecular weight of polypropylene.

Epoxy resin is known in the marine industry for incredible toughness and strength for bonding. Epoxy resin has a greater ability to flex and strain with the fibers without micro-fracturing and good resistant to water absorption. Epoxy bonds to all sorts of fibers very well and also offers excellent results in repair ability when it is used to bond two different materials together.

Alkali treatment

The quality of a fiber reinforced composite depends considerably on the fiber-matrix interface because the interface acts as a binder and transfers stress between the treatments of fibers using chemical agent like sodium hydroxide (NaOH). For treatment process water by volume is taken along with 2% of NaOH. The fibers are soaked in the water for 24 hours as shown in Figure 3, and then the fibers are washed thoroughly with distilled water to remove the final residues of alkali. Good bonding is expected due to improved wetting of fibers with the matrix. In order to develop composites with better mechanical properties and good environmental performance, it is necessary to impart fibers by chemical treatments [9,10]. The extracted fibres treated untreated and chopped fibres as shown in Figures 3 and 4.

Laminate preparation for specimens

Compression moulding is a well-known technique to develop variety of composite products. It is a closed moulding process with high pressure application. In this method the matched metal moulder

used to fabricate composite product. In compression moulder the base plate is stationary while upper plate is movable. Reinforcement and matrix are placed in the metallic mold and the whole assembly is kept in between the compression moulder (Figures 5 and 6).

Testings and Observations

Flexural test specimen

Specimens for the flexural test are cut on a jig saw machine as per the ASTM standards. The dimensional details of each type of specimen are presented in respective diagrams [10]. Specimens are cut from laminas on a jig saw machine as per ASTM D790 standards. Rhe standard regulated shaped specimens are used for testing. The dimensions of the flexural test specimen are shown in the Figure 7.

Figure 4: Showing treated, untreated fibers.

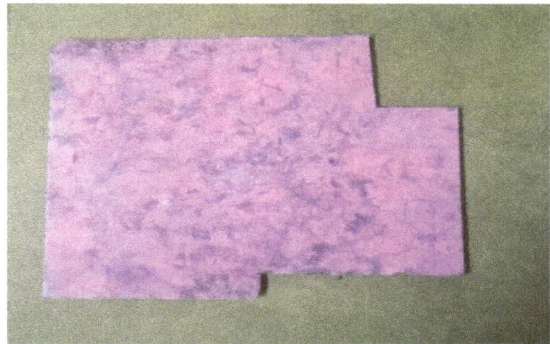

Figure 5: Kenaf + Polypropylene laminate (FRPMC1).

Figure 2: Hibiscus Cannabinus (Kenaf Plant).

Figure 3: Treatment of Fiber in 2% NaOH solution.

Figure 6a: Polmera + Epoxy laminate (FRPMC2).

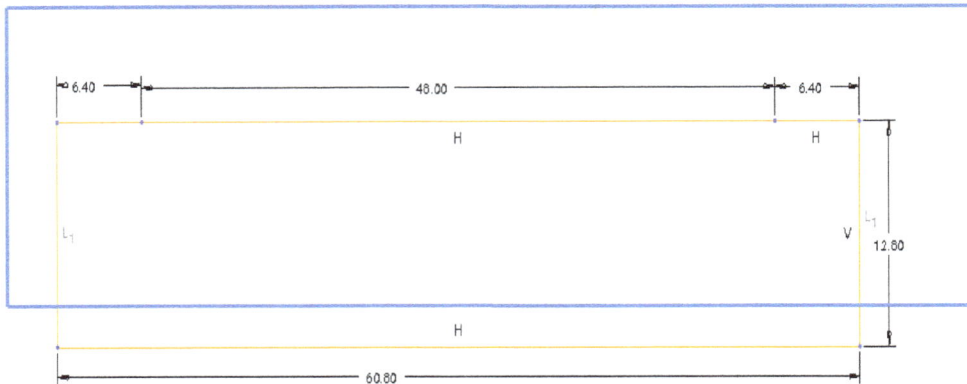

Figure 6b: Dimensions of a flexural test specimen.

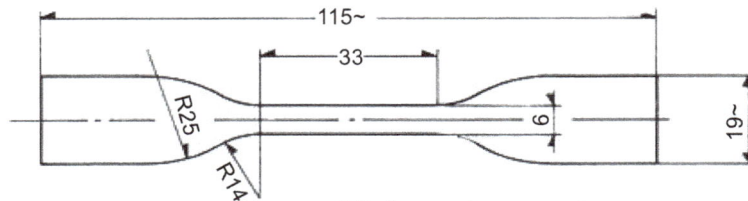

All dimensions are in mm

Figure 7: Dimensions of tensile test specimen.

Tensile test specimen

Specimens for the tensile test are cut on a jig saw machine as per the ASTM standards the dimensional details of each type of specimen are presented in respective diagrams [9]. Specimens are cut from laminas on a jig saw machine as per ASTM D638 standards (Tables 1 and 2). The standards type 4 dumbbell shaped specimens are used for testing. The dimensions of the tensile test specimen are shown in the Figures 8 and 9.

Results and Conclusion

For the Tensile properties at the peak load, the tensile strength on polmera fiber with epoxy is 15.529 N/mm² which is less than the kenaf fiber reinforced with the polypropylene is 25.17 N/mm² while in compared with Md. Roshnal Hossain [2], the strength is 844 N/mm² but as an anisotropic material, jute fiber has a large scatter in tensile properties depending on test specimen span length, test machine slippage and presence of inherent and surface of defects, according to Buenaventurada P. Calabia [11] the tensile strength is 30 N/mm² cotton reinforced with poly(butylene succinate).

For the flexural properties at the peak load, the flexural strength is 125 N/mm² on polmera fiber with epoxy whereas in the kenaf fiber reinforced with the polypropylene is 88 N/mm², while according to Byoung-Ho Lee [12] the flexural strength of 31N/mm² for the kenaf and polypropylene composite (Figures 10 and 11).

From the analysis it is observed that stress variation in the material shown by the different colors in the above image. The maximum stress is at red colour that is

σ max = 381.08MPa

The minimum stress is at the green color section that is

σ min = 57.331MPa

Specimen Description	Polmera with Epoxy	Kenaf with Polypropelene
Load at Peak (N)	3.345	4.5
Tensile Strength(N/mm²)	15.529	21.17
Load at Break(N)	3.27	5.8
% of Elongation	3.985	5.22
Tensile modulus(N/mm²)	528.5	796.2

Table 1: Tensile test observations.

Specimen Description	Polmera with Epoxy	Kenaf with Polypropelene
Flexural modulus(N/mm²)	4580	2322
Flexural strength(N/mm²)	125	88
Deflection(mm)	4	5
Peak load(N)	76	41

Table 2: Flexural test observations.

Figure 8: Flexural test specimens ASTM D-670.

Figure 9: Tensile test specimen ASTM D-638.

Figure 11: Tensile testing machine.

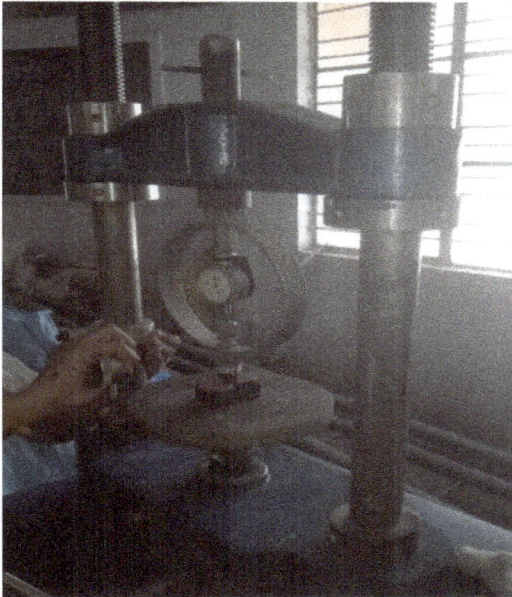

Figure 10: Flexural testing machine.

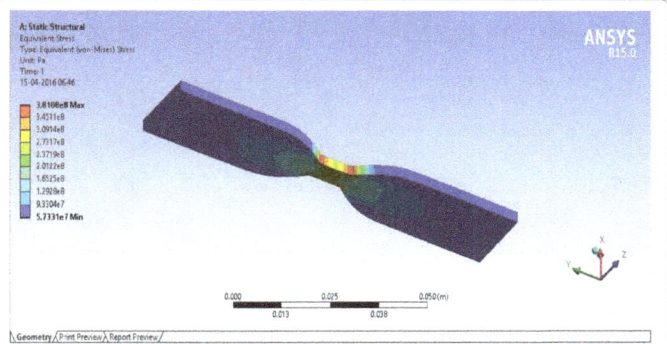

Figure 12: Equivalent stress of the composite specimen.

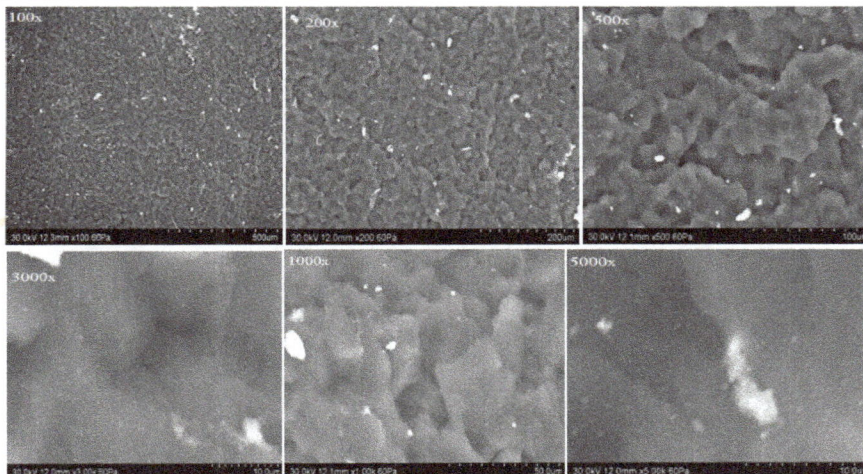

Figure 13: SEM analysis on the surfaces of the polypropylene resin with Kenaf fiber shows the fine distribution of fiber in resin in matrix with the required resolutions [8].

The actual value obtained by the calculations is less than the analysis value so the material Properties are accurate and accepted. It is concluded that the epoxy has far more to offer in its ability to flex, prevent delamination and ease of use for repair work. Using epoxy leads to better quality marine products (Figures 12 and 13).

The kenaf fibres were successfully used to fabricate composites with 30% fiber and 70% resin these fibres are bio-degradable and highly crystalline with well aligned structure. So it has been known that they also have higher tensile strength than other natural and synthetic composites and intern it would not induce any serious environmental problem like in synthetic fibers [12].

Future Work

In the present investigation a compression moulding technique was used to fabricate the composites. However the results provided in this project can act as a base for the utilization of these fibers. Further work may be carried out to know the Dielectric properties of composites. Orientation of fibers may be considered to get the better mechanical properties. Design of experiment (DOE) can be used to get the optimum results.

References

1. Zgoul MH, Habali SM (2009) An investigation into plastic pipes as hot water transporters in domestic and industrial applications. Jordan Journal of Mechanical and Industrial Engineering 2: 191-200.

2. Hossain MR, Islama MA, Vuureab AV, Verpoestb I (2013) Tensile behavior of environment friendly jute epoxy laminatedcomposite. Procedia Engineering 56: 782-788.

3. Xu Y, Kawata S, Hosoi K, Kawai K, Kuroda S (2009) Thermomechanical properties of the silanized-kenaf/polystyrene composites. Express Polymer Letters 3: 657-664.

4. Randjbaran E, Zahari R, Jalil A, Abang Abdul Majid DL (2014) Hybrid composite laminates reinforced with kevlar/carbon/glass woven fabrics for ballistic impact testing. The Scientific World Journal 2014: 1-7.

5. Adekunle KF (2015) Surface treatments of natural fibres-a review: part 1. Journal of Polymer Chemistry 5: 41-46.

6. Tewari M, Singh VK, Gope PC, Chaudhary AK (2012) Evaluation of mechanical properties of bagasse glass fiber reinforced composites. Journal of materials and environmental science 3: 171-184.

7. Ashik KP, Sharma RS (2015) A review on mechanical properties of natural fiber reinforced hybrid polymer composites. Journal of minerals and materials characterization and engineering 3: 420-426.

8. Arrzkhiz FZ, Achaby ME, Malha M (2015) Mechanical and thermal properties of natural fibers reinforced polymer composites: Doum/low density polyethylene. Journal of Materials & Design 43: 200-205.

9. Ku H, Wang H, Pattarachaiyakoop N, Trada M (2011) A review on the tensile properties of natural fibre reinforced polymer composites. Composites Part B: Engineering 42: 856-873.

10. Udaykumar PA, Ramalingaiah RS (2014) Studies on effects of short coir fiber reinforcement on flexural properties of polymer matrix. International Journal of Research in Engineering and Technology 3: 37-41.

11. Calabia BP, Ninomiya F, Yagi H, Oishi A, Taguchi K, et al. (2013) Biodegradable poly(butylene succinate) composites reinforced by cotton fiber with silane coupling agent. International journal of Polymers 5: 128-141.

12. Lee BH, Kim HJ, Yu WR (2009) Fabrication of long and discontinuous natural fiber reinforced polypropylene biocomposites and their mechanical properties. Journal of Fibers and Polymers 10: 83-90.

22

Prediction of Hardness, Yield Strength and Tensile Strength for Single Roll Melt Spinning of 5083 Al-alloy Ribbons

Jassim AK[1]* and Hammood AS[2]

[1]*Department of Materials Engineering, University of Basra, Iraq*
[2]*Department of Materials Engineering, University of Kufa, Iraq*

Abstract

In this paper, an empirical model is applied to predict the hardness, yield strength, and tensile strength of rapid solidified ribbons. The discovered empirical equation is obtained depends upon the experimental results of rapid solidification process for 5083 Al-alloys. The empirical equations predict values and describe the behavior of ribbon with consideration of ribbon thickness, grain size, hardness, yield strength, and tensile strength. The experimental work involves difference operation conditions and the results indicate that orifice diameter, nozzle roll wheel gap, and melting temperature have direct impact on the quality of alloy. Additionally, the results showed that there is a good agreement between experimental and predicted values where the correlation coefficient is 0.99. The experimental show that there is a possibility to produce very thin ribbons with thickness in micrometer by reducing the distance between nozzle and roll wheel, and reduce the orifice diameter of casting. The hardness, and yield strength increased due to increasing the number of small grain size in the ribbons structure and rapidly heat transfer of the small ribbons thickness. Moreover, the optimal melting temperature of this alloy is 925°C which produces high ribbon hardness compared with other melting temperature that used in this research.

Keywords: Rapid solidification; Melt spinning; Ribbons; Empirical model; Mechanical properties

Introduction

Most metallic materials are produced from their liquid state. It is very important to understand their solidification path and the resulting microstructures which can be significant affected by melt undercooling [1]. Rapid solidification technology is an important process for the development of the advance metallic materials. It is a means to produce new alloys with superior properties and develop unusual even novel microstructure, which frequently exhibit beneficial properties. The process has direct impact on the properties of alloys. It is improved the mechanical properties and refining the microstructure [2].

Rapid solidification process is a tool for modifying the microstructure of alloy which is applied to produce a metallic glass with better mechanical and microstructure properties compared with those of conventional casting. Yield stress and hardness of bulk metallic glasses can be twice as these of steel [3]. The hardness of rapid solidified ribbons increases with decreasing thickness of ribbons which becomes more than twice of the original hardness of alloy before rapid solidification process because it produces homogenous structure with massive Nano grain size. It has one of the highest melt cooling rate among all continuous casting process because the solidification time and movement of the melt and substrate are very short [4].

Rapid solidification technique with one roller melt spinning technique is an important method that used to produce bulk metallic glasses materials nowadays. In this technique, a stream of molten metal is directed at a rapidly moving substrate. The final product is in the form of ribbon with thickness in micrometer. Quenching on the inside or outside of a rapidly rotating cylinder now appears to be a widely utilized technique and production rate of 2000 m/min obtainable [5].

Bogno have been studied characterization of rapidly solidified metallic alloy using combination of experiments and modelling [1]. Kramer et al. investigated various melt pool characteristics and their influence on the melt spun ribbon using high speed digital imaging. They showed that ribbon thickness can be computed depending on

the density, viscosity, and melt stream diameter [6]. Tayebeh et al. have been studied the effect of cooling rate during melt spinning of fine ribbons. They use curve fitted with a correlation factor of 0.985 and reveals that the thickness of ribbon is inversely proportional to the wheel speed with the power of -1.231 which is in reasonable agreement with calculated and reported values. They found that as shorter and the melt drop has to a longer distance before solidification and the hardness of as spun ribbon decrease with increasing the wheel speed [7].

Adam L. Woodcraft predicted the thermal conductivity of Al-alloys in the cryogenic to room temperature based on a measurement of the thermal conductivity or electrical resistivity at single temperature [8]. Egami et al. proposed a correlation between the minimum solute concentrations that require for glass formation [9]. Wannaparhan used two common classes of numerical methods which are finite element method FEM and finite difference method FDM to study several important parameters such as the solid/liquid interface velocity and the cooling rate. All of these parameters can be correlated the final microstructure and properties of the casting. In addition, used numerical methods for the prediction of temperature distribution during solidification [10].

Zhang et al. applied thermo kinetic model to study the re-calescence characteristic in rapid solidification of copper. The effect of the heat transfer coefficient, the melt thickness and the nucleation temperature were investigated. Results showed that lower nucleation temperature and thinner melt lead to a longer re-calescence effect

*Corresponding author: Jassim AK, Department of Materials Engineering, University of Basra, Iraq, E-mail: ahmadkj1966@yahoo.com

while larger heat transfer coefficient results in a weaker re-calescence effect. A dimensionless number was derived to measure the extent of re-calescence [11].

The empirical model proposed in this study is a useful prediction tool help to analysis the relationship between rapid solidification parameters (orifice diameter, nozzle-roll wheel gap, and melting temperature) and the thickness, hardness, and tensile strength of rapid solidified ribbons. It was applied without any materials, energy, and laborious time consuming, and without machining trials. The discovered empirical equations were feasible to make the prediction of mechanical properties. The constants and coefficients of these equations were calculated by multiple regression method using data-fit software.

The main advantages of data fit curve model is that it is easy to use and simple program which can be used as an aid for data visualization where no data are available and summarize the relationship among two or more variables. It is the process of constructing a curve or mathematical function that has the best fit to a series of data points; it is involved when ab exact fit to the data is required.

Experimental Work

Ribbon thickness is one of the most important factors that affect the heat transfer and the microstructure refining. Therefore, the effect of operation parameters that include orifice diameter, nozzle-roll wheel gap, and melting temperature on the ribbons thickness are studied. In this work, Al-Mg alloy type 5083 was used to product rapid solidified ribbons with very thin thickness in the range of micrometer. A series of rapid solidification process with single roll melt spinning technique were carried out to produce ribbons with different operation parameters [2-4]. The results were used to discover the empirical equations for prediction the mechanical properties of rapid solidified ribbons.

Discovered Empirical Equations

The indirect method represented by empirical equations is more particles for measurements. The multiple regression analysis method used to develop the empirical models to predict the ribbons thickness, grain size, hardness, yield strength, and tensile strength under difference conditions is curve-fitting treatment of data tools which is called data fit software or LAB Fit curve fitting software-V7.2.48-(1999-2011). Difference case study was used to discover the empirical equations for computing the mechanical properties for rapid solidified Al-Mg alloys type 5083.

Data fit for thickness

The discovered empirical equation that obtained in this work is shown in Equation (1). The correlation coefficient of this equation is equal to 0.93. The discovered equation was used to predict the thickness of ribbons TH depending on the melting temperature MT, and the distance between crucible nozzles and roll wheel surface G at constant orifice diameter OD equal to 2.5 mm. Therefore, the discovered empirical equation describes the behavior of rapid solidified ribbons thickness under the effect of melting temperature and distance between nozzle and roll wheel as shown in Figure 1 and the effect of distance on the ribbon thickness is shown in Figure 2. The constant values (A, B, C, and D) that used in the empirical equation of ribbon thickness are shown in Table 1 which was generated by the data fit program.

$$TH = \left(A*G^{(B+(C*MT))}\right) + \left(\frac{D}{MT}\right) \quad (1)$$

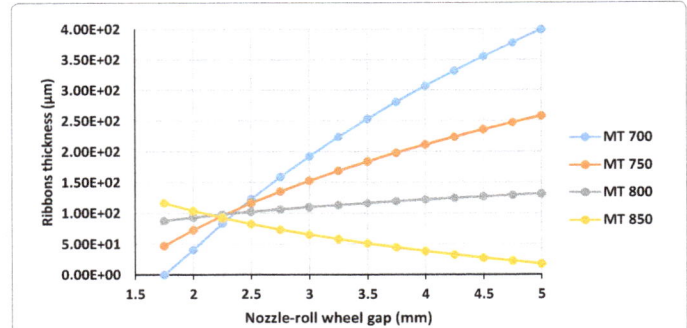

Figure 1: Effect of G and MT on the thickness of rapid solidified ribbons at constant OD equal to 2.5 mm.

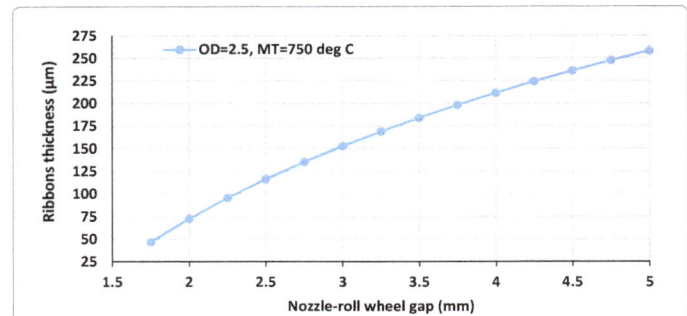

Figure 2: Effect of G on the thickness of rapid solidified ribbons at MT of 750°C and OD of 2.5 mm.

A	B	C	D
1885	1.21	-0.00149	-1455988

Table 1: Constant values.

A	B	C
-116.4	-0.135	95508

Table 2: Constant values.

Case No.	Variable Parameters		Constant Parameter
1	OD	G	MT 750
2	G	MT	OD 2.5

Table 3: Operation conditions.

Data fit for grain size

The discovered empirical equation for prediction the grain size GS of rapid solidified ribbons was discovered depending on the thickness of ribbons TH as shown in Equation (2). The constant values (A, B, and C) that used in the equation of ribbon grain size are shown in Table 2 which were generated by the data fit program. The regression value of this equation is 0.968.

$$GS = \frac{(A + TH)}{(B + (C*TH^2))} \quad (2)$$

Data fit for hardness

The hardness of rapid solidified ribbons H computes using three parameters; one of them is constant and two are variants. The discovered empirical equations were obtained depending on the orifice diameter OD, nozzle-roll wheel gap G, and melting temperature MT as shown in Table 3. Equations 3 and 4 showed the empirical equations and their constant values (A1, B1, C1, A2, B2, C2, and D2) are shown in Table 4 which was generated by data fit program. Correlation coefficient of these equations is 0.97 and 0.89. The results from computations are shown in Figures 3-5 which demonstration the relationship between gap, ribbon thickness, and melting temperature. The better agreement between prediction and experimental data are shown.

$$H = \left(A1*OD\right) + \left(B1*G\right) + \left(\frac{C1}{OD^4}\right) \tag{3}$$

$$H = (A2 + B2*G)/(1 + C2*MT + D2*MT^2) \tag{4}$$

Data fit for tensile strength

The hardness H that obtained by empirical equation can be used instead of the measured hardness value to compute the tensile strength TS value [12]. Equations (5, 6 and 7) were derived to get new empirical equation to predict the tensile strength of Al-Mg alloys ribbons without measuring the hardness of ribbons. The new discover empirical equation use to compute tensile strength depending on the orifice diameter (OD), and distance between nozzles and roll wheel (G) without doing any experimental work. Table 5 present the constant values (A, B, and C) of this equation which were generated by data fit program.

$$TS = K * hardness \tag{5}$$

A1	B1	C1	
15.44	-0.162	1799	
A2	B2	C2	D2
1.262	-0.0997	-0.0022	0.0000012

Table 4: Constant values.

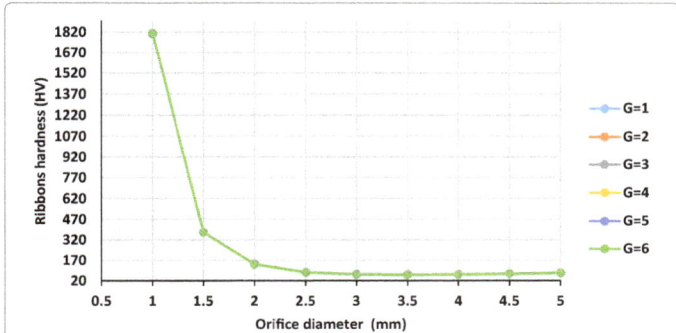

Figure 3: Effect of OD and G on the hardness of rapid solidified ribbons at constant MT equal to 750°C.

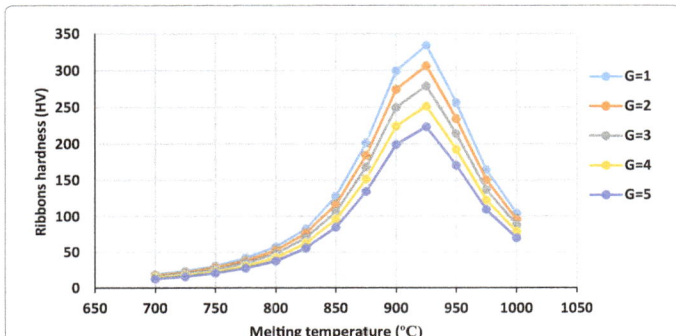

Figure 4: Effect of G and MT on the hardness of rapid solidified ribbons at constant OD equal to 2.5 mm.

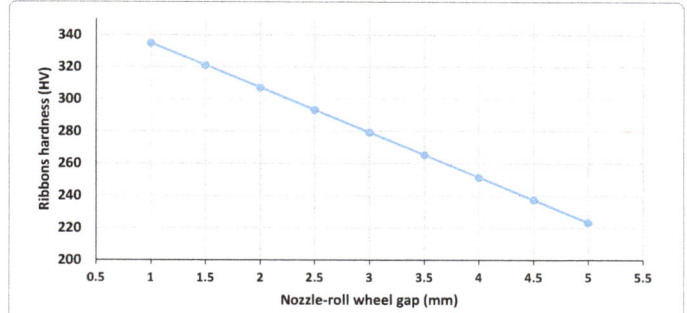

Figure 5: Effect of G on the hardness of rapid solidified ribbons at MT of 925°C and OD of 2.5 mm.

A	B	C
126316	55.4	-3.0994
A1	B1	C1
1884.68	1.21253	-0.001489
A2	B2	C2
-116.39	-0.1353	95508
A3	B3	C3
126315.5	55.43	-3.0994
D1		
-1455988		

Table 5: Constant values.

$$H = \left(A*OD\right) + \left(B*G\right) + \left(\frac{C}{OD^4}\right) \tag{6}$$

$$TS = K*\left[\left(A*OD\right) + \left(B*G\right) + \left(\frac{C}{OD^4}\right)\right] \tag{7}$$

Data fit for yield strength

Yield strength YS of rapid solidified ribbons compute depending on the grain size of ribbons. Discovered empirical equation that use to predict yield strength is shown in Equations (8, and 9) and constant values (A, A1, A2, A3, B, C, B1, B2, B3, C, C1, C2, and C3) are presented in Table 5 which was generated by the data fit program. The relationship between yield strength and grain size is shown in Figure 6.

$$YS = A*\left(Ln\left(GS + B\right)\right)^C \tag{8}$$

$$YS = A3*\left(Ln\left(\frac{A2 + \left(\left(A1*G^{(B1+(C1*MT))}\right) + \left(\frac{D1}{MT}\right)\right)}{B2 + \left(C2*\left(\left(A1*G^{(B1+(C1*MT))}\right) + \left(\frac{D1}{MT}\right)\right)^2\right)} + B3\right)\right)^{C3} \tag{9}$$

Results and Discussion

The discovered empirical equations that obtained in this work can be used to predict the mechanical properties of rapid solidified ribbons types 5083 without doing any experimental work. It is depending on the major operation conditions of rapid solidification process include orifice diameter, nozzle-roll wheel gap, and melting temperature. A predict mechanical properties include ribbons thickness, hardness, and tensile strength. The result curve showed that when the distance between crucible nozzle and roll wheel surface increased the ribbon thickness increased which agree with the experimental work. This relationship depends on the melting temperature. For instant, when the melting temperature equal to 700, 750, and 800°C, the above relation is correct. But, when the melting temperature increase more than 800°C, ribbons

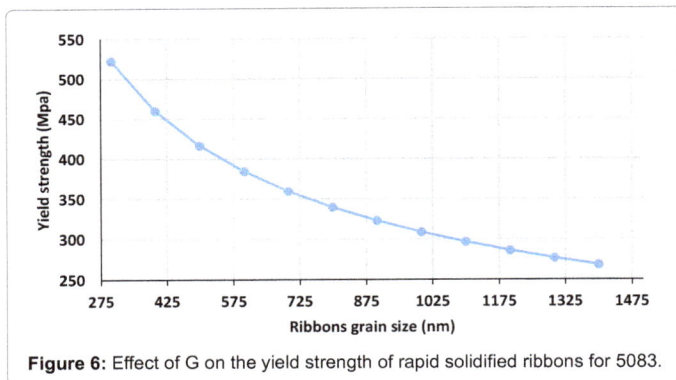

Figure 6: Effect of G on the yield strength of rapid solidified ribbons for 5083.

thickness decrease with increasing the distance between nozzle and roll wheel as shown in Figures 1 and 2. The dependence of the ribbon dimensions on single roll melt spinning parameters has been presented in terms of mathematical expressions. Due to increasing the thickness of ribbons with increasing the distance between nozzle and roll, the hardness of ribbon was decreased which agree with the relationship that obtained by empirical modelling as shown in Figure 5. Additionally, Figure 4 showed that the hardness of ribbons increases with increasing the melting temperature up to 925°C, then decrease with increasing the melting temperature. On the other hand, the hardness increase with increasing the distance between nozzle and roll gap especially when the melting temperature increased. From Figure 3 can be indicated that the hardness of ribbons decreased with increasing the orifice diameter. This relationship is constant for all distance between nozzle and roll. There is inverse proportional relationship between yield strength and orifice diameter as shown in Figure 6. When the grain size increase, the yield strength decreased. However, hardness is directly proportional to tensile strength. With the hardness increase, the tensile strength increased too.

The rapid solidification technique was found be effective process for refining structural particles in 5083 alloy, where highly refined intermetallic components as well as grain size were found to be beneficial to the modification of mechanical properties.

Conclusions

It can be concluded that the hardness, yield strength and tensile strength of 5083 Al-alloy ribbons increased by rapid solidification process. There is a good agreement between experimental results and empirical results, where the correlation reached to 0.99. Empirical modelling and equation are economical and useful for predicting of mechanical properties for rapidly solidified alloys which forming by single roll melt spinning process. The maximum ribbon hardness was obtained at melting temperature equal to 925 degrees centigrade with orifice diameter of 2.5 mm and gap between orifice nozzle and roll surface equal to 1 mm. The major factors that influence the mechanical properties of rapidly solidified ribbons are the shape and grain size of ribbons, because hardness and yield strength of rapidly solidified ribbons were increased with degreased the ribbon grain size and ribbons thickness.

References

1. Bogno AA, Maitre A, Henein H, Gandin CA (2014) Characterization of rapidly solidified metallic alloys using combination of experiments and modeling. COM 2014 Conference metallurgists Proceedings.

2. Jassim AK, Hammood AS (2014) single roll melt spinning technique applied to produce micro thickness rapid solidified ribbons type 5083 Al-Mg alloy. International Parallel Conference on Researches in Industrial and Applied Science 1-7.

3. Jassim AK, Hammood AS(2014) Sustainable manufacture process for bulk metallic glasses production using rapid solidification with melt spinning technique. 2014 international conference on materials science and materials engineering (MSME2014).

4. Jassim AK, Hammood AS (2014) Single roll melt spinning technique applied as a sustainable forming process to produce very thin ribbon and 5052 and 5083 Al-Mg alloys directly from liquid state. 13th Global conference on sustainable manufacturing-decoupling growth from recourse use.

5. Karpe B, Kosec B, Bizjak M (2011) Modeling of heat transfer in the cooling wheel in the melt-spinning process, Journal of Achievements in Materials and Manufacturing Engineering 46: 88-94.

6. Kramer M, Mecco H, Dennis K, Vargonova E, McCallum R, et al. (2007) Rapid solidification and metallic glass formation-experimental and theoretical limits. Journal of Non-crystalline solids 353: 3633-3639.

7. Gheiratmand T, Madaah Hosseini HR, Davami P, Ostadhossein F, Song M, et al. (2013) On the effect of cooling rate during melt spinning of FINEMET ribbons. The Royal Society of Chemistry 5: 7520-7527.

8. Woodcraft AL (2005) Predicting the thermal conductivity of aluminum alloys in the cryogenic to room temperature range. Cryogenics 45: 421-431.

9. Jinna M (2009) Titanium-based Bulk Metallic Glasses: Glass Forming Ability and Mechanical Behavior. University Joseph Fourier.

10. Wannaparhan S (2005) Roles of supercoiling and cooling rate in the microstructural evolution of copper- cobalt alloys. University of Florida, USA.

11. Zhang X, Atrens A (1992) Rapid solidification characteristics in melt spinning. Materials Science and Engineering A159: 243-251.

12. Bolton W(1988) Engineering Materials Technology. 2nd, Butterworth-Heinemann, British Library, UK.

23

The Effect of Grain Boundaries on the Elastic, Acoustical and Thermo-Physical Properties of Metal-Ceramics Composites

Abramovich A*

St. Petersburg State University of Technology and Design, Higher School of Technology and Design, 198095 St. Petersburg, Russia

Abstract

The purpose of this work was searching of the formation of grain boundaries in metal-ceramics composites at various metal concentrations and sintering temperatures, influence of these boundaries on elastic moduli, coefficient absorption ultrasonic waves (USW) and thermo-conductivity to find the coupling of these properties and to estimate the optimal value of the metal concentration for achieve high quality of ready composites "corundum-stainless steel". These boundaries are formed in the sintering process. In this work, cermets sintered in a high vacuum at different temperatures are investigated. Cermets (metal-ceramic composites) are modern construction materials used in different branches of industry. Their toughness and heat resistance are determined by their elastic and thermo-physical properties. In addition, these properties are significantly dependent on the grain boundaries in the material. The elastic moduli and absorption coefficient were measured by the ultrasonic method at room temperature; measurement of the thermal conductivity coefficient was carried out at temperature 200°C. In addition the samples structure was investigated by optical and scanning electron microscopy (SEM), cermets composition was determined by energy-dispersive X-ray spectroscopy method (EDS). We found two extremes for the concentration dependence of the elastic moduli (E and G) on the stainless steel concentrations, the nature of which is unknown. Similar dependence is observed also for the thermal conductivity coefficient and coefficient absorption ultrasonic waves. A discussion of the results is based on the structure cermet model as multiphase micro heterogeneous media with isotropic physical properties is also presented.

Keywords: Metal-ceramic composite; Sintering; Grain boundaries; Elastic and thermo-physical properties

Introduction

Composite materials, based on ceramics and metals, are widely used in different branches of industry. These materials are often created for the production of constructions with the need for high strength, thermal stability, thermal shock resistance and resistance to aggressive media. Investigations of the physical and technical properties of these materials have been published in periodical issues and monographs [1,2]. Thermal shock resistant composites have to possess high thermal conductivity and mechanical stability, which are determined by the elastic moduli. Thus it is supposed that the composite have to possess these properties simultaneously. However ceramics have small thermo conductivity and high stability while metal possess high thermo conductivity and low stability. Obviously that the mixing of row metal and ceramic powders gives so-called "mechanical mixture" possessed low stability even if the powder mix was compacted by pressure. In this connection to prepare composites usually the sintering process uses. The sintering temperatures can be higher or lower than metal melt temperature. This process leads to activation of the thermo-chemical reactions on grain boundary. The new phase of composite is resulting of sintering process, moreover just her properties determinate the stability and thermo conductivity of ready composite. As of now, there is no uniform understanding about the connection between the elastic and thermo-physical properties of sintered dispersed materials [3].

The purpose of this work was searching of the formation of grain boundaries in metal-ceramics composites at various metal concentrations and sintering temperatures; influence these boundaries on elastic moduli and thermo conductivity to find the coupling of these properties, to estimate the optimal value of the metal concentration for achieve high quality of ready composites "corundum-stainless steel". The present report is development of the authors preceding work complimented by new investigations of cermets microstructure

and their acoustical properties based on elastic moduli and thermo conductivity.

Materials and Methods

We investigated cermet samples based on α-Al_2O_3 in combination with commercial quality stainless steel (18 wt.% Cr, 9 wt.% Ni, 1 wt.% Ti and 72 wt.% Fe). To fabricate a cermet, an initial fine grained mixture was prepared by milling α-Al_2O_3 powder (2-25 μm) in a ball mill in the presence of balls of 1-2 cm in diameter made from stainless steel. The milling was terminated when the steel content in the α-Al_2O_3 powder became equal to 2.2, 4.0, 5.5, 11.0 and 21.0 vol%. Then, the mixture obtained was doped with a plasticiser and subjected to dry compaction under a pressure of 100 MPa, followed by sintering in a high vacuum at either 1500 or 1600°C. Finally, the samples were cooled in a furnace at an average rate of 100°C/h and no further treatment was made. The prepared cermets ranged in volume porosity from 3 to 7% and had steel contents from 2.2 to 21.0 vol%. The cermet samples (two series for two temperatures of sintering) were then cut into smaller parts of the required dimensions (10 mm in diameter and 8-15 mm in length), which were further ground and polished, depending on the measuring method. To investigate the elastic properties of the cermets, we measured their density using the hydrostatic method and the velocity of longitudinal and transverse ultrasound waves with a frequency of

***Corresponding author:** Abramovich A, St. Petersburg State University of Technology and Design, Higher School of Technology and Design, 198095 St. Petersburg, Russia, E-mail: a.abramovich@spbu.ru

2 MHz. The measurement of mechanical stability requires special preparation of many samples to provide reliable results, whereas for the measurement of dynamic elastic moduli, only one sample is required. Using the pulsed phase-interferometer method [4] made it possible to determine the velocity of the ultrasound waves within an accuracy of 0.1-0.2% and 5% for the elastic moduli. Using the data obtained, we calculated Young's modulus, E, and the shear modulus, G, from the well-known formulas of the elasticity theory for an isotropic medium [4]. Absorption coefficient was measured at 5 MHz frequency of USW within accuracy ~10%. The thermal conductivity coefficient λ was measured using an adiabatic calorimeter, ИТ-λ-400 (Thermometer Company, Russia). The measuring of λ was carried out by a method of comparison with the standard sample. The measurement accuracy of λ was 2-3% for all samples. Examination of the surface structures (cleaved facet) of the cermet samples was made with a JSM-840 scanning electron microscope (JEOL Ltd. Company, Japan), the distribution of particles with size be measured by Mastersizer 3000 (Malvern Instruments Ltd, UK), metallographic and X-ray spectroscopy (EDS) analysis by Thixomet. MACRO (Thixomet, Russia).

Results and Discussion

In Figure 1, a typical microstructure of the studied cermets, obtained by scanning electron microscopy (SEM), is presented. It can be seen there are three phases in the samples, corundum grains, metal drops and pores. The metal drops have sizes of 2-3 µm and are disposed usually on the joints of the corundum grains, smaller drops

are disposed on the grain faces. It should be noted that inter phase and inter grain boundaries are not observed and the size of the corundum grains did not change after sintering.

Figure 2 shows the curve of the distribution of corundum particles with size in initial powder. From the curve it follows that on the whole the particles have average size ~1.5-2.0 µm. More large particles (10-20 µm) also are presented. Comparison this between with curve and calculation of the distribution of corundum particles in Figure 3 showed good accordance. Hence at sintering the recrystallization of corundum particles don't observed. Pore and steel particles are distributed of the cermet volume evenly.

From Figures 1-5 it is clear that the grain boundaries phase (probably spinel $FeAl_2O_4$ [5-9]) are invisible. Thus it can suppose that grain boundary phase in cermets is thin layer equal some nm. If the size of this phase is considerably more she became visible. It can see from Figure 6 (new phase cluster size equals 10-20 µm).

Minimum of absorption coefficient α can be seen in Figure 7 at steel concentration ~5 vol%. Earlier in the work [6] concentrations dependences of elastic moduli and thermal conductivity coefficient of the same cermets were investigated (Figures 8 and 9). It can be seen that these dependences have minima also at ~5 vol %. Detailed explanation of this effect was presented in work [6]. Briefly it caused by kinetics of grain boundaries formation: she stabilized at this concentration. It should be noted that observed connection α and λ is describe in acoustical physics of solid state [3,4] as thermo-elastic effect (where

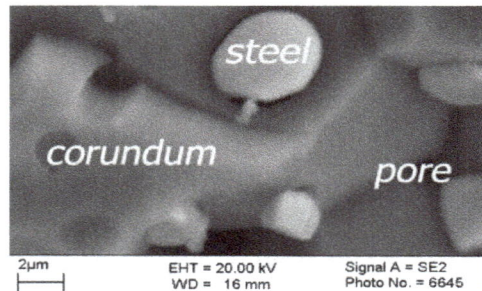

Figure 1: Microstructure of the cermets sample N1sintered in vacuum at 1500°C by SEM method. It is clear that the grain boundaries are no visible.

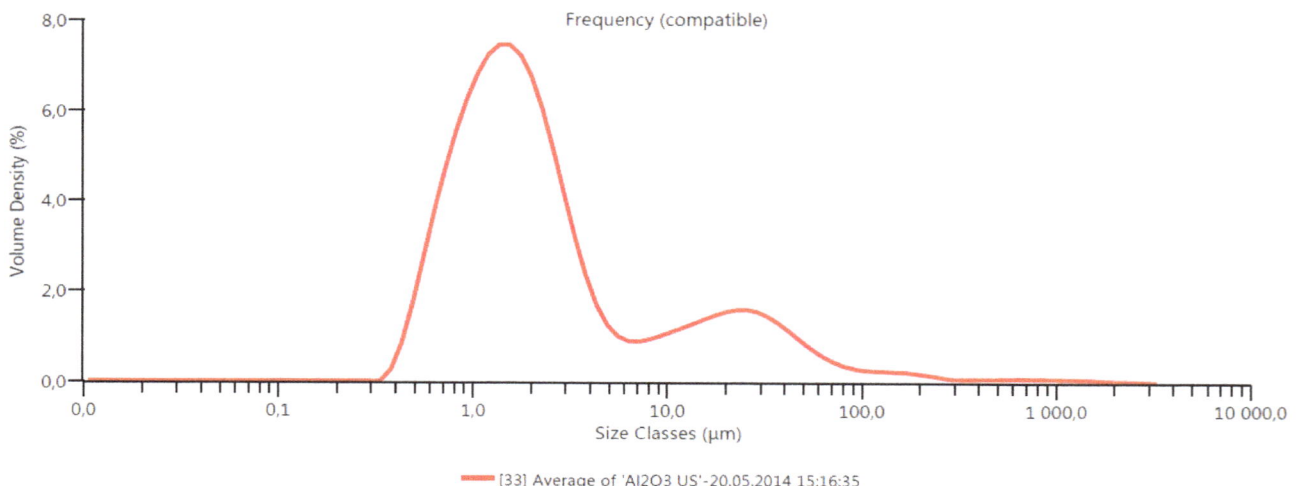

Figure 2: The distribution of corundum particles with size in initial corundum powder.

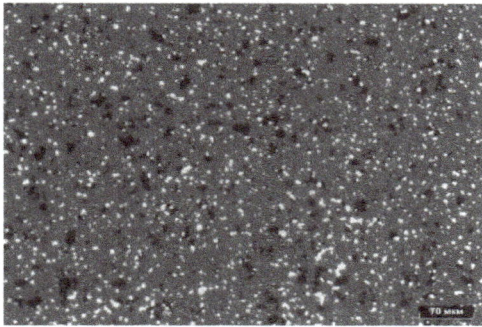

Figure 3: Optical image of cermet N1 shows different cermet areas: black-pores, white-steel, grey-corundum.

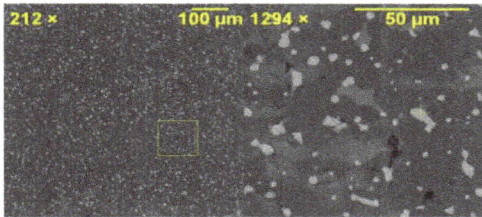

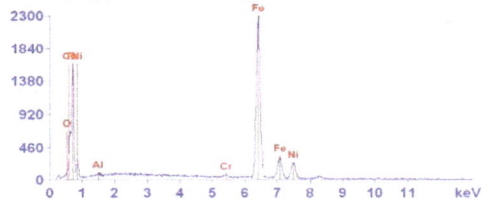

Figure 4: Optical image of cermet N1 (different scale, top) and its corresponding EDS spectrum showing inclusion of steel (bottom).

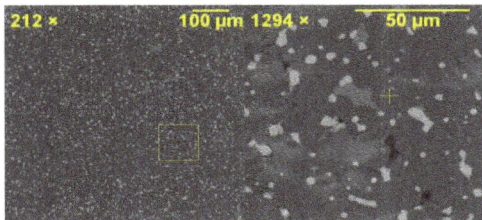

Figure 5: Optical image of cermet N1 (different scale, top) and its corresponding EDS spectrum showing corundum matrix Al_2O_3 (bottom).

Figure 6: Optical image of cermet N1 (different scale, top) and its corresponding EDS spectrum showing inclusion of new grain boundaries phase made possible spinel $FeAl_2O_4$ (bottom).

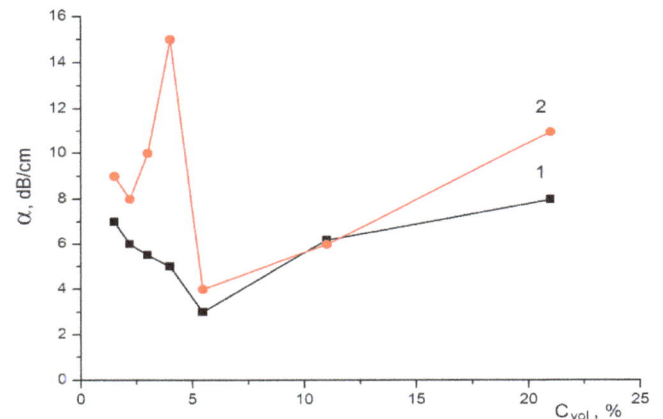

Figure 7: Concentration dependences of the absorption coefficient USW (α) in cermets N1, N2 sintered at temperatures of 1500°C (curve1) and 1600°C (curve 2), respectively.

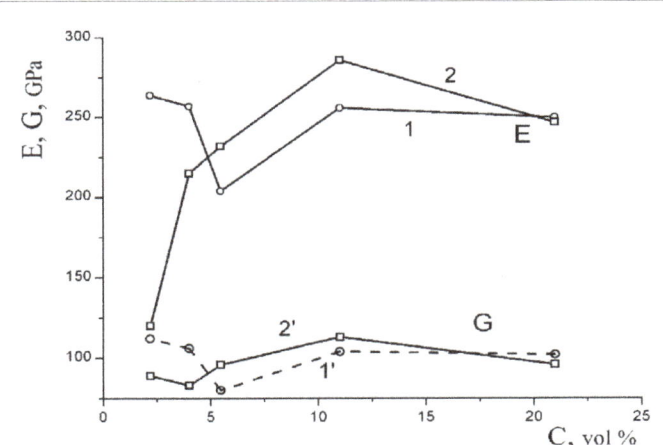

Figure 8: Concentration dependences of elastic moduli, E and G, for cermets, sintered at various temperatures: 1500°C – curves 1 and, 1' and; 1600°C – curves 2 and, 2' [6].

$\alpha \sim \lambda$). Thus it is believed that the structure of sintered composite is similar to structure of polycrystalline solid body [7-10].

Conclusion

In the oxide cermets the concentration dependences of elastic moduli, absorption coefficient of USW and the thermal conductivity coefficient have been obtained and discussed; relation between of

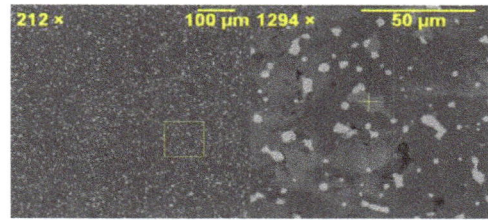

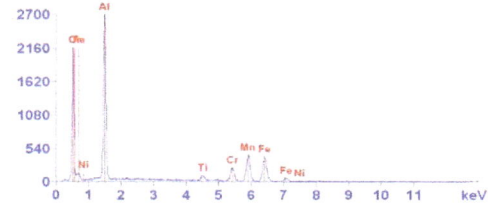

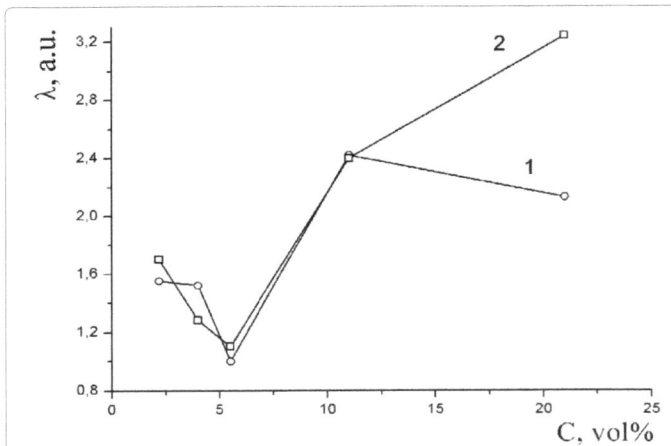

Figure 9: Concentration dependences of the thermal conductivity coefficient (arbitrary units) in cermets sintered at temperatures of 1500 (1) and 1600°C (2), $\lambda=1$ corresponds to the standard sample value [6].

elastic moduli, absorption coefficient and the thermal conductivity for metal concentrations 2-11 vol.% was established; the optimal value of metal concentration for investigated composites "corundum-stainless steel" were determined. To summarize, this work shows us that the thermo-physical and ultrasonic spectroscopy methods are simple and informative in the technology of synthesis of sintered composites.

Acknowledgements

Part of scientific research was performed at the Center for Innovative Technologies of Composite Nanomaterials of Research park of St. Petersburg State University.

References

1. Kingery WD, Bowen HK, Uhlmann DR (1976) Introduction to Ceramics. Wiley Interscience, New York.

2. Kislyi PS, Bodnaruk MS, Borovikova MS (1985) Cermets. Naukova dumka, Kiev, Ukraine (in Russian).

3. Klemens PG (1969) Thermal Conductivity, RP. Academic Press, London, UK, 1.

4. Truell R, Elbaum C, Chick BB (1972) Ultrasonic Methods in Solid State Physics. Academic Press, New York.

5. Karban OV, Abramovich AA, Khazanov EN, Taranov AV, Konygin GN, et al. (2013) Investigation of structure and kinetics of the thermal phonons cermets based alumina with iron supplementation synthesized by different methods. The Chemical Physics and Mesoscopy 15: 758-763.

6. Abramovich A (2016) Acoustical and thermo physical properties of metal-ceramics composites in dependence on few of metal volume concentration. IOP Conf. Series: Materials Science and Engineering 123: 012053.

7. Gusev AI, Rempel AI (2001) Nanocrystalline materials. Physmathlit, Moscow, Russia (in Russian).

8. Glezer AM, Stolyarov VL, Tomchuk AA, Shurygina NA (2016) Grain boundary engineering and superstrength of nanocrystals. Tech Phys Lett 42: 51-54.

9. Slepnev AG (2007) Evaluating the mechanical properties of grain-boundary phases in nano- and submicrocrystalline materials using a model of an elastic multilayer periodic medium. Tech Phys Lett 33: 936-938.

10. Ivanov SN, Khazanov EN, Taranov AV, Mikhailova IS, GropyanovVM, et al. (2001) Grain Boundaries and Elastic Properties of Aluminum-Oxide and Stainless-Steel-Based Cermets. J Physics Solid State 43: 665-669.

24

Synthesis and Characterization of Polyaniline/Ignimbrite Nano-Composite Material

Ertekin B[1], Çimen Z[2], Yilmaz H[2,*] and Yilmaz UT[2]

[1]Department of Milk Technology, Faculty of Agriculture, Adnan Menderes University, 09100 Aydın, Turkey
[2]Department of Chemistry, Polatli Faculty of Arts and Science, Gazi University, 06900 Polatli, Ankara, Turkey

Abstract

In this study, Polyaniline-ignimbrite (PAN-IB) which is a novel nanocomposite material consisting of electrically conductive Polyaniline (PAn) polymer and Ignimbrite (IB) natural insulating material was synthesized chemically using KIO_3 as a radical initiator in aqueous media with conventional radical polymerization method. The synthesized nanocomposites including ignimbrite with various monomer-ignimbrite percentages were monitored with scanning electron microscopy (SEM) and structural characterizations were examined with FTIR spectroscopy. Thermogravimetric analysis (TGA), particle size analysis with dynamic light scattering, magnetic susceptibility and conductivity measurements were also examined for all synthesized composites. With making a composite with aniline, the conductivity of ignimbrite (3×10^{-7} Scm^{-1}) reached to 2.7×10^{-5} Scm^{-1}. However increasing the ratio of monomer which is added to ignimbrite did not make a significant change on conductivity of the resulting composite.

Keywords: Nanocomposites; Polyaniline; Ignimbrite; Magnetic susceptibility

Introduction

Nowadays, polymeric nano-composite materials have been gaining substantially growing importance in industrial field. Fascinating features of these polymers such as solidity, lightness, flexibility, plasticity, having high chemical and corrosion resistance, thermal stability when combined with adding a filling substance to the ultimate material exhibit mechanical features like flame resistance or thermal stability and give rise to wide range of products. Sports equipment, rocket parts and the applications in the automobile industry are some examples of the use of these materials.

There is an aroused interest in nanotechnology in composite applications because of new superior features that the use of nanotechnology offers. Due to the reason that the filling substance is in nanoscale; there might be some features that cannot be observed in macro scale and only be realized in nanoscale. This phenomenon can be explained with the increased surface area of the substance in nanoscale with respect to its initial state when it is in macro scale. Increased surface area differentiates the interactions between nano particles; and as a result of that there might be a change in the material weight, stiffness, chemical and thermal features [1]. Furthermore, as the size of the particles decreases, the bonds between atoms and molecules in particles differentiates, the number of atoms and molecules on the surface of the particles increases; and consequently these facts explain why particles in nanoscale demonstrate different behavior [2]. In addition, polymer nanocomposites are the materials that are used as matrix polymers. Polymer nanocomposites exhibit different behaviors depending upon some configuration properties such as; the ratio of polymer and filling substances they include, their chemical properties, and the crystallinity of polymer matrix.

In general, ceramic matrix composites are manufactured to improve crack resistance. As distinct from ceramic matrix composites, in the materials derived from polymer matrix composites; either endurance and solidity or flexibility and processability are the concepts that are worked on. For example, in polymer matrix composites, material features such as elasticity and thermal expansion are dependent on the filler's structure; however the strength of the composite will vary depending on the particle size of the filling substance. One of the first

studies on this subject is, polymer nanocomposite material which is conducted in Toyota Central Research Laboratories in the early 1990s [3]. As the little amount of filling material added to polymer, the positive changes on thermal and mechanical properties of the product have been observed in the research.

Conductive polymers have been the subject of many studies in recent years due to their electrical conductivity, mechanical strength, corrosion resistance, and ability to be synthesized both chemically and electrochemically. Among conducting polymers; polyaniline (PAN) is a unique substance due to its excellent electrical, magnetic, and optical properties. It is thought to have a high potential in commercial applications regarding conductive polymers because of its low cost of synthesis and raw material. Its insolubility in many commonly used solvents limits its processability with solution process or melting process methods [4,5]. Improvement in the processability of polyaniline can be achieved by preparing copolymer, composite or blends with other polymers or inorganic materials [6-9].

Conductive polymers have a wide range of application areas such as; conductive coloring [10], optical devices [11], membranes [12], biomedical applications [13], the removal of heavy metals [14,15], and solar cells [16]. In present study, a natural rock ignimbrite that widely spreads around Cappadocia region in Turkey is milled and turned into a composite inside polyaniline matrix, there by obtaining both conductive and light material with high mechanical strength.

Ignimbrite is a natural insulating material that widely spreads around Cappadocia region in Turkey. It is a light and easily processable rock

---

*Corresponding author: Yilmaz H, Polatli Faculty of Arts and Science, Department of Chemistry, Gazi University, 06900 Polatli, Ankara, Turkey
E-mail: hasim@gazi.edu.tr

which is the deposit of a pyroclastic density current and traditionally used in the construction and chemical industries. It is made of a very poorly sorted mixture of volcanic ash (or tuff when lithified) and pumice lapilli, whose color may be white, grey, pink, beige, brown or black depending on composition and density. Due to its lightness it can be used in insulation and siding applications of buildings.

In the present study ignimbrite is mechanically milled up to nanometer scale, defused to a polyaniline matrix during polymerization process and a semiconducting Pan-IB nanocomposite material have been produced. Thermal, morphological, electrical and magnetic characterizations have been examined and this new material has been offered as a suitable material for conducting polymer technology applications.

Materials and Methods

Materials

Ignimbrite was supplied from Nevşehir province's natural area and ball milled up to nanometer scale before used. KIO_3 (99.995%), diethyl ether and HCl were purchased from Sigma Aldrich and used without purification. Aniline (≥99.5%) was purchased from Sigma Aldrich and vacuum distilled before used, then stored at -18°C.

Characterization

For the characterizing of electrical conductivity of composites the four point probe technique was used with a Nippon NP-900 multimeter (Osaka, Japan). Fourier transform infrared (FT-IR) spectra were recorded with a Spectrum 100 model spectrophotometer (Perkin Elmer, USA) to had structural characterization. Magnetic susceptibility measurements were carried out with a Sherwood Scientific model MKI Gouy scale with a procedure reported elsewhere [17]. The physical morphologies of the composites were monitored by Zeiss LS-10 field emission SEM instrument equipped with an Inca Energy 350 X-Max spectrometer (Oxford Instruments). Samples were sputter-coated with Au (60%) and Pd (40%) alloy using a Q150R instrument (Quorum Technologies). Images were obtained at 3×10^{-4} Pa working pressure and 15 keV accelerating voltage using In Lens detection mode (2 mm working distance). Dynamic light scattering analyses were performed at room temperature using a 90Plus Nanoparticle Size Analyzer (Brookhaven Instruments, Holtsville, NY). Thermogravimetric analysis was performed with a Seteram SETSYS thermal analyzer at temperature range of 25-1200°C at a heating rate of 10°C min^{-1} under argon atmosphere with a gas flow rate of 20 ml min^{-1}.

Synthesis of Pan-IB nanocomposite

A certain amount of aniline monomer was dissolved in 0,5M HCl solution and bubbled with N_2 15 minutes with stirring at 5°C. Then the KIO_3 solution was added dropwise to the medium under N_2, and the solution was stirred. 1 hour later, 1 gram of ignimbrite dispersed in HCl solution and added dropwise to the reaction mixture. The reaction was continued under nitrogen atmosphere for 24 h. Then the composite was collected and washed with 1.5 M HCl, water, and diethyl ether, respectively. We synthesized the PAn/Ignimbrite composites, including ignimbrite at different percentages, by varying the amount of aniline monomer at a constant $n_{aniline}/n_{oxidant}$ ratio. Reaction conditions and amount of components were given in Table 1.

Results and Discussions

Electrical conductivity

The electrical conductivity of composites was measured on pressed

Sample Coding	Composite Ratio	IB (gram)	Aniline (ml)	KIO_3 (mol)
IB	-	1	-	
K1	1 × 1	1	1	5.09×10^{-6}
K2	1 × 2	1	2	1.02×10^{-5}
K3	1 × 3	1	3	1.53×10^{-5}
K4	1 × 4	1	4	2.04×10^{-5}

Table 1: Reaction conditions and amount of components. T: +5°C, $d_{aniline}$: 1.0217 gmL^{-1}.

pellets of composite powders. The average thickness of the compressed pellets was 1.0 mm. As is known the most common green polyaniline emeraldine salt has conductivity on a semiconductor level of the order of 10^0 Scm^{-1} [18]. In this study addition of Ignimbrite, decreased the conductivity and conductivities of composites ranged from 2.64×10^{-5} to 2.70×10^{-5} Scm^{-1} and were close to each other as it can be seen from the Table 2.

Magnetic susceptibility

The measurement of magnetic properties of conducting polymer is important to know what the charge carrier is and how the electrical conduction occurs. Of these properties, magnetic susceptibility is a significant factor in determining the type of magnetism and density of states at Fermi level [19]. Magnetic susceptibility data of ignimbrite and PAn-IB nanocomposites were measured. As seen in Table 2, magnetic susceptibility values were all positive and range from 1.73×10^{-6} to 10.25×10^{-6}. The positive values of Gouy scale measurements reveal that the conductivity mechanisms of PAn-IB composites are polarone in nature [20].

Dynamic light scattering measurements

Average particle size of ignimbrite was measured to be 34.7 and PAn-IB composites were changed between $47.5 \geq d_{0.5} \geq 61.3$ nm (Table 2).

FTIR results

Figure 1 displays the FTIR spectrum of ignimbrite (a), aniline (b) and PAn-IB (K3) (c) nanocomposite. There is about 70% SiO_2 component in the chemical composition of ignimbrite stone. Characteristic peaks at 1028 cm^{-1} and 911 cm^{-1} were assigned to the stretching and bending vibration of O-Si-O bonds respectively. Aniline and polyaniline showed the main characteristics bands about 3351 cm^{-1}-3686 cm^{-1} attributed to the stretching from amine groups. Two broad bonds at 3351 cm^{-1} and 3360 cm^{-1}, indicated the NH_2 groups of aniline. It was observed that these peaks of aniline were shifted to 3686 cm^{-1} on the spectrum of PAn-IB. This sharp and single bond indicate the secondary N-H groups of polyaniline. And also by comparing the peaks of aniline and PAn-IB, it was observed that peaks from amine groups were shifted due to the presence of SiO_2 particles in polymer matrix.

TGA analysis

The amount of weight loss and the thermal stability of PAn-IB composites were determined using TGA at temperature range of 25-1200°C at a heating rate of 10°C min^{-1} under argon atmosphere with a gas flow rate of 20 ml min^{-1}. TGA curves of Pan-IB (K3) are shown in Figure 2. As can be seen from TGA curves, nanocomposite shows decomposition with two steps. The first step weight loss at 105°C indicates the loss of small units such as solvents and monomers in the composites. The second weight loss at 557°C shows degradation of the polymer [21].

Sample Coding	Conductivity (σ, Scm^{-1}) × 10^{-5}	Magnetic Susceptibility (Xg, cmg^{-1}) × 10^{-6}	Average Particle Size $d_{0.5}$, nm
IB	0.03	+1.73	34.7
K1	2.64	+4.72	50.3
K2	2.66	+7.58	61.3
K3	2.67	+7.90	50.8
K4	2.70	+10.25	47.5

Table 2: Conductivity, magnetic susceptibility and dynamic light scattering data of ignimbrite and PAn-IB nanocomposites.

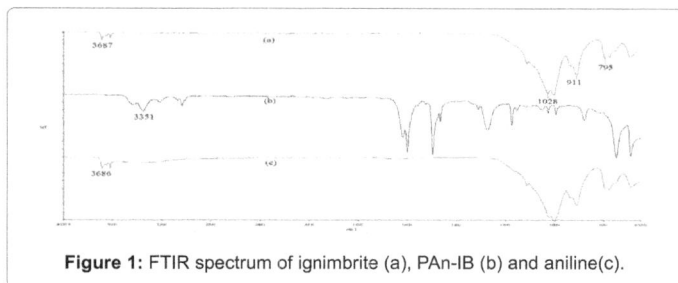

Figure 1: FTIR spectrum of ignimbrite (a), PAn-IB (b) and aniline(c).

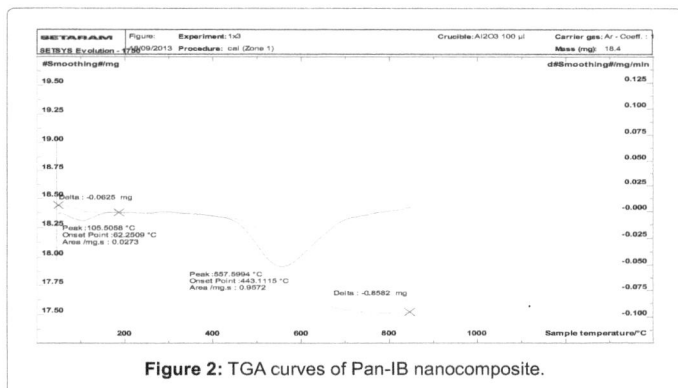

Figure 2: TGA curves of Pan-IB nanocomposite.

SEM analysis

Figures 3a-3f shows the SEM micrographs of PAn-IB nanocomposites and pure ignimbrite powder. All composites reveal granular, nonporous, aggregated surface morphologies with diverse sizes. As shown in Figures 3a-3d, percentage of composites doesn't affect the morphological image considerably. But it is obvious that ignimbrite and Pan-IB nanocomposites exhibit different view in a closer look. As shown in Figures 3e and 3f pure ignimbrite powder exhibits sharp, layered and rod like structure and transforms granular and bulk structure when polymerized with aniline.

Conclusions

In this study, a novel nanocomposite material synthesis and structural, electrical, magnetic and morphological characterizations are present. Pan-IB nanocomposites at 52 nm average particle diameter with 2.7×10^{-5} Scm^{-1} electrical conductivity were synthesized. All magnetic susceptibility measurements were obtained as positive and this situation indicated that the conductivity mechanisms were polarone. According to the FTIR spectra existence of Si-O and N-H bonds in the same spectrum implied that the formation of PAn-IB composite. It was appeared that thermal stability of composite was observed up to 557°C that showed the degradation of the polymer from TGA curves. Scanning Electron Microscopy (SEM) was used for microstructures analysis. Finally it was concluded that the Pan-IB semiconducting nanocomposite is a novel material and suitable for many future applications on science and technology.

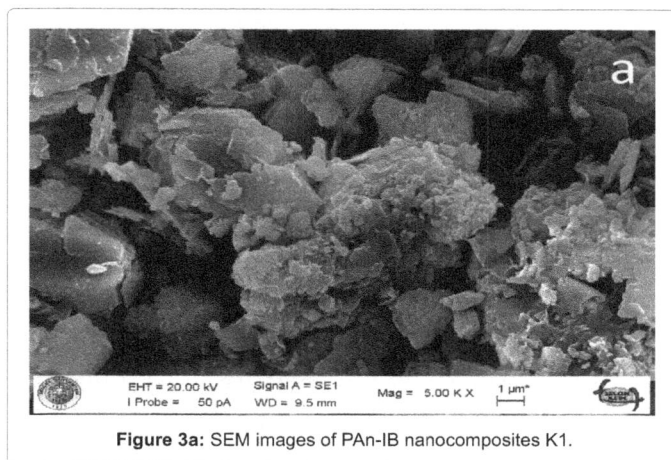

Figure 3a: SEM images of PAn-IB nanocomposites K1.

Figure 3b: SEM images of PAn-IB nanocomposites K2.

Figure 3c: SEM images of PAn-IB nanocomposites K3.

Figure 3d: SEM images of PAn-IB nanocomposites K4.

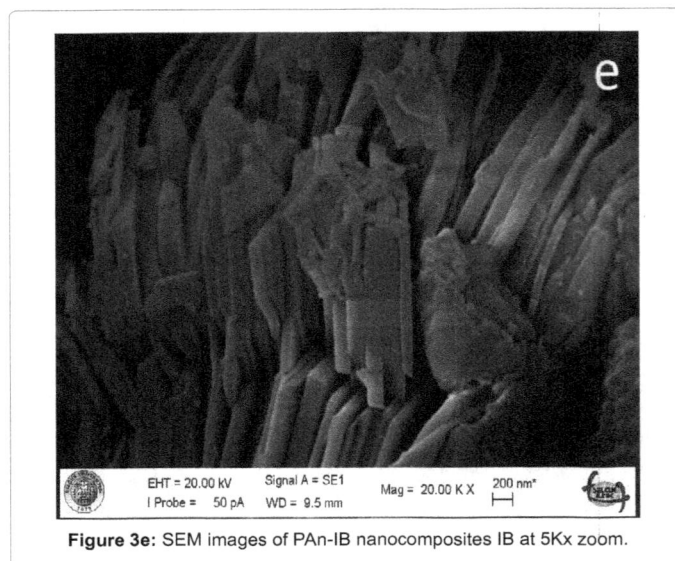

Figure 3e: SEM images of PAn-IB nanocomposites IB at 5Kx zoom.

Figure 3f: SEM images of PAn-IB nanocomposites K1 at 20Kx zoom.

References

1. Saptarshi SR, Duschl A, Lopata AL (2013) Interaction of nanoparticles with proteins: Relation to bioreactivity of the nanoparticle. J Nanobiotech 11: 26-37.

2. Maeda H (1992) The tumor blood vessel as an ideal target for macromolecular anticancer agents. J Controlled Release 19: 315-324.

3. Usuki A, Kawasumi M, Kojima Y, Okada A, Kurauchi T, et al. (1993) Swelling behavior of montmorillonite cation exchanged for ω-amino acids by ε-caprolactam. J Mater Res 8: 1174-1178.

4. Yin W, Li J, Li Y, Wu J, Gu T (2001) Conducting composite film based on polypyrrole and crossliked cellulose. J Appl Polym Sci 80: 1368-1373.

5. Machado JM, Karasz FE, Lenz RW (1988) Electrically conducting polymer blends. Polymer 29: 1412-1417.

6. Paoli MAD, Waltman RJ, Diaz AF, Bargon J (1984) Conductive composites from poly(vinyl chloride) and polypyrrole. J Chem Soc Chem Commun. 15: 1015-1016.

7. Lindsey SE, Street GB (1984) Conductive composites from poly(vinyl alcohol) and polypyrrole. Synth Met 10: 67-69.

8. Cassignol C, Cavarero M, Boudet A, Ricard A (1999) Microstructure conductivity relationship in conducting polypyrrole/epoxy composites. Polymer 40: 1139-1151.

9. Bhat NV, Gadre AP, Bambole VA (2001) Structural, mechanical and electrical properties of electropolymerized polypyrrole composite film. J Appl Polym Sci 80: 2511-2517.

10. Eisazadeh H, Spinks G, Wallace GG (1993) Conductive electroactive paint containing polypyrrole colloids. Mater Forum 17: 241-245.

11. Falco EHL, Azevedo WM (2002) Polyaniline-poly (vinyl alcohol) composite as an optical recording material. Synth Met 128: 149-154.

12. Misoska V, Ding J, Davey JM, Price WE, Ralph SF, et al. (2001) Polypyrrole membranes containing chelating ligands: synthesis characterization and transport studies. Polymer 42: 8571-8579.

13. Benabderrahmane S, Bousalem S, Mangeney C, Azioune A, Vaulay MJ, et al. (2005) Interfacial physicochemical properties of functionalized conducting polypyrrole particles. Polymer 46: 1339-1346.

14. Eisazadeh H (2007) Removal of mercury from water using polypyrrole and its composites. Chin J Polym Sci 25: 393-397.

15. Eisazadeh H (2007) Removal of chromium from waste water using polyaniline. J App Polym Sci 104: 1964-1967.

16. Zafer C, Kus M, Turkmen G, Dincalp H, Demic S, et al.(2007) New perylene derivative dyes for dye-sentisized solar cells. Sol Energy Mater Sol Cells 91: 427-431.

17. Gok A, Sari B (2002) Chemical synthesis and characterization of some conducting polyaniline derivatives: Investigation of the effect of protonation medium. J Appl Polym Sci 84: 1993-2000.

18. Stejskal J, Gilbert RG (2002) Polyaniline, preparation of a conducting polymer. Pure Appl Chem 74: 857-867.

19. Ryu KS, Chang SH, Jeong SK, Oh EJ, Yo CH (2000) Magnetic properties of conducting polyanilines induced by organic acids and solvents. Bull Korean Chem Soc 21: 238-240.

20. Road A (1993) Magnetic Susceptibility Balance, Instruction Manuel, Christian Scientific Equipment, East Gateshead Industrial Estate.

21. Alves WF, Venancio EC, Leite FL, Kanda DHF, Malmonge LF, et al. (2010) Thermo-analyses of polyaniline and its derivatives. Thermochim Acta 502: 43-46.

25

To Develop a Biocompatible and Biodegradable Polymer-Metal Composite with Good; Mechanical and Drug Release Properties

Najabat Ali M¹*, Ansari U¹, Sami J¹, Qayyum F² and Mir M¹

¹Biomedical Engineering and Sciences Department, School of Mechanical and Manufacturing Engineering, National University of Sciences and Technology (NUST), Islamabad, Pakistan

²Mechanical Engineering Department, University of Engineering and Technology, Taxilla, Pakistan

Abstract

For achieving additional benefits and improving the material characteristics two or more materials are often combined together in the form of composites. Composites are important because of their light weight, high strength and flexibility of design. Composite materials provide various advantages based on their particulate or fibrous nature and on the basis of individual qualities of the constituting elements of the composites. Besides the multiplied benefits achieved with the composite materials, they being composed of two different materials exhibit greater challenges and biocompatibility threats which need to be addressed while developing a composite material. A structural composite of bioabsorbable nature is developed using a polymeric material and metal particles. The composite material so developed would provide altered strength and flexibility, better than the individual constituting materials for use in various biomedical devices and would eventually degrade on subject to exposure to the physiological environment. The two different varieties of the composite have been developed using metal particles and metal salt and they have been tested for their tensile, degradation and drug release properties, which have been found satisfactory for use of the composite in various biomedical devices and drug release applications.

Keywords: Composites; Organo-metallic; Polymer-metallic; PVA

Introduction

Several devices being used in the biomedical industry like artificial joints, implants, templates for tissue repair, external and internal fixation devices; require the use of biocompatible and biodegradable materials, so for this purpose researchers are focusing on developing new biomaterials that could be used for aforementioned applications.

Biocompatibility of materials is of prime importance to avoid inflammatory reactions and cyto-toxic responses. With metallic implants, biocompatibility is difficult to attain because of the corrosive nature of metals. Moreover metals exhibit mechanical properties which are not comparable to the natural tissues and thus their mechanical properties need to be altered. This can be done by making composite materials [1].

When particles (as flakes or powder) of the reinforcement material are embedded or distributed in the matrix material, the particulate types of composite is formed, the other types being the fiber reinforced composites and laminar composites [2]. In this article a particulate reinforced composite has been developed by embedding metal particles in a polymeric matrix.

Polymers are used as matrix material for being light in weight, more economical, easily process able, widely available and for their environmental acceptability. The type of polymer matrix and its interaction with the reinforcement material greatly affect the physical and mechanical properties of the composite [3].

A novel composite material of biodegradable and biocompatible nature with good drug release properties can be developed by using polyvinyl alcohol and magnesium. Magnesium is a biodegradable metal with an established biocompatibility and is added to polyvinyl alcohol (PVA) to develop a novel composite with properties better than those of the constituting metal (magnesium) and polymer (polyvinyl alcohol). Aspirin serves as a model drug, to establish the drug release properties of the composite.

Poly-vinylalcohol (PVA)

Poly-vinyl alcohol (PVA) is a very promising functional polymer. Its hydroxyl groups make it easy to be modified chemically which can ultimately result in modifications in itsphysical structure and properties. Owing to it's good chemical and physical properties, excellent film forming nature and hydrophillicity, Poly-vinyl alcohol (PVA) is widely used for making polymeric films [4]. Poly-vinyl alcohol (PVA) blends have better properties in terms of physical characteristics, biological compatability and film forming behavior [5]. Poly-vinyl alcohol (PVA) is the polymer of choise for many bio-medical devices and implants because of its hydrohilic nature which makes it a highly permiable material. PVA membranes are commonly used in medical devices which are in contact with the blood, like hemodyalysis membranes. For improving compatability with blend PVA is often blended with polyelectrolytes, heparin and poly-ethylene glycol [6].

Acetyl-salicylic acid (ASA)/Asprin

It improves the blood compatibility of the device because of its anti-coagulant, blood thinning effect which it creates through its anti-platelet nature, therefore prevents arterial and cerebral thrombosis. It is a known anti-inflammatory and analgesic drug with worldwide acceptance. It also prevents the adhesion of platelets to artificial devices and implants [6].

*Corresponding author: Najabat Ali M, Biomedical Engineering and Sciences Department, School of Mechanical and Manufacturing Engineering, National University of Sciences and Technology, Sector H-12 Islamabad, Pakistan, E-mail: drmurtaza@smme.nust.edu.pk

Keeping in mind the wide range of benefits of polyvinyl alcohol (PVA) and magnesium (Mg) metal, a particulate composite material has been developed in two different forms; using metal particles and by metal salt. The composite so formed has been tested and compared with the mother polymer (PVA) for its tensile strength, degradation rate and drug release behavior, with aspirin as a model drug, which can be replaced and tested with any other drug according to use.

Materials and Methodology

The poly-vinyl alcohol used for composite formation had a molecular weight of approximately 72000 g/mol. Polyvinyl alcohol 72000 BioChemica was used. Simple polymer membranes were developed using 5% and 10% PVA solutions. The solutions were made by dissolution of polyvinyl alcohol (PVA) in water. Clear solutions were made and allowed to cure on stainless steel plate over an area of 3*3 inches square to obtain 5% and 10% PVA membranes for comparison purposes with the composite membranes.

Magnesium particles of 100-200 μm. were used for composite formation. The poly-metal composite film was prepared by solvent casting method. Magnesium (Mg) particles are homogeneously suspended in the 10% poly-vinylalcohol (PVA) solution through magnetic stirring. The poly-vinylalcohol-magnesium (PVA-Mg) suspension was then cured to obtain the metalopolymer composite film. The composite formed this way was compared for its properties with a poly-metal blend of Magnesium sulphate salt ($MgSO_4$) and poly-vinyl alcohol (PVA) which is prepared by homogeneously dissolving the magnesium salt particles in 10% poly-vinylalcohol (PVA) solution and casting the resulting solution at room temperature.

For casting purposes 10 ml of each solution (both the composite and the simple poly-metal blend) was spread over an area of 9 inches square on a stainless steel plate. The surface of the casting plate was uniform and polished to achieve smooth composite membranes. Also by controlling the quantity of solution and the area of its spread, uniform thickness of the membranes was ensured. With the aforementioned quantities and dimensions, 0.1 mm thick membranes were achieved. The thickness of the membranes can be increased by increasing the quantity or decreasing the area of spread and vice versa. With the given quantities of solution, the casting time is 16-18 hours, which might increase with the increased quantity of solution.

Same procedure was adopted for developing composite-drug membranes. Aspirin was used as a model drug here. 75 mg aspirin was added into the polymer-metal composite solution and polymer-metal salt solution and mixed properly before casting. Similar curing parameters were adopted as for simple composite membranes.

The membranes developed included 5% PVA membrane, 10% PVA membrane, 1% mg fin 10% PVA composfifte membrane, 2.5% mg in 10% PVA composite membrane, 1% $MgSO_4$ in 10% PVA membrane, 2.5% $MgSO_4$ in 10% PVA membrane, 1% mg and 75 mg aspirin in 10% PVA composite membrane and 1% $MgSO_4$ and 75 mg aspirinin 10% PVA membrane.

Characterization

To establish the physico-mechanical properties of the composite, tensile testing and degradation testing were done. These tests were performed to determine the stress-strain behavior and degradation time and pattern of the composite and to have a comparative review of the properties of the composite with respect to the parent polymer. Later the drug release tests were conducted to have an idea about the pattern of drug release from the composite material.

Tensile testing was performed on a universal testing machine (UTM), SHIMADZU AGX Plus at a strain rate of 5 mm/min. which was selected by hit and trial method. The gauge length for the samples was kept 25 mm.Tensile testing was done for all the composite membranes and for the membranes of the parent polymer i.e., polyvinyl alcohol (PVA).

For degradation rate analysis 1*1 inches square samples of the membranes were weighed and placed over a plastic mesh for support and immersed in 1ml. normal saline solution at 37°C. Membrane samples of the composite dipped in the saline medium were weighed after every 15 minutes (subtracting the weight of the plastic mesh from the total weight) by taking them out of the medium and then immersing into the saline again. This was repeated till the membrane got fully degraded and only the plastic mesh was left behind. The weight of the membranes was plotted graphically with time to determine the degradation rate and pattern of the composite membranes.

To establish the kinetics of drug release from the composite, the drug containing samples of composite membrane were degraded by a similar procedure but instead of weighing the membranes as in degradation testing, the degradation medium was collected for UV spectrophotometry, by UV-2800 UV/VIS Spectrophotometer to determine the quantity of drug released into the medium. The quantity of medium taken out was replaced with the same quantity of fresh medium everytime.

Results

The results of tensile testing are a direct indicator of mechanical properties of the material. Firstly two different membranes of polyvinyl alcohol (PVA) containing 5% and 10% polyvinyl alcohol (PVA) were tested for their tensile strengths. 10% PVA membrane (maximum force of 13.059 N and maximum stress of 5.22 N/mm²) was found to have more strength as compared to 5% PVA membrane (maximum force of 10.369 N and maximum stressof 4.147 N/mm²) as indicated by their stress-strain curves. Thus 10% concentration of polyvinyl alcohol was selected to develop a composite with magnesium (Mg) metal powder. The results of tensile testing for 5% and 10% polyvinyl alcohol membranes in the form of stress-strain graphs are shown in Figures 1a and 1b respectively.

The degradation rate of 5% polyvinyl alcohol membrane was found to be 0.593 mg/min. It was calculated by measuring the slope of the graph plotted between time on X-axis and the changing weight on

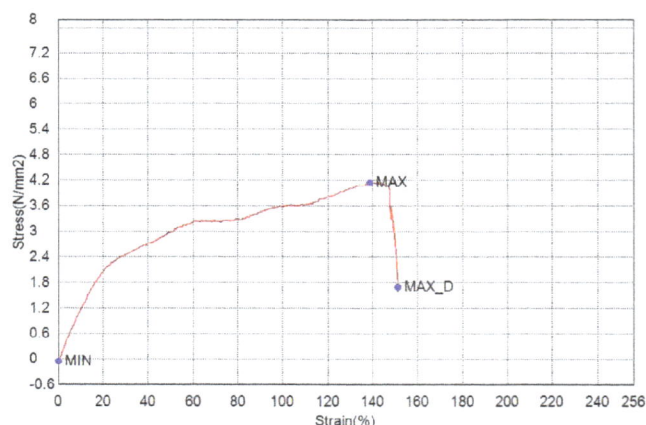

Figure 1a: Stress-strain graph for 5% PVA membrane.

Y-axis. 10% polyvinyl alcohol membrane on the other hand degraded faster at a rate of 0.350 mg/min. which can be attributed to greater concentration of solute (polyvinyl alcohol) present in it.

Composites of polyvinyl alcohol containing 1% and 2.5% magnesium powder when tested for their mechanical strength showed a considerable reduction in strength, the strength was further reduced by increasing the concentration of magnesium powder from 1% to 2.5%. Maximum force for 1% magnesium containing composite was found to be 10.4618 N and for 2.5% magnesium containing composite it was 8.4177 N. Similarly the maximum stress at entire area was calculated to be 4.184 N/mm² and 3.3671 N/mm² for 1% and 2.5% magnesium containing composites respectively. The results of tensile testing for 1% and 2.5% magnesium particles containing 10% polyvinyl alcohol composite membranes in the form of stress-strain graphs are shown in Figures 2a and 2b respectively.

When these composites were studied for their degradation rates 1% magnesium containing composite degraded quickly than the composite containing 2.5% magnesium with the degradation rates of 0.5 mg/min and 0.3 mg/min respectively.

Performing the tensile tests and degradation experiments, with 1% MgSO$_4$ (maximum force of 38.3314 N and maximum stress of 15.332 N/mm²)and 2.5% MgSO$_4$ (maximum force of 24.792 N and maximum stress of 9.9169 N/mm²) containing 10% polyvinyl alcohol (PVA) membranes, indicated more strength than the composite films of similar concentrations and the membranes of parent polymer, and a uniform degradation rate of 0.5 mg/min. for both concentrations of MgSO$_4$ containing membranes. The results of tensile testing for 1% and 2.5% magnesium sulphate containing 10% polyvinyl alcohol composite membranes in the form of stress-strain graphs are shown in Figures 3a and 3b respectively.

For drug release studies the better compositions (as shown by the results of tensile and degradation tests) of the composite and the polymer-salt blend were selected for drug incorporation into them. Thus 1% magnesium particle containing composite and 1% MgSO$_4$ containing polyvinyl alcohol-magnesium sulphate blends were added with the drug and studied for the pattern of drug release from them. The composite showed a sustained pattern of drug release while the magnesium salt containing polymer membrane showed a conventional inclined pattern of drug release.

Discussion

The fact that 10% PVA membrane is mechanically stronger than 5% PVA membrane shows that increasing the concentration of polyvinyl alcohol has a positive effect on the mechanical strength of both the polymer membrane and the metalo-polymer membrane. 10% PVA membrane degrades quickly because the larger amount of PVA in the membrane provides greater area for water interaction and eventual hydrolysis and degradation.

For polyvinyl alcohol-magnesium composite films, addition of magnesium powder decreases the overall strength thus beneficially reducing the flexibility of the polyvinyl alcohol by the hinging effect created between the polymer (PVA) and the magnesium metal particles. This reduction in the flexibility of polyvinyl alcohol is required for certain biomedical applications like stent grafts where a limited/quantified amount of expansion is required.

The reduction in tensile strength with the addition of magnesium particles in the composite and it's direct relation with the quantity of magnesium particles is because of the fact that magnesium particles

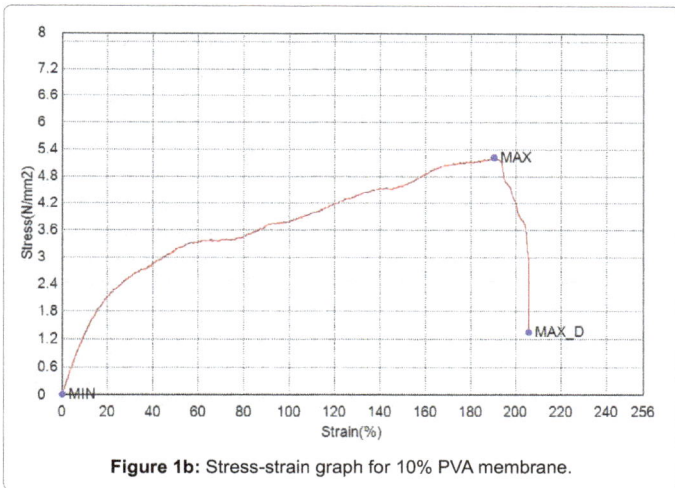

Figure 1b: Stress-strain graph for 10% PVA membrane.

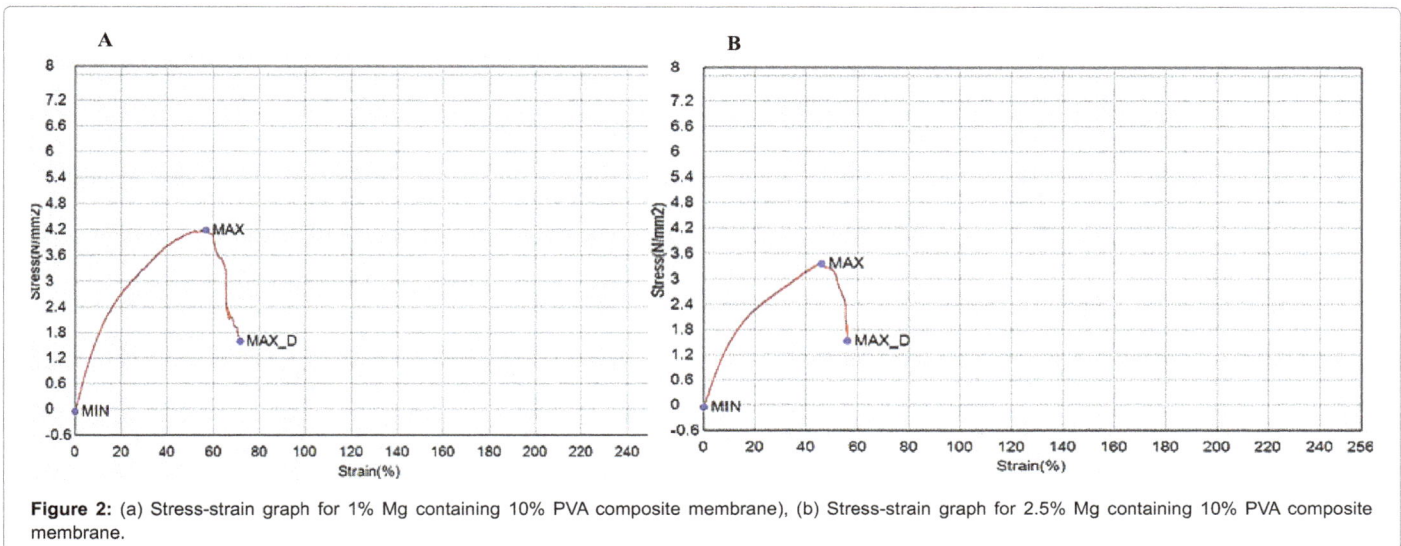

Figure 2: (a) Stress-strain graph for 1% Mg containing 10% PVA composite membrane), (b) Stress-strain graph for 2.5% Mg containing 10% PVA composite membrane.

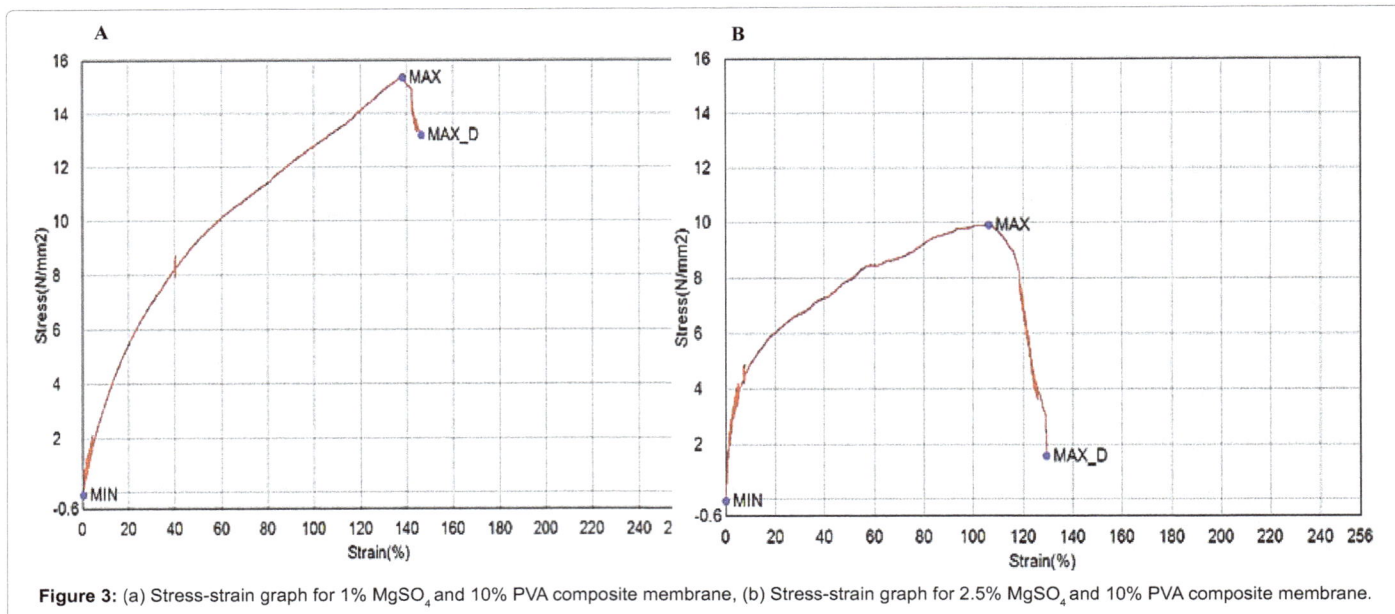

Figure 3: (a) Stress-strain graph for 1% MgSO$_4$ and 10% PVA composite membrane, (b) Stress-strain graph for 2.5% MgSO$_4$ and 10% PVA composite membrane.

provide crack propagation sites in the composite membrane, higher the number of magnesium particles in the composite, greater are the chances of crack initiation and propagation and lesser is the flexibility and strength of the composite membrane in comparison to the parent polymeric membrane.

1% magnesium containing composite degrades quickly than the composite with 2.5 magnesium powder in it. This could be because of the reason that the magnesium particles hold around portions of polyvinyl alcohol (PVA) with them, thus creating islands of PVA around magnesium particles, these islands degrade slowly with time, with the rest of the membrane being degraded quickly. Since the composite with 2.5% magnesium particles has more number of such islands so it degrades slowly as compared to the composite containing 1% magnesium particles.

The magnesium sulphate containing PVA membranes have better tensile strength because they do not undergo crack initiation and propagation in an accelerated manner as in the composites containing magnesium metal particles. Moreover the presence of foreign solute material (MgSO$_4$) in the polymer membrane increases its flexibility in comparison to the mother polymer sheet because of the possible hydrogen boding between the sulphate ions and hydroxyl groups of solvent (water). Lesser amount of foreign solute (1% MgSO$_4$) containing polymer membrane therefore has relatively more strength because of more active hydrogen bonds than in the other with higher concentration (2.5% MgSO$_4$).

As far as the same degradation rate for both the concentrations of MgSO$_4$ (1% and 2.5%) is concerned, the homogeneously dissolved MgSO$_4$ forms a uniform blend with the polymer, without creating islands of salt particles with the polymer around them, thus enabling degradation at similar rates regardless of the quantity of magnesium salt present.

The sustained release of drug from the composite membrane is because of the polymer surrounded magnesium islands which keep on releasing the drug at regular intervals, on the other hand the polymer blend of magnesium sulphate releases the drug based on the concentration of drug in the reservoir (the membrane) which decreases with time following first order kinetics of drug release.

Conclusion

Adding magnesium in small quantities to polyvinyl alcohol alters the tensile, degradation and drug release properties of the mother polymer. Increasing the quantity of reinforcement material to larger levels however doesn't lead to ideal improvement in properties. Different results are obtained by changing the form of the reinforcement material (pure metal powder or metal salt) with decrease in mechanical strength by the addition of metal particles and an increase in mechanical strength by the addition of metal salt; so any of these altered composite materials of biodegradable nature can be used based on the area of application for which they are intended to be used.

References

1. Hossain KMZ, Felfel R, Rudd CD, Ahmed I (2014) Biocomposites: Natural and Synthetic Fibers. In: M.K. Mishra (ed.) Encyclopedia of Biomedical Polymers and Polymeric Biomaterials. Taylor and Francis. pp: 585-601.

2. Kumar R, Singh T, Singh H (2015) Natural Fibers Polymeric Composites with Particulate Fillers – A Review Report. International Journal of Advanced Engineering Research and Applications 1: 21-27.

3. Ali M, Chee C, Hock C (2014) Effect of Fiber Loading and Chemical Treatment on Properties of Kenaf Fiber Reinforced Polyvinyl Alcohol Biocomposites. ISCASE-2014 Malaysia.

4. Elharati M (2009) Composite Membranes.

5. Kamoun EA, Kenawy ERS, Tamer TM, El-Meligy MA, Mohy Eldin MS (2015) Poly (vinyl alcohol)-Alginate Physically Cross-linked Hydrogel Membranes for Wound Dressing Applications: Characterization and Bio-evaluation. Arab J Chem 8: 38-47.

6. Paul W, Sharma CP (1997) Acetylsalicylic Acid Loaded Poly(vinyl alcohol) Hemodialysis Membranes: Effect of Drug Release on Blood Compatibility and Permeability. J Biomater Sci Polym Ed 8: 755-764.

Reduction of Sulphur Content of AGBAJA Iron Ore Using Sulphuric Acid (H₂SO₄)

Reduction of Sulphur Content of AGBAJA Iron Ore Using Sulphuric Acid (H_2SO_4)

Ocheri C[1]* and Mbah AC[2]

[1]Department of Metallurgical and Materials Engineering, University of Nigeria, Nsukka, Nigeria
[2]Department of Metallurgical and Materials Engineering, Enugu State University of Science and Technology, Nigeria

Abstract

Iron ores are used in blast furnace for the production of pig iron; AGBAJA Iron ore has an estimated reserve of over I billion metric tonnes. Unfortunately, this large reserve cannot be utilized for the production of pig iron due to its high sulphur contents. This work studied the reduction of sulphur content of AGBAJA iron ore. Acid leaching methods were used to reduce sulphur contents of the ore sulphuric acid of different concentrations were used at various leaching times, acid concentrations and particle sizes. Atomic Absorption Spectrophotometer, X-ray fluorescence spectrophotometer, Digital muffle furnace and Absorbance-concentration technique were used for experimentation and chemical analysis. The reduction of the sulphur content of AGBAJA Iron Ore using Acid leaching process experiments were carried out at the National Metallurgical Development Centre (NMDC), Jos in Plateau State of Nigeria. Sulphur is one of the main harmful elements in ferrous metallurgy and it affects the quality of iron and steel produced. At present, Nigeria has some large iron ore deposits including AGBAJA which bear tremendous iron ore with high sulphur content of 0.12%. Central composite design technique was applied to obtain optimum conditions of the processes. Surface response plots were also obtained. The percentage degrees of reduction of sulphur content of AGBAJA Iron ore were found to increase with increase in acid concentration and leaching time and a decrease in particle size for the three acids. The experimental results for percentage removal of sulphur are 87.77% the optimum % removal of sulphur is 87.73%. The result of this work has shown that AGBAJA Iron Ore if properly processed can be used in our metallurgical plants and also can be exported since sulphur contents of the ore have been reduced drastically.

Keywords: Reduction; Sulphur content; AGBAJA; Iron Ore and Sulphuric Acid (H_2SO_4)

Introduction

The AGBAJA ore is a fairly lean, acidic iron ore with high sulphur content. The ore is an earthy, friable material containing magnetite and goethite, together with minor aluminosilicates and phosphates of iron and aluminium. The ore contains approximately 54% iron and shows thermal effects associated with the elimination of water. The texture and chemical composition of the AGBAJA ore suggests that despite its magnetic character it cannot be easily beneficiated for use in a direct, non-conventional iron making process. The various slag-forming constituents (silica, alumina, lime and magnesium oxide) are so closely associated with the iron-bearing constituents that separation is impossible to achieve by simple physical means. Furthermore, the high sulphur content (about 0.12%) would probably give rise to problems in steel production unless a conventional, oxidizing; liquid-metal process (such as basic oxygen steelmaking) is used following blast-furnace production of liquid iron. For the AGBAJA iron ore the basicity value is very low (approximately 0.035) and hence the ore would need significant additions of lime, limestone or a lime-rich ore to make a self-fluxing sinter or pellet suitable for iron production. The reduction/removal of the high sulphur content can be achieved through the process of leaching using nitric acid.

Statement of the problem

AGBAJA iron ore is the largest iron deposit in Nigeria with an estimated reserve of over 1 billion tonnes. This iron ore has high relative high sulphur content. Consequently, the iron ore deposit is abundant in both research work and exploitation. The high sulphur content in steel making cause brittleness or crackability depending on the type of Steel products. The sulphur content in the AGBAJA iron ore has a detrimental effect on the steel making process using the ore as raw materials in steel making. It is therefore, necessary to drastically remove/reduce the sulphur content.

Purpose and goals

The purpose of this work is to carry out experiments on the reduction of sulphur content of AGBAJA iron ore using acid leaching. Leaching of lean ores or complex ore in different acids has proved successful for several years. However, the leaching of sulphur contaminated iron ore has made a very limited progress. This underscores the ongoing intense research in the area for several decades. Depending upon the degree of association of sulphur with the minerals in the iron body, iron ore can therefore the sulphur content can be reduced using the following mineral acids which are often used as leaching agent of hydrochloric acid (HCl). The purpose therefore is to use all the three mentioned acid leaching with a view to reducing/removing of sulphur content in order to making it useable for steel making process in the Blast Furnace.

In Nigeria today, it is a known fact that most of the steel industries are not functional due to lack of infrastructure. Despite all the endowment of the abundant mineral resources that are readily available due to natural blessings. This project work was carried out

*Corresponding author: Ocheri C, Department of Metallurgical and Materials Engineering, University of Nigeria, Nsukka, Nigeria
E-mail: cyrilocheric@gmail.com

with a view to making use of this minerals in which the iron ore falls into the materials used in the production of iron and steel. Nigeria is endowed with large deposits of iron ores; however, the ores are not suitable for the production of Direct Reduced Iron (DRI). The ore closest to specification in terms of sulphur content is the Itakpe iron ore deposit which has a reserve of about 300 million tonnes. The ore is beneficiated from its natural occurrence of 36.8% Fe and 45.8% acid gangue (SiO_2)+Al_2O_3 to 63% Fe and 8% acid gangue by the National Iron Ore Mining Company (NIOMCO). The latter parameters are only suitable for the Blast Furnace (BF) plant located at Ajaokuta Steel Company Limited which is yet to be completed. However, a sister plant located at Aladja which has been in operation since 1982 requires some 1.5 million tonnes of high quality iron ore (66% Fe and <3.5% of SiO_2+Al_2O_3).

The supply of the high grade ore has so far been met only through importation which imposes high financial constraint on the operation of the Aladja plant. It can be found that at a landing cost conservatively put at 104 dollar/tonne of the ore and at even 50% capacity utilization. More than 10 billion Naira will be expended on the procurement of the ore alone. In order to meet the challenges posed by the importation of ore to meet the local demands at the steel industries in Nigeria, it therefore becomes imperative to investigate the suitability of the AGBAJA Iron ore with a view to carry out the reduction of the sulphur content of AGBAJA iron ore using acid leaching. High sulphur content found in the AGBAJA iron ore has been verified to have adverse effects for use in the Blast Furnace of the Ajaokuta Steel Company Limited. Nigeria is one of the richest countries of the world in terms of mineral deposits. Among these deposits is iron ore, located at AGBAJA, Kogi State, Nigeria. AGBAJA iron ore deposit, the largest in Nigeria is about 1.3 billion tonnes. AGBAJA iron ore is of low silicon modulus (SiO_2/Al_2O_3), fine texture and contains about 1.4-2.0% sulphur.

Iron is one of the most abundant elements in the earth's crust, being the fourth most abundant element at about 5% by weight [1]. Astrophysical and seismic evidence indicate that iron is even more abundant in the interior of the earth and is apparently combined with nickel to make up the bulk of planets core.

Iron ores are mainly composed of iron oxides, and oxyhydroxides, with other accessory gangue phases. These iron ores cannot be used in the production of steel in their raw states. For them to be maximally used in the production of quality steel, they must be upgraded or beneficiated.

Although the terms coarse-grained, intermediate size and fine grained are not assigned definite or specific dermacative values in mineral processing, a fine grained iron ore is often construed as one in which mineral matter is so finely disseminated within the gangue matrix that crushing and grinding, to effect liberation, produce minute particles that respond poorly to conventional beneficiation equipment and/or processes (froth flotation, magnetic separation gravity separation etc) [2]. Uwadiale and Whewell [3] observed that the utilization of AGBAJA iron ore is hampered by its poor response to established industrial beneficiation techniques, this is as a result of fine grained texture of the iron ore.

Sulphur may be incorporated either into the crystal lattice of iron oxides or into the gangue minerals [4]. This element has a deleterious effect on the workability of steel. For that reason, in most places only premium low sulphur ores are extracted leaving many iron mines around the world enriched in which are unsaleable, high-sulphur iron ore [4].

If steel is produced with high level of sulphur, that steel will be brittle and therefore not ideal for industrial application hence the need for reduction. Depending on the degree of association of sulphur with the minerals in the ore body, iron ore can be reduced either physically or chemically [5,6].

Comminution followed by wet magnetic separation or froth flotation is generally employed when the sulphates gangue minerals appear as discrete inclusions in the iron oxide matrix (primary mineralization) [5]. However, when sulphur is disseminated in the iron oxide structure, possibly forming cryptocrystalline phosphates or forming solid solutions with the iron oxide phases (secondary mineralization), the reduction can only be processed by chemical routes [4-6].

The chemical reduction involves the hydrometallurgical processing of the ore, that is, the selective leaching of sulphur in the ore with a reagent usually acid or base. Since early in the 19th century, suggested the use of sulphuric acid to remove sulphur compounds from lumps of iron ore. Nevertheless, a real scientific interest in hydrometallurgical processing of high sulphur iron ores can only be noticed after the last third of the 20th century, when several papers and patents were published [4-6]. Ever since, traditionally low prices of iron ore products had impeded the large-scale industrial application of chemical reduction. At the present time, an increase in world steel production has increased demand for iron ore with a consequent increase in the price for this commodity, making hydrometallurgical sulphate removal viable [7]. In the last eight years, the situation of iron ore markets has changed dramatically due to an increase in the world steel consumption, pushed up mainly by the economic growth of China and other Asian emerging markets.

The mechanism and process analysis of reduction of sulphur content of AGBAJA iron ore concentrate using powdered potassium trioxochlorate (v) ($KClO_3$) as an oxidant has been reported [8]. Concentrates were treated at a temperature range 500°C-800°C.

The prevalence of this amount of sulphur is a major setback to its utilization in the blast furnace or direct reduction process [9]. The removal of sulphur from iron and steel presents problems because of similarity of the standard free energies of formation of iron oxide and sulphur pentoxide [9]. Consequently, in the reducing conditions of the blast furnace to recover some 99.5% of the iron charged, near complete reduction of sulphur pentoxide from the acid blast furnace occurs. As the sulphur in the ore impregnates the pig iron, there occur two distinct processes of tackling the problem: pyrometallurgical route and hydrometallurgical route. The first route employs basic slag during the conversion to steel. This technique covers the activity coefficient of sulphur pentoxide in the slag [9,10]. The second route delves into ways of reducing sulphur in the iron at relatively low temperatures. Leaching of lean ores or complex ore in different acids has proved successful for several years. However, the leaching of sulphur contaminated iron ore has made a very limited progress. This underscores the ongoing intense research in the area for several decades. As is well known, smelting process is effective for dispersion but with very high cost, and it is still under fundamental research. For physical separation, comminution followed by wet magnetic separation or froth flotation is generally employed when the sulphur in the gangue mineral appears as discrete inclusion in the iron bodymatrix (primary mineralization) [5-7]. Low sulphur extraction, high grinding cost and iron loss are the major disadvantages of the method. However, when sulphur is disseminated in the iron structure, possibly forming cryptocrystalline sulphates or solids solutions with the iron oxide phases (secondary mineralization), and the beneficiation can only proceed by chemical routes [5,7]. He investigated reduction with acid leaching. In their studies, the acid concentrations were very high and low sulphur extraction was obtained. In this study, the feasibility of reduction of sulphur content of AGBAJA iron ore by acid leaching at various dilution ratios, dwell time and particle size will be investigated. Also, the Design of Experiment for the evaluation of interactive factorial effects on the sulphur removal will be investigated. AGBAJA iron ore deposit is the largest in Nigeria with a reserved deposit of One Billion metric tonnes. However, when compared with other iron ore, it is evidently shown that the sulphur content of the AGBAJA iron ore is high (0.12%) with low silicon modules of (SiO_2/Al_2O_3=0.89) and fine grain texture which constitute major problem for utilization in the Blast Furnace (BF) direct reduction process.

Although it is Nigerian's largest known ore deposit estimated at over one billion tonnes, its utilization is hampered by its poor industrial beneficiation techniques. It is in the light of the above that the work seeks to find an alternative way of effectively reducing the sulphur content of AGBAJA iron ore apart from the conventional method of roasting sulphur. Sulphur which is one of the main harmful elements to ferrous metallurgy affects the quality of iron and steel products. At present Nigeria has some large iron ore deposits including AGBAJA which bear tremendous iron ore with high sulphur content. The removal of sulphur from iron ores involves smelting process, physical separation and chemical leaching. Smelting process is effective for reduction but very expensive and still under research. High sulphur extraction, high grinding cost and iron loss are the major disadvantages of the ore. The ore is leached with a suitable solution in a relatively simple process as it can directly treat the fines without strict requirements for the particles sizes. The cost of chemical leaching process mostly depends on the consumption of the acid leaching. In the study, the feasibility of reduction by acid leaching will be investigated. The study is to evaluate the effect acid leaching of Sulphuric Acid (H_2SO_4), this with a view of using aid leaching agent for the purpose of reducing the Sulphur content so that the ore could be used for the production of iron and steel through the process of the Blast Furnace located at the Ajaokuta Steel Company Limited. The experiment therefore to bring all these potentials to bear in the Iron and steel industry.

Materials and Methods

The experiments on the research works were carried out in the course of studying the reduction of sulphur content of AGBAJA iron ore using acid leaching agent H_2SO_4 for, reduction. Mineral processing involves collection of ore samples, the prepared ore samples were carried out by comminution and the chemical analysis of raw and scrubbed ore. Furthermore, modeling of the observed variables and data generated were used to carry out Central Composite Design (CCD) modeling technique.

The materials and equipment that were used in this study include the following: Hand trowel, pan, mesh screens, blender, Electronic weighing balance, Mechanical shaker, porcelain pot, pH indicator device, Oleic acid, sodium silicate, distilled water, kerosene, beakers, crushers, X-ray fluorescence, spectrometer, atomic absorption spectrophotometer, LF6484A flotation machine, aero froth, cylinders, autoclave hot air oven, inoculating loop, petridishes, conical flask, binocular microscope, staining rack, glass, slides, test tubes, deionized water, and the acid leaching agent includes H_2SO_4.

General description AGBAJA iron ores

The ore samples were compacted bonded crystalline. The AGBAJA sample consisted mainly of aggregates of brown, compact, fine-grained

material with some larger, extremely friable particles. This ore is strongly magnetic. The lump ore samples were crushed mechanically and sieved to give particles in the size range 1-1.7 mm (16-10 mesh). Care was taken to ensure that the size fractions were the representative of the lump material. The ore particles were crushed further for certain experimental techniques. Analyses for calcium, magnesium, iron, aluminium sulphur were made by atomic absorption spectrometry; silica was determined by a combination of gravimetric and colorimetric method, x ray diffraction analysis that were performed (Figure 1).

Reduction of sulphur content of AGBAJA iron ore using sulphuric acid

The scrubbed iron ore was further pulverized and sieved with a view to obtaining particle sizes which range from 10 microns, 20 microns, 30 microns, 40 microns, 60 microns and 80 microns. Sulphuric acid (H_2SO_4), solutions of different moles of 0.2, 0.4, 0.8, 1.0, 2.0, 4.0 8.0, 12.0 and 16.0 were prepared. 100g of particle size of 10 microns of the scrubbed iron ore was weighed and subsequently poured into a conical flask. 100 mL of 0.2 M of sulphuric acid was poured into the conical flask which contained the ore.

The mixture was vigorously stirred with a shaker to ensure proper homogeneity. The contents were allowed to leach for 5 min, 10 min, 20 min, 30 min, 60 min and 120 min. At the end of each period, the solutions were cooled and thereafter the solutions were filtered. The residues were collected and washed to ensure that the solutions were neutralized with distilled water; the residues were air dried and oven dried. The experiment were repeated for 0.4, 0.8, 1.0, 2.0 and 4.0 moles and particle sizes of 20 microns, 30 microns, 40 microns, 60 microns and 80 microns. Photographic Display of the Leaching Procedure is shown in Figures 2-5.

Hand specimen identification and specific gravity calculation

The physical analysis of the specimen indicates that the specimen has the following characteristics:

➤ Color   dark brown

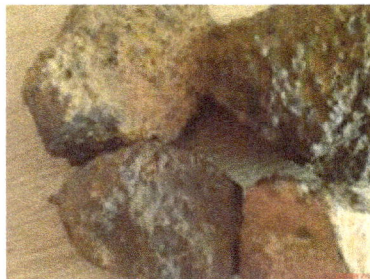

Figure 1: Sample of the Iron ore as obtained from Agbaja mines.

Figure 2: Weighing balance.

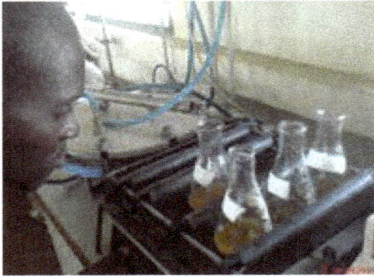

Figure 3: Mechanical Shaker.

Figure 4: Blast Furnace at Ajaokuta Steel.

Figure 5: Direct Reduction Iron (DRI) Plant, Aladja Delta Steel.

➢ Streak   brown

➢ Specific gravity calculation

Mass of iron ore   =297.58 g

Mass of equal volume of water   =69.9 g

Sp. Gravity=Mass of substance/Mass of equal volume of water

           =297.58 g/69.9 g

           =4.26

The micrograph in plate 6 shows the yellowish-brown colour indicates rings of iron mineral without clear grain boundaries at 250 magnifications, suggesting very fine grains. The visible bright yellow seen on deep brown patches of silica signifies the presence of sulphur in the ore. Moreover, other elements which are not visible from the micrograph occur probably in their apatite phase which makes them difficult to be seen visibly.

The above result when compared with standard values indicates that the color, the streak and the specific gravity showed that it is goethite FeO(OH) ore.

Hand specimen identification and specific gravity calculation

The physical analysis of the specimen indicates that the specimen has the following characteristics:

➢ Color   dark brown

➢ Streak   brown

➢ Specific gravity calculation

Mass of iron ore   =297.58g

Mass of equal volume of water   =69.9g

Sp. Gravity=Mass of substance/Mass of equal volume of water

           =297.58g/69.9g

           =4.26

Specific gravity test: The specific gravity of a body is the ratio of the weight of the body to that of an equal volume of water.

Sieve analysis test

Sample preparation: The samples were properly mixed and large residual sample lumps were broken up. The samples were reduced to test sizes by means of Roll Crusher. The test samples were dried to a constant weight in an oven regulated at $80 \pm 5°C$. The mass of the dry test sample was measured in gram.

Procedure: The sieve chosen for the test were arranged in a stack, with the coarsest sieve on the top and the finest at the bottom. A tight-fitting pan or receiver was placed below the bottom sieve to receive the final undersize, and a lid (cover) was placed on top of the coarsest sieve to prevent escape of the sample.

The micrograph in Figure 6 shows the yellowish-brown colour indicates rings of iron mineral without clear grain boundaries at 250 magnifications, suggesting very fine grains. The visible bright yellow seen on deep brown patches of silica signifies the presence of sulphur in the ore. Moreover, other elements which are not visible from the micrograph occur probably in their apatite phase which makes them difficult to be seen visibly.

The above result when compared with standard values indicates that the color, the streak and the specific gravity showed that it is goethite FeO(OH) ore.

Also from the result obtained from the sieve analysis, a screen analysis graph of cumulative weight percentage retained against particle size was plotted. It can be observed that the 60 μm sieve retained the most quantity of iron ore particles. This shows that the AGBAJA iron ore has fine grain size.

Result of ED-XRF analysis

The chemical composition of AGBAJA iron ore as revealed by ED-X-ray florescence is shown in Table 1 below which implies that

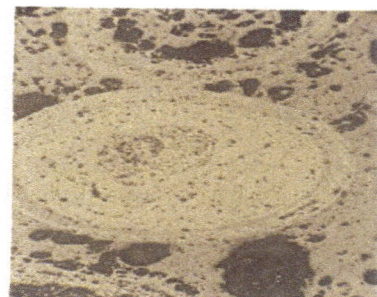

Figure 6: Micrograph of the ore as obtained after the thin section analysis.

the ore is most likely to be a mixture of hemeatite and magnetite. The various compositions of minerals have different effects on the ore; the presence of silicon promotes the formation of gray cast iron and as seen from Table 1. AGBAJA iron ore contains considerably high amount of silicon. It is noted that high sulphur causes hot shortness which is brittleness of steel at high temperatures. Moreover the aluminium in the ore does not pose any major problem; but makes it difficult to tap off the liquid slag. Also from the result obtained from the sieve analysis, a screen analysis graph of cumulative weight percentage retained against particles size was plotted. It can be observed that the 60 μm sieve retained the most quantity of iron ore with fine grain size.

Experimental Results and Discussion

Reduction of sulphur of AGBAJA iron ore using sulphuric acid

Influence of sulphuric acid concentration on reduction of sulphur of AGBAJA iron ore: The influence of sulphuric acid concentrations on the reduction of sulphur content of AGBAJA iron ore are illustrated in Figures 7-11. In Figure 7, the percentage degree of reduction of sulphur content remained fairly constant between 0.2 M and 1 M. From 1 M, there was slight increase in percentage reduction of sulphur content. The highest percentage reduction was achieved at 46.66% for leaching time of 120 min and for particle size of 80 microns while the least value was obtained at 45.00% at 0.2 M concentration.

Between 0.2 M and 0.4 M, there was a slight increase in percentage reduction of sulphur content as shown in Figure 8. From 0.4 M, the percentage degree of reduction of sulphur content was slightly increases, attaining the highest value of 50.00%.

The percentage of reduction remained constant from 0.2 M to 1.0 M for leaching time of 120 min, 60 min and 30 min and slightly increased after1.0 M as depicted in Figure 9. For leaching time of 20 min, 10 min and 5 min, for all the concentrations, the increase in percentage reduction of sulphur content was gradually achieved. The

highest value of percentage reduction was 67.77% for 4.0 M at particle size of 40 microns. The percentage degree of reduction of sulphur content was constantly achieved between 0.2 M and 0.8 M. There was a slight increase between 0.8 M and 1.0 M. As from 1.0 M, the percentage of reduction of sulphur content became fairly constant. 80.00% of sulphur was reduced as the highest value for concentration of 4.0M at leaching time of 120 min as shown in Figure 10. Figure 11 shows that the percentage of reduction was constantly achieved except for 120 min, where there was little increase from 2 M. The highest value of 87.77% was obtained for 10 microns at concentration of 4.0 M. It could be inferred that the percentage degree of reduction of sulphur content is directly proportional to H_2SO_4 concentration.

Effect of leaching time on the percentage degree of reduction of sulphur content of AGBAJA iron ore

The leaching time effects are represented in Figures 12-16. In Figure 12, the percentage degree of reduction of sulphur increases between 5 min and 30 min significantly but from 30 min there was steady increase in percentage of reduction of sulphur content. The highest percentage reduction of sulphur was achieved at 46.66% for 4 M at leaching time of 120 min. The least value was obtained at 34.44% for 4 M at leaching time of 5 min.

There was a significant increase in the percentage degree of reduction of sulphur between 5 min and 60 min. But from 60 min there was slight increase in percentage reduction of sulphur as depicted in Figure 13. At exactly 30 min, for 4 M concentration 43.33% of sulphur content was reduced while the highest value of 50.00% was reduced at 120 min at the same conditions. This shows that time is of essence in sulphur reduction from the iron ore. Between 5 min and 60 min, the percentage degree of reduction of sulphur content was significant while from 60 min to 120 min there was an average slight increase in percentage reduction of 1.1% as shown in Figure 14. Figure 15 shows that between 5 min and 30 min, there was appreciable increase in percentage reduction of sulphur content. There was a slight percentage

Mineral	K$_2$O	CaO	TiO$_2$	MnO	Fe$_2$O$_3$	MgO	Al$_2$O$_3$	SiO$_2$	P$_2$O$_5$	S*	Al$_2$O$_3$
Composition	0.04	0.13	0.37	0.40	89.40	0.40	9.60	2.28	1.20	0.12	3.38
Mineral	K$_2$O	CaO	TiO$_2$	MnO	Fe$_2$O$_3$	MgO	Al$_2$O$_3$	SiO$_2$	P$_2$O$_5$	S*	Al$_2$O$_3$
Composition	0.04	0.13	0.37	0.40	89.40	0.40	9.60	2.28	1.20	0.12	3.38

Table 1: Chemical composition of Agbaja Iron ore.

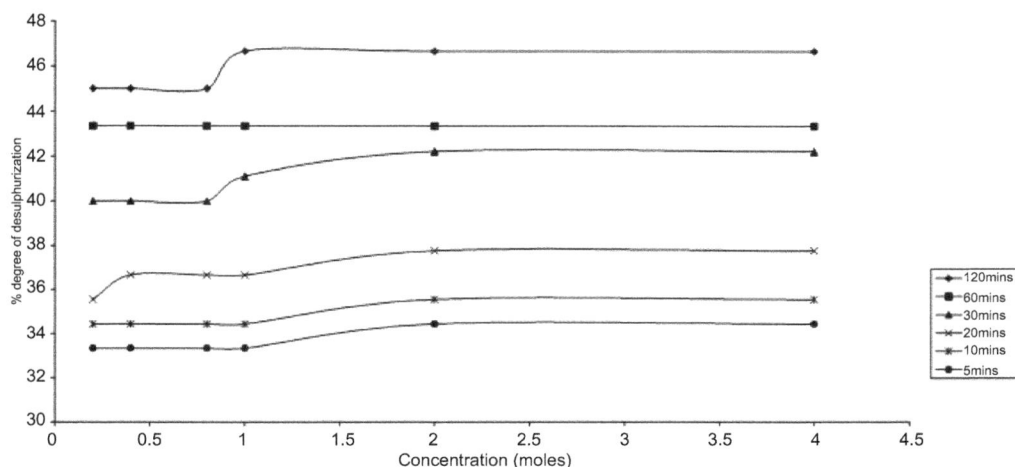

Figure 7: Influence of H_2SO_4 concentration on the %degree of reduction of sulphur for 80microns.

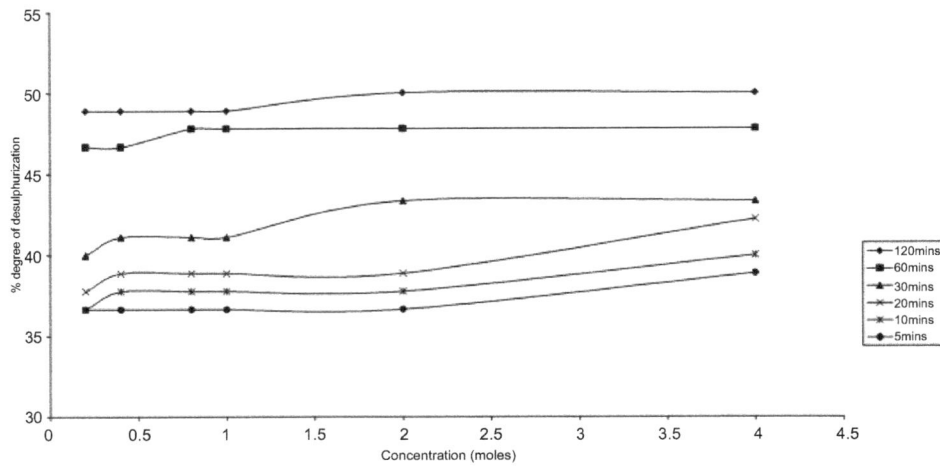

Figure 8: Influence of H_2SO_4 concentration on the %degree of reduction of sulphur for 60 microns.

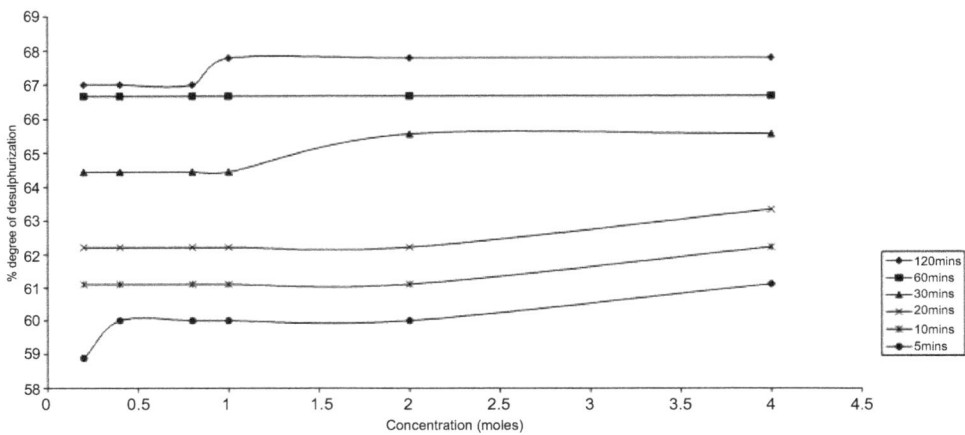

Figure 9: Influence of H_2SO_4 concentration on the %degree of reduction of sulphur for 40 microns.

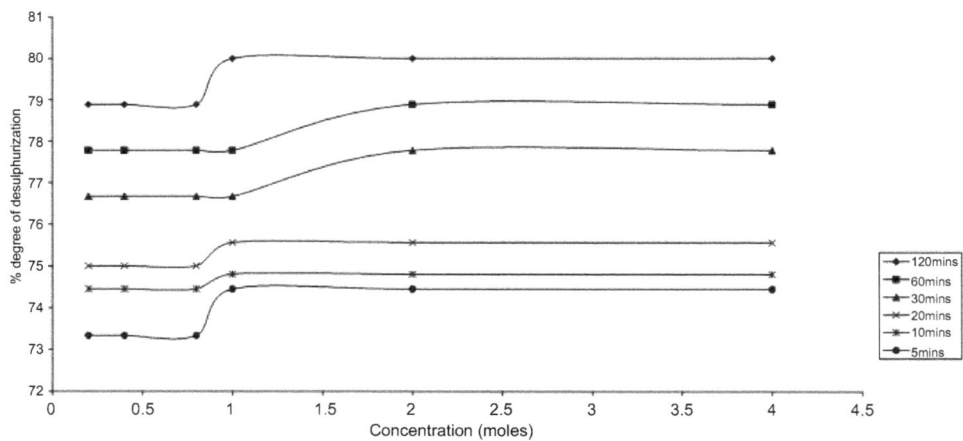

Figure 10: Influence of H_2SO_4 concentration on the %degree of reduction of sulphur for 20 microns.

of reduction of sulphur content from 30 min to 120 min for particle size of 20 microns. The highest value obtained was at 80.00% for 4.0 M concentration. The percentage degree of reduction of sulphur content significantly increased between 5 min and 60 min while from 60 min there was a slight increase in sulphur reduction as shown in Figure 16. As the leaching time increases the percentage degree of sulphur content equally increased.

Effect of particle size on the percentage degree of reduction of sulphur content of AGBAJA iron ore: The effects of particle size on the percentage degree of reduction of sulphur content of AGBAJA iron ore are shown in Figures 17-22. In Figure 17, the percentage degree of reduction of sulphur content decreased sharply from 10 microns to 60 microns while from 60 microns to 80 microns there was a fairly significant percentage reduction of sulphur content. The highest value of 84.44% was obtained for the particle sizes of 10 microns at 4.0 M.

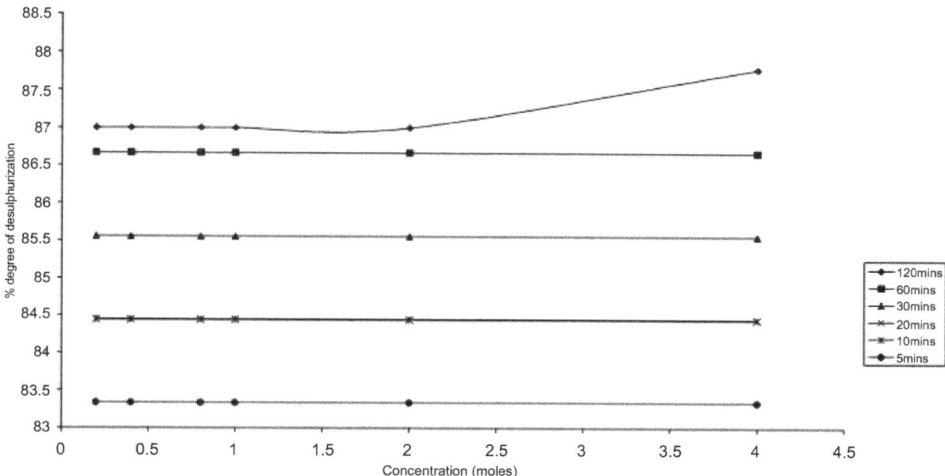

Figure 11: Influence of H_2SO_4 concentration on the %degree of reduction of sulphur for 10 microns.

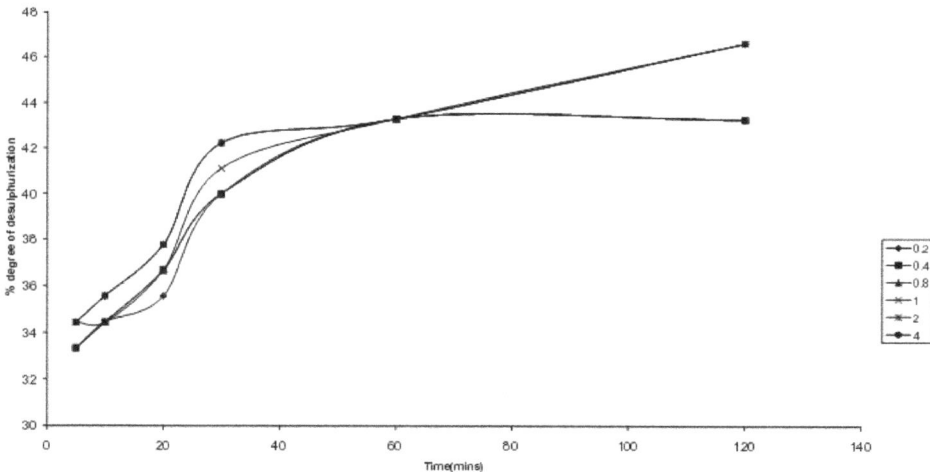

Figure 12: Effect of leaching time on the % degree of reduction of sulphur for 80microns using H_2SO_4.

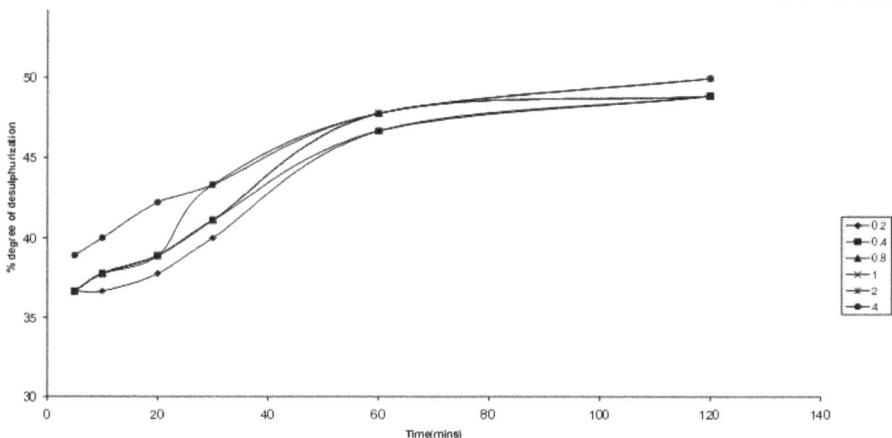

Figure 13: Effect of leaching time on the %degree of reduction of sulphur for 60microns using H_2SO_4.

Figures 18 and 19 showed similar trend. More than 40.0% reduction of sulphur content was achieved between particle sizes of 10 microns and 60 microns. Between 60 microns and 80 microns about 2.0% reduction was achieved implying that large particle sizes do not favour the reduction of sulphur content.

The percentage degree of reduction of sulphur content increased significantly between 10 microns and 60 microns. But between 60 microns and 80 microns there was fairly significant percentage of reduction of sulphur content as depicted in Figure 20. The highest percentage of reduction of sulphur content was achieved at 85.55% for concentration of 4.0 M at particle size of 10 microns.

Between 10 microns and 60 microns the percentage of reduction of sulphur content was largely significant as shown in Figures 21 and 22. It slightly increased between 60 microns and 80 microns. The highest value obtained for Figure 21 was obtained at 86.66% while that of Figure 22 was obtained at 87.77% at particle size of 10 microns at the same concentrations. From the forgoing analysis, it could be inferred that percentage degree of reduction of sulphur content was inversely proportional to particle diameter. In other words, as particle size decreases, the percentages of reduction of sulphur content increases.

Surface Response Plots

Surface response as controlled by leaching variables. The surface responses were plotted using MATLAB. The surface response plot for percentage reduction of sulphur content using time and concentration as variables is represented in Figure 23. The time effect is more pronounced than concentration as shown in the figure. Optimum value is located on the surface at yellow-dark red colour.

In Figure 24, the surface response plot for percentage reduction of sulphur content using time and particle size as variable is depicted. The time responded better than particle size. Linear relationship is represented in this plot. Figure 25 represents the surface response plot for percentage reduction of sulphur using particle size and concentration. The influence of concentration was more than that of particle size. It also had a linear relationship.

Summary

AGBAJA ore is an acidic oolitic ore consisting of goethite, magnetite and major amount of aluminous and siliceous minerals. It cannot be used directly in a blast furnace or other reduction process without further treatment, sintering, pelletizing or briquetting. Acid leaching is an effective method of reducing sulphur content from iron ores.

The recycle of sulphuric acid solutions is incorporated processes which obtained results for the reduction of sulphur content of the AGBAJA Iron ore as a by-product which makes the whole process for

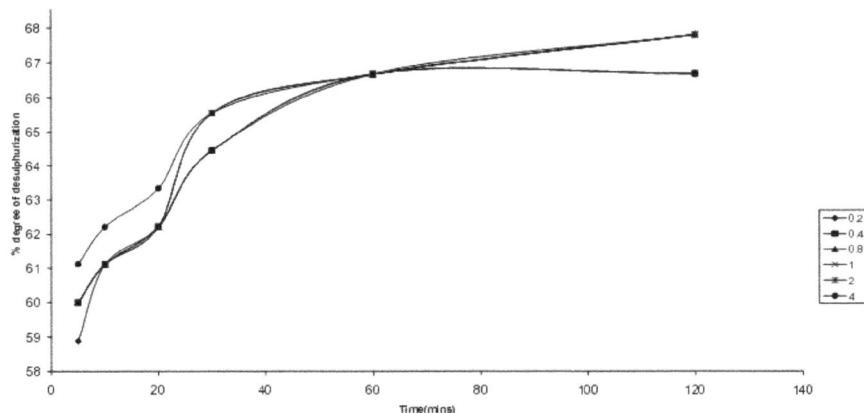

Figure 14: Effect of leaching time on the %degree of reduction of sulphur for 40microns using H_2SO_4.

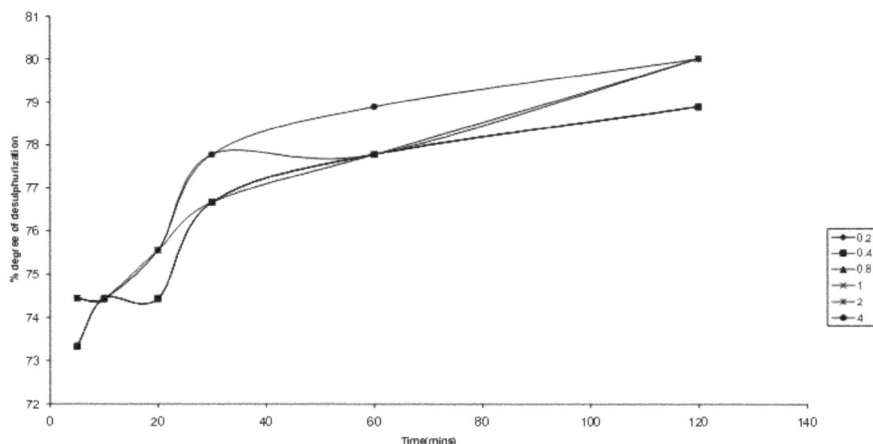

Figure 15: Effect of leaching time on the %degree of reduction of sulphur for 20microns using H_2SO_4.

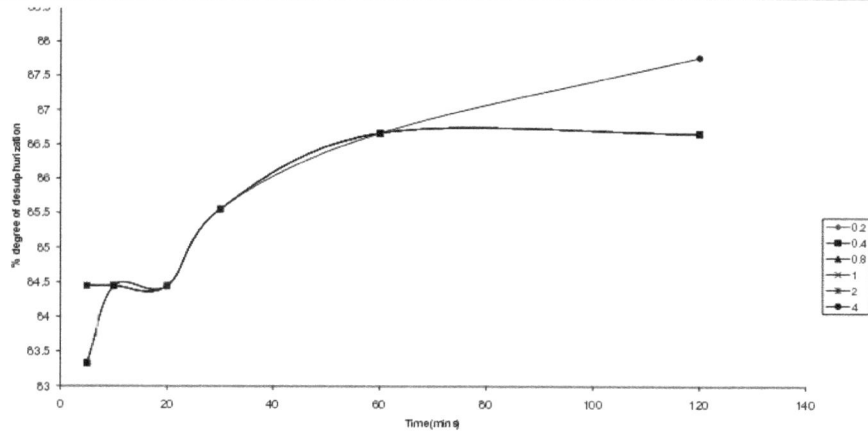

Figure 16: Effect of leaching time on the % degree of reduction of sulphur for 10microns using H_2SO_4.

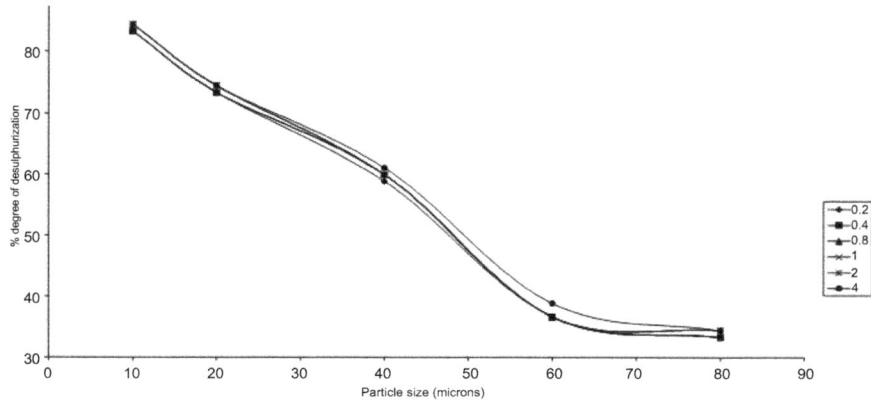

Figure 17: Effect of particle size on the %degree of reduction of sulphur for 5mins using H_2SO_4.

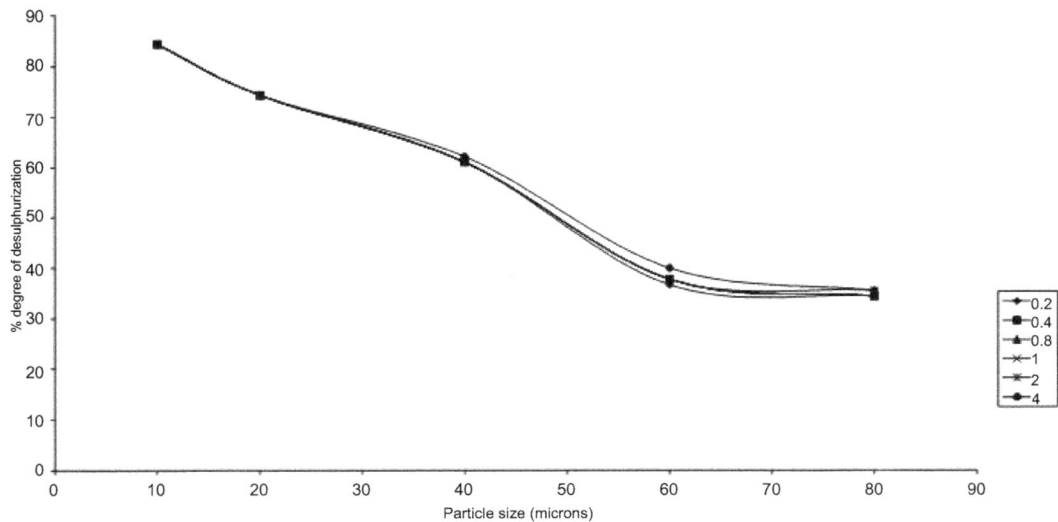

Figure 18: Effect of particle size on the %degree of reduction of sulphur for 10mins using H_2SO_4.

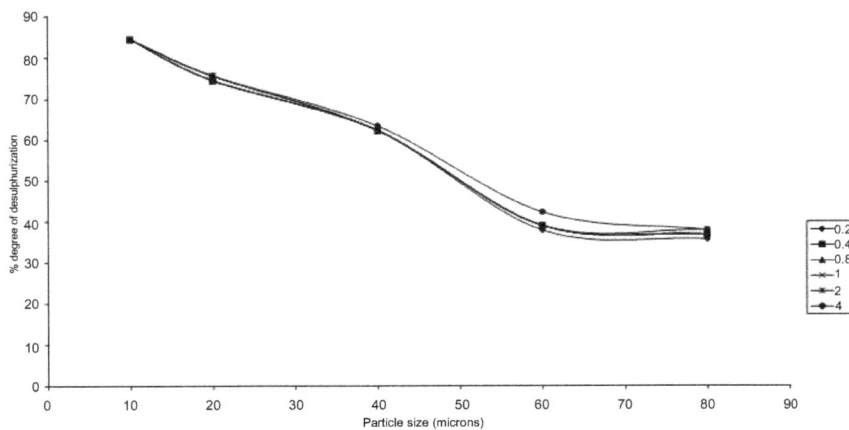

Figure 19: Effect of particle size on the %degree of reduction of sulphur for 20mins using H_2SO_4.

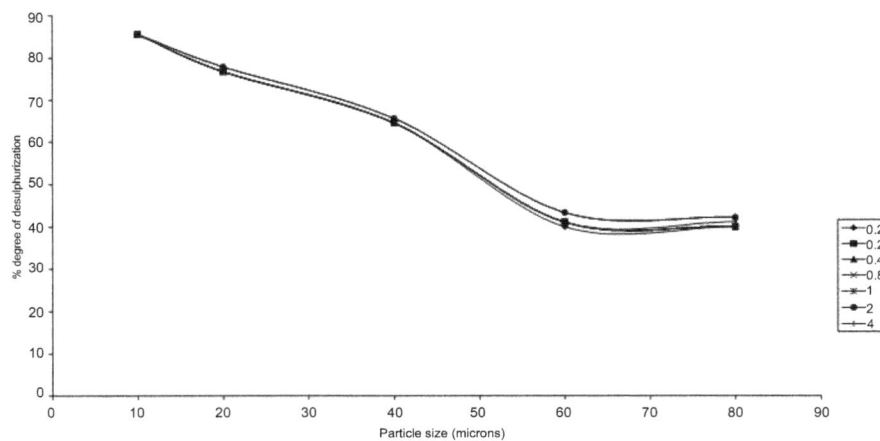

Figure 20: Effect of particle size on the %degree of reduction sulphur for 30mins using H_2SO_4.

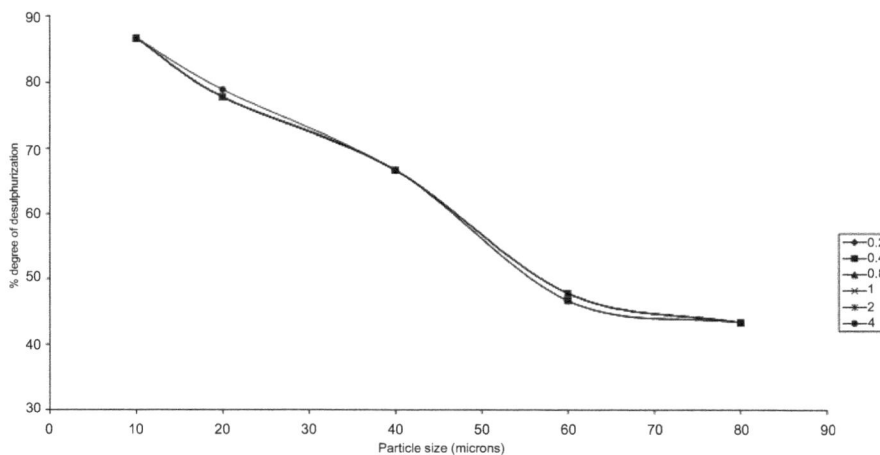

Figure 21: Effect of particle size on the %degree of reduction of sulphur for 60mins using H_2SO_4.

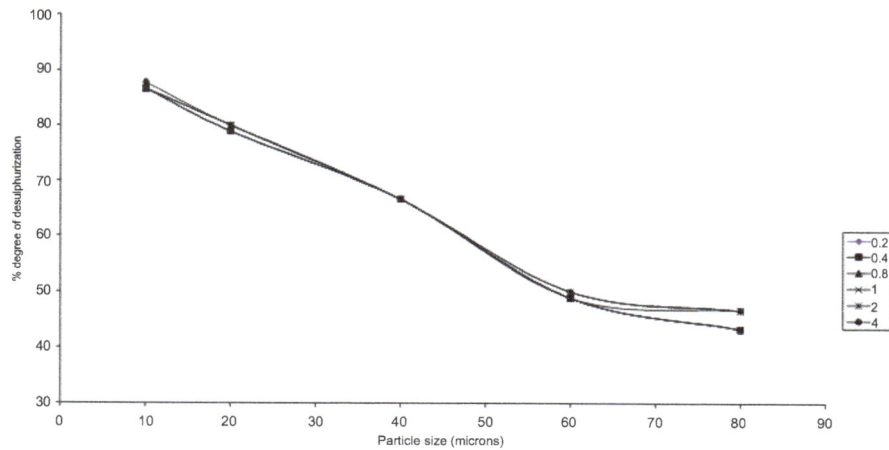

Figure 22: Effect of particle size on the %degree of reduction of sulphur for 120 mins using H_2SO_4.

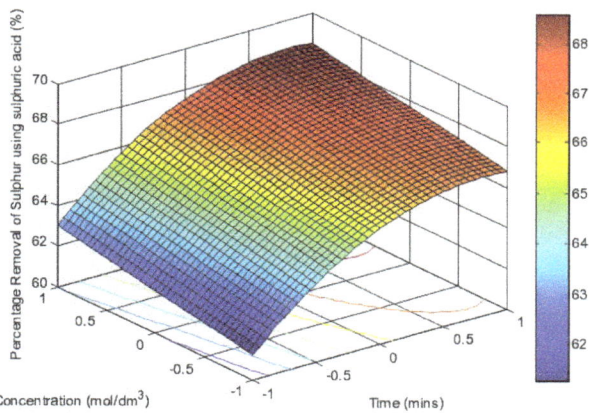

Figure 23: Percentage reduction of sulphur content using H_2SO_4 (time and concentration as variables).

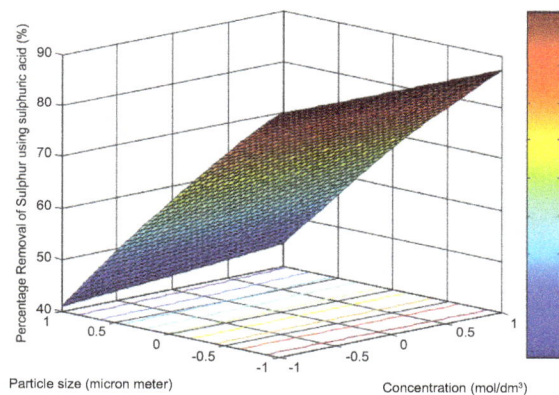

Figure 25: Percentage reduction of sulphur content using H_2SO_4 (particle size and concentrations as variables).

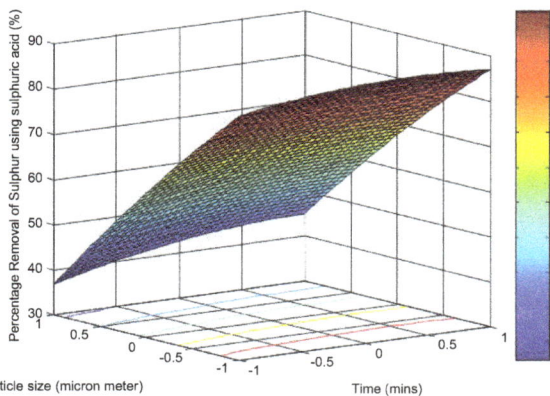

Figure 24: Percentage reduction of sulphur content using H_2SO_4 (time and particle as variable).

reduction of sulphur content more economical. It is therefore, noted that the percentage degrees of reduction of sulphur content using the acid leaching gave remarkable results. The Central Composite Design (CCD) used relevant parameters and data to develop the models. The developed model was tested and the results were quite adequate in the reduction of sulphur content of AGBAJA iron ore. The experimental results for percentage degree of the reduction of sulphur content are 87.77%. It is quite significant to note that the optimum values for percentage reduction of sulphur content obtained from the model for H_2SO_4 is 87.73%.

The experimental results also showed that as the leaching time and concentration increases, the percentage degree of reduction of sulphur content also increases while the particle size decreased as the leaching agent's increases. It is significantly observed that the level of reduction of sulphur content indicates that all the necessary processes were put in the right perspectives as being carried out in this research work, the final aims and objectives of the research work has been achieved.

Conclusion

The research work has availed the researcher the opportunity to explore all the experimental processes to understand the techniques involved in the reduction of sulphur content of AGBAJA iron ore using the leaching process. It should therefore be known that the large deposit of the iron ore which was initial adjudged as not suitable for the production of Direct Reduced Iron (DRI) and for the usage at the Blast Furnace at Ajaokuta Steel Company Limited due to the harmful nature of the ore because of high sulphur of 0.12% could be drastically be reduced.

This will further indicate that the fear of harmful impurities/effects that were initially envisaged to cause some harmful effects will be reduced to a minimal level through the reduction of the sulphur content using acid leaching process as indicated in this research work.

References

1. Alafara AB, Adekola FA, Lawal AJ (2009) Investigation of Chemical and Microbial Leaching of Iron ore in Sulphuric acid. J Appl Sci Environ Manage 11: 39-44.

2. Uwadiale GGOO (1990) Upgrading fine-grained Iron Ores: General review (ii) Agbaja Iron Ore. Elsevier Science Publishing Co Inc, USA.

3. Uwadiale GGOO, Whewel RJ (1988) Effect of Temperature on magnetizing reduction of Agbaja Iron Ore. Metallurgical Transaction B 19B: 731-735.

4. Dukino R, England B, Kneeshaw M (2000) Phosphorus distribution in BIF derived iron ores of Hamersley province, Western Australia. Transactions of Institute of Mining and Metallurgy (Section B: Applied Earth Science) 109: 168-176.

5. Kokal HR (1990) The Origin of Sulphur in iron making raw materials and methods of removal. Proceedings of the 63rd Annual meeting- the Minesota section AIME and 51st international symposium Duluth Minesota.

6. Fonsea D, Souza C, Araujo A (1994) Hydrometallurgical routes for the reduction of sulphur in iron ores. Hydrometallurgy 2003 I Fifth International Conference in Honor of Professor Ian Ritchie.

7. Kokal HR, Singh MP, Naydyonov VA (2003) Removal of Sulphur Lisakovsky iron ore by roast-leach process. The Minerals, Metals & Materials Society, Warrendale, USA.

8. Nwoye CI (2009) Process Analysis and mechanism of Reduction of AGBAJA iron Oxide Ore. J MME 4: 27-32.

9. Alafara AB, Adekola FA, Folashade AO (2005) Quantitative leaching of a Nigerian Ore in Hydrochloric acid. J Appl Sci Environ 9: 15- 20.

10. Alafara AB, Adekola FA, Bale RB (2003) Correlation between global thermodynamic functions and experimental data in multicomponent heterogeneous systems. J Chem Soc of Nigeria 28: 40-44.

Prediction of Geometry of Loop Formed on Terry Fabric Surface Using Mathematical and FEM Modelling

Singh JP[1]* and Behera BK[2]

[1]*Department of Textile Technology, U.P. Textile Technology Institute, Kanpur, India*
[2]*Department of Textile Technology, Indian Institute of Technology Delhi, New Delhi, India*

Abstract

The objective of this study is to understand the loop formation phenomenon of yarn by considering their non-linear bending behaviour and the effect of loop shape factor on properties of terry fabric. The yarn is modelled as a continuum thin solid beam, and the governing buckling equation is derived using Timoshenko's elastic theory and the Bernoulli-Euler theorem. Since the formation of loop is effected by large deformation caused by the weight of yarn too, geometric non-linearity is also considered and Runge Kutta method of numerical technique is used to solve the governing equation. Further, finite element modelling technique is also used to see the accuracy of the prediction which is further verified by the actual experimental results. The results of the research prove that the finer yarn produce loops which are having more circularity i.e., higher loop shape factor, as compared to the loops produced from coarse yarn. It is also being proved that the increasing the loop length increases circularity of the loop i.e., higher loop shape factor.

Keywords: Loop geometry; Terry fabric; Non-linear material; Numerical analysis; Finite element simulation

Introduction

The functional and aesthetic characteristics of terry fabric are predominantly governed by the geometrical profile of the loop [1-3]. The geometrical configuration of loop is primarily determined by yarn characteristics. Substantial amount of research work has been carried out to reveals the relationship between yarn properties and fabric properties. Loop geometry which is the peculiarity of the terry fabric has been ignored by most researchers. It seems to be interesting to know the relation between yarn properties and loop geometry.

Formation of loop on fabric surface is effected by buckling of yarn which is largely influenced by its bending behaviour. The yarn bending rigidity can be evaluated experimentally by defining it as a Bernoulli-Euler beam and differentiating the moment curvature relationship. Similar to the fabric, in the early stage of yarn bending process, a higher moment to overcome the interfibre friction is required to bend the unit curvature, and then after less moment is need for further bending. In both friction couple theory [4-7] and bilinear model [8,9] this early stage phenomenon is neglected, assuming a linear moment curvature relationship. In some latest research moment-curvature relationship was explained by an exponential function [10,11]. Therefore, we considered yarn bending as a highly non-linear phenomenon without any assumption [12,13].

The objective of this study is to understand the loop formation phenomenon of yarn by considering their non-linear bending behaviour. The yarn is modelled as a continuum thin solid beam, and the governing buckling equation is derived using Timoshenko's elastic theory and the Bernoulli-Euler theorem. Since the formation of loop is effected by large deformation caused by the weight of yarn too, geometric non-linearity is also considered and Runge Kutta method of numerical technique is used to solve the governing equation. Further, FEM modelling technique is also used to see the accuracy of the prediction which is further verified by the actual experimental results.

Formation of Loop during Terry Weaving

Formation of terry loop during terry weaving (Figure 1) is a process

of buckling. Buckling is a mode of failure in which the structure experiences sudden failure when subjected to a compressive stress. During formation of loop, the structural behaviour has been found markedly non-proportional to the applied load which suggests the high non-linearity of the system of loop formation. These nonlinearities have to be considered to obtain the correct solution. Instead of one step solution found in linear problems, the non-linear problem is solved by incremental method [14,15].

Loop Shape Factor

Aspect ratio and shape factor are two important quantities that explain the geometry of any shape. Aspect ratio is the ratio of major

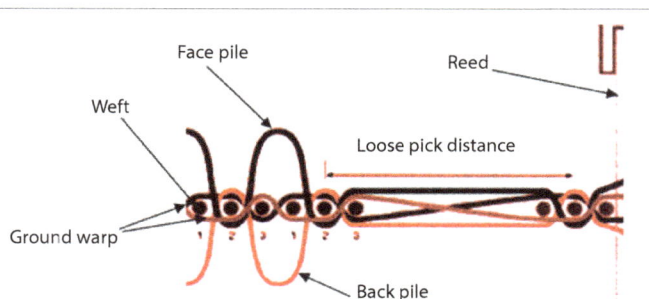

Figure 1: Loop formation showing the yarn segment between fell of cloth and reed.

*Corresponding author: Singh JP, Department of Textile Technology, UP Textile Technology Institute, Kanpur, India
E-mail: jpsingh.iitd@gmail.com

axis length to the minor axis length. It is suitable for the regular shapes while shape factor is suitable for all kind of shapes. Loop shape factor is a measure of circularity of loop. Here loop shape factor is calculated as shown below.

Loop shape factor=loop area/perimeter of the loop

=loop area/(l+d)

Where l=loop length, d=distance between two legs (Figure 2).

Theoretical Analysis

Loop geometry has been described here by the loop shape factor. Loop shape factor is a measure of circularity of the terry pile loop which is defined as the ratio of maximum loop height to the maximum loop width.

Mathematical modelling

Using KES-FB2 bending tester, moment curvature relationships and consequently regression equation (1) were obtained:

$$m(k) = c_0 k + c_1 \left\{1 - e^{-\beta k}\right\} \tag{1}$$

where m=moment, k=curvature, and c_0, c_1, β are constants having a value of 0.024, 0.015, and 2.89, respectively.

The yarn bending properties were successfully modelled here using exponential function giving standard error compared to the KES-FB2 m-k relationship below 0.002 and R^2 higher than 0.96. Differentiating Equation (1) gives the bending rigidity- curvature relationship, Equation (2):

$$b(k) = c_0 + \beta c_1 e^{-\beta k} \tag{2}$$

where b=bending rigidity, k=curvature, and c_0, c_1, β are constants. Equation is plotted in Figure 3 for 100% cotton yarn, which shows that the bending rigidity is non-linear in nature.

Mathematical model and governing equation: According to large deformation beam theory, the curvature for a large deformation problem can be defined as:

$$k = \frac{d\theta}{ds} \tag{3}$$

Where θ = tangent angle at some point of the beam, s = arc length.

According to elastic beam theory bending rigidity can be expressed as b (k). So Bernoulli-Euler theory for m-k relationship can be expressed as:

$$\frac{d\theta}{ds} = \frac{m}{b(k)} \tag{4}$$

Elastica model and governing equation: Yarn had been modelled after Timoshenko's elastic assuming it as bent bean with an identical cross-section. The moment equilibrium of the elastic model shown in Figure 4, gives the governing Equation (5).

$$b(k)\frac{d\theta}{ds} = b(s)\frac{d\theta}{ds} = -Py + me + Rx - \int_0^s w(x - x')ds' \tag{5}$$

Where P=compression load applied at the ends of beam, R=reactive force for the weight of beam, me= external moment, w=weight per unit length.

Differentiating Equation (5) by gives

$$b_d(s)\frac{d^2\theta}{ds^2} = -P\sin\theta + (R - ws)\cos\theta \tag{6}$$

Where:

$$b_d(s) = \left\{c_0 + \beta c_1\left(1 - \beta\frac{d\theta}{ds}\right)\exp\left(-\beta\frac{d\theta}{ds}\right)\right\} \tag{7}$$

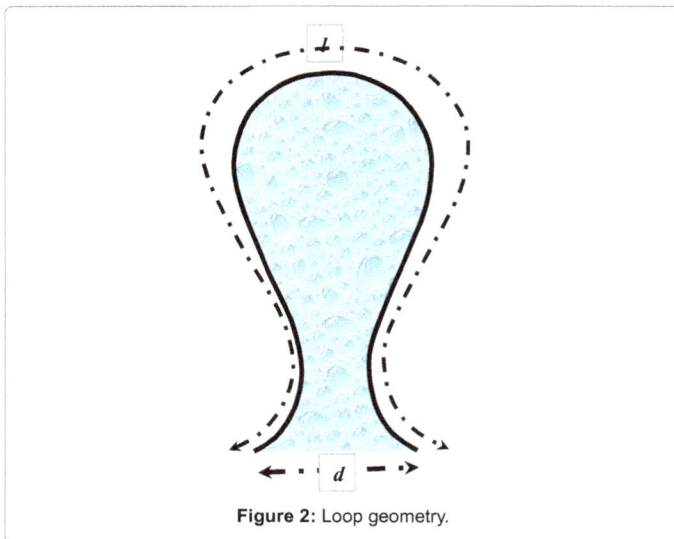

Figure 3: Relationship between bending rigidity and curvature of yarn.

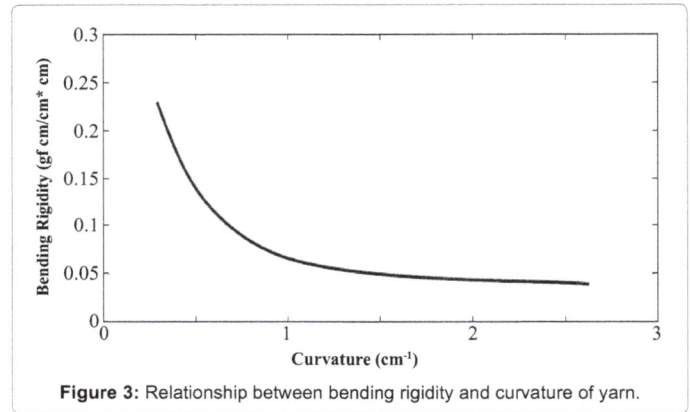

Figure 4: Yarn Elastic loop model.

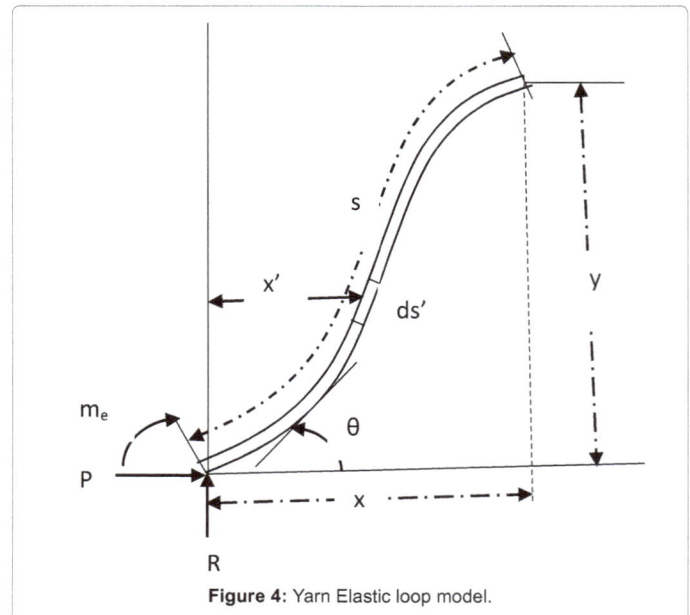

Figure 2: Loop geometry.

Numerical analysis: Highly non-linear differential Equation (6) can be solved by fourth-order Runge-Kutta method. The second order governing equation is modified into a system of two first order equations and normalised as in Equation (7):

$$\frac{d\theta}{ds} = \varphi \tag{7a}$$

$$\frac{d\varphi}{ds} = -\overline{P}\sin\theta + \left(\overline{R} - \overline{W}s\right)\cos\theta \tag{7b}$$

Formation of yarn loop which is governed by the highly nonlinear differential Equation (7) is a two point boundary problem which can be effectively solved by shooting method considering the following condition.

$$0 \le \overline{s} \le 1 \tag{8a}$$

$$\theta\big|_{\overline{s}=0,0.5,1} = 0 \tag{8b}$$

$$\overline{R} = 0.5\overline{W} \tag{8c}$$

Finite element modelling

The functional and aesthetic characteristics of terry fabric are predominantly governed by the geometrical profile of the loop. The geometrical configuration of loop is primarily determined by yarn characteristics. Need of precise study on how the yarn properties control the geometrical profile of the loop motivated us to explore the outcome using FEM.

According to principle of virtual work, it possible to represent the distributed displacement field {u} of any solid body by

$$\{u\} = [N]\{r\} \tag{9}$$

where [N]=set of known function of coordinates; {r}=set of constants

It follows that the algebraic equations for the axially loaded bar analysis are of the form

$$[K]\{r\} = \{R\} \tag{10}$$

Where [K]=stiffness matrix; {r}= nodal displacement; {R}=equivalent nodal loads

The expression for $[K^d]$ supposes that N is known. In the case of simple beams with a single axial load at one end, we have N= − P and

$$\left[K^d\right]\{r\} = -P\int[N']^T[N']\{r\}\,dx \tag{11}$$

Adding shape function (equation 3) to equation 2

$$\left([K]+\left[K^d\right]\right)\{r\} = \{R\} \tag{12}$$

Where $[K^d]$=differential stiffness;

$$[K] = \int[B^T][G][B]\,dv = \int EI[N'']^T[N'']\,dx$$

$$[R] = \int[N]^T f(x)\,dx + \sum[N(x_i)]^T F_{Zi} + W\left[N'(x)_j\right]^T M_{yj}$$

Equation (12) is a matrix equation for a beam- column. In case of buckling

$$\left([K]+\left[K^d\right]\right)\{r\} = \{0\}$$

Subsequently, critical load can be found by setting:

$$\left|[K] - P\left[K^d\right]\right| = 0 \tag{13}$$

Expanding the determinant in Equation (13) will produce a polynomial in P and the lowest root of this polynomial is critical value of P. For large deformation we take shape function as:

$$[n] = \frac{1}{L^3}\left[L^3 - 3L^2s + 2s^3, L^3s - 2L^2s^2 + Ls^3, 3Ls^2 - 2s^3, -L^2s^2 + Ls^3\right] \tag{14}$$

Where L= length of element, s=distance measured from a node at one end of the element and is positive in the direction of the other node. The element stiffness matrix is:

$$[K_i] = \frac{EI}{L_i^3}\begin{bmatrix} 12 & -6L_i & -12 & -6L_i \\ -6L_i & 4L_i^2 & 6L_i & 2L_i^2 \\ -12 & 6L_i & 12 & 6L_i \\ -6L_i & 2L_i^2 & 6L_i & 4L_i^2 \end{bmatrix} \tag{15}$$

Inserting shape function into the integral for the element differential stiffness matrix gives us--

$$[K^D]_i = \int_0^{L_i}[n']_i^T[n]_i\,ds\,\frac{1}{30L_i}\begin{bmatrix} 36 & -3L_i & -36 & -3L_i \\ -3L_i & 4L_i^2 & 3L_i & L_i^2 \\ -36 & 3L_i & 36 & 3L_i \\ -3L_i & -L_i^2 & 3L_i & 4L_i^2 \end{bmatrix} \tag{16}$$

Yarn model: Yarn path and cross- section for designing yarn model, its path is represented by the yarn centre line in three dimensional spaces. SolidWorks 2010 software package is used to build the model. The yarn cross-section is a 2D shape of the yarn when cut by a plane perpendicular to the yarn path tangent. The yarns are treated as solid volumes with circular cross-section. So a circle is swept along the pre designed yarn path to build the yarn geometry. The final outcome is a bend elastica. The bend elastica (Figure 5) is the true representation of the yarn segment between cloth fell and the reed as this yarn segment is in bend condition. Keeping the same spline for the path of the yarn different yarn model has been created by changing the circle diameter for different count of yarn.

Material model: Yarns are modelled as continuum solid bent beam with identical cross-section. The yarn is treated as a non-linear orthotropic material. The longitudinal direction is defined by 11, which is parallel to fibres; the transverse plane is described by the directions 22 and 33, which are characterized by a plane of isotropy at every point in the material. The orthotropic behaviour of the yarn is typically described using a 3D stiffness matrix containing nine independent constants [16]. Since the yarn is transversally isotropic, $E_{22} = _{33}$, $\upsilon_{12} - \upsilon_{13}$ and $G_{12} = G_{13}$. The longitudinal modulus E_{11} is approximated as a linear function of fibre volume fraction V_f of a yarn and fibre modulus E_f by the following equation:

$$E_{11} = \frac{E_f}{V_f} \tag{17}$$

It is assumed that all fibres within a yarn are perfectly parallel and hence no stiffening of the yarn will occur due to fibre straightening at low strains. For simplicity a constant E_{11} ($E_{11} = E_f/V_{f0}$, V_{f0} is initial fibre volume fraction of the yarn) was used in the simulations.

The transverse stiffness, E_{33} ($E_{22} = E_{33}$) can be expressed as a function of strain to express the nonlinearity of the material because the material matrix is no longer constant [17]. The transverse stiffness reduces during the loop formation as the gap between fibres increases.

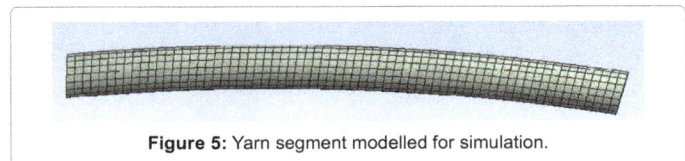

Figure 5: Yarn segment modelled for simulation.

$$E_{33}(\varepsilon_{33}) = \frac{\sigma_{33}}{\varepsilon_{33}} = \frac{-a\left(\frac{V_{f0}}{e^{\dot{a}_{33}}}\right)^{b} + a\left(V_{f0}\right)^{b}}{\varepsilon_{33}} \qquad (18)$$

The initial value of E_{33} is

$$E_{33}(0) = \lim_{\varepsilon_T \to 0} \frac{\sigma_T}{\varepsilon_T} = \frac{d\sigma_T}{d\varepsilon_T} = ab\left(\frac{V_{f0}}{e^{\varepsilon_{33}}}\right)^{b} \qquad (19)$$

Where V_{f0} is initial fibre volume fraction; $a = 1151$, $b = 12.24$

Due to transverse isotropy, the transverse shear behaviour is characterizes by [18]:

$$G_{23} = \frac{E_{33}}{2(1 + v_{23})} \qquad (20)$$

Material property used for simulation are E_{11} (MPa)=390, E_{33} (MPa)=0.75, G_{12} (MPa)=0.2, G_{23} (MPa)=3.13, v_{12}=0.32, v_{23}=0.32, density (kg/m³) =1530.

FE implementation: According to the proposed algorithm yarn structure model was constructed. The Solidworks 2010 and Ansys 14 software package were used to model the yarn and predict the behaviour and loop shape factor. The yarn was discredited using solid-45, 4-noded tetrahedral three-dimensional elements.

Boundary condition: Keeping yarn length 15 mm constant, yarn diameter varied up to four levels. Fixed support is applied at one end of the yarn, displacement is applied in negative x-direction only (keeping other direction frozen) at the other end. Static structural analysis has been performed keeping large deflection active so that the loop can be formed. In another simulation, yarn length was kept 18 mm and yarn count varied up to two levels.

Actual loop profile

To see the actual loop profile and set up the precise boundary condition for modelling, an experimental setup is done according to Figure 6 which shows the process of loop formation as in actual practice on loom.

Results and Discussion

Loop shape by mathematical modelling

Numerical analysis of the mathematical model gives the loop shape shown in Figure 7 for different weight per unit length/bending rigidity (w/b) ratio. It is clear from Figure 7 that the yarn having lower w/b ratio forms loop of higher shape factor.

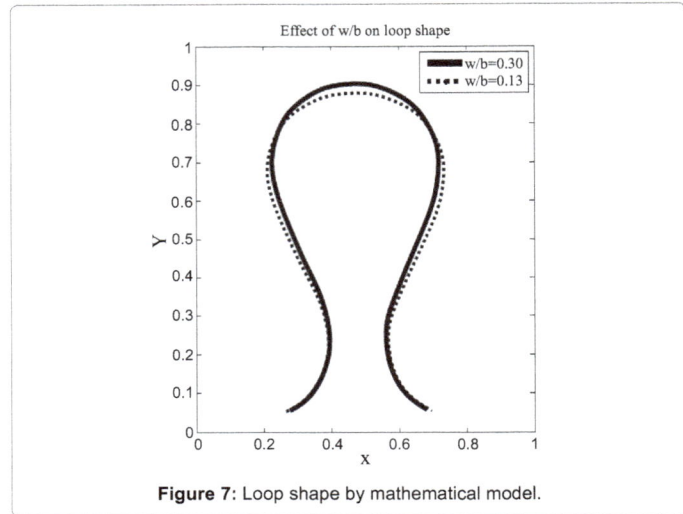

A=Slider, B=Yarn, C=Pulley, D= Dead

Figure 6: Experimental set up for the process of loop formation.

Loop shape by FEM model

Considering non-linear bending behaviour of yarn i.e., non-linearity in geometrical and bending properties, loop geometry of different yarn count was studied. Results of FEM model were given in the Figures 8-13. These figures the variation in loop shape with the change in yarn count and loop length keeping twist level same. The loop shape factor increases with increase in yarn fineness and this phenomenon is attributed to the reduction in w/b ratio of the yarn.

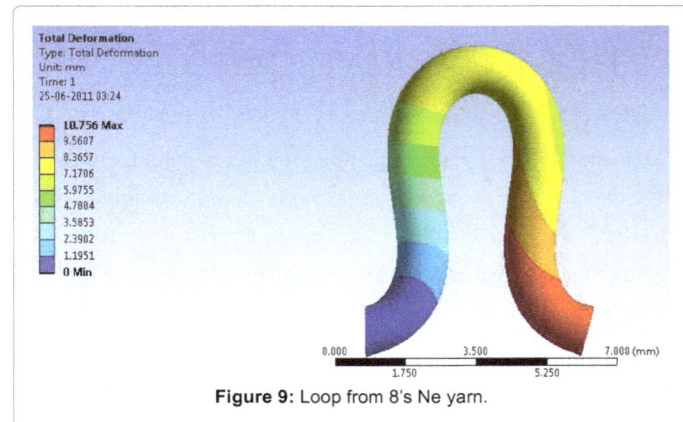

Figure 7: Loop shape by mathematical model.

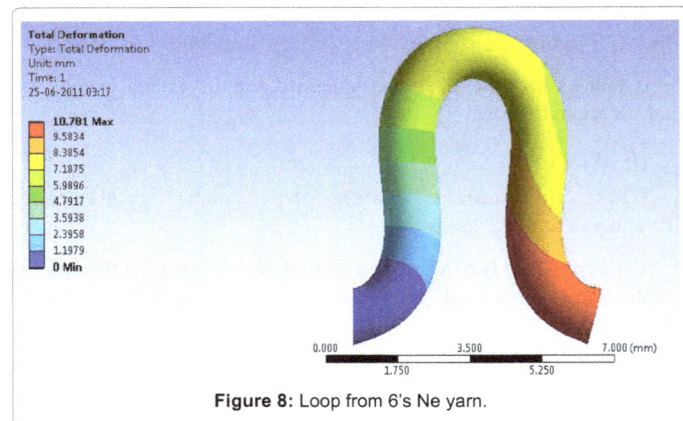

Figure 8: Loop from 6's Ne yarn.

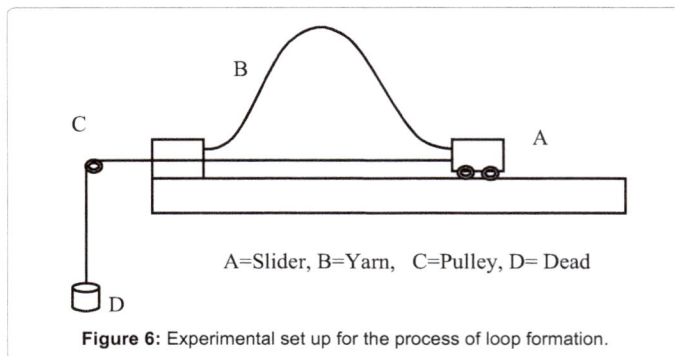

Figure 9: Loop from 8's Ne yarn.

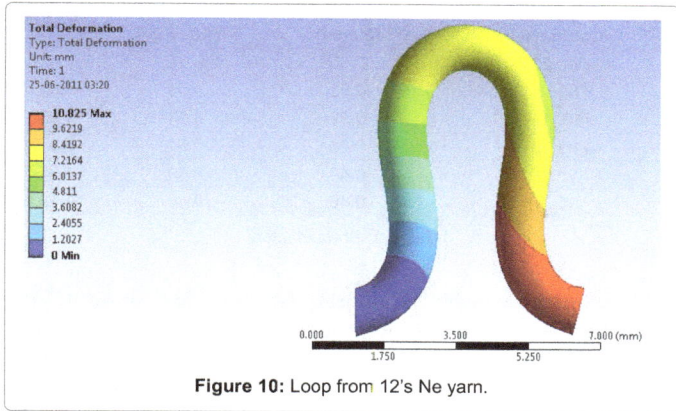

Figure 10: Loop from 12's Ne yarn.

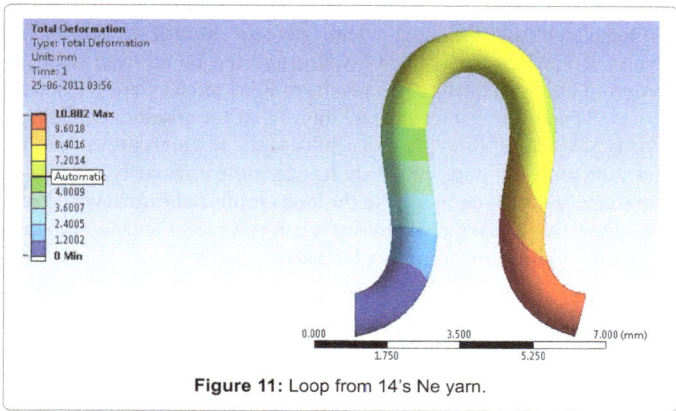

Figure 11: Loop from 14's Ne yarn.

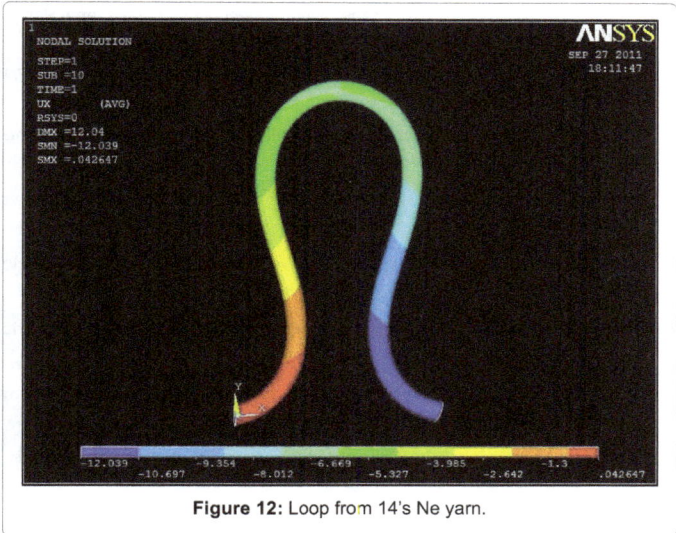

Figure 12: Loop from 14's Ne yarn.

These results are very well supported by the results of our mathematical model as well as earlier research [19,20].

Figures 12 and 13 show the effect of loop length on loop shape factor, the loop shape factor increases with increase in loop length keeping yarn twist level constant.

Actual loop shape

Images of the actual loop shape from different loop length and yarn count produced on the experimental set up has been shown in

Figures 14-17. These figures show the similar effect of yarn count and loop length on loop shape factor as the FEM modelled loops.

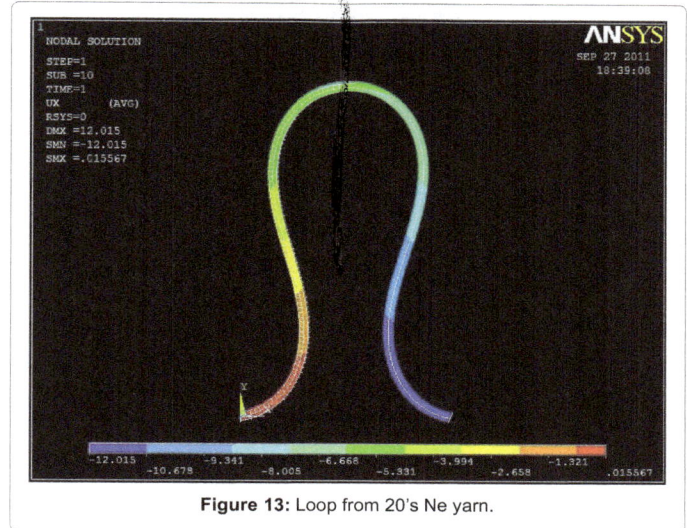

Figure 13: Loop from 20's Ne yarn.

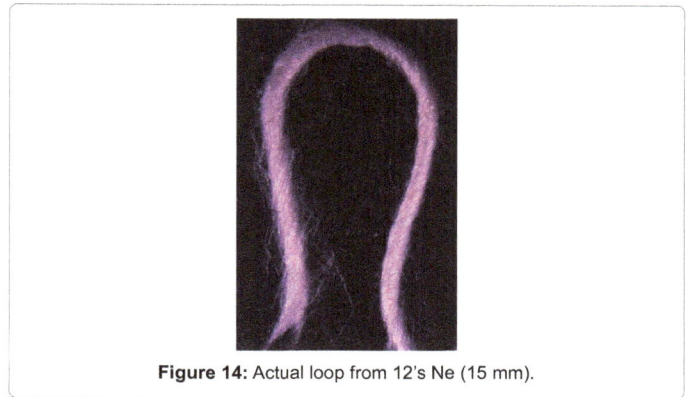

Figure 14: Actual loop from 12's Ne (15 mm).

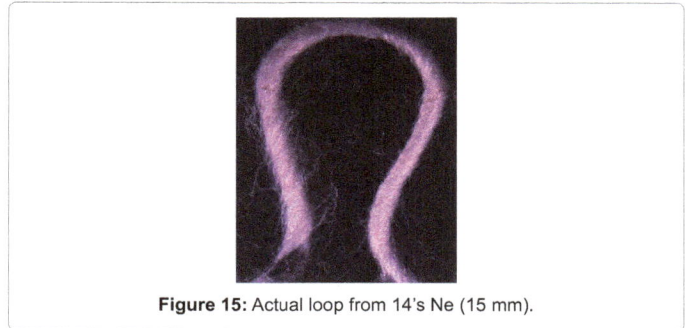

Figure 15: Actual loop from 14's Ne (15 mm).

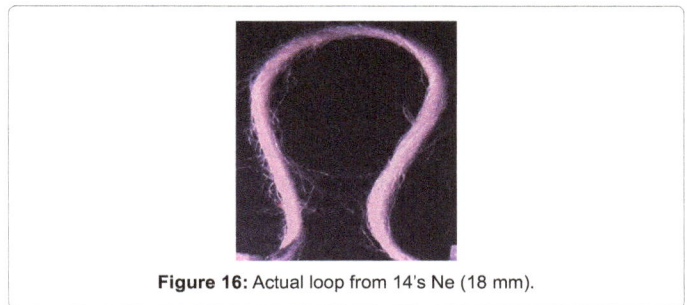

Figure 16: Actual loop from 14's Ne (18 mm).

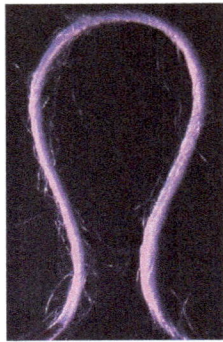

Figure 17: Actual Loop from 20's Ne (18 mm).

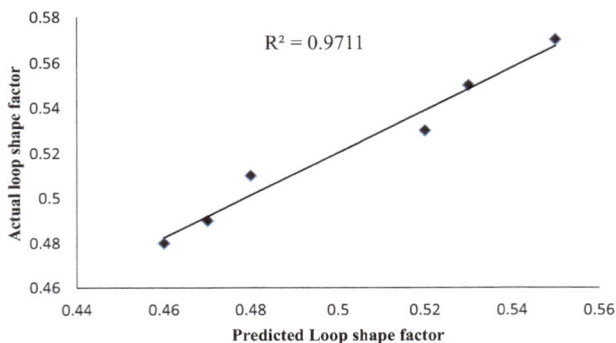

Figure 18: Predicted v/s Experimental loop shape factor.

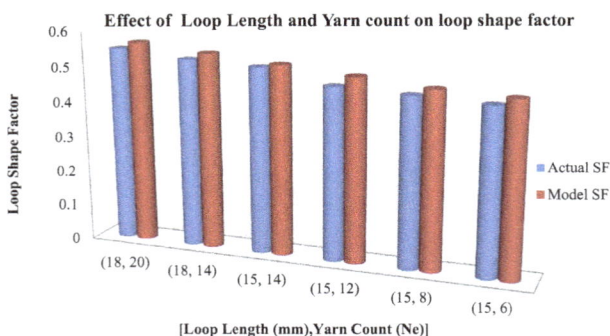

Figure 19: Effect of loop length and yarn count on loop shape factor.

Loop Length (mm), Yarn Count (Ne)	Shape Factor (Model)	Shape Factor (Actual)	Error %
(18, 20)	0.55	0.57	3.52
(18, 14)	0.53	0.55	3.65
(15, 14)	0.52	0.53	1.93
(15, 12)	0.48	0.51	5.91
(15, 8)	0.47	0.49	4.08
(15, 6)	0.46	0.48	4.16

Table 1: Shape factor.

of yarn which is highly non-linear. Geometric non-linearity and bending non-linearity both governs the yarn buckling process and consequently the shape of the loop. As mentioned in earlier research, geometric non-linearity is important for modelling large deformation like buckling and non-linear bending rigidity is important to get the real fabric behaviour from model. Considering these two non-linearity, mathematical modelling and finite element modelling was done. Results of FEM model was very well supported by the results of the numerical analysis. Further, the results of FEM model were verified by the actual experimental results and found that the absolute percentage error is 3.86 and R^2 is 0.951. The results of the research prove that the finer yarn produce loops which are having more circularity i.e., higher loop shape factor, as compared to the loops produced from coarse yarn. It is also being proved that the increasing the loop length increases circularity of the loop i.e., higher loop shape factor.

Prediction accuracy of FEM model

Shape factor of the loops and Area under the loop along with the RMSE between predicted and actual results are given in the Table 1. Figure 18 shows that the prediction of loop geometry using finite element method is good which gives R^2 value of 0.951.

Effect of loop length and yarn count on loop shape factor

It is clear from Figure 19 that the loop shape factor affected by loop length and yarn count. Finer yarn and higher loop length gives higher loop shape factor.

Conclusions

Formation of loop on terry fabric surface is affected by buckling

References

1. Behera BK, Singh JP (2013) Investigating absorbency behaviour of terry fabric. Res J Text Appal.

2. Behera BK, Singh JP (2014) Objective evaluation of aesthetic characteristics of terry pile structures using image analysis technique. Fibers and Polym 15:2633-2643.

3. Singh JP, Behera BK (2013) Compression behaviour of terry fabric. Proceedings of 13th Autex, Dresden, Germany.

4. Grosberg P (1966) The mechanical properties of woven fabrics part ii: the bending of woven fabrics. Text Res J 36: 205-211.

5. Grosberg P, Swani NM (1966) The mechanical properties of woven fabrics part III: The buckling of woven fabrics. Text Res J 36: 332.

6. Clapp TG, Peng H (1990) Buckling of woven fabrics part I: Effect of fabric weight. Text Res J 60:228-234.

7. Clapp TG, Peng H (1990b) Buckling of woven fabrics part II: effect of fabric weight and frictional couple. Text Res J 60: 285.

8. Ghosh T (1987) Computational model of the bending behaviour of plain woven fabrics. North Carolina State University.

9. Leaf GAV, Anandjiwala RD (1985) A generalized model of plain woven fabric. Text Res J 55: 92-99.

10. Kang TJ, Joo KH, Lee KW (2004) Analyzing fabric buckling based on nonlinear bending properties. Text Res J 74: 172-177.

11. Cornelissen B, Akkerman R(2009) Analysis of yarn bending behaviour.

12. Ivančo V (2006) Nonlinear finite element analysis. University of Applied Sciences-Technology, Business and Design, Wismar.

13. Deshpande S (2010) Buckling and post buckling of structural components.

14. Bao L, Takatera M, Shinohara A (2002) Analysis of large non-linear elastic deformation of fabrics. J Textile Inst 93: 410-419.

15. Zienkiewiez OC, Taylor RL (1991) The finite element method. (4th Ed) McGraw-hill, USA.

16. Hull D, Clyne CW (1996)An Introduction to Composite Materials. (2nd Ed) Cambridge University Press, Cambridge.

17. Sherburn M (2007) Geometric and mechanical modeling of textiles. Nottingham University, USA.

18. Lin H, Sherburn M, Crookston J, Long AC, Clifford MJ, et al. (2008) Finite element modelling of fabric compression. Model and Simul Mater Sci Eng 16: 035010.

19. Yu WR, Kang TJ, Chung K (2000) Drape simulation of woven fabrics by using explicit dynamic analysis. J Textile Inst 91: 285-301.

20. Zhou N, Ghosh TK (1998) On-Line measurement of Fabric bending behaviour. Text Res J 68: 533.

28

Transition Temperature Behaviour in Pan Based Composite Materials with and without SIC Filler

Venkateswara Rao CH[1]*, Usha Sri P[1] and Ramanarayanan R[2]

[1]Mechanical Engineering Department, UCE, Osmania University, Hyderabad, India
[2]TO'D', ACC, Advanced Systems Laboratory, DRDO, Hyderabad, India

Abstract

This article focus on the study of Transition or Back wall Temperature behavior in PAN based composite materials with and without Silicon Carbide filler was investigated. PAN fabric is one of the most useful reinforcement materials in the composites, its major use being the manufacture of components in aerospace, automotive and missile technology. In this study, PAN carbon fabric laminates were prepared by Hand lay-up technique with and without Silicon Carbide filler. The experimental work was done on studying the Transition or Back wall Temperature characterization and ablative studies through Oxy-acetylene testing. The study results revealed that it exhibits the better thermal protection shield in PAN based Silicon Carbide filler laminate compared without adding filler material. However, the ablation rate study results were exhibits better erosion rate of CP laminate with Sic filler for thermal insulation.

Keywords: PAN carbon fabric; Phenolic resin; Silicon carbide (SIC) filler; Hand lay-up technique; Oxy-acetylene flame

Introduction

In aerospace application the re-entry vehicle structures are generally made of a structural member and thermal protection member made by the combination of Carbon/ Epoxy (CE) and over above Carbon/Phenolic (CP). For optimum performance of electronic packages mounted inside the CE + CP shell, the temperature inside the shell should be ideally be less than 100°C. The major objective of this study the back wall temperature of Carbon-Phenolic laminate with and without Silicon Carbide filler laminates.

The study of back wall temperature represents an understanding of that material which was selected when subjected to high temperature under a specific time period, what would be its back wall temperature so as to assess its suitability for its intended use. Since, advanced composite materials are very widely used in aerospace applications because of its exotic properties. The final materials would be able to form complex shapes and should be as light as possible.

This paper describes the preparation of laminate, cutting of samples and Oxy-acetylene flame test. For this study carbon/phenolic material is selected and silicon carbide powder (SIC) as filler material. Carbon fabric (PAN based) is used as reinforcement and phenolic resin is used as matrix, Sic filler is mixed with resin.

Materials and Methods

The experimental study was carried out to know the back wall temperature of the Carbon Phenolic Laminate with and without Silicon carbide filler. The PAN based carbon fabric impregnated with Phenolic resin with and without Sic as filler. Raw materials used in this work are PAN carbon plain wave woven fabric as 0/90 fiber orientation. The adhesion made from Phenolic resin grade ABRON-PR100 (WS). Thickness of cloth is measuring 0.28-0.38 mm. In this experiment PAN carbon fabric along with Phenolic resin laminates are prepared with Silicon Carbide powder used as filler. The Laminates are prepared with and without filler material and subjected to Oxy-acetylene testing. The following Table 1 describes the raw material constituents used for this experiment.

Carbon fabric

Poly Acrylo Nitrile (PAN) based Phenolic Composites are gradually becoming the most advanced materials having better properties to meet the Thermal requirements. The aerospace or missiles structures are require high strength, light weight and should withstand the high temperatures. Phenolic based composites are widely used as thermosetting resins having the better Thermal properties [1]. Composite materials are used to replace the conventional materials almost in every field of application. PAN based phenolic composites are widely used in air frame structures for aerospace and missile application due to their exotic properties such as low weight to high strength and also withstanding higher temperatures [1,2]. The thermal properties of carbon phenolic reinforced composites depending on the properties of ingredients used in the material.

Phenolic resin

Phenolic resins are the oldest synthetic polymers used commercially available of ABRON-PR100 (WS) Phenolics to meet the requirement for low smoke and toxicity. Phenolics are formed by the reaction of phenol (carbolic acid) and formaldehyde and catalyzed by an acid or base. Phenolics generally dark in colour and therefore used for applications in which colour does not matter [3]. The phenolic products are usually red, blue, brown or black in colour. These thermo set resins have typically been cured at high temperatures and usually high pressures [2,4,5]. Phenolic resin provides intermolecular hydrogen bonding as a domain driving force to interact with hydroxyl, carbonyl, amide, ester, and other hydrogen bonding functional groups.

Silicon carbide filler

Silicon Carbide (Sic) Fillers affect the tribological behaviour of the polymers and change the properties of the composite materials [4]. The present investigation has been concentrated to determine the back wall

---

*Corresponding author: Venkateswara Rao CH, Ph.D Scholar, Mechanical Engineering Deptartment, UCE, Osmania University, Hyderabad, India, E-mail: hvenkateswararao@rediffmail.com

temperature on the carbon phenolic laminate with and without Sic filler by using Oxy-acetylene flame.

Preparation of Laminate

For preparation of laminate the carbon fabric is impregnated with phenolic resin (equivalent weight of reinforcement) and the phenolic resin is applied uniformly on the reinforcement and it is allowed for some time for solvent to evaporate. Then this prepregs are cut into size (250 × 250 mm) and 10 layers of these prepregs are cut and kept ready. Then the mould is cleaned thoroughly and release agent (wax polish/petroleum jelly) is applied uniformly over the mould for easy removal of laminate from the mould after curing. Then the cut prepregs are placed in the mould one over the other till all the 12 layers are placed in the mould refer the Figure 1a-1c. The mould is kept in a vacuum bag and cured in the autoclave by a cure cycle of temperature Vs time under vacuum and pressure.

Two laminates are prepared as per the above procedure. First laminate carbon fabric + phenolic resin without any filler is made and it is designated as Laminate L_1. The second laminate carbon fabric + phenolic resin with Sic filler (20%) added is made and it is designated as Laminate L_2.

Preparation of test specimen

The laminates are taken out of the autoclave are to be cut into required sizes as per standard i.e 250 × 250 × 3 mm square were cut using a tile saw. Two specimens were cut in each category stated with and without Sic filler. The thermocouple is bonded in center of the test specimen from the rear surface with help of the high temperature adhesive cerma bond and subjected to Oxy-acetylene flame test.

Testing of specimen on oxy-acetylene test bed

The oxyacetylene test bed (OTB) is a small scale experimental setup to know the transition or back wall temperature of the said test laminates. The oxy-acetylene flame capable of producing a flame temperature up to 3000°C using a calibrated oxyacetylene welding torch. This type of experimental setup is used for testing the composite materials at relatively low costs while still simulating extreme conditions in real time applications [4,5].

OTB setup contains a data acquisition system to measure the in situ temperature of the test specimens using embedded thermocouples. Test sample of 4' × 4' is held on the fixture, and oxy acetylene torch is held at a predetermined distance (d =10 cm) in front of the laminate focusing at the center. The torch is lit and the sample is subjected to exposure for more than 1 minute and the back wall temperature recorded for the said laminates refer the Figure 2a and 2b.

Results

The comparative study has been carried out between the two laminates designated as Carbon Phenolic laminate without filler and Carbon Phenolic laminate with Silicon carbide filler. The test results were tabulated in the Tables 2 and 3 respectively.

Laminate 1

Type of laminate: Carbon phenolic prepeg without filler (L_1)

Filler: No filler

Flame temperature: 3000 centigrade

Laminate 2

Type of Laminate: Carbon phenolic prepeg with filler (L_2)

Filler: Silicon Carbide **220 mesh

Percentage of filler: %20 Sic

Flame temperature: 3000 centigrade

Ablation properties

The Erosion was measured after the tests refer the Figure 3a and 3b, and it is tabulated as given below in Table 4.

Discussion

The Transition or back wall temperature of the samples are measured, by focusing the thermal beam set up located perpendicular to the sample surface. The graphical representation of the two laminates shown in Figure 4.

The investigation revealed that Carbon Phenolic laminate with Silicon carbide filler exhibits the better transition temperature compared without adding the Sic filler material. The graphical representation of the ablation rate study of the Carbon Phenolic laminate with and without Sic filler. The ablation rate test results are revealed that Carbon Phenolic Laminate without Sic filler (L_1) shows the 0.8 mm erosion rate (erosion rate can be defined as rate at which the wearing of rocks and other deposits on the earth surface by the action of water, ice, wind take place.) is less than with Carbon Phenolic Laminate with Sic filler (L_2) shown as 0.1 mm erosion (Figure 5).

Constituent	Grade
Reinforcement	PAN carbon fabric T-300
Phenolic resin	ABRON-PR100(WS)
Filler material	Silicon Carbide powder MW40.09

Table 1: Details of materials used.

Sl No.	Initial reading (ambient temperature t_0)	Time (Sec)	Temperature achieved (T_A)
1	33	15	34
2	-	30	57
3	-	45	118
4	-	60	192
5	-	75	253

Table 2: Represents carbon phenolic prepeg without filler.

Sl No.	Initial reading (ambient temperature T_0)	Time (Sec)	Temperature achieved (T_A)
1	33	15	52
2	-	30	88
3	-	45	96
4	-	60	100
5	-	75	105

Table 3: Represents carbon phenolic prepeg with filler.

Sl. No	Nomenclature	Thickness before test (mm)	Thickness after test (mm)	Erosion (mm)
1	Laminate L_1	3.00	2.2	0.8
2	Laminate L_2	3.00	2.9	0.1

Table 4: Shows erosion rate.

Conclusion

The Test samples were successfully manufactured with Phenolic resin, PAN based carbon fabric with and without Silicon carbide filler by hand layup technique. The test samples were subjected to Oxy-acetylene flame test and test results are revealed that the Carbon Phenolic laminate with Silicon carbide filler exhibits the better transition temperature behavior compared without Silicon carbide filler laminate comparatively withstood the high temperature during ablation test. The CP laminate with Sic filler exhibits the better erosion rate than other laminate without filler. The addition of Sic filler improved the better temperature withstanding ability in the laminate and it is the best composite laminate for thermal insulation.

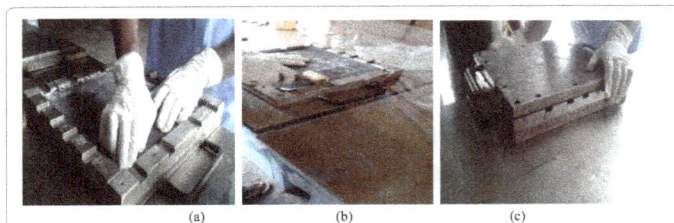

Figure 1: (a-c) Represents the preparation of laminate by hand layup technique.

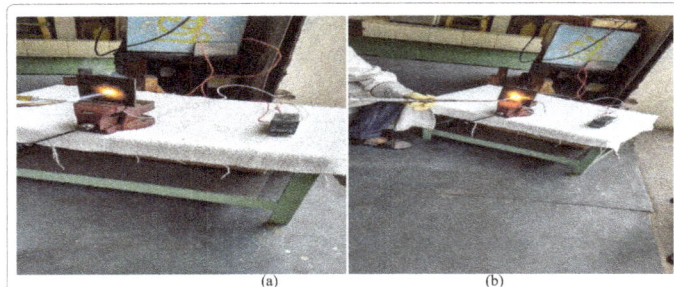

Figure 2: (a and b) Represents the oxy-acetylene test bed setup.

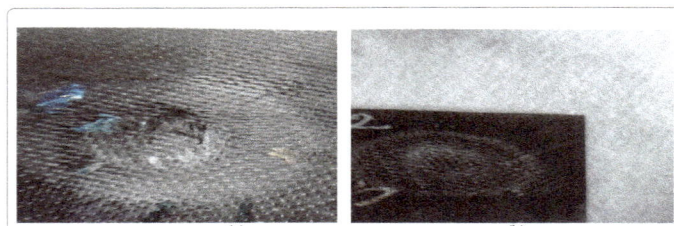

Figure 3: (a and b) Represents the post test samples.

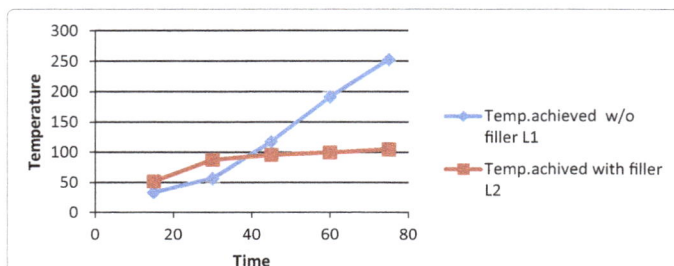

Figure 4: Graphical representation of comparative transition study of two laminates.

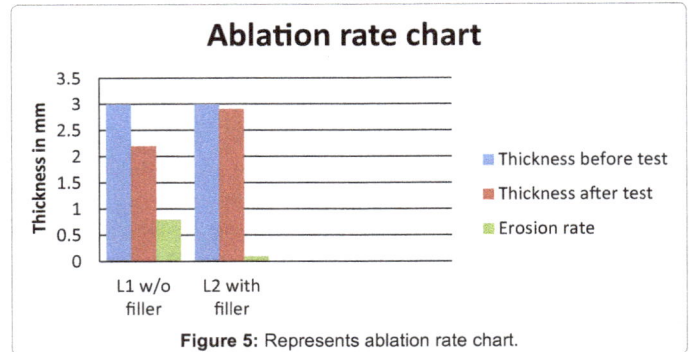

Figure 5: Represents ablation rate chart.

References

1. Mallick PK (1993) Fiber Reinforced Composites: Materials, Manufacturing and Design. (3rdedn) Marcel Dekker Inc, New York.

2. Tate JS, Gaikwad S, Theodoropoulou N, Trevino E, Koo JH (2013) Carbon/ Phenolic Nano composites as Advanced Thermal Protection Material in Aerospace Applications. Journal of Composites.

3. Torre L, Kenny JM, Maffezzoli AM (1998) Degradation behavior of a composite material for thermal protection systems. Journal of Materials Science 33: 3137-3143.

4. Devendra K, Rangaswamy T (2012) Determination of mechanical properties of Al_2O_3, $Mg(OH)_2$ and Sic. filled E-glass epoxy composites. Int J Engg Res App 2: 2028-2033.

5. Pulci G, Tirillò J, Marra F, Fossati F, Bartuli C, et al. (2010) Carbon-phenolic ablative materials for re-entry space vehicles: manufacturing and properties. Composites A 41: 1483-1490.

29

Structural and Hydriding Properties of the LaZr2Mn4Ni5-AB3 Type Based Alloy Prepared by Mechanical Alloying from the LaNi5 and ZrMn2 Binary Compounds

Elghali M* and Abdellaoui M

Laboratoire des matériaux utiles, Institut National de Recherche et d'Analyse Physico-chimique, pole technologique de Sidi Thabet, 2020 Sidi Thabet, Tunisia

Abstract

In this study, we report on the synthesis of a new AB_3-type compound $LaZr_2Mn_4Ni_5$ (with a content of 40 wt%) at room temperature during 5h of mechanical alloying. This compound was synthesized from the two binary compounds $LaNi_5$ ($CaCu_5$-type structure, P6/mmm space group) and $ZrMn_2$-Laves phase ($MgZn_2$-type structure, $P6_3$/mmc space group) in order to take advantage of the hydrogen absorption properties of these two types of intermetallic compounds. Structural properties were investigated using X-ray diffraction (XRD). The surface morphology of the cycled electrode was observed by a scanning electron microscope (SEM). The electrochemical properties of the $LaZr_2Mn_4Ni_5$–based alloy were determined using the chrono-potentiometry method. The experimental results indicate that the discharge capacity reaches a maximum value of 300 mAh/g. Solid-gaz reaction shows that this compound is able to form the $LaZr_2Mn_4Ni_5H_{13}$ hydride at room temperature at an absorption plateau pressure of about 7 bar.

Keywords: Intermetallic compound; Mechanical alloying; AB3-type compound; Hydrogen storage properties; Electro-chemical discharge capacity

Introduction

Intermetallic compounds AB_n (A = Y; rare earth, M = transition metal, $1 \leq n \leq 5$) are able to store reversibly large amount of hydrogen and are therefore potential materials for energy storage. The H_2 absorption-desorption reaction can be performed either by solid-gas or electrochemical routes [1]. Much research have been performed in order to improve the overall properties of the hydrogen storage alloys and to develop new types of hydrogen storage alloys, used as negative electrode materials for the Ni-MH battery. Applications of Ni-MH batteries require continuously an increase in the energy density [2,3]; weight capacities are still low for practical applications and many efforts are conducted worldwide to develop materials with improved performances regarding energy density [4]. It is well known that element substitution is one of the effective methods for improving the overall properties of the hydrogen storage alloys [5], by the replacement of part of the rare earths or the transition metals of the intermetallic compounds by lighter atoms [3].

Kadir [6] have presented a new series of ternary alloys, AMg_2Ni_9 (where A = La, Ce, Pr, Nd, Sm and Gd), whose structures are built up from alternating $MgNi_2$ Laves-type phases and rare-earth based AB_5 layers. They found that these compounds crystallize in an ordered variant of the $PuNi_3$-type rhombohedral structure (R-3 m space group) [1]. Moreover, it is reported that some of the A-Mg-Ni-based AB_3-type alloys also exhibited promising electrode properties. For example, the La-Mg-Ni-Co system AB_x ($x = 3.0$-3.5) type quaternary alloys were found to have large discharge capacities of 387-410 mAh/g, higher than those of the commercially used AB_5-type alloys [7].

In this work we will try to synthesize a new quaternary AB_3-type compound $LaZr_2Mn_4Ni_5$ starting from $ZrMn_2$ Laves phase and $LaNi_5$ assuming the reaction $LaNi_5 + 2 ZrMn_2 \rightarrow LaZr_2Mn_4Ni_5$, La (A atom type) and Ni (B atom type) are then partially substituted respectively by Zr and Mn. According to thermodynamic properties reported for the binary compounds, the intergrowth between $LaNi_5$ and $ZrMn_2$ should lead to a compound having intermediate properties between those of the two starting binary compounds. The compound $LaNi_5$, which crystallizes in the hexagonal $CaCu_5$-type structure (Haucke phase), exhibits exceptional thermodynamical properties toward hydrogen absorption storing up to 6.6 H per formula unit (f.u.) [8]. However, its equilibrium pressure (P = 1.7 bar at room temperature) is too high for practical applications and the molecular mass of La implies a weight capacity limited to 370 mAh/g [1]. In another hand, the use of Zr-based AB_2 alloys as electrodes in nickel-metal hydride batteries has been intensively studied in recent years. This is due to their large hydrogen reversible capacity compared with AB_5-type compounds presently used in commercial devices [9]. Furthermore, it has been found that the Mn element is beneficial in many respects for the rare earth-based hydrogen storage alloys [10-12]. In this work, as it is not possible to obtain the $LaZr_2Mn_4Ni_5$ compound by high-temperature melting synthesis due to the non-miscibility of the two elements La and Zr [3], this compound will be elaborated by mechanical alloying (MA) since it is a process suitable for alloying non-miscible materials and producing powders having a fine microstructural scale [3,13-17]. In this paper, the structure and the hydrogen storage properties of the mechanically alloyed compound $LaZr_2Mn_4Ni_5$ will be investigated.

Experimental Details

The MA process was carried out using a Fritsch 'Pulverisette 7'

---

*Corresponding author: Elghali M, Laboratoire des matériaux utiles, Institut National de Recherche et d'Analyse Physico-chimique, Pole technologique de Sidi Thabet, 2020 Sidi Thabet, Tunisia, E-mail: mouna.elghali@gmail.com

planetary ball mill, starting from a mixture of two binary compounds $LaNi_5$ and $ZrMn_2$ crushed into powder. These two alloys were prepared by Ultra High Frequency (UHF) induction melting of the pure elements La, Ni and Zr, Mn respectively. The purities of those starting metallic elements were 99.9%, 99.98%, 99.8% and 99.8% respectively. The ingots were melted five times to ensure good homogeneity. For the $LaNi_5$ compound, the synthesis was carried out under secondary vacuum in a water-cooled copper crucible from a stoichiometric mixture of La and Ni elements with atomic composition 1:5. The synthesis of the $ZrMn_2$ compound was made under vacuum and then under argon atmosphere when some losses of Mn by sublimation were observed. Binary compounds were analyzed by electron probe microanalysis (EPMA) using a CAMECA SX-100 to ensure their compositions and homogeneity. A mixture of 2 g of these two alloys was then placed in a cylindrical container which was tightly closed under argon atmosphere in a glove box filled with purified argon. The container made by tungsten carbide was loaded with 5 balls (Ø = 12 mm, m = 6.7851 g) with a balls to powder weight ratio equal to 17:1.

The MA conditions correspond to $0.6645.10^{-1}$ J/Hit kinetic shock energy, 93.35 Hz shock frequency and 3.1025 W/g injected shock power [17,18]. The alloy was milled for 5 hours at a disc rotation speed of 450 rpm. The sample obtained was named as S5.

The structural properties of this sample was obtained by X-ray powder diffraction (XRD) that was performed using Cu; Kα radiation at room temperature on a (θ-2θ) Panalytical X'Pert Pro MPD diffractometer. Diffracted intensities were measured in the range 10°-100° with a two-theta step of 0.04° and the diffractogram were fitted using the Rietveld method with the Fullprof software [19-21]. The Rietveld method was used to refine the lattice parameters, the atomic parameters (site occupancy and atomic coordinates) and to calculate the weight contents of the existing phases.

Electrochemical properties were tested using a composite negative electrode made of the intermetallic compound as active material, carbon black as electronic conductor and poly-tetra-fluoro-ethylene (PTFE) as a binder. The "latex" technology has been used for the electrode preparation; ninety percent of the alloy powder is mixed with 5% of the black carbon and 5% of the polytetrafluoroethylene (PTFE). Two 0.5 cm² pieces of this latex are pressed on each side of a nickel grid current collector to prevent the electrode plate from breaking into pieces during the charge-discharge cycling [22]. This grouping forms the negative electrode of Ni-MH accumulator. The counter electrode was formed by the Ni-oxyhydroxide/dihydroxide ($NiOOH/Ni(OH)_2$), whereas the reference electrode was the Hg/ HgO (1 M KOH). The electrolyte was 1 M KOH solution, prepared with deionised water. All electrochemical measurements were conducted at 25°C in a conventional three-electrode open-air cell using VMP system. The discharge capacity of the electrode was determined by a galvanostatical charging–discharging for 30 cycles at C/3 and D/6 regime, respectively between -0.6V and -1.3 V versus Hg/HgO. Every cycle was carried out by charging fully at 150 mA/g for 3 h (this time was majored by 50% due to efficiency of the charging reaction) and discharging at 75 mA/g.

The surface morphology of the cycled electrode was observed by a scanning electron microscope (SEM). The pressure-composition-temperature curve (P-C-T) was measured using a volumetric Sievert's method for pressure between 0.1 and 1 MPa at room temperature [23]. The sample was first activated by 3 hydriding-dehydriding cycles to reduce the grain size and to increase the kinetic (absorption at 25°C under 9 bar and desorption under primary vacuum at 35°C).

In addition, the hydride formation is confirmed by X-ray diffraction measurements carried out after the hydrogen absorption.

Result and Discussions

Structural characterization of the mechanically alloyed compound $LaZr_2Mn_4Ni_5$

The results of the structural characterization of the two starting compounds $LaNi_5$ and $ZrMn_2$ shows that $LaNi_5$ compound is single phase with the composition $LaNi_{4.98(3)}$ and crystallizes in the hexagonal $CaCu_5$-type structure (P6/mmm space group) with a = 5.0115(1) Å and c = 3.9850(1) Å. For the $ZrMn_2$ compound, the corresponding XRD pattern can be indexed in the $P6_3/mmc$ structure with parameters a = 5.0425(1) Å and c = 8.2835(3) Å, more details are given in our previous work [24].

MA starting from $LaNi_5$ and $ZrMn_2$ leads to the formation of a nanocrystalline AB_3-type phase with the hexagonal $PuNi_3$-type structure (S.G: R-3m) (Z = 3) [6,25]. This phase is formed in coexistence with $ZrMn_2$ phase ($MgZn_2$-type structure, $P6_3/mmc$ space group) and a cubic nanocrystalline AB_2-type phase Zr-Mn-Ni (C15-type) (S.G: Fd-3 m) [26].

Figure 1 shows the XRD pattern refinement of S5 sample for which Rietveld structural parameters are reported in Table 1. As can be seen in Figure 1, the peaks of $LaNi_5$ compound totally disappear after 5 h of MA whereas those of $ZrMn_2$ phase are still present. For the formed AB_3 phase with $PuNi_3$-type structure, the results show that La atoms are located not only at the 3a site of the $PuNi_3$-type structure, but also at the 6c site [24]. Both 3a and 6c sites are then occupied by both La and Zr elements indicating that the alloy is not a fully ordered compound having the same structure as previously reported for AMg_2Ni_9 compounds [6,25,27-29].

In our previous work, we have showed by transmission electron microscopy (TEM) examinations that the particles size of the mechanically alloyed sample during 5 h is rather inhomogeneous, ranging from 0.5 to 5 μm and the La + Zr/ Mn + Ni ratio have an average about 3 [24]. These results confirm the XRD analysis and show that that the particles produced by MA of $LaNi_5$ and $ZrMn_2$ formed a quaternary compound with AB_3 composition.

Electrochemical measurements

Figure 2a presents a typical discharge curve of the $LaZr_2Mn_4Ni_5$ alloy electrode at 25°C after being activated. Obviously, this curve has

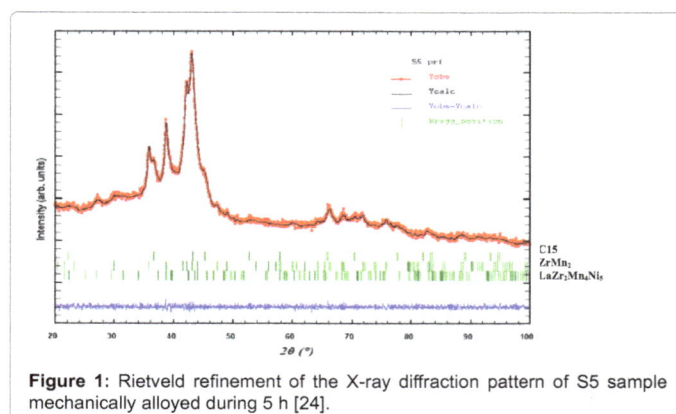

Figure 1: Rietveld refinement of the X-ray diffraction pattern of S5 sample mechanically alloyed during 5 h [24].

Figure 2: Rietveld refinement of the X-ray diffraction pattern of S5 sample mechanically alloyed during 5 h [24].

Figure 3: Variation of the discharge capacity of (a) the ZrMn2 compound and (b) the C15 Laves phase type Zr-Mn-Ni, measured at atmospheric pressure and room temperature as function of the cycle number.

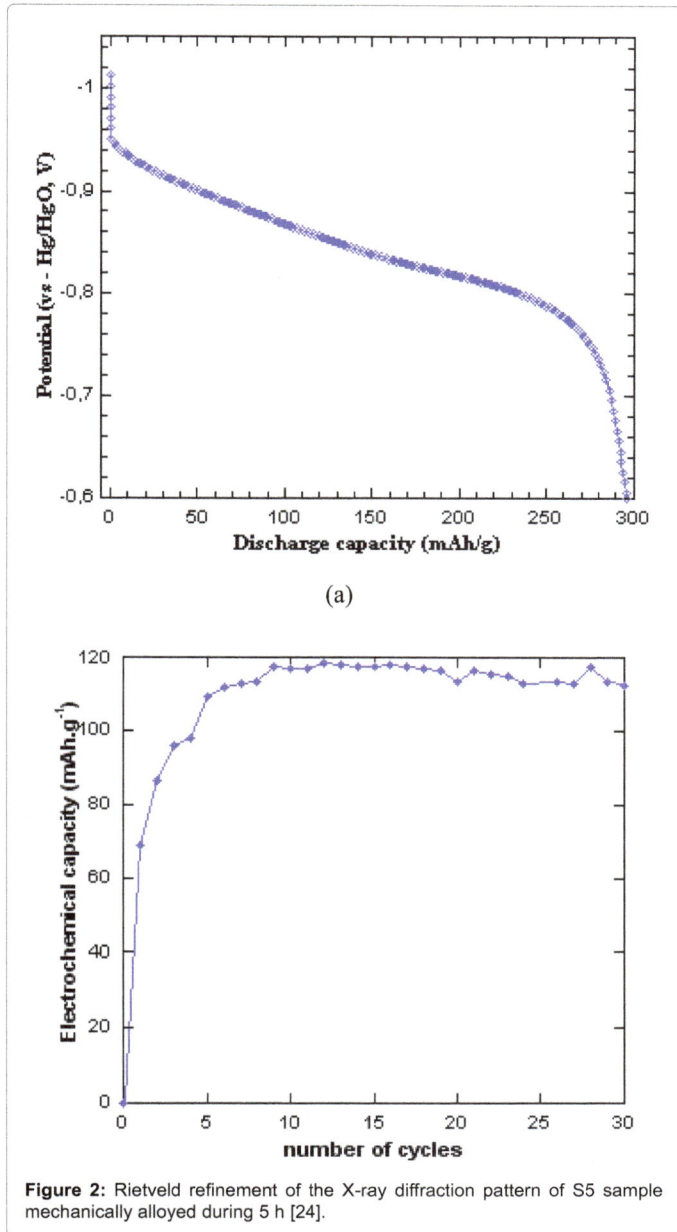

a wide discharge potential plateau based on the oxidation of desorbed hydrogen from the hydride. The mid-discharge potential is about -0.849V.

Figure 2b shows the variation of the electrochemical discharge capacity of the AB_3-based alloy (S5) as function of the number of cycles. However, as this alloy contains additionally to the AB_3 phase, the $ZrMn_2$ compound and the Zr-Ni-Mn Laves phase (C15-type) with significant mass proportions, it was, therefore, necessary to characterize their reactivity with hydrogen and measure their individual electrochemical discharge capacity to determine their contribution to the total discharge capacity.

Figure 3a shows the discharge capacity of the $ZrMn_2$ compound measured for 30 cycles of charging/discharging by galvanostatic cycling at room temperature and atmospheric pressure to deduce the discharge

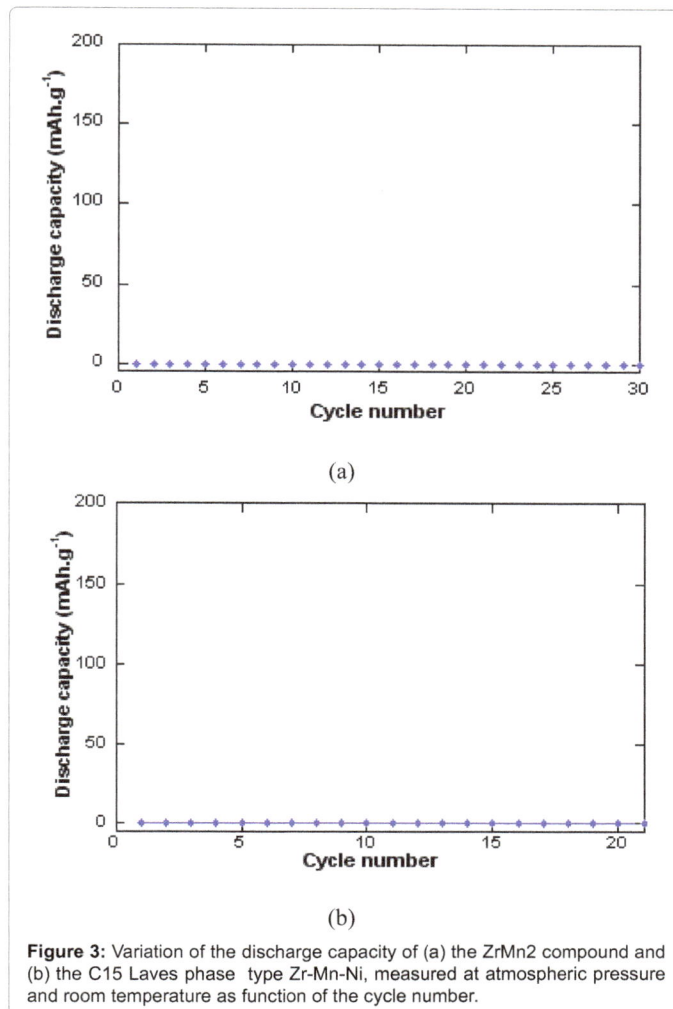

capacity of the AB_3-type compound. Therefore $ZrMn_2$ compound, in these conditions does not contribute to the total discharge capacity determined in Figure 2b for S5 sample based on $LaZr_2Mn_4Ni_5$ compound. According to the literature, the electrochemical discharge capacity of the $ZrMn_2$ compound is practically zero at 30°C and atmospheric pressure [30]. Nevertheless, it absorbs nearly 4 H/f.u, for maximum pressure of 8 bar, at room temperature, by solid-gas reaction [31]. The electrochemical discharge capacity of the C15 Laves phase Zr-Mn-Ni was also measured (Figure 3b) in order to determine its contribution to the discharge capacity of the S5 alloy. This latter compound was synthesized by UHF induction melting and it's a single phase which structure can be described in the cubic Fd-3 m space group (Figure 4). Figure 3b shows that similarly to the $ZrMn_2$ compound, the Zr-Mn-Ni Laves phase does not absorb hydrogen and thus does not contribute to the total discharge capacity of the synthesized AB_3-type based alloy (S5). Thus, the discharge capacity of the prepared alloy depends only of the mass proportion of the $LaZr_2Mn_4Ni_5$ AB_3-type compound. The corresponding capacity obtained can be therefore expressed in AB_3-type phase wheight (mAh/g AB_3), its variation versus the cycle number is given in Figure 5.

According to Figure 5, the $LaZr_2Mn_4Ni_5$ has a good cycle life since the discharge capacities have approximately stable values even after 30 cycles of charging/discharging. The maximum reversible capacity is

Phases	Content (wt%)	a (Å)	c (Å)	R_f (%)	R_{Bragg} (%)	X^2
$ZrMn_2$	38(1)	5.0231(f)	8.2475(f)	0.705	0.508	
AB_2 (C15)	21.5(0.6)	6.955(2)	—	0.229	0.104	
$AB_3(LaZr_2Mn_4Ni_5)$	39.9(0.9)	4.815(2)	27.269(18)	0.482	0.348	1.18

(f) : fixed parameter

Table 1: Structural parameters determined by Rietveld refinement of the XRD pattern of the S5 sample.

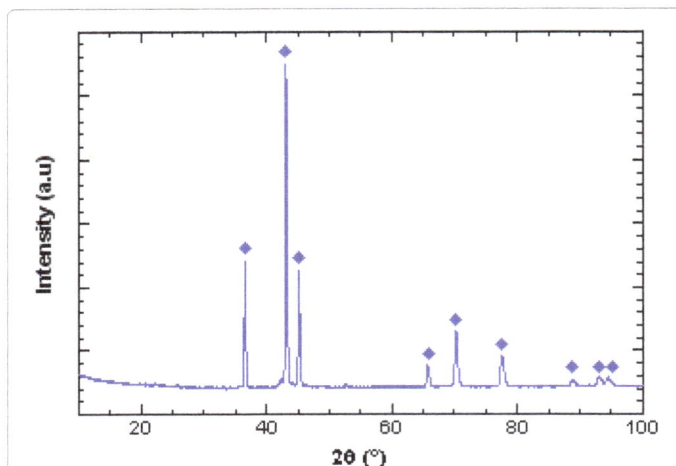

Figure 4: X-ray diffraction pattern of the Zr-Mn-Ni Laves phase (C15 type) synthesized by UHF induction melting, indexed in Fd-3m space group.

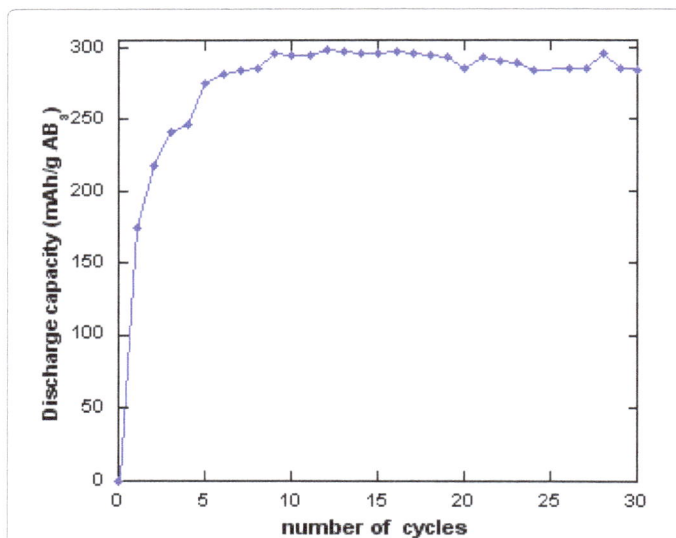

Figure 5: Variation of the electrochemical capacity of LaZr2Mn4Ni5 versus the cycle number.

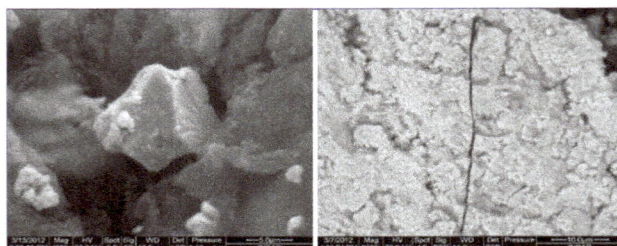

Figure 6: SEM micrographs of electrode surface morphologies of S5 sample (a) before cycling and (b) after 30 cycles of charging discharging.

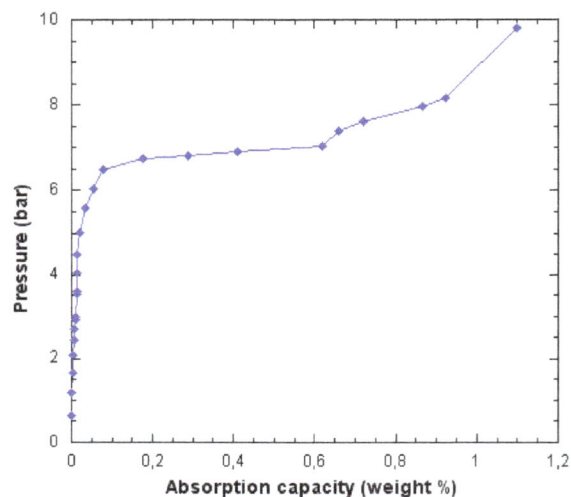

Figure 7: PCT curve measured at 25°C for the sample elaborated by mechanical alloying from the two binary compounds LaNi5 and ZrMn2.

about 298.56 mAh.g^{-1} equivalent to 9.12 H/f.u. The discharge capacity of the alloy powder increases to reaches 298.56 mAh/g after 12 cycles and then decreases to 293.6 mAh/g after a few charge-discharge cycles.

The capacity decay of the hydrogen storage alloy electrode is mainly due to the pulverization and oxidation of the active material components to form oxides or hydroxides [32,33].

Figure 6 shows the electrode surface morphologies by SEM of the S5 sample before and after 30 charge-discharge cycles. The alloy consists of large particles and clusters of fine particles. In fact the formed oxide or hydroxide layer acts as a barrier to the hydrogen diffusion and leads to the decrease of the number of hydrogen atoms which can be absorbed by the material. In fact, during the cycling, the rare earth elements such as La or the transition metal such as Mn segregate to the grain boundaries, where they were subject to corrosion. The corrosion products are disposed on the surface of grain particles as $La(OH)_3$ or as Mn_3O_4 and act as a barrier to the hydrogen diffusion in the particle volume which in turn decrease the alloy discharge capacity. It was shown that the thickness of the corrosion layer increases generally with the number of charging/discharging cycles [34-36].

Gas hydrogenation P-C-T isotherm

The pressure-composition-isotherm (P-C-T) measured at 25°C is presented in Figure 7. The PCT curve of this alloy has two absorption plateau pressures, showing the presence of two types of hydrides $A_xB_yH_z$ and $A'_xB'_yH'_z$. The first one is characterized by the formation of the first plateau at a pressure of about 7 bar, corresponding to an absorption capacity of 0.6 wt%, the second is characterized by the second plateau at a maximum pressure of about 8 bar corresponding to an absorption capacity of 0.32 wt%. Indeed, according to the literature, $ZrMn_2$

compound, one of the constituents of the synthesized composite, can absorb nearly 4 H/f.u by solid-gas reaction for maximum pressure of 8 bar, at room temperature [31]. Thus, the second plateau can be attributed to the $ZrMn_2$ compound and the first one to the $LaZr_2Mn_4Ni_5$ compound. Therefore, by converting the absorption weight capacities to H/f.u capacities, the hydride $A_xB_yH_z$ is $LaZr_2Mn_4Ni_5H_{13}$. We can then conclude that the $LaZr_2Mn_4Ni_5$ compound absorbs hydrogen at room temperature; its absorption capacity in these conditions is in the range of 13 H/uf.

The hydride formation as a result of the solid-gaz absorption of hydrogen by the alloy has been ascertained by XRD analysis. Figure 8 presents the XRD diffractogram of the hydrided alloy. This Figure 8 shows no diffraction peaks of other phases and/or none other of pure metals are observed in the XRD diffractogram of the formed hydride and all the initial phases or compounds before hydrogenation are identified. It is also found that all the hydrided phases still preserve their structures. The hexagonal PuNi3-type structure, is preserved for the $LaZr_2Mn_4Ni_5$ compound like in some RMg_2Ni_9 type alloys [25,27,28].

The lattice parameters and the cell volumes, determined by the the Fullprof program are summarized in Table 2. It is seen that the cell volume has not changed significantly. The reason for this observation is associated with a partial desorption of the solid-gas stored hydrogen in the course of the ex-situ XRD measurement.

This unavoidable desorption, in our case, is a result of hydrogen under pressure driven decomposition initiated upon taking the alloy out of the sample holder, clearly proving that the formed hydride is not stable, it should therefore have an equilibrium plateau pressure greater than atmospheric pressure at room temperature which is explaining the value of 7 bar of the absorption plateau pressure measured for the $LaZr_2Mn_4Ni_5$ compound. Some other AB3-type compounds exhibit plateau pressures greater than 1 atm: the $LaMg_2Ni_9$ compound has a plateau pressure of H absorbing about 22 bar at 80°C [37], this compound according to Liao [7] shows a plateau pressure of 4 bar at 25°C. Denys [38] have determined the enthalpy value of the hydride formation for this latter compound it is about -22.5 KJ $(mol_{H_2})^{-1}$ at 293 K, this proves that the studied $LaMg_2Ni_9$-AB3 type compound is also stable with an absorption equilibrium plateau pressure at 122 bar and a desorption plateau pressure at 18 bar at 20°C. These values are in agreement with that of $LaZr_2Mn_4Ni_5$ compound reported in this work.

Conclusion

The hydrogen storage alloy $LaZr_2Mn_4Ni_5$ was successfully

Phase	a (Å)	c (Å)
$ZrMn_2$	5.0297(7)	8.216(1)
$LaZr_2Mn_4Ni_5$	4.821(3)	27.29(3)

Table 2: Lattice parameters determined for the hydrided $LaZr_2Mn_4Ni_5$-based alloy after hydrogenation at room temperature.

synthesized at room temperature, for only 5 h of mechanical alloying, from the mixture of the two binary compounds $LaNi_5$ and $ZrMn_2$ at 40 wt.%. The following conclusions can be drawn concerning structural characterization and the hydrogen absorption properties of the mechanically alloyed compound $LaZr_2Mn_4Ni_5$ determined either by solid-gas or electrochemical routes:

- XRD analysis show the formation of a composite that is mainly consist of $LaZr_2Mn_4Ni_5$ compound with hexagonal PuNi3-type structure coexisting with the $ZrMn_2$ binary compound (MgZn2-type structure C14, P63/mmc space group) and a cubic AB2-type Laves phase Zr-Mn-Ni (MgCu2-type structure C15, Fd-3 m space group).

- The partial substitution of La and Ni by respectively Zr and Mn does not change the main phase structure but affects the lattice parameters. The overall effect is an increase of the cell volume and of c/a ratio as compared to the $LaMg_2Ni_9$ compound.

- Rietveld analysis proves that $LaZr_2Mn_4Ni_5$ is not fully ordered since both of La and Zr atoms are located simultaneously in the two types of allowed sites 3a and 6c contrarily to the $LaMg_2Ni_9$ compound where La atoms occupy only 3a site and Mg atoms the 6c site.

- The electrochemical measurements show that the discharge capacity of the synthesized alloy depends only of the mass proportion of the $LaZr_2Mn_4Ni_5$ compound. The corresponding capacity obtained can be therefore expressed in AB3-type phase wheight (mAh/g AB3), it is of about 300 mAh/g AB3.

- The $LaZr_2Mn_4Ni_5$ compound can absorb hydrogen at room temperature; its absorption capacity in these conditions is in the range of 13 H/uf at a plateau pressure of about 7 bar.

Acknowledgments

We are very grateful to M. Latroche and V.P. Boncour (CMTR/ICMPE-CNRS) for all the facilities given by to prepare the initial alloys by UHF and also for the useful discussion. We thank E. Leroy for technical assistance in the EPMA and TEM analysis. We are also thankful to the CMCU (10G1208) for the financial support of our collaboration.

References

1. Latroche M, Baddour-Hadjean R, Percheron-Guégan A (2003) Crystallographic and hydriding properties of the system La1-xCexY2Ni9 (xCe = 0, 0.5 and 1). Journal of Solid State Chemistry 173: 236-243.

2. Kohno T, Yoshida H, Kawashima F, Inaba T, Sakai I, et al. (2000) Hydrogen storage properties of new ternary system alloys : La2MgNi9, La5Mg2Ni23, La3MgNi14. Journal of alloys and compounds 311: L5-L7.

3. Msika E, Latroche M, Cuevas F, Percheron-Guégan A (2004) Zr-substitution in LaNi5-type hydride compound by room temperature ball milling. Material Science Engineering B 108: 91-95.

4. Latroche M, Percheron-Guégan A (2003) Structural and thermodynamic studies of some hydride forming RM3-type compounds (R = lanthanide, M = transition metal). Journal of Alloys and Compounds 356: 461-468.

5. Liao B, Lei YQ, Chen LX, Lu GL, Pan HG, et al. (2004) A study on the structure and electrochemical properties of La2Mg(Ni0.95M0.05)9 (M = Co, Mn, Fe, Al, Cu, Sn) hydrogen storage electrode alloys. Journal of Alloys and Compounds 376: 186-195.

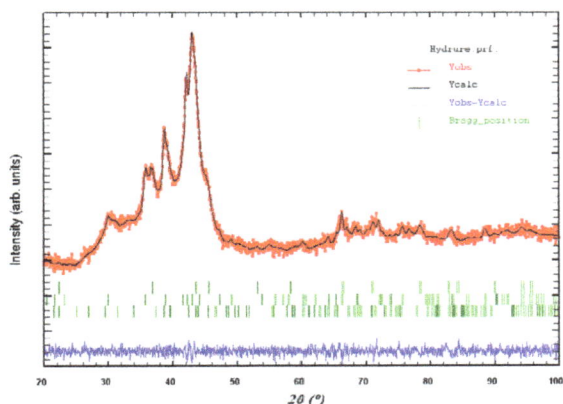

Figure 8: Rietveld refinement of the X-ray diffraction pattern of the hydrided alloy.

6. Kadir K, Sakai T, Uehara I (1997) Synthesis and structure determination of a new series of hydrogen storage alloys; RMg$_2$Ni$_9$ (R = La, Ce, Pr, Nd, Sm and Gd) built from MgNi$_2$ Laves-type layers alternating with AB$_5$ layers. Journal of Alloys and Compounds 257: 115-121.

7. Liao B, Lei YQ, Lu GL, Chen LX, Pan HG, et al. (2003) The electrochemical properties of La$_x$Mg$_{3-x}$Ni$_9$ (x = 1.0–2.0) hydrogen storage alloys. Journal of Alloys and Compounds 356-357: 746-749.

8. Van Vucht JHN, Kuijpers FA, Bruning HCAM (1970) Reversible room-temperature absorption of large quantities of hydrogen by intermetallic compounds. Philips Research Reports 25: 133-140.

9. Joubert JM, Sun D, Latroche M, Percheron-Guégan A (1997) Electrochemical performances of ZrM$_2$ (M = V, Cr, Mn, Ni) Laves phases and the relation to microstructures and thermo-dynamical properties. Journal of Alloys and Compounds 253-254: 564-569.

10. Lartigue C, Percheron-Guégan A, Achard JC (1980) Thermodynamic and structural properties of LaNi$_{5-x}$Mn$_x$ compounds and their related hydrides. Journal of the Less-Common Metals 75: 23-29.

11. Reilly JJ, Adzic GD, Johnson JR, Vogt T, Mukerjee S, et al. (1999) The correlation between composition and electrochemical properties of metal hydride electrodes. Journal of Alloys and Compounds 293-295: 569-582.

12. Sakai T, Oguro K, Miyamura H, Kato A, Ishikawa H (1990) Some factors affecting the cycle lives of LaNi$_5$-based alloy electrodes of hydrogen batteries. Journal of the Less-Common Metals 161: 193-202.

13. Liu BH, Li ZP, Chen CP, Liu WH, Wang QD (1995) The effects of mechanical grinding on the hydrogen storage properties of MlNi$_{4.7}$Al$_{0.3}$ alloy. Journal of Alloys and Compounds 231: 820-823.

14. Corré S, Bououdina M, Kuriyama N, Fruchart D, Adachi GY (1999) Effects of mechanical grinding on the hydrogen storage and electrochemical properties of LaNi$_5$. Journal of Alloys and Compounds 292: 166-173.

15. Abdellaoui M, Cracco D, Percheron-Guégan A (1998) Structural characterization and reversible hydrogen absorption properties of Mg$_2$Ni rich nanocomposite materials synthesized by mechanical alloying. Journal of Alloys and Compounds 268: 233-240.

16. Mokbli S, Abdellaoui M, Zarrouk H, Latroche M, Percheron-guégan A (2008) Hydriding and electrochemical properties of amorphous rich Mg$_x$Ni$_{100-x}$ nanomaterial obtained by mechanical alloying starting from Mg$_2$Ni and MgNi$_2$. Journal of Alloys and Compounds 460: 432-439.

17. Abdellaoui M, Gaffet E (1995) The physics of mechanical alloying in a planetary ball mill: Mathematical treatment. Acta Metal Materialia 43: 1087-1098

18. Abdellaoui M, Gaffet E (1994) A mathematical and experimental dynamical phase diagram for ball-milled Ni10Zr7. Journal of Alloys and Compounds 209: 351-361.

19. Rietveld HM (1967) Line profiles of neutron powder-diffraction peaks for structure refinement. Acta Crystallographica 22: 151-152.

20. Rietveld HM (1969) A profile refinement method for nuclear and magnetic structures. Journal of Applied Crystallography 2: 65-71.

21. Carvajal JR (1993) Recent Advances in Magnetic Structure Determination by Neutron Powder Diffraction. Journal of Physics 192: 55-69.

22. Chuab-Jan L, Feng-Rong W, Wen-Hao C, Wei L, Wen-Tong Z (2001) The influence of high-rate quenching on the cycle stability and the structure of the AB$_5$-type hydrogen storage alloys with different Co content. Journal of Alloys and Compounds 315: 218-223.

23. Sieverts A (1907) Occlusion and diffusion of gases in metals. Z Phys Chem 60: 129-201.

24. Elghali M, Abdellaoui M, Boncour VP, Latroche M (2013) Synthesis and structural characterization of mechanically alloyed AB$_3$-type based material: LaZr$_2$Mn$_4$Ni$_5$. Intermetallics 41: 76-81.

25. Kadir K, Kuriyama N, Sakai T, Uehara I, Eriksson L (1999) Structural investigation and hydrogen capacity of CaMg$_2$Ni$_9$: a new phase in the AB$_2$C$_9$ system isostructural with LaMg$_2$Ni$_9$. Journal of Alloys and Compounds 284: 145-154.

26. Suzuki A, Nishimiya N (1984) Thermodynamic properties of Zr (Ni$_x$Mn$_{1-x}$)$_2$-H$_2$ systems. Material Research Bulltin 19: 1559-1571.

27. Kadir K, Sakai T, Uehara I (1999) Structural investigation and hydrogen capacity of YMg$_2$Ni$_9$ and (Y$_{0.5}$Ca$_{0.5}$)(MgCa)Ni$_9$: New phases in the AB$_2$C$_9$ system isostructural with LaMg$_2$Ni$_9$. Journal of Alloys and Compounds 287: 264-270.

28. Kadir K, Sakai T, Uehara I (2000) Structural investigation and hydrogen storage capacity of LaMg$_2$Ni$_9$ and (La$_{0.65}$Ca$_{0.35}$)(Mg$_{1.32}$Ca$_{0.68}$)Ni$_9$ of the AB$_2$C$_9$ type structure. Journal of Alloys and Compounds 302: 112-117.

29. Chen J, Takeshita HT, Tanaka H, Kuriyama N, Sakai T, et al. (2000) Hydriding properties of LaNi$_3$ and CaNi$_3$ and their substitutes with PuNi$_3$-type structure. Journal of Alloys and Compounds 302: 304-313.

30. Kown I, Park H, Songa MY (2002) Electrochemical properties of ZrMnNi$_{1+x}$ hydrogen-storage alloys. International Journal of Hydrogen Energy 27: 171-176.

31. Cuevas F, Joubert JM, Latroche M, Percheron-Guégan A (2001) Intermetallic compounds as negative electrodes of Ni/MH batteries. Applied Physics A 72: 225-238.

32. Geng M, Wen Han J, Feng F, Northwood DO (2000) Electrochemical measurements of a metal hydride electrode for the Ni/MH battery. International Journal of Hydrogen Energy 25: 203-210.

33. Pan H, Liu Y, Gao M, Lei Y, Wang Q (2003) A Study of the Structural and Electrochemical Properties of La$_{0.7}$Mg$_{0.3}$(Ni$_{0.85}$Co$_{0.15}$) x (x=2.5 5.0) Hydrogen Storage Alloys. Journal of Electrochemical Society 150: A565-A570.

34. Abdellaoui M, Mokbli S, Cuevas F, Latroche M, Percheron-Guégan A, et al. (2003) Structural and electrochemical properties of amorphous rich Mg$_x$Ni$_{100-x}$ nanomaterial obtained by mechanical alloying. Journal of Alloys and Compounds 356-357: 557-561.

35. Liu W, Wu H, Lei Y, Wang Q, Wu J (1997) Amorphization and electrochemical hydrogen storage properties of mechanically alloyed Mg-Ni. Journal of Alloys and Compounds 252: 234-237.

36. Nohara S, Yamasaki K, Zhang SG, Inoue H, Iwakura C (1998) Electrochemical characteristics of an amorphous Mg$_{0.9}$V$_{0.1}$Ni alloy prepared by mechanical alloying. Journal of Alloys and Compounds 280:104-106.

37. Jin G, Dan H, Guangxu L, Shuyuan M, Wenlou W (2006) Effect of La/Mg on the hydrogen storage capacities and electrochemical performances of La-Mg-Ni alloys. Materials Science and Engineering B 131: 169-172.

38. Denys RV, Yartys VA (2011) Effect of magnesium on the crystal structure and thermodynamics of the La$_{3-x}$Mg$_x$Ni$_9$ hydrides. Journal of Alloys and Compounds 509S: S540-S548.

30

Recommendations for Compatibility of Different Types of Polymers with Potassium/Sodium Formate-Based Fluids for Drilling Operations: An Experimental Comparative Analysis

Kakoli M, Davarpanah A*, Ahmadi A and Jahangiri MM

Department of Petroleum Engineering, Islamic Azad University, Science and Research Branch, Tehran, Iran

Abstract

A formate-based fluid has been successfully used in many high pressure High temperature (HPHT) well operations since they were introduced in field practice. The laboratory research was carried out to determine composition of formate-base drilling fluid. It was formulated using sodium and potassium formate salts, Carboxymethyl Cellulose (CMC), Polyanionic Cellulose (PAC) and other types of polymers. In this research, the compatibility of different polymers with fluids, including potassium/sodium formate salts is being studied. Having said this, however, polymers, when taken to high temperatures, lose their properties. Therefore, this experimental procedure has been done at a temperature of 250°F over a period of 16 h. For doing these tests, six types of potassium/sodium formate fluids were made by different polymers. All samples formulation regarding their type and amount of water used in preparing fluids and the volume of salts used are simultaneously kept constant. On the contrary, the only differences were the types of polymers used in the different formulations. Formulation of formate-base fluids gives the best rheological properties in terms of AV/PV, YP and shale recovery than other fluids.

Keywords: CMC; PAC; Formate fluids; Compatibility

Abbreviations: CMC: Carboxymethyl Cellulose; HPHT: High Pressure High Temperature; OBM: Oil Based Mud; PAC: Polyanionic Cellulose; PV: Plastic Viscosity; AV: Apparent Viscosity; WBM: Water Based Mud; YP: Yield Point; API: American Petroleum Institute; PAC-UL: Polyanionic Cellulose-Ultra low Viscosity; PAC-R: Polyanionic Cellulose-Regular Viscosity; HT-PM: High Technology Polymer Materials; PHPA: Partial Hydrolyzed Polyacrylamid; $CaCO_3$: Calcium Carbonate; Na_2CO_3: Sodium Carbonate

Introduction

To produce hydrocarbons wells need to be drilled. The main objective when drilling a well is to drill a hole as fast as possible without accidents. Drilling is an important part when producing hydrocarbons and drilling fluids represent one fifth (15-18%) of the total cost of well drilling [1]. Therefore it is of interest to develop better solutions for a less costly operation. Better techniques have been made to improve the production, such as horizontal wells, directional drilling and managed pressure drilling. Drilling fluid, sometimes referred to as drilling mud, is used in drilling operations [2]. It is circulated down the drill string, through the bit and back to the surface through the annulus. The particle improves the mechanical strength and reduces the filtrate loss.

The development of deep offshore operations gives new and more technical challenges due to the harsh conditions encountered at these water depths. The extreme conditions that exist require an adaption and a particular design of the drilling and cementing fluids. One of the most challenging problems of deep offshore drilling is the range of temperatures and pressures. The temperature of the drilling fluid when circulating in the well may range from 0°C to 150°C and it is important that the drilling fluid maintain acceptable rheological properties within the whole range. The rheological properties of the mud will strongly depend on the temperature and the pressure variations.

To solve the problems with the mud properties it is often necessary to use several additives either separately or concurrently. The major drawback of the conventional additives is that they are generally unstable at high temperatures normally encountered in deep wells. The rheological properties of traditional water- and oil-based muds may change, often dramatically, when regions of high temperature and/or pressure are encountered in a deep well. Alternatively, there are several natural and synthetic polymers available which exhibit better resistance to thermal, bacterial and even mechanical degradation. For these reasons the new polymers are increasingly replacing conventional additives as rheology modifiers in the drilling industry [3].

When drilling a well it is very important to know the exact pressure drop for many reasons. Some of the reasons might be [4].

- To optimize the pressure drop on the drilling bit in order to get a maximum impact on the formation, and thereby increase the rate of penetration.

- For optimizing the flow rate in the annulus, the area between the borehole wall and the drill pipe, to get a better transport of drilled cuttings to the surface as well as to maintain a proper hole cleaning.

- To avoid fracture of the formation crossed due to the underestimation of the annular pressure drop.

- To detect any unexpected changes of the standpipe pressure, due to changes in the hydraulic drilling circuit (i.e., washout, plugged nozzles and fluid kick) and make opportune decisions to restore the original conditions.

---

*Corresponding author: Davarpanah A, Department of Petroleum Engineering, Islamic Azad University, Science and Research Branch, Tehran, Iran
E-mail: Afshindpe@gmail.com

- To better design the mud pumps available on the drilling rig.

In addition to the reasons mentioned above might the drilling of ultra-deep wells with high temperatures and pressures influence the rheological properties of the drilling fluids in several ways [4]. Physically, decreases in temperature and increases in pressure both affect the mobility of the system and lead to an increase of apparent viscosities and viscoelastic relaxation times [5]. The effect of pressures is expected to be greater with oil-based systems due to the oil phase compressibility [6].

Electrochemically, an increase in temperature will increase the ionic activity of electrolytes and the solubility of any partially soluble salt that may be present in the mud. This could change the balance between the inter-particle attractive and repulsive forces and also the degree of dispersion and flocculation in the mud systems. In some occasions this can also affect the emulsion stability of oil-based muds [7]. All these phenomena have a big impact on rheological properties, especially as far as viscoelasticity and thixotropy are concerned.

Chemically, all hydroxides react with clay minerals at temperatures above 90°C and for many kinds of muds this can result in a change in the structure and also a change in the rheological properties [4].

Due to the large number of variables involved, the behavior of the drilling mud at high pressures and temperatures may be very difficult to explain because of the complexity. It can be very difficult to set general guidelines for each group of muds (oil-based muds (OBM), water based muds (WBM), etc.) or even for the same kind of mud as small differences in composition may result in large differences in the rheological behavior [4].

Methodology of Work

Field description

The vertical well to be drilled was an exploration well that could provide information on potential reservoirs and lithological information of the field. No offset data were available on the well and the nearest well information was 80 km away. Geologist forecast from this well required drilling through reactive shales. Table 1 shows the lists the interval parameters for drilling.

The objective was to drill a 8 1/2-in. hole section from 10900 ft, to the casing point at a measured depth (MD) of 12500 ft. A 7-in. casing string was then to be run and cemented. The FBM optimized for member (A) was expected to provide maximum shale stabilization and inhibition to achieve maximum ROP without any incidents such as tight hole, pipe stuck and hole filling.

Laboratory tests

Material: Initially, optimized base fluid systems are developed at different concentrations as shown in Table 2. Fann V-G meter 35SA model (Fann Instrument Company, Houston, Texas) was used to

Formation Type	Member A compose of shale
Thickness interval depth	400 feet
Interval Hole size	8 1/2 inches
Fluid Type	Formate based mud
Bit type	Mill Tooth bit
Nozzle size	3*16/32 inches
String Rotation speed (rpm)	100-130 rpm
Weight on Bit (WOB)	20-25klb

Table 1: Interval well Parameters.

Material	Sample No	A	B	C	D	E	F
1	Water (sea) (cc)	350	350	350	350	350	350
2	Na_2CO_3 (g)	0.3	0.3	0.3	0.3	0.3	0.3
3	Poly drill (g)	-	-	-	-	-	6
4	Poly thin (g)	-	-	-	-	-	1
5	PAC-UL (mg)	3	-	-	-	-	-
6	PAC-R (mg)	1	-	-	-	-	-
7	HT-PM (g)	-	4	-	-	-	-
8	STARCH (g)	-	-	4	-	-	-
9	CMC(g)	-	-	-	4	-	-
10	Polysal (g)	-	-	-	-	4	-
11	PHPA(PolyPlus) (g)	1	1	1	1	1	1
12	Duo Tec (mg)	1	1	1	1	1	1
13	KCl (g)	21	21	21	21	21	21
14	Formate Na (g)	50	50	50	50	50	50
15	Formate K (g)	50	50	50	50	50	50
16	$CaCO_3$ (g)	50	50	50	50	50	50

Table 2: Materials that was used in formulation of formate sample muds for evaluating the compatibility of different polymers.

S/n	Additive(s)	Function(s)
1	Water	Base fluid
2	Polyanionic Cellulose	Filtration Control
3	PHPA	maintaining the rheology of the fluid inside the wellbore
4	$CaCO_3$	Weighting agent- bridging material
5	Na_2CO_3	Calcium precipitant and pH reducer in cement contaminated mud

Table 3: Shows the additive and its functions.

measure the dial readings which were further empirically correlated to determine rheological properties like plastic viscosity, apparent viscosity, yield point, fluid loss; and also the initial and 10 min gel strength of the prepared homogenous solutions.

Drilling fluid design: Materials that was used in formulation of formate sample muds for evaluating the compatibility of different polymers. Component of each formate sample are illustrated more in Table 2.

The materials used for this study are:

✓ Shale sample

✓ Fresh water

✓ Hamilton Beach Mixer

✓ Mud balance API filter press

✓ Variable speed rheometer

✓ Marsh funnel

✓ Fann V-G meter

✓ PH meter.

Table 3 shows the additive and its functions.

Fann V-G meter: The V-G meter is a rotational type viscometer in which the fluid is contained between coaxial cylinders. The outer cylinder rotates at a constant speed and the viscous drag of the fluid on the inner cylinder or bob exerts a torque that is indicated on a calibrated dial. The torque is proportional to shear stress and the rotational speed is proportional to shear rate. The indicated dial reading times 1.067 is equivalent to shear stress in lb/100 sq. ft. And the rotational speed in rpm times 1.703 is equivalent to shear rate in recipical seconds.

Two models of the V-G meter in common use are the Fann 35 and 34. The Model 35 is a six-speed model (600, 300, 200, 100, 6, and 3 rpm) and the Model 34 is a two-speed model (600 and 300 rpm). These instruments provide measurements of the actual flow parameters of shear rate and shear stress and also provide a means of making gel strength measurements. With this information we are better equipped to diagnose flow behavior and prescribe mud treatment than with the funnel viscosity. After the shear stress/shear rate data are collected, they can be handled and reported in a number of ways. Traditionally, these data have been used to calculate plastic viscosity and yield point in the Bingham plastic rheological model, and these parameters have been reported on the mud check sheet. The difference in the V-G Meter dial readings at 600 and 300 rpm is the plastic viscosity, and the plastic viscosity subtracted from the 300 rpm reading is the yield point.

Since the Bingham plastic model does not truly represent the shear rate/shear stress behavior of most muds, the calculated yield point is not equivalent to the true yield stress and the plastic viscosity is not a true viscosity. However, the wealth of experience we have acquired in the use of these parameters make them quite useful in predicting mud performance and diagnosing mud problems.

Rheological properties measured with a rotational viscometer are commonly used to indicate solids buildups flocculation or de-flocculation of solids, lifting and suspension capabilities, and to calculate hydraulics of a drilling fluid.

A rotational viscometer is used to measure shear rate/shear stress of a drilling fluid - from which the Bingham Plastic parameters, PV and YP, are calculated directly. Other rheological models can be applied using the same data. The instrument is also used to measure thixotropic properties, gel strengths. The following procedure applies to a Fann Model 35, 6-speed VG Meter.

Plastic Viscosity (PV) and Yield Point (YP): There are following steps:

1. Obtain a sample of the mud to be tested. Record place of sampling. Measurements should be made with minimum delay.

2. Fill thermal cup approximately 2/3 full with mud sample. Place thermal cup on viscometer stand. Raise cup and stand until rotary sleeve is immersed to scribe lie on sleeve. Lock into place by turning locking mechanism (Figure 1).

3. Place thermometer in thermal cup containing sample. Heat or cool sample to desired test temperature of 115° ± 2°F.

4. Flip VG meter toggle switch, located on right rear side of VG meter, to high position by pulling forward.

5. 6/94 3-2

6. Position red knob on top of VG meter to the 600-rpm speed. When the red knob is in the bottom position and the toggle switch is in the forward (high) position -this is the 600-rpm speed.

7. With the sleeve rotating at 600-rpm, wait for dial reading in the top window of VG meter to stabilize (minimum 10 seconds). Record 600-rpm dial reading.

8. With red knob in bottom position, flip the VG meter toggle switch to low position by pushing the toggle switch away from you. Wait for dial reading to stabilize (minimum 10 seconds). Records 300-rpm dial reading.

9. The Plastic Viscosity and Yield Point are calculated from the 600-rpm and 300-rpm dial readings (Figure 2).

Gel Strength (10-sec/10-min): There are following steps to be followed:

1. With red knob in bottom position, flip toggle switch to 600-rpm position (forward position). Stir mud sample for 10 seconds.

2. Position red knob to the 3-rpm speed. When the red knob is in the middle position and the toggle switch is in low (rear) position - this is the 3-rpm speed. Flip toggle switch to off position. Allow mud to stand undisturbed for 10 seconds.

3. After 10 seconds, flip toggle switch to low (rear) position and note the maximum dial reading. This maximum dial deflection is the 10-second (initial) gel strength in lb/100 ft2. Record on the mud check sheet.

4. Pull toggle switch to high and position red knob to 600-rpm speed. Stir mud for 10 seconds.

5. After 10 seconds, and while mud is still stirring, position red knob to the 3-rpm speed. Flip toggle switch to off position and allow mud to stand undisturbed for 10 min.

6. After 10 min, flip toggle switch to low (rear) position and note the maximum dial reading. This maximum dial deflection is the 10-min gel strength in lb/100 ft2. Record on the mud check control of filtration properties of a drilling fluid can be useful in reducing tight hole conditions.

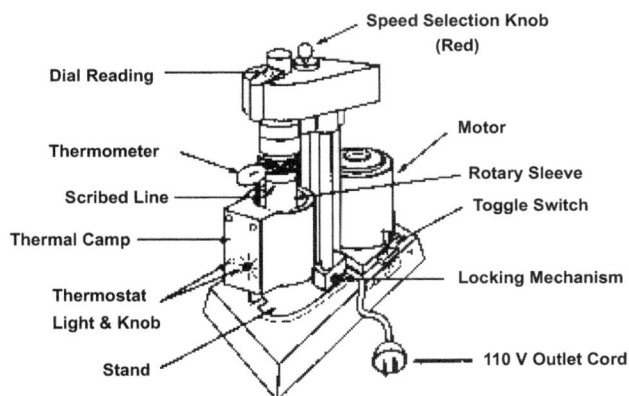

Figure 1: Fann Model 35 6-Speed Viscometer.

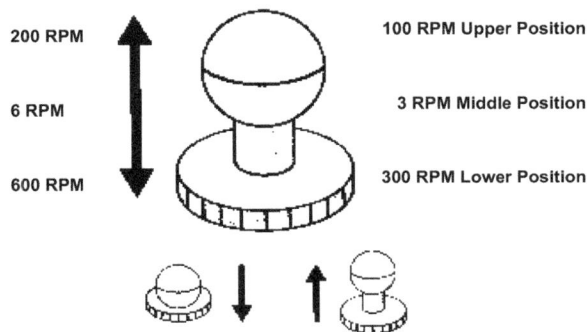

Figure 2: Speed Selection Knob (Caution: Change gears only when motor is running).

Marsh funnel for measuring apparent viscosity: The Marsh funnel is a crude method for measuring the consistency of a fluid. Although listed on the mud check sheet as viscosity, the Funnel viscosity is not in the true sense a viscosity at all. The test consists of filling the funnel to the bottom of the screen with mud (1500 ml) and timing how long it takes for one quart to flow out of the funnel. The time in seconds is reported as the funnel viscosity. Fresh water at 70°F will have a funnel viscosity of 26 seconds. This test has the advantage of being quick, simple, and requiring very little equipment. It is useful in showing gross changes in the overall "viscosity" of a fluid, but it does not measure specific flow parameters. It can be changed by changes in plastic viscosity, yield point, gel strength, or density. For this reason it should be used only to monitor a mud and not to diagnose problems or to prescribe treatment (Figure 3).

Figure 4 shows apparent viscosity for formate fluids in presence of different polymers. As it can be seen in the Figure 4, the amount of apparent viscosity for fluids including CMC and PAC polymers are more than the other samples. Apparent viscosity for other fluids are approximately remains constant (25 centipoises). Furthermore, viscosity loss after applying temperature are extremely low and it has the minimum amount within formate fluids contained starch.

Figure 5 shows plastic viscosity for different fluids. In the comparison with Figure 4, the amount of plastic viscosity for fluids including CMC and PAC polymers is far high is high. Hence, it isn't appropriate for drilling operations. The amount of plastic viscosity for other samples is in the proper limited and optimum value for drilling operations. As it can be seen in Figure 5, the plastic viscosity loss after applying temperature to the fluids considered be negligible that consequently cause of increasing thermal stability of polymers by formate salts. The amount of plastic viscosity loss for CMC and starch are minimum than others.

Figure 6 shows Yield point for potassium/sodium formate fluids in combining of different polymers. In the same vein, to other Figure 7, fluids that included CMC and PAC polymers have higher yield point than the other samples. The amount of yield point for other samples was relatively remained unchanged and this value could be properly limited for drilling operations and mud pumps required less initial pressure for flowing the fluid.

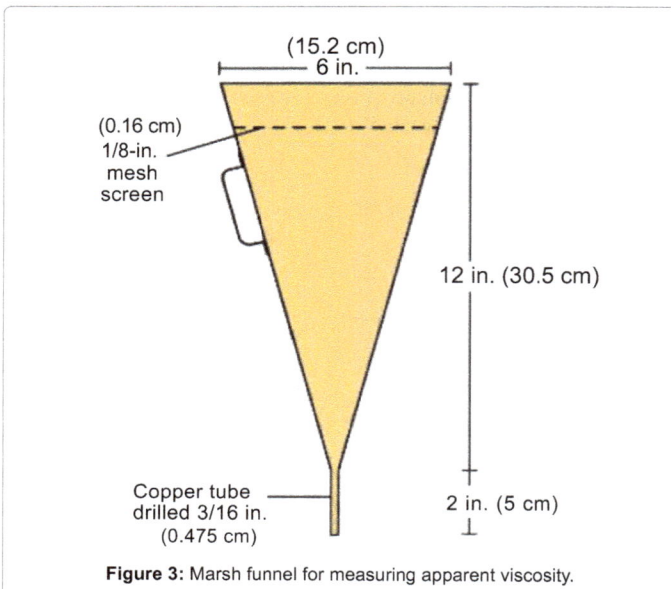

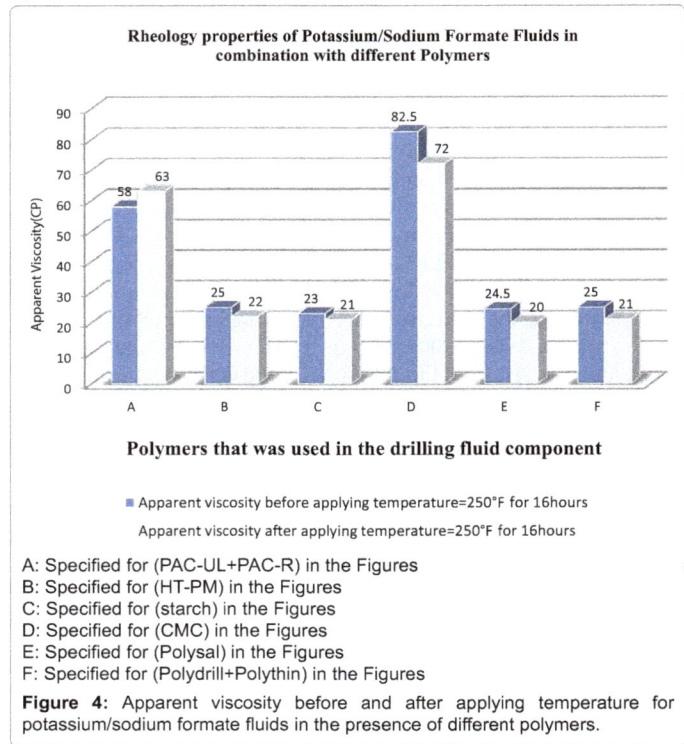

Rheology properties of Potassium/Sodium Formate Fluids in combination with different Polymers

- Apparent viscosity before applying temperature=250°F for 16hours
- Apparent viscosity after applying temperature=250°F for 16hours

A: Specified for (PAC-UL+PAC-R) in the Figures
B: Specified for (HT-PM) in the Figures
C: Specified for (starch) in the Figures
D: Specified for (CMC) in the Figures
E: Specified for (Polysal) in the Figures
F: Specified for (Polydrill+Polythin) in the Figures

Figure 4: Apparent viscosity before and after applying temperature for potassium/sodium formate fluids in the presence of different polymers.

Figure 7 shows the fluid loss for different types of polymers. As it can be seen, the amount of fluid loss before and after applying temperature for fluids contain HT-PM, CMC and starch are nearly similar and the rise of fluid loss after applying pressure could be ignored. The maximum amount of fluid loss after and before applying temperature are the fluids than contain poly-drill and poly-thin. Samples produced mud cakes are thin and flexible.

Figure 8 shows shale recovery for formate fluids in presence of different polymers. The amount of shale recovery are being measured by and experimental test according to API-13I. This test didn't propose the accurate amount of shale recovery. Besides, the accuracy of steps of test procedure and specially washing shale grains by saturated salty water has noticeable effect on the experimental results. The result of this test is also extremely affected by fluid viscosity. As a result, by increasing fluid viscosity, shale grains washing might be more difficult and mud with high viscosity are separated from shaly grains strictly. Thereby, in this occasion some amount of mud didn't separate from shaly grains. Subsequently shaly grains weight after applying temperature are more than the real amount. So, the amounts of shale recovery are being shown more than the real amount. In this experiment, as it can be seen in Figure 8 the amount of shale recovery for fluids contain CMC and PAC polymers is more than the other samples. In addition, these fluids have more viscosity than the others. Therefore, cause of increasing in the amount shale recovery depended directly to the high viscosity. As a consequence, overall in most cases combination of formate fluids with several polymers, show noticeable shale recovery.

Figure 9 shows the Effect of Xanthan polymer concentration on the formate fluid yield point with mud weight of 78 PCF. Figure 10 shows the Effect of Xanthan polymer concentration on the formate fluid Gel strength with mud weight of 78 PCF.

Figure 3: Marsh funnel for measuring apparent viscosity.

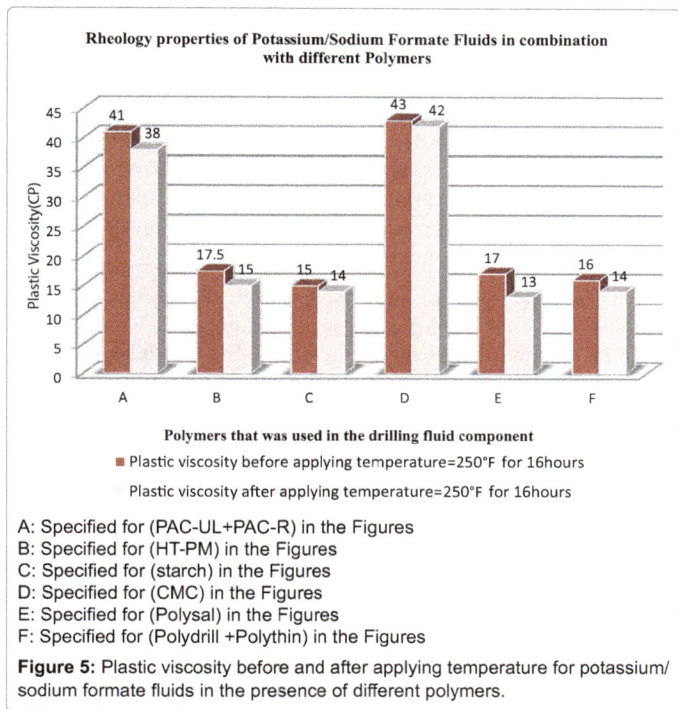

A: Specified for (PAC-UL+PAC-R) in the Figures
B: Specified for (HT-PM) in the Figures
C: Specified for (starch) in the Figures
D: Specified for (CMC) in the Figures
E: Specified for (Polysal) in the Figures
F: Specified for (Polydrill +Polythin) in the Figures

Figure 5: Plastic viscosity before and after applying temperature for potassium/sodium formate fluids in the presence of different polymers.

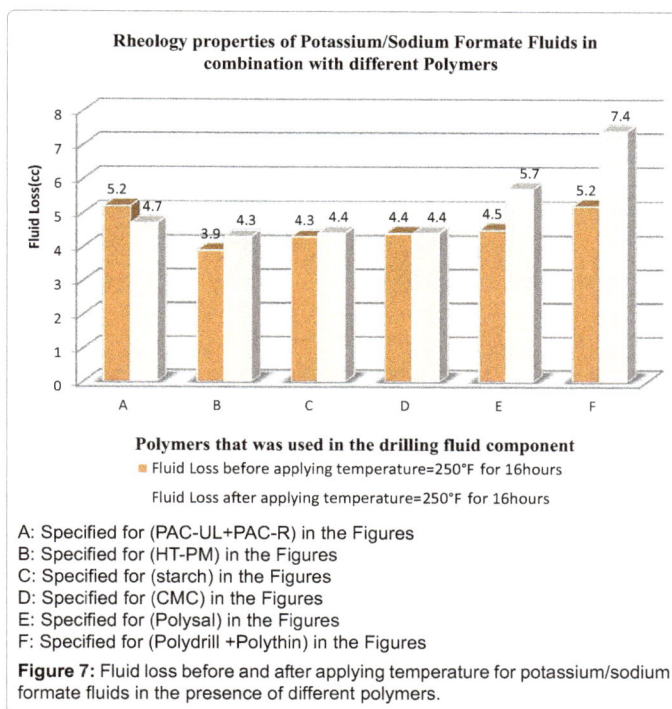

A: Specified for (PAC-UL+PAC-R) in the Figures
B: Specified for (HT-PM) in the Figures
C: Specified for (starch) in the Figures
D: Specified for (CMC) in the Figures
E: Specified for (Polysal) in the Figures
F: Specified for (Polydrill +Polythin) in the Figures

Figure 7: Fluid loss before and after applying temperature for potassium/sodium formate fluids in the presence of different polymers.

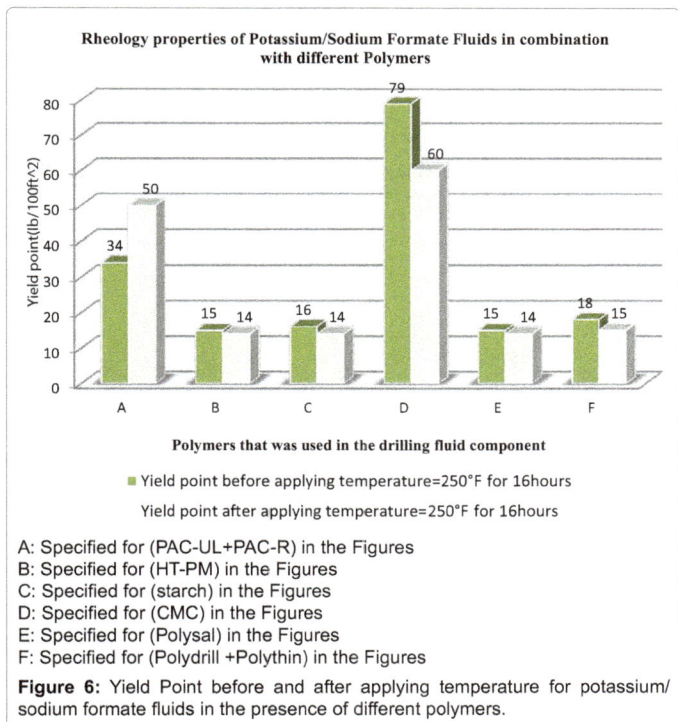

A: Specified for (PAC-UL+PAC-R) in the Figures
B: Specified for (HT-PM) in the Figures
C: Specified for (starch) in the Figures
D: Specified for (CMC) in the Figures
E: Specified for (Polysal) in the Figures
F: Specified for (Polydrill +Polythin) in the Figures

Figure 6: Yield Point before and after applying temperature for potassium/sodium formate fluids in the presence of different polymers.

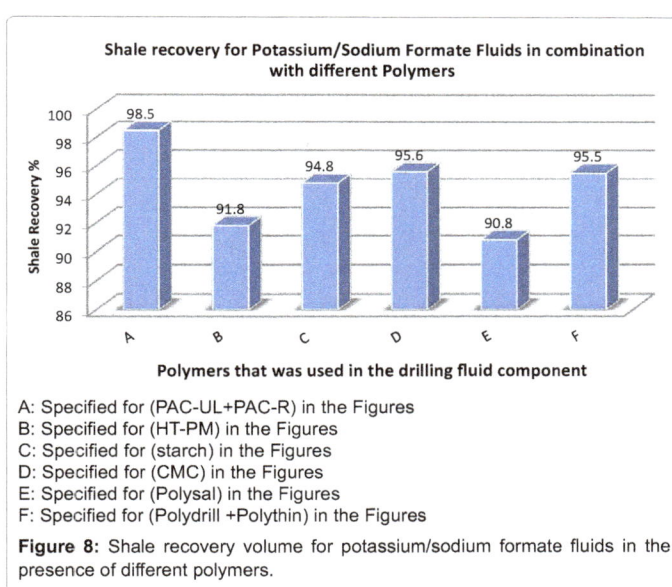

A: Specified for (PAC-UL+PAC-R) in the Figures
B: Specified for (HT-PM) in the Figures
C: Specified for (starch) in the Figures
D: Specified for (CMC) in the Figures
E: Specified for (Polysal) in the Figures
F: Specified for (Polydrill +Polythin) in the Figures

Figure 8: Shale recovery volume for potassium/sodium formate fluids in the presence of different polymers.

Results and Conclusions

Results

Regarding to the polymers and several bio-polymers compatibility tests with potassium/sodium formate fluids, it can be observed that formate fluids with all used polymers and bio-polymers showed proper and reasonable compatibility.

Rheological properties (AV, PV and Yp) for formate fluids in combination of PAC and CMC polymers are more than other used polymers. Reduction in rheological properties for different polymers through the formate fluids are extremely low. As for, formate salt have impacts on increasing thermal stability of polymers and bio-polymers. Thereby, the amount of fluid loss before and after applying temperature has no differential effects and the fluid loss for different polymers such as HT-PM polymer, CMC bio-polymers and starch are minimum and it is in the optimum level for drilling operations of a reservoir.

The amount of shale recovery in combination with several polymers is optimum and the amount of shale recovery for all samples is more than 90%. It should be noted that shaly sample that was used for shale recovery tests including high percent of clay materials is in type of montmorillonite that has high water absorption and will easily rescued.

Yield point after Hot rolling for 78 PCF mud vs Xanthan gum Concentration

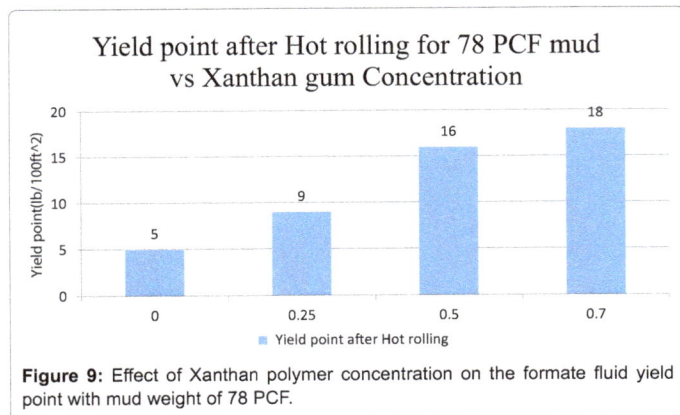

Figure 9: Effect of Xanthan polymer concentration on the formate fluid yield point with mud weight of 78 PCF.

Gel 10sec/10 min after Hot rolling for 78 PCF mud vs Xanthan gum Concentration

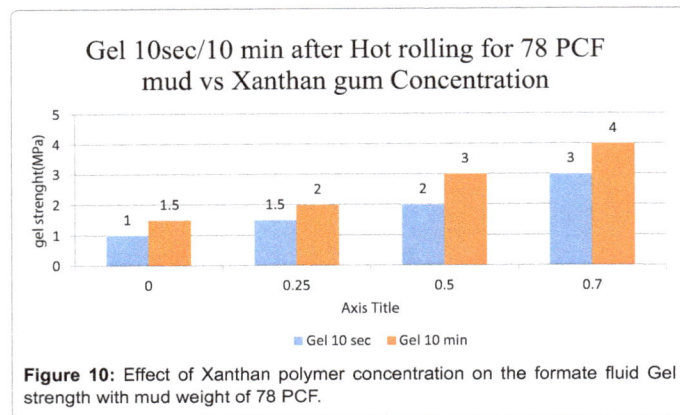

Figure 10: Effect of Xanthan polymer concentration on the formate fluid Gel strength with mud weight of 78 PCF.

Conclusions

The results of the comparison of several fluid tests compared to each other, such as silicate, glycol and potassium chloride fluids with formate fluids illustrated that formate fluids maintain rheological properties and fluid loss after applying temperature better than other fluids. Moreover, these muds have higher thermal stability in 250°F during 16 h rather than silicate, glycol and potassium chloride fluids. The amount of shale recovery in these muds is always more than silicate, glycol and potassium chloride fluids.

Experimental tests of mud pollution with several pollutants illustrated that mud rheological properties changes after it is polluted by several pollutants like cement and acids didn't have a large volume. Potassium/sodium formate fluids maintained their properties properly and did not cause a strict reduction in mud rheological properties.

References

1. Khodja M, Khodja-Saber M, Canselier JP, Cohaut N, Bergaya F (2010) Drilling Fluid Technology: Performances and Environmental Considerations. InTech 227-257.

2. Gao P, Yin D (2006) Simulation study on the conditions of converting injection wells to production wells in low permeability reservoirs. Journal of Daqing Petroleum Institute 6: 12-22.

3. Maglione R, Robotti G, Romagloni R (1996) In-Situ Rheological Characterization of Drilling Mud. Society of Petroleum Engineers.

4. Ferry JD (1980) Dependence of Viscoelastic Behaviour on Temperature and Pressure. Viscoelastic Properties of Polymers. John Wiley & Sons Inc, New York City.

5. Briant J, Denis J, Parc G (1989) Variation in Viscosity with Pressure," Rheological Properties of Lubricants. IFP Publications, Paris.

6. Schramm LL (1992) Emulsions: Fundamentals and Applications in the Petroleum Industry. Advances in Chemistry Series, USA.

7. Patil RC, Deshpande A (2012) Use of Nanomaterials in Cementing Applications. SPE-155607 SPE International Oilfield Nanotechnology Conference and Exhibition, Noordwijk, The Netherlands.

31

Synthesis of Graphene Oxide by Modified Hummers Method and Hydrothermal Synthesis of Graphene-NiO Nano Composite for Supercapacitor Application

Narasimharao K*, Venkata Ramana G, Sreedhar D and Vasudevarao V
Sreenidhi Institute of Science and Technology, Ghatkesar, Hyderabad, India

Abstract

Graphene oxide was synthesized by using modified Hummers method and also Graphene-NiO nanocomposite prepared by hydrothermal method with the use of graphene oxide solution, $Ni(NO_3)2 \cdot 6H_2O$ and urea as raw materials. The synthesized nanocomposite was characterized by the XRD, Raman, SEM, TGA and energy dispersive spectrometer analysis. The results demonstrate that NiO nano particles uniformly covered on the surface of the graphene layer and Raman spectroscopy states that well formation of the Graphene-NiO Nano composite. CV curves of Graphene-NiO electrodes at different scan rates conveys that capacitance characteristic is very different from that of traditional electric double-layer capacitance in which the shape is normally an ideal rectangular shape. Therefore, cyclic voltammetry analysis is the evidence of these materials are having the ability of supercapacitor electrode material properties. TGA Analysis is used for the estimation of how much of graphene is exist in the Graphene-NiO nanocomposite.

Keywords: Graphene; Nanocomposite; Material; Capacitance; Pseudocapacitors

Introduction

Graphene is a one-atom thick structure of sp^2-bonded carbon atoms. These carbon atoms are densely packed in a honeycomb crystal lattice structure. Since the detection in 2004, graphene has attracted fabulous research interest in energy-storage technologies due to its unusual properties, like great mechanical strength, large specific surface area and high electrical conductivity [1]. A lot of scientists said that graphene would be a competitive material for energy storage applications like batteries, solar cells and super capacitors etc. [2,3]. Now days, a lot of research has been launched into the development of graphene-based nanocomposites for supercapacitor applications, particularly for nano-composites that consist of graphene and transition-metal oxides such as MnO_2, ZnO, NiO, Co_3O_4 because they combine the advantages of both components and may offer special properties through the reinforcement or modification of each other [4]. Several graphene and transition-metal oxide nanocomposites such as graphene-NiO, graphene-MnO_2, graphene-Mn_3O_4, graphene-Bi_2O_3, graphene-Co_3O_4 and graphene- ZnO, have been exploited and improved performance of pseudo supercapacitor has been found in these type of composite systems [5].

Electrochemical capacitors, which are also called as supercapacitors, are widely investigated due to their interesting properties in terms of fast recharge capability and high power density [6-8]. Depends upon the charge-storage mechanism, electrochemical supercapacitors can be divided into (a) Electrical Double-layer capacitors (EDLCs), in which capacitance arises from charge separation between electrode-electrolyte interface, and (b) pseudocapacitors concerns reversible redox reactions, in which the dominant process is of pseudo capacitive origin. Till now, carbon materials are widely used as electrode materials for EDLCs due to their good cycle life, large surface area, good processing ability and low cost [9-12]. However, carbon materials have some drawback like it suffers from low specific capacitance [13-16]. To improve energy and power densities, much effort has been dedicated to investigate pseudo capacitive transition-metal oxides which give higher capacitance due to multi electron transfer during fast faradaic reactions [17-21].

Transition-metal oxides can be separated into two groups: one is noble-metal oxides and another one is cheap-metal oxides. One example for the first one is RuO_2, whereas Na_xMnO_2, K_xMnO_2, NiO, and Co_3O_4 are well-known as examples of the second one. Various forms of RuO_2 exhibit superior electrochemical response [22-26]. Unluckily, the cost of RuO_2 has limited its technological viability. So, NiO is considered to be a promising alternative for electrode materials in redox electrochemical capacitors because of its low cost, high capacitance, ease of synthesis [27-30].

The addition of graphene in electrochemical supercapacitors requires a graphene film synthesized on a large-scale with decent uniformity. There are numerous deposition techniques available for the synthesis of graphene like chemical reduction of graphite oxide, chemical vapour deposition, epitaxial growth and vaccum filtration [31-36]. Up to now, the graphene is prepared by the GO film reduction. These are still of great value due to its low-cost, ease of process and low-temperature [37]. Hence nickel oxide (NiO) among many transition metal oxides, shows outstanding cycle-life stability and high specific capacitance resulting from fast electron and ion transport, excellent structural stability and large Electro active surface area [2]. But these superior properties vary tremendously depending upon preperation methods [38,39].

*Corresponding author: Narasimharao K, Sreenidhi Institute of Science and Technology, Ghatkesar, Hyderabad 500130, Telangana, India
E-mail: knr367@gmail.com

Experimental Procedure

Required chemicals

Graphite Flakes (acid treated 99%, Asbury Carbons), Potassium permanganate (99%, RFCL), Sodium nitrate (98%, Nice chemicals), Hydrogen peroxide (40% wt, Emplura), Sulphuric acid (98%, ACS), Hydrochloric acid (35%, RANKEM), $Ni(NO_3)_2*6H_2O$ and Urea.

Graphene oxide preparation by using modified Hummer's method

Graphene oxide (GO) was prepared from graphite flakes by using modified Hummer's method. The step by step synthesis is as follows :

1. 2 g of Graphite flakes and 2 g of $NaNO_3$ and 50 ml of H_2SO_4 (98%) were mixed in a 1000 ml volumetric flask kept under at ice bath (0-6°C) with stirring continuously.

2. The sample mixture was stirred for 2 hrs at the same temperature and 6 g of potassium permanganate ($KMnO_4$) was added to the suspension very slowly. The addition rate was controlled carefully to preserve the reaction temperature lower than 14°C.

3. Then the ice bath was removed, and the sample mixture was stirred at 30°C til it became pasty brownish and kept under stirring for 2 hrs. For every half an hour, increase the temperature.

4. Then it was weakened with the slow addition of 100 ml of water. The reaction temperature was increased quickly to 96°C with effervescence, and the color changes to brown type of color.

5. Further, this solution mixture was weakened by the addition of 200 ml of water stirred continuously.

6. The solution mixture was finally treated with 8 ml H_2O_2 to terminate the reaction by the form of yellow colour.

7. For purification, the mixture was washed by centrifugation and rinsing with 8% HCL and then deionized (DI) water for various times.

8. After filtration and then it dried in hot air oven, the graphene Oxide (GO) was obtained as a powder.

Preparation of graphene-NiO nano composite by hydrothermal method

1. Firstly, 2.98 g nickel nitrate ($Ni(NO_3)_2.6H_2O$) was dissolved in 8 ml of deionized water and added into 20 ml graphene oxide with stirring for 60 min.

2. The pH of the solution was maintained to around 8.0 by the addition of 20 ml urea drop by drop.

3. The solution mixture was stirred forcefully for 60 min to mix equally and transferred into a 100 ml steel autoclave. Then the autoclave was sealed, moved into muffle furnace and kept at 150°C for 5 h.

4. After that the autoclave cools down to room temperature, the solution mixture was vacuum-filtered to get the precursor then washed with ethanol and DI water 2 or 3 times to remove the probable absorbed ions, metal salts and remained raw material, then dried at 80°C for 12 hrs.

5. At last, the obtained precursor was calcinated at 400°C for 4 h in a muffle furnace. Then the Graphene–NiO nanocomposite was obtained after calcination and collected for characterization.

6. For comparison, pure NiO was also prepared by the same process without GO.

7. The graphene was synthesized by pyrolyzing dried GO in N_2 atmosphere at the same temperature with hybrid material (400°C for 4 h).

Characterization

The obtained precursors were characterized by X-ray powder diffraction (XRD), Field-emission scanning electron microscopy (SEM), EDS and Raman spectroscopy.

Electrochemical tests

Electrochemical tests were investigated by Electrochemical Work station. The three-electrode cell consists of Ag/AgCl as reference electrode and Platinum (Pt) foil as counter electrode. The working electrode was prepared by mixing the Active material (1 mg), Ethanol (1 ml) and Nafion solution (1 ml) and sonicated it for 30 min to form gel. And this Gel was dropped onto the Working electrode and dried it overnight. The electrolyte was 1 M H_2SO_4 aqueous solution.

Results and Discussions

X-ray diffraction analysis was done to know the crystal size and crystal phase and structure of the synthesized samples (Figure 1). The diffraction peaks related to pure Cubic NiO occurs at 2Θ=37.1°, 43.1°, 62.7°, 75.3°, 79.2° can be readily written as (1 1 1), (2 0 0), (2 2 0), (3 1 1) and (2 2 2) crystal planes, and the widening of the diffraction peaks advises a very small size of NiO nanoparticles. The average grain size of pure NiO and G–NiO are 80 nm and 12 nm calculated by using Scherer's formula (D =0.89λ/B cosΘ). It is clearly show that the particle size of pure NiO is larger than that of G–NiO and it is mostly due to the presence of graphene. In the G–NiO sample, there is no distinctive peak of Graphene oxide (2Θ=10.9°) or graphite Flakes (2Θ=24.6°), advising that the graphene oxide was well condensed.

Figure 2 shows the Raman spectrum of Graphene-NiO and NiO. Four Raman peaks positioned at about 383, 521, 710 and 1092 cm^{-1} are detected in both spectra, equivalent to the shaking peaks of NiO. The first two peaks could be recognized to the 1st order transverse optical and longitudinal optical phonon types of NiO, respectively. The peaks at 710 and 1092 cm^{-1} could be allotted to 2nd order transverse optical and longitudinal optical phonon modes of NiO. For graphene-NiO Nanocomposite, the Raman peaks of D line and G line is allotted to the E_{2g} phonon of Carbon sp^2 atoms, and the D line is a living mode of k-point phonons of A_{1g} symmetry. The remaining peaks are in agreement with the pure nickel oxide. Such a result further confirms that the crystalline structure of graphene-NiO nanocomposite has been obtained.

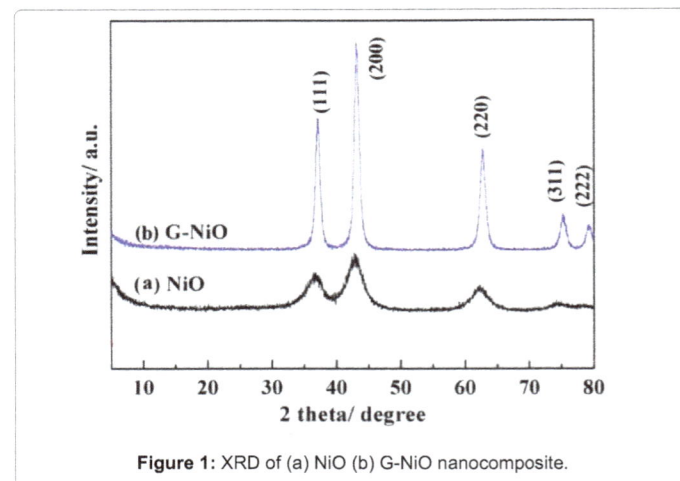

Figure 1: XRD of (a) NiO (b) G-NiO nanocomposite.

The microstructure, morphology and particle size of as-synthesized pure nickel oxide and Graphene-NiO Nanocomposites were understood by SEM and TEM analysis. The SEM image of pure nickel oxide (Figure 3a) shows that the particle size is around 80 nm and these nanoparticles combined with each other. Figure 3b shows a SEM image of the Graphene-NiO nanocomposite consists of thin, randomly aggregated and wrinkled graphene sheets closely related with each other and forming a chaotic solid with particle size of 1-10 nm. The higher magnification SEM image of Figure 3b further tells that tiny NiO nanoparticles are spread on the curly graphene nanosheets. Unusually, the particle size of the NiO grown on graphene is much lesser than pure NiO synthesized by the same process, because of the confining effect of disordered graphene nanosheets. More likely, the anchoring NiO nanoparticles could act as a inserter to prevent the re-stacking of separate graphene nanosheets.

Figure 3c we can tell only elements C, O and Ni are present. The percentage weight of NiO is calculated to be approaching the theoretical value, advising that all the Ni^{2+} are completely dropped as the hydrolysis of urea.

Figure 4 shows the CV curves of the Graphene-NiO electrode at different scan rates. The shapes of the CV disclose that the capacitance characteristic is very different from that of traditional electric double-layer capacitance in which the shape is generally an ideal rectangular shape. A NiO supercapacitor in an alkaline solution depend on charge storage in the electric double layer at the electrode/electrolyte boundary and charge storage in the host product through redox reactions on the surfaces and hydroxyl ion diffusion in the host product. As scan rate increases from 10 mv/sec to 200 mv/sec the current subsequently increases while the shape of CV curves changes little and much rapid current responses on voltage reversal at each end potential, which indicates good electrochemical capacitive nature for graphene-NiO porous structures. Figure 5 shows the Tga analysis of the G-NiO nanocomposite. By the help of Tga, we can estimate how much of graphene is present in the G-Nio nano composite.

From the result shown in the above graph, the weight loss of water is approximately equal to 7.6% when the temperature varies from the room temperature to 160°C. Next the weight loss regarding functional group decomposition is equal to 1.6% and its temperature rises from the 160°C to 250°C. And 12% of weight loss is due to the reaction of carbon and oxides and its temperature changes from 250°C to 620°C. Last but not least the weight loss of graphene is 13%.

Conclusion

In summary, synthesis of Graphene oxide was done by using modified hummers method and Graphene-NiO nanocomposites have been effectively prepared by a hydrothermal method. SEM analysis tells that graphene sheets were well ornamented by the NiO nanoparticles to form a complex ordered nanostructures with rich pore size distribution and large specific surface area. XRD analysis specified that

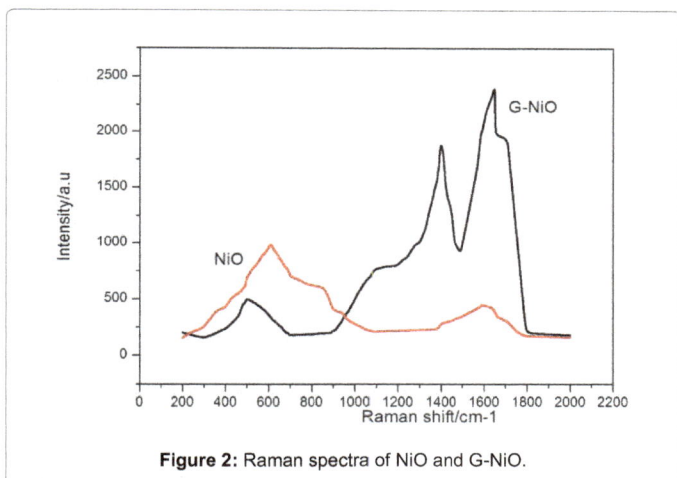

Figure 2: Raman spectra of NiO and G-NiO.

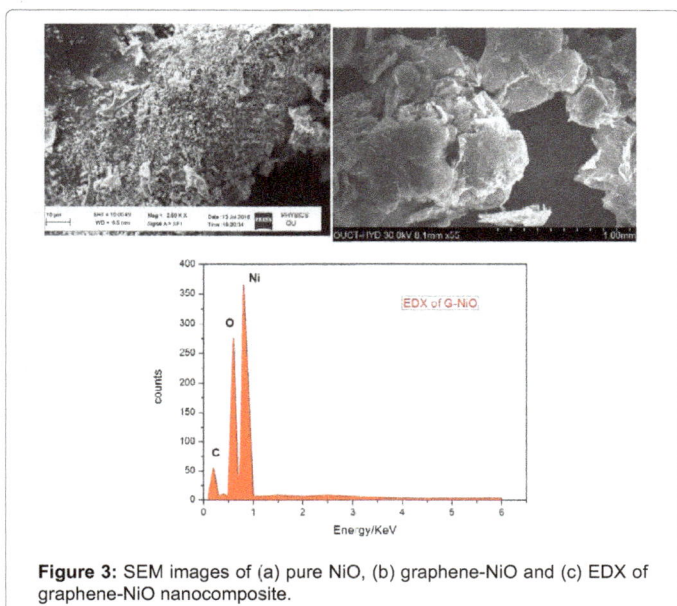

Figure 3: SEM images of (a) pure NiO, (b) graphene-NiO and (c) EDX of graphene-NiO nanocomposite.

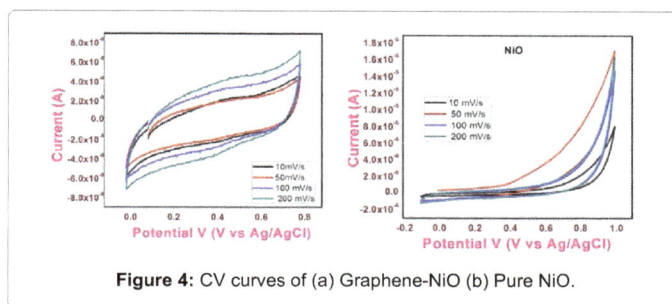

Figure 4: CV curves of (a) Graphene-NiO (b) Pure NiO.

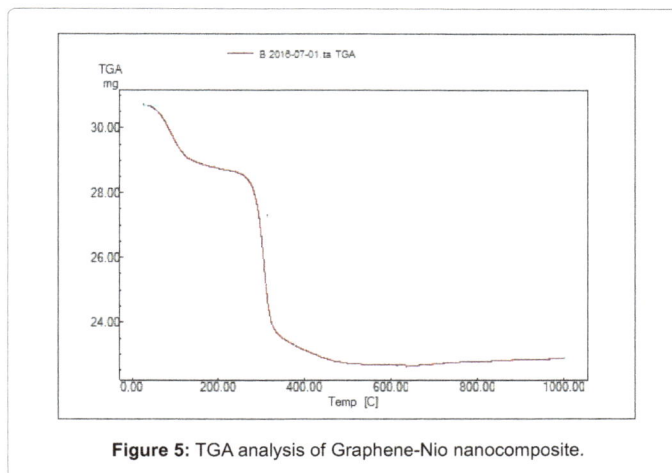

Figure 5: TGA analysis of Graphene-Nio nanocomposite.

GO and G-NiO composite phases and the average grain size of pure NiO and porous G-NiO are 82 nm and 12 nm. Raman spectrum of GO and Graphene-NiO displays that well formation of GO and composite. Cyclic voltammetry analysis is the evidence of these materials are having the ability of supercapacitor electrode materials. And also TGA Analysis gives the information about how much of graphene is present in the Graphene-NiO) nanocomposite.

References

1. Lee JK, Smith KB, Hayner CM, Kung HH (2010) Silicon nanoparticles-graphene paper composites for Li ion battery anodes. Chem Commun 46: 2025-2027.

2. Zhao B, Song J, Liu P, Xu W, Fang T, et al. (2011) Monolayer graphene/NiO nanosheets with two-dimension structure for supercapacitors. J Mater Chem 21: 18792-18798.

3. Lee JW, Ahn T, Soundararajan D, Ko JM, Kim JD (2011) Non-aqueous approach to the preparation of reduced graphene oxide/a-Ni(OH)$_2$ hybrid composites and their high capacitance behaviour. Chem Commun 47: 6305-6307.

4. Zhang Y, Zhu J, Song X, Zhong X (2008) Controlling the synthesis of CoO nanocrystals with various morphologies. J Phys Chem C 112: 5322-5327.

5. Wei W, Yang S, Zhou H, Lieberwirth I, Feng X, et al. (2013) 3D graphene foams cross-linked with pre-encapsulated Fe$_3$O$_4$ nanospheres for enhanced lithium storage. Adv Mater 25: 2909-2914.

6. Futaba DN, Hata K, Yamada T, Hiraoka T, Hayamizu Y, et al. (2006) Shape-engineerable and highly densely packed single-walled carbon nanotubes and their application as super-capacitor electrodes. Nat Mater 5: 987-994.

7. Zhu Y, Murali S, Stoller MD, Ganesh KJ, Cai W, et al. (2011) Carbon-based supercapacitors produced by activation of grapheme. Science 332: 1537-1541.

8. Jiang C, Zhan B, Li C, Huang W, Dong X (2014) Synthesis of three-dimensional selfstanding graphene/Ni(OH)$_2$ composites for high-performance supercapacitors. RSC Adv 4: 18080.

9. Chen S, Duan J, Tang Y, Zhang Qiao S (2013) Hybrid hydrogels of porous grapheme and nickel hydroxide as advanced supercapacitor materials. Chem Eur J 19: 7118-7124.

10. Chen K, Chen L, Chen Y, Bai H, Li L (2012) Three-dimensional porous graphene based composite materials: electrochemical synthesis and application. J Mater Chem 22: 20968-20976.

11. Xu MW, Bao SJ, Li HL (2007) Synthesis and characterization of mesoporous nickel oxide for electrochemical capacitor. J Solid State Electrochem 11: 372-377.

12. Stankovich S, Dikin DA, Dommett GHB, Kohlhaas KM, Zimney EJ, et al. (2006) Graphene-based composite materials. Nature 442: 282-286.

13. Kim E, Son D, Kim TG, Cho J, Park B, et al. (2004) A mesoporous/crystalline composite material containing tin phosphate for use as the anode in lithium-ion batteries. Angewandte Chemie International Edition 43: 5987-5990.

14. Lou XW, Deng D, Lee JY, Feng J, Archer LA (2008) Self-supported formation of needlelike co$_3$o$_4$ nanotubes and their application as lithium-ion battery electrodes. Advanced Materials 20: 258-262.

15. Hou Y, Cheng Y, Hobson T, Liu J (2010) Design and synthesis of hierarchical MnO$_2$ nanospheres-carbon nanotubes-conducting polymer ternary composite for high performance electrochemical electrodes. Nano Letters 10: 2727-2733.

16. Zhu X, Zhu Y, Murali S, Stoller MD, Ruoff RS (2011) Nanostructured reduced graphene oxide-Fe$_2$O$_3$ composite as a high-performance anode material for lithium ion batteries. ACS Nano 5: 3333-3338.

17. Hosogai S, Tsutsumi H (2009) Electrospun nickel oxide-polymer fibrous electrodes for electrochemical capacitors and effect of heat treatment process on their performance. J Power Sources 194: 1213-1217.

18. Wu NL (2002) Nanocrystalline oxide supercapacitors. Mater Chem Phys 75: 6-11.

19. Brezesinski T, Wang J, Tolbert SH, Dunn B (2010) Ordered mesoporous a-MoO3 with iso-oriented nanocrystalline walls for thin-film pseudocapacitors. Nat Mater 9: 146-151.

20. Rudge A, Davey J, Raistrick I, Gottesfeld S, Ferraris JP (1994) Conducting polymers as active materials in electrochemical capacitors. J Power Sources 47: 89-107.

21. Snook GA, Kao P, Best AS (2011) Conducting-polymer-based supercapacitor devices and electrodes. J Power Sources 196: 1-12.

22. Yuan C, Zhang X, Su L, Gao B, Shen L (2009) Facile synthesis and self-assembly of hierarchical porous NiO nano/micro spherical superstructures for high performance supercapacitors. J Mater Chem 19: 5772-5777.

23. Ding S, Zhu T, Chen JS, Wang Z, Yuan C, et al. (2011) Controlled synthesis of hierarchical NiO nanosheet hollow spheres with enhanced supercapacitive performance. J Mater Chem 21: 6602-6606.

24. Li X, Zhu Y, Cai W, Borysiak M, Han B, et al. (2009) Transfer of large-area graphene films for high-performance transparent conductive electrodes. Nano Lett 9: 4359-4363.

25. Lin CC, Yen CC (2008) Manganese oxide precipitated into activated carbon electrodes for electrochemical capacitors. J Appl Electrochem 38: 1677-1681.

26. Cheng MY, Hwang BJ (2010) Mesoporous carbon-encapsulated NiO nanocomposite negative electrode materials for high-rate Li-ion battery. J Power Sources 195: 4977-4983.

27. Wang YG, Xia YY (2006) Electrochemical capacitance characterization of NiO with ordered mesoporous structure synthesized by template SBA-15. Electrochim Acta 51: 3223-3227.

28. Yang S, Wu X, Chen C, Dong H, Hu W, et al. (2012) Spherical a-Ni(OH)$_2$ nanoarchitecture grown on graphene as advanced electrochemical pseudocapacitor materials. Chem Commun 48: 2773-2775.

29. Gao F, Wei Q, Yang J, Bi H, Wang M (2013) Synthesis of graphene/nickel oxide composite with improved electrochemical performance in capacitors. Ionics 19: 1883-1889.

30. Rolison DR, Long JW, Lytle JC, Fischer AE, Rhodes CP, et al. (2009) Multifunctional 3D nanoarchitectures for energy storage and conversion. Chem Soc Rev 38: 226-252.

31. Xu Y, Huang X, Lin Z, Zhong X, Huang Y, et al. (2013) One-step strategy to graphene/Ni(OH)$_2$ composite hydrogels as advanced three-dimensional supercapacitor electrode materials. Nano Res 6: 65-76.

32. Zhang F, Zhu D, Chen XA, Xu X, Yang Z, et al. (2014) A nickel hydroxide-coated 3D porous graphene hollow sphere framework as a high performance electrode material for supercapacitors. Phys Chem Chem Phys 16: 4186-4192.

33. Hummers WS, Offeman RE (1958) Preparation of graphitic oxide. J Am Chem Soc 80: 1339.

34. Pham VH, Dang TT, Hur SH, Kim EJ, Chung JS (2012) Highly conductive poly(methyl methacrylate) (PMMA)-reduced graphene oxide composite prepared by self-assembly of PMMA latex and graphene oxide through electrostatic interaction. ACS Appl Mater Interfaces 4: 2630-2636.

35. Pei S, Cheng HM (2012) The reduction of graphene oxide. Carbon 50: 210-3228.

36. Lee JW, Ahn T, Soundararajan D, Ko JM, Kim JD (2011) Non-aqueous approach to the preparation of reduced graphene oxide/a-Ni(OH)2 hybrid composites and their high capacitance behaviour. Chem Commun 47: 6305-6307.

37. Ferrari AC, Robertson J (2000) Interpretation of raman spectra of disordered and amorphous carbon. Phys Rev B 61: 14095-14107.

38. Wang B, Park J, Wang C, Ahn H, Wang G (2010) Mn$_3$O$_4$ nanoparticles embedded into graphene nanosheets: preparation, characterization, and electrochemical properties for supercapacitors. Electrochim Acta 55: 6812-6817.

39. Tuinstra F, Koenig JL (1970) Raman spectrum of graphite. J Chem Phys 53: 1126-1130.

Permissions

All chapters in this book were first published in JMSE, by OMICS International; hereby published with permission under the Creative Commons Attribution License or equivalent. Every chapter published in this book has been scrutinized by our experts. Their significance has been extensively debated. The topics covered herein carry significant findings which will fuel the growth of the discipline. They may even be implemented as practical applications or may be referred to as a beginning point for another development.

The contributors of this book come from diverse backgrounds, making this book a truly international effort. This book will bring forth new frontiers with its revolutionizing research information and detailed analysis of the nascent developments around the world.

We would like to thank all the contributing authors for lending their expertise to make the book truly unique. They have played a crucial role in the development of this book. Without their invaluable contributions this book wouldn't have been possible. They have made vital efforts to compile up to date information on the varied aspects of this subject to make this book a valuable addition to the collection of many professionals and students.

This book was conceptualized with the vision of imparting up-to-date information and advanced data in this field. To ensure the same, a matchless editorial board was set up. Every individual on the board went through rigorous rounds of assessment to prove their worth. After which they invested a large part of their time researching and compiling the most relevant data for our readers.

The editorial board has been involved in producing this book since its inception. They have spent rigorous hours researching and exploring the diverse topics which have resulted in the successful publishing of this book. They have passed on their knowledge of decades through this book. To expedite this challenging task, the publisher supported the team at every step. A small team of assistant editors was also appointed to further simplify the editing procedure and attain best results for the readers.

Apart from the editorial board, the designing team has also invested a significant amount of their time in understanding the subject and creating the most relevant covers. They scrutinized every image to scout for the most suitable representation of the subject and create an appropriate cover for the book.

The publishing team has been an ardent support to the editorial, designing and production team. Their endless efforts to recruit the best for this project, has resulted in the accomplishment of this book. They are a veteran in the field of academics and their pool of knowledge is as vast as their experience in printing. Their expertise and guidance has proved useful at every step. Their uncompromising quality standards have made this book an exceptional effort. Their encouragement from time to time has been an inspiration for everyone.

The publisher and the editorial board hope that this book will prove to be a valuable piece of knowledge for researchers, students, practitioners and scholars across the globe.

List of Contributors

Daik R, Ezzaouia H, Amor SB and Meddeb H
Photovoltaic Laboratory Research and Technology Centre of Energy, Borj-Cedria Science and Technology Park, BP 95, 2050 Hammam-Lif, Tunisia

Lajnef M
Sfax Preparatory Engineering Institut, Route Menzel Chaker, 0.5 km, BP 1172, 3080 Sfax, Tunisia

Elgharbi S and Férid M
National Research Center in Sciences of Materials, Borj-Cedria Science and Technology Park, B.P. 95 Hammam-Lif, 2050, Tunisia

Abdessalem K
L3M, Department of Physics, Faculty of Sciences of Bizerte, 7021 Zarzouna, Tunisia

Merah N and Khan Z
Mechanical Engineering Department, King Fahd University of Petroleum and Minerals, Dhahran, KSA

Naik MK
The Petroleum Institute, Abu Dhabi, UAE

Al-Sulaiman F
National Company of Mechanical Systems, Riyadh, Saudi Arabia

Vishwanathan K and Springborg M
Physical and Theoretical Chemistry, University of Saarland, Germany

Shah MMK, Ismail A and Sarifudin J
Faculty of Engineering, Universiti Malaysia Sabah, Sabah, Malaysia

You PL
Institute of Quantum Electronics, Guangdong Ocean University, Zhanjiang, China

Chtaini A and Touzara S
Molecular Electrochemistry and Inorganic Materials Team, Sultan Moulay Slimane University, Faculty of Science and Technology Béni Mellal, Morocco

Cheikh Ould S'Id E
Faculty of Science and Technology, University of Science, Technology and Medicine, B.P. 5026, Nouakchott, Mauritania
Faculty of Sciences, Laboratory of water and environment (team Biomaterials and Electrochemistry), University Chouaib Doukkali, El Jadida, Morocco

Chamekh M
Faculty of Science and Technology, University of Science, Technology and Medicine, B.P. 5026, Nouakchott, Mauritania

Mabrouki M
Laboratory of Industrial Engineering, Sultan Moulay Slimane University, Faculty of Science and Technology Béni Mellal, Morocco

Kheribech A
Faculty of Sciences, Laboratory of water and environment (team Biomaterials and Electrochemistry), University Chouaib Doukkali, El Jadida, Morocco

Jibowu T, Rohani S and Ragab D
University of Western Ontario, London

Pereira da Silva S, Costa de Moraes D and Samios D
Laboratory of Instrumentation and Molecular Dynamics, Department of Physical Chemistry, Chemistry Institute, Federal University of Rio Grande do Sul Av. Bento Gonçalves 9500, Porto Alegre, Brazil

Kalra A, Lowe A and Al-Jumaily AM
Institute of Biomedical Technologies, Auckland University of Technology, New Zealand

Muzaffar MU
PMAS Arid Agriculture University Attock Campus, Pakistan

Khan NA and Rehman UU
Materials Science Laboratory, Quaid-i-Azam University, Islamabad, Pakistan

Ali SA
Department of Physics, Government College, University Lahore, Pakistan

Talwar DN
Department of Physics, Indiana University of Pennsylvania, USA

Wan L and Feng ZC
Laboratory of Optoelectronic Materials and Detection Technology, Guangxi Key Laboratory for Relativistic Astrophysics, School of Physical Science & Technology, Guangxi University, China

Tin CC
Department of Physics, Auburn University, Auburn, USA

Chowdhury AMS
Department of Applied Chemistry and Chemical Engineering, Faculty of Engineering and Technology, University of Dhaka, Bangladesh

Poddar P
Department of Applied Chemistry and Chemical Engineering, Faculty of Engineering and Technology, University of Dhaka, Bangladesh
Office of the Chief Chemical Examiner, CID, Bangladesh Police, Mohakhali, Dhaka, Bangladesh

Islam MS, Sultana S and Nur HP
Bangladesh Council of Scientific and Industrial Research, Dhanmondi, Dhaka, Bangladesh

Singh VK, Bansal G, Agarwal M and Negi P
G.B.P.U.A.T, Pantnagar, Uttarakhand, India

Bogis Haitham
Center of Excellence for Industrial Design and Manufacturing Research (CEIDM) Mechanical Engineering, King Abdulaziz University Jeddah, Saudi Arabia

Khasawneh FA
Industrial and Technological Development Center Mechanical and Aerospace Engineering, University of Missouri-Columbia Columbia, Missouri-65211, USA

El-Gizawy A. Sherif
Center of Excellence for Industrial Design and Manufacturing Research (CEIDM) Mechanical Engineering, King Abdulaziz University Jeddah, Saudi Arabia
Industrial and Technological Development Center Mechanical and Aerospace Engineering, University of Missouri-Columbia Columbia, Missouri-65211, USA

Richetta M, Medaglia PG, Mattoccia A, Varone A and Pizzoferrato R
Department of Industrial Engineering, University of Rome "Tor Vergata", Rome, Italy

EL-Shihy AM, Shabaan HF and Al-Kader HM
Faculty of Engineering, Zagazig University, Zagazig, Egypt

Hassanin AI
Faculty of Engineering, Egyptian Russian University, Cairo, Egypt

Elshafie S and Boulbibane M
Faculty of Engineering, Sports and Science, University of Bolton, Bolton, UK

Whittleston G
Department of Civil Engineering, University of Salford, UK

El-Eskandarany MS
Nanotechnology and Advanced Materials Program, Energy and Building Research Center, Kuwait Institute for Scientific Research, Kuwait

Ocheri C
Department of Metallurgical and Materials Engineering, University of Nigeria, Nigeria

Mbah AC
Department of Metallurgical and Materials Engineering, Enugu State University of Science and Technology, Nigeria

Sharma S, Dua A and Malik A
Dyal Singh College, University of Delhi, Lodhi Road, New Delhi, India

Guduru KK and Banothu S
Christu Jyothi Institute of Technology Science, Jangaon, Warangal, Hyderabad, India

Pandu R
Marri Laxman Reddy Institute of Technology & Management, Hyderabad, India

Vinaya K
ACE Engineering College, Telangana, India

Jassim AK
Department of Materials Engineering, University of Basra, Iraq

Hammood AS
Department of Materials Engineering, University of Kufa, Iraq

Abramovich A
St. Petersburg State University of Technology and Design, Higher School of Technology and Design, 198095 St. Petersburg, Russia

Ertekin B
Department of Milk Technology, Faculty of Agriculture, Adnan Menderes University, 09100 Aydın, Turkey

Çimen Z, Yilmaz H and Yilmaz UT
Department of Chemistry, Polatli Faculty of Arts and Science, Gazi University, 06900 Polatli, Ankara, Turkey

Najabat Ali M, Ansari U, Sami J and Mir M
Biomedical Engineering and Sciences Department, School of Mechanical and Manufacturing Engineering, National University of Sciences and Technology (NUST), Islamabad, Pakistan

Qayyum F
Mechanical Engineering Department, University of Engineering and Technology, Taxilla, Pakistan

Singh JP
Department of Textile Technology, U.P.Textile Technology Institute, Kanpur, India

Behera BK
Department of Textile Technology, Indian Institute of Technology Delhi, New Delhi, India

Venkateswara Rao CH and Usha Sri P
Mechanical Engineering Department, UCE, Osmania University, Hyderabad, India

Ramanarayanan R
TO'D', ACC, Advanced Systems Laboratory, DRDO, Hyderabad, India

Elghali M and Abdellaoui M
Laboratoire des matériaux utiles, Institut National de Recherche et d'Analyse Physicochimique, pole technologique de Sidi Thabet, 2020 Sidi Thabet, Tunisia

Kakoli M, Davarpanah A, Ahmadi A and Jahangiri MM
Department of Petroleum Engineering, Islamic Azad University, Science and Research Branch, Tehran, Iran

Narasimharao K, Venkata Ramana G, Sreedhar D and Vasudevarao V
Sreenidhi Institute of Science and Technology, Ghatkesar, Hyderabad, India

Index